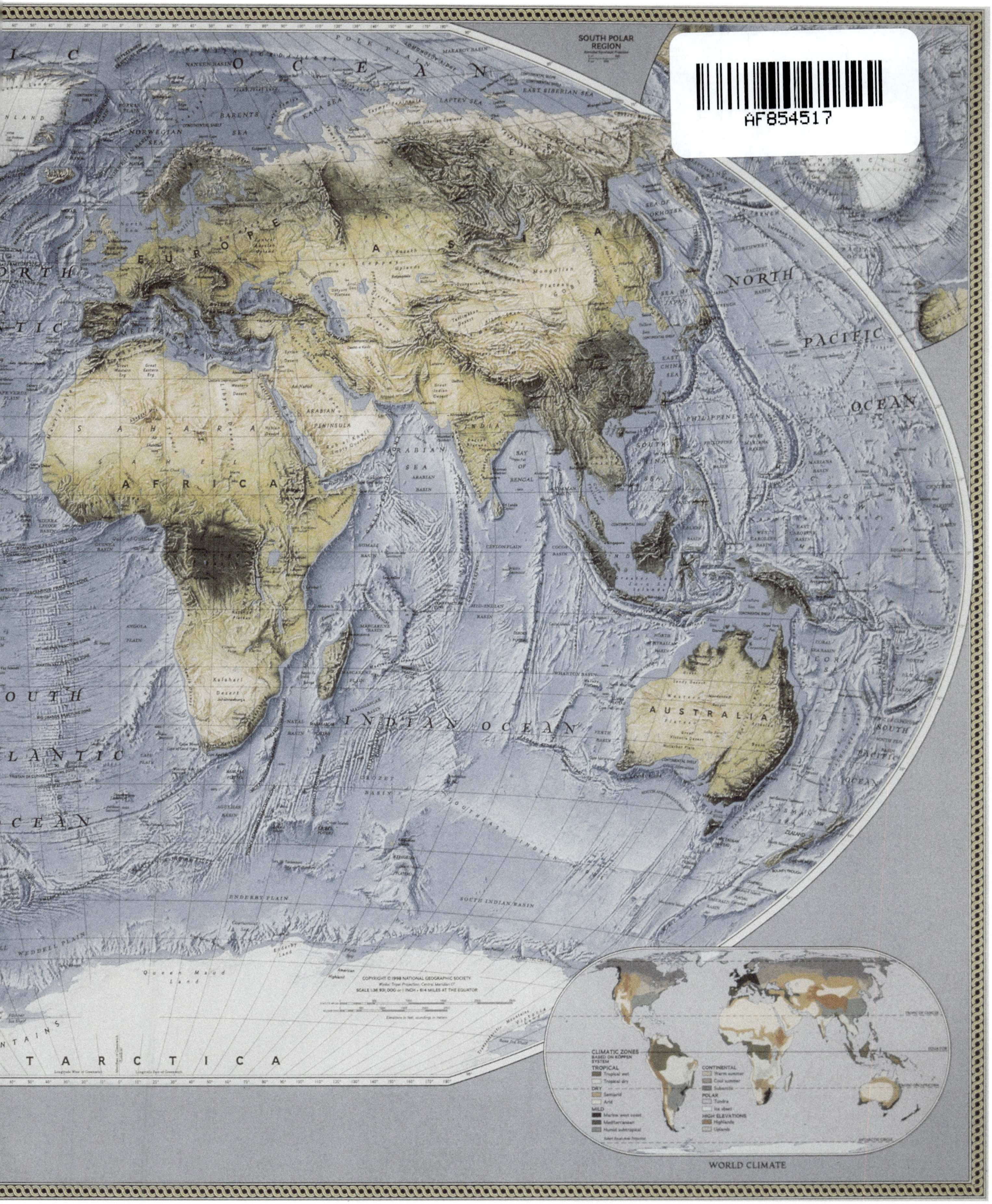
SOUTH POLAR REGION
O C E A N
EUROPE
A S I A
NORWEGIAN SEA
BARENTS SEA
KARA SEA
LAPTEV SEA
EAST SIBERIAN SEA
SEA OF OKHOTSK
NORTH PACIFIC OCEAN
S A H A R A
A F R I C A
ARABIAN PENINSULA
ARABIAN SEA
BAY OF BENGAL
INDIA
PHILIPPINE SEA
SOUTH CHINA SEA
EAST CHINA SEA
SOUTH ATLANTIC OCEAN
INDIAN OCEAN
AUSTRALIA
SOUTH PACIFIC OCEAN
ENDERBY PLAIN
SOUTH INDIAN BASIN
WEDDELL PLAIN
Queen Maud Land
American Highland
A N T A R C T I C A
COPYRIGHT © 1998 NATIONAL GEOGRAPHIC SOCIETY
SCALE 1:38,931,000 or 1 INCH = 614 MILES AT THE EQUATOR
Transantarctic Mountains
Ross Ice Shelf
WORLD CLIMATE
CLIMATIC ZONES BASED ON KÖPPEN SYSTEM
TROPICAL
Tropical wet
Tropical dry
DRY
Semiarid
Arid
MILD
Marine west coast
Mediterranean
Humid subtropical
CONTINENTAL
Warm summer
Cool summer
Subarctic
POLAR
Tundra
Ice sheet
HIGH ELEVATIONS
Highlands
Uplands

Fundamentals of World Regional Geography

Radford University
GEOG 102: Africa & Asia

Joseph J. Hobbs

CENGAGE Learning®

Australia • Brazil • Japan • Korea • Mexico • Singapore • Spain • United Kingdom • United States

CENGAGE
Learning®

Fundamentals of World Regional Geography: Radford University, GEOG 102: Africa & Asia

Senior Project Development Manager:
Linda deStefano

Market Development Manager:
Heather Kramer

Senior Production/Manufacturing Manager:
Donna M. Brown

Production Editorial Manager:
Kim Fry

Sr. Rights Acquisition Account Manager:
Todd Osborne

Printed in the United States of America

Fundamentals of World Regional Geography, Third Edition
Joseph J. Hobbs

ISBN-13: 978-1-285-55218-7

ISBN-10: 1-285-55218-0

Cengage Learning
5191 Natorp Boulevard
Mason, Ohio 45040
USA

Cengage Learning is a leading provider of customized learning solutions with office locations around the globe, including Singapore, the United Kingdom, Australia, Mexico, Brazil, and Japan. Locate your local office at:
international.cengage.com/region.

Cengage Learning products are represented in Canada by Nelson Education, Ltd.
For your lifelong learning solutions, visit **www.cengage.com/custom.**
Visit our corporate website at **www.cengage.com.**

Brief Contents

Maps

Preface

To appreciate how the world works today, you will benefit from a solid grounding in the environmental, cultural, historic, economic, and geopolitical contexts of the world's regions and nations. *Fundamentals of World Regional Geography* establishes that foundation for you and offers you an opportunity to explore in more detail the events, issues, and landscapes of your world.

My aim in writing the book is to make the world feel *accessible* to you. I hope that no place or issue will seem too far away or strange. I encourage you to think and talk more about what you read here, including through social media. I've started a Twitter site for us all to communicate with one another about what's happening in the world: follow me on Twitter: *@worldprof*

Chapters 1 through 3 provide the basic concepts, tools, and vocabulary for world regional geography. In the first chapter, geography's uniquely spatial approach to the world is introduced, along with some of the discipline's milestone concepts and its considerable career possibilities—especially those growing from the "geospatial revolution." The second chapter covers the essential characteristics of the world's physical processes and how some of them have been altered by human activity. Climate change and the treaties to control it have a prominent role in that chapter. Chapter 3 traces the modification of landscapes by human actions, describes trends and projections of population growth, and considers efforts to slow destructive trends in resource use.

Then come eight chapters exploring the world's regions through a consistent, thematic approach focusing in turn on five elements: "Area and Population," "Physical Geography and Human Adaptations," "Cultural and Historical Geographies," "Economic Geography," and "Geopolitical Issues." The final section of each chapter, entitled "Regional Issues and Landscapes," contains a selection of short studies of critical problems in global affairs—for example, the Palestinian-Israeli conflict—and exemplary or important problems in human or physical geography, such as ethnicity and petroleum in Nigeria, and the sources and impacts of deforestation in the Amazon Basin. The book is built for a one- or two-semester course. Instructors and students have the freedom to use some or all of content in a given chapter. If time is limited, you can concentrate on the five thematic elements of the chapter and limit the case studies selected from the Regional Issues and Landscapes section. It does not matter what order you read the book in; after the three introductory chapters, no chapter presumes that you have read any other chapter.

New to the Third Edition

Both longtime and first-time users of *Fundamentals of World Regional Geography* should be pleased with this text. I have tried to make the writing more *accessible* than ever. I use first person fairly often so that students can recognize the geographer behind the writing.

- I have written some new *Perspectives from the Field* essays, describing a meeting with the leader of a "Cargo Cult" in Vanuatu and observing landscape changes in Southeast Asia over a period of decades, for example.
- The visual layout is appealing. Cartographer Andrew Dolan builds on his reputation as the best cartographer of the world regional geography texts. For this edition's world maps he uses the Winkel-Trippel projection for the first time, to excellent effect. Also for the first time, I have dipped into my extensive collection of digital photographs. Since the second edition, I have researched, travelled, and photographed across Oceania, South and East Asia, and Latin America.
- There is more *geospatial* content throughout, with extensive use of remotely-sensed imagery and practical applications of GIS.
- For the first time I will correspond regularly with students and professors, especially to track issues related to the book. With the twitter handle *@worldprof*, I will share events and insights with readers.
- The book has, of course, been thoroughly updated. Since the second edition was published, the world has experienced the Great Recession. The economic fallout is a common thread throughout the book. The continued ascendance of China is another pervasive theme.
- I have made it easier to understand and use the important "margin cross-reference" feature which stitches common strands together all across the world.
- A new feature called "Try It" encourages students to use their own critical thinking and problem-solving skills to address a wide range of issues. The student may do these individually or in group exercises assigned by the instructor.
- More attention has been given to solutions, jumping from the principles of sustainable development to specific applications, especially in the developing world.
- I work purposefully to be objective, always trying to present pros and cons of controversial issues (e.g., of large dams) and multiple perspectives of complex political problems (e.g., of the Arab/Israeli conflict).
- Major metropolitan area populations are in tables rather than in the text, where they would interrupt the flow of prose.
- The Arctic Ocean appears to be opening up to trade and minerals exploration as seasonal and permanent sea ice coverage shrinks. In several chapters, we follow the race for Arctic routes and resources.
- I continue using the pattern of five geographic themes for each region: Area and Population; Physical Geography and Human Adaptations; Cultural and Historical Geographies; Economic Geography; and Geopolitical Issues. The book's signature emphasis on geopolitics is as strong as ever. This

edition has an enhanced focus on Economic Geography, introducing students to such blocs as the "PIGS" of Europe and the emerging market "BRICS." Conventional and alternative energy sources receive considerable attention in each region.

Chapter-Specific Changes

- Chapter 1, "Objectives and Tools of World Regional Geography," introduces "the Geospatial Revolution," with discussion of GIS and Remote Sensing. Students are encouraged to use Google Earth habitually. The AAG's exciting forecasts about job growth in Geography are shared.
- Chapter 2, "Physical Processes and World Regions," introduces the concept of the Anthropocene and includes a useful new map on the human "footprint." The term *global warming* is largely replaced by the more nuanced and variable *climate change*. Measures for adaptation to and mitigation of climate change are reviewed. There is more information on marine resources. I have removed the detailed information on weather fronts; profs can introduce these with other materials if they wish.
- Chapter 3, "Human Processes and World Regions," digs more deeply into the paradoxes of population, including how "underpopulation" as well as overpopulation can threaten economic well-being.
- Chapter 4 on Europe helps students make sense of the "Euromess:" the continent's sovereign debt crisis, problems with the Euro, and threats to the integrity of the European Union itself. Europe's growing immigration problems, especially in the south, are examined in detail. How can Europe restrict movement while widening Schengenland?
- Chapter 5, "Russia and the Near Abroad," sees Russia deepening its dependence on petroleum while Vladimir Putin consolidates his hold on power. The two are closely tied in what is described as "Putinomics." Russia continues to use oil and natural gas as bludgeons in foreign policy, through a geographic cat's cradle of pipelines.
- Chapter 6 is about the Middle East and North Africa. The major innovation here is a thorough analysis of the Arab Spring, including both the broad underlying themes and fallout, and an individual country survey showing how the revolution played out differently. We examine the near-death and resurrection of Iraq's southern marshes. Iran's prospective nuclear weapons development and the potential for preemptive military action are described. By popular demand, Afghanistan has been moved out of the Middle East and into Asia.
- Chapter Seven, the book's longest (because it covers the most populous region), is now "South and East Asia" rather than "Monsoon Asia." Villagers still outnumber urbanites here, but in anticipation of a rapid change in that equation there is attention to Asia's megacities. I discuss the phenomenon of "language death" due partly to computerized communication, and also the potential of computers to save endangered languages.
- Chapter 8 takes us across Oceania and Antarctica. The "Lucky Country" of Australia continues on it lucky streak on the back of ever-growing commodities demand in China. The Alliance of Small Island States is more alarmed than ever over rising sea levels.
- Chapter 9 on Sub-Saharan Africa portrays the continent at its most hopeful state is decades. Conflicts have subsided. Ironically, some of the greatest threats come from having too much wealth in oil and other natural resources: this is the so-called "resource curse." Technologies are growing quickly in Africa. Mobile phones are providing a shortcut to the information superhighway, and African software developers have contributed an important crowdsourcing tool, Ushahidi.
- Chapter 10 on Latin America describes the unspeakable tragedy of Haiti's 2010 earthquake and the world's other seismically prone cities like Port-au-Prince that are "rubbles in waiting." There are new discussions of religious movements, including those of Santeria and Maria Lionza. Brazil's impressive recent economic growth, based in part on newly discovered offshore oil reserves, is described.
- Chapter 11's disccusion of United States and Canada depicts two developed countries that still have their doors open to immigrants. In the United States, we had a 2010 census data form for 63 distinct ethnic/racial groups. The recent booms and busts of the Great Recession are depicted. We look at the phenomenon of "birth tourism" and the slowdown in illegal immigration due to economic recession. We see the Arid West becoming more arid, with some authorities saying this is the old and the new normal. Meteorological natural hazards of many kinds seem to be on the rise. Canada's tar sands and natural gas released by "fracking" have given North America more energy independence than they have enjoyed in decades. The *Deepwater Horizon* accident and its impacts on the Gulf are here. Outsourcing has begun to give way to insourcing (onshoring) and new life in American manufacturing.

Acknowledgments

I am grateful to everyone who encouraged and helped me with the book, especially Cindy, Katie, Lily, and Mom. Andy Dolan is the best cartographer there is. Andy wrote much of glossary and, as always, provided a lot of guidance on what he thought should be added to and removed from the text. The Cengage Learning team was headed up by Acquisitions Editor Aileen Berg and included Development Editor Liana Sarkisian, Project Manager Teresa Christie of MPS North America, Editorial Assistant Margaux Cameron, Media Editor Alexandria Brady, Marketing Manager Jack B. Cooney, Marketing Coordinator Lianne Ramsauer, Marketing Communications Manager Darlene Macanan, Rights Acquisitions Specialist Roberta Broyer, Photo Researcher Jeremy Glover of Bill Smith Group, Text Permissions Researcher Sue C. Howard, and Art Director Pam Galbreath.

Ancillaries

This text is accompanied by a number of ancillary publications to assist instructors and enhance student learning.

Instructor Resources

PowerLecture with JoinIn and ExamView

This one-stop digital library and presentation tool includes preassembled Microsoft® PowerPoint® lecture slides. In addition to a full Instructor's Manual and Test Bank, PowerLecture also includes an image library, animations, videos, blank maps, Google Earth activities, JoinIn™ Student Response System, and ExamView® testing software with all the test items from the printed Test Bank in electronic format, enabling you to create customized tests in print or online.

Online Instructor's Manual and Test Bank

By Chad Garick of Jones County Junior College. Streamline and maximize the effectiveness of your course preparation using such resources as *chapter summaries and outlines, learning objectives, lecture suggestions, key terms, and answers to in-text questions*. This time-saving resource also includes a Test Bank that offers over 70 questions. Available for download on the instructor's companion Web site.

ExamView Computerized Testing Software

Featuring automatic grading, ExamView® allows you to create, deliver, and customize tests and study guides (both print and online) in minutes.

WebTutor Toolbox™

WebTutor™ Toolbox for WebCT™ or Blackboard® provides access to all the content of this text's rich Book Companion Web site from within your course management system. Robust communication tools—such as course calendar, asynchronous discussion, real-time chat, a whiteboard, and an integrated e-mail system—make it easy for your students to stay connected to the course.

Student Resources

Geography CourseMate

The more you study, the better the results. Make the most of your study time by accessing everything you need to succeed in one place. Geography CourseMate includes:

- An interactive eBook with highlighting, note taking, and an interactive glossary
- Interactive learning tools, including:
  - Quizzes
  - Flashcards
  - Videos
  - Animations
  - ... and more!

About the Author

Katie Hobbs

Joe Hobbs received his B.A. at the University of California Santa Cruz and his M.A. and Ph.D. at the University of Texas Austin. He is a professor and chairman of the Department of Geography at the University of Missouri. He is a geographer of the Middle East with many years of field research on Bedouin peoples and natural environments in the deserts of Egypt. Joe's interest in the region grew from his boyhood in Saudi Arabia. His profession in Geography has roots in all the travels and time abroad with his mom and dad. His research in Egypt has been supported by grants from Fulbright, the American Council of Learned Societies, the American Research Center in Egypt, and the National Geographic Society Committee for Research and Exploration. In the 1990s, he served as the team leader of the Bedouin Support Program, a component of the St. Katherine National Park project in Egypt's Sinai Peninsula. In 2007, he led an effort to establish a national plan for environmental management in the United Arab Emirates. From 2009–2012 he worked with a team funded by the Norwegian Research Council, studying the interactions between nomadic pastoralists and acacia trees in Egypt and the Sudan. His current research interests include indigenous peoples' participation in national parks and other protected areas, and educational and diplomatic relations between Vietnam and the United States.

Joe is the author of the books *Bedouin Life in the Egyptian Wilderness* and *Mount Sinai* (both University of Texas Press), co-author of *The Birds of Egypt* (Oxford University Press), and co-editor of *Dangerous Harvest: Drug Plants and the Transformation of Indigenous Landscapes* (Oxford University Press).

Joe has taught graduate and undergraduate courses in world regional geography, environmental geography, the geography of the Middle East, the geography of caves, and the geography of global current events, the geographies of drugs and terrorism, and a field course on the ancient Maya geography of Belize. He has received the University of Missouri's highest teaching award, the Kemper Fellowship, as well as MU awards for leadership in international education. He has led adventure tours to remote areas in Latin America, Africa, the Indian Ocean, Asia, Europe, and the High Arctic. Joe lives in Missouri with his wife Cindy, daughters Katie and Lily, and lots of animals.

> ***"If a man takes no thought about what is distant, he will find sorrow near at home."***
>
> —Confucius

The Earth at night. In general, the prosperous countries are lit up, while poor countries are darker. The differences between prosperous and poor countries are discussed often in the book.

C. Mayhew & R. Simmon (NASA/GSFC), NOAA/ NGDC, DMSP Digital Archive

Joe Hobbs

1 Objectives and Tools of World Regional Geography

Before you begin reading, there is an important feature of the book that you need to know about: *cross-referencing*. The book is written with global interconnections in mind. Globalization is understandable as a concept, but how exactly does it work? The page and figure numbers in the book's margins serve to tie the diverse strands of global issues together. For example, when you read in Chapter 4 that Europeans generally dislike genetically modified foods, page numbers in the margin lead you to discussions of other countries' policies on GM foods and how genetic modification can affect ecosystems. Likewise, the discussion of how genetic modification of food crops can affect ecosystems has a page number in the margin, leading you back to the discussion of Europeans and GM foods. As you read about China's economic growth and appetite for raw materials, margin numbers will take you to places all around the world where these forces come into play. I hope you will use this feature often and learn a lot from it.

chapter outline

chapter objectives

This chapter should enable you to

- Appreciate the richness of geography
- Identify the six essential elements and the five themes of geography
- Learn some key concepts in geography
- Appreciate the book's overall objectives
- Learn the basic language of maps
- Understand the "geospatial revolution," geographic information systems (GIS), and remote sensing
- See how geographic knowledge is put to work in the job market

1.1 Welcome to World Regional Geography

About every five years, the National Geographic Society does a survey to see how much Americans between the ages of 18 and 24 know about the world. The findings can be discouraging, especially when it comes to the big issues; take knowledge of Afghanistan, for example. From a base inside Afghanistan, al-Qa'ida planned its attacks against the United States on September 11, 2001. The United States responded by opening a war in Afghanistan that would become the longest in American history. This conflict often dominated the news, and almost every American knew or had seen someone deployed to Afghanistan. But where is Afghanistan? In National Geographic's most recent survey, 90 percent of the Americans polled could not locate Afghanistan on a map.

So what? Does it matter if you can't find Afghanistan on a map? Does it matter if you can't even find the United States on a map? Years ago, geography earned a reputation for forcing students to memorize lists of places and facts (•**Figure 1.1**). And the truth is, by themselves, those pieces of knowledge probably mean little. But geography is all about context and connections. Understanding *where* things are makes it much easier to appreciate and answer the *who*, *what*, *when*, *why*, and *how* questions in life, at every scale—from your daily activities to world affairs. Geography always starts with the "where" question, but it is far more interesting and important than its old reputation for memorizing places suggests. Helping you to understand contexts and relationships, geography can help you make better informed judgments and decisions. See the Try It feature on page 3.

Geographic knowledge has the power to transform our lives and contribute to the welfare of our communities and our countries. By the end of this chapter, you will know what geography is, recognize the benefits you might gain from learning world regional geography, understand the organization and objectives of this book, and learn some of the key concepts and tools of geography.

What Is Geography?

Geography, a term first used by the Greek scholar Eratosthenes in the third century B.C.E.,[1] literally means "description of the Earth" but is probably best characterized as "the study of the Earth as the home of humankind." Focusing on interactions between people and the environments in which we live, the modern academic discipline of geography has its roots in the Greek and Roman civilizations and the Scientific Revolution in Europe.

• **Figure 1.1** Geography used to be associated with memorizing mind-numbing facts. Not any more!

Geography has an important niche in modern science. Its role is well summarized by the National Geographic Society's **"six essential elements of geography."** They are as follows:

1. *The World in Spatial Terms.* Geography studies the relationships among people, places, and environments by mapping information about them into a spatial context (*spatial* means "of or relating to space").
2. *Places and Regions.* The identities and lives of individuals and peoples are rooted in particular places and in human constructs called regions.
3. *Physical Systems.* Physical processes shape the Earth's surface and interact with plant and animal life to create, sustain, and modify ecosystems.
4. *Human Systems.* People are central to geography; human activities, settlements, and structures help shape the Earth's surface, and humans compete for control of the Earth's surface.
5. *Environment and Society.* The physical environment is influenced by the ways in which human societies value and use the Earth's physical features and processes.
6. *Uses of Geography.* Knowledge of geography enables people to develop an understanding of the relationships among people, places, and environments over time—that is, of the Earth as it was, is, and might be.

Think about U.S. Involvement in Afghanistan

Try It

Why has it been difficult for the United States to be involved in Afghanistan? Suggest two or three reasons.

In trying to answer this important question (important partly because of the sacrifices, costs, and risks that came with U.S. involvement), it may be difficult for you *not* to use a map. You will probably consider Afghanistan in its regional context, looking at its neighbors and how it is positioned relative to major oceans and mountain ranges. You may think about some of the connections and interactions between places, cultures, beliefs, natural resources, politics, and other factors related to Afghanistan.

If it is a tough question to answer now, be patient; this book will help you answer it (on pages 240–244)—and many other questions that ask you to think critically about the world today.

After you have considered this question, you may have more empathy for a man who had to make tough choices about U.S. engagement in Afghanistan. Here is what the former U.S. Secretary of Defense Robert Gates told cadets at the U.S. Military Academy at West Point in 2011:

> Any future defense secretary who advises the president to again send a big American land army into Asia or into the Middle East or Africa should "have his head examined," as General [Douglas] MacArthur so delicately put it . . . Just think about the range of security challenges we face right now beyond Iraq and Afghanistan: terrorism and terrorists in search of weapons of mass destruction, Iran, North Korea, military modernization programs in Russia and China, failed and failing states, revolution in the Middle East, cyberpiracy . . . , natural and man-made disasters, and more. And I must tell you, when it comes to predicting the nature and location of our next military engagements, since Vietnam, our record has been perfect. We have never once gotten it right.

You may have to read those last two sentences more than once to appreciate Gates' purposeful irony. He might have wished that American presidents were advised by geographers! Geography is all about "getting it right" when it comes to understanding how the world works. Geographic knowledge of the *where, who, what, when, why,* and *how* can help guide informed decision making at all scales, from how you might get from point A to point B in town, to whether the United States should commit troops to a ground war in Asia.

Source: Geography for Life: National Geography Standards 1994, National Geographic Research and Exploration, Washington, DC, pp. 34–35.

Another summary of what geography is all about is known as the **Five Themes of Geography.** The National Council for Geographic Education and the Association of American Geographers developed this list. Because of its effective simplicity, some geographers prefer using it:

1. Location
2. Place
3. Human–Environment Interaction
4. Movement
5. Region

Although brief, these six elements and five themes cover a lot of ground. Geography is probably the most all-encompassing of the social sciences (a point of pride for geographers). Broadly, the discipline has two major branches, **physical geography** and **human geography,** each of which has roots and relationships with other disciplines in the social and physical sciences (•**Figure 1.2**). Geographers often bridge the social and natural sciences in their research, publication, and teaching (another point of pride for them). As you can see in the center of Figure 1.2, where all the components of the discipline converge, geography is almost always concerned with the theme of **human–environment interaction.** This concern has put geographers at the cutting edge of science and policy in the twenty-first century, because so many of the Earth's most pressing problems—climate change, population growth, and hunger, for example—involve the coupling of human and environmental systems.

Geographers' interests in human–environment interaction, and especially in the ways in which people are changing the face of the Earth, go way back. The great German geographer Alexander von Humboldt (1769–1859) began geography's modern era in a series of classic studies on this theme. From field observations in Venezuela, he concluded, "Felling the trees which cover the sides of the mountains provokes in every climate

Try It The Geography of Anyplace

Try using the "five themes of geography" to characterize any place. Staying with the U.S.–Afghanistan connection for now, here is an example to work from, using Ground Zero in Manhattan.

Geographic Characteristics of Ground Zero

Location: Lower Manhattan, New York City (with an exact location of north latitude: 40 degrees, 42 minutes, 43 seconds, and west longitude: 74 degrees 00 minutes, 49 seconds (later in the chapter, we will look at latitude and longitude).

Place: Formerly, office buildings and firms at the heart of one of the world's great financial centers (a reason it was targeted for destruction); now, a place of historical significance and collective grief for people of the United States.

Human–Environment Interaction: Lower Manhattan occupies low-lying ground that once was marshy swampland. Construction of the twin towers of the World Trade Center, as well as the buildings erected after the 9/11 attacks, required special foundations to keep the Hudson River's water from pouring in.

Movement: Before 9/11, the daily comings and goings of office workers in the World Trade Center; on 9/11, the diversion of airplanes to target the buildings; after 9/11, the flow of mourners, tourists, and construction crews to the site.

Region: Situated in region of the United States known as the Northeast, in a humid continental climate region (in the next chapter we will look at such physical regions).

You can use these themes to appreciate any place geographically, from the Great Pyramids of Egypt to where you are now. Try it.

two disasters for future generations: a want of fuel and a scarcity of water."[2] A century and a half later, we are von Humboldt's future generations. Look at some of the most pressing global environmental issues that concern us today: they include deforestation and shortages of fresh water.

In Humboldt's wake, other geographers in Europe and the United States wrote about environmental changes due to deforestation and the expansion of agriculture and industry. The American geographer Carl Sauer (1889–1975) wrote, "We have accustomed ourselves to think of ever expanding productive capacity, of ever fresh spaces of the world to be filled with people, of ever new discoveries of kinds and sources of raw materials, of continuous technical progress operating indefinitely to solve problems of supply. Yet our modern expansion has been affected in large measure at the cost of an actual and permanent impoverishment of the world."[3] These words have a modern ring to them, but Sauer, a geographer at the University of California–Berkeley, wrote them in 1938. Sauer is credited with founding the **landscape perspective** in American geography, based on the method of studying the transformation through time of a **natural landscape** to a **cultural landscape.** Essentially, Sauer challenged us 16
to think of what the world would look like without people and then understand what people have done to reshape the world through time. Geographers are interested in how the forces of nature and culture have been at work in the creation of the **landscape**—the collection of physical and human geographic features on the Earth's surface—and in particular the roles that human ideas, activities, and cultures play in modifying the landscape. **Culture**—the system of values, beliefs, and attitudes that shapes and influences perception and behavior[4]—underlies many of our decisions about how to use and modify the landscape. That is why geographers are so concerned with cultural features such as ethnicity, language, and religion, and why you will learn much about these issues in this book.

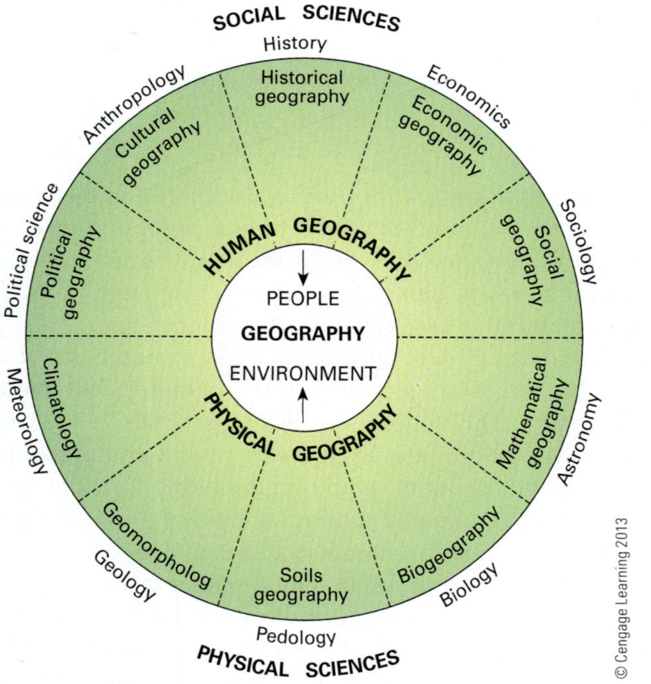

• **Figure 1.2** Selected subfields of geography. These are the main subject areas in human geography and physical geography and their links with the most closely related disciplines in the social and natural sciences.

The World Regional Approach to Geography

The world regional approach to geography ranges across the human and physical subfields of geography, synthesizing, simplifying, and characterizing the human experiences of Earth as home. It is impossible to deal with something as large and diverse as our planet without an organizing framework. World regional geography simplifies the task by dividing the world into **regions** (•**Figure 1.3** and •**Table 1.1**). These subdivisions of space are human constructs, not "facts on the

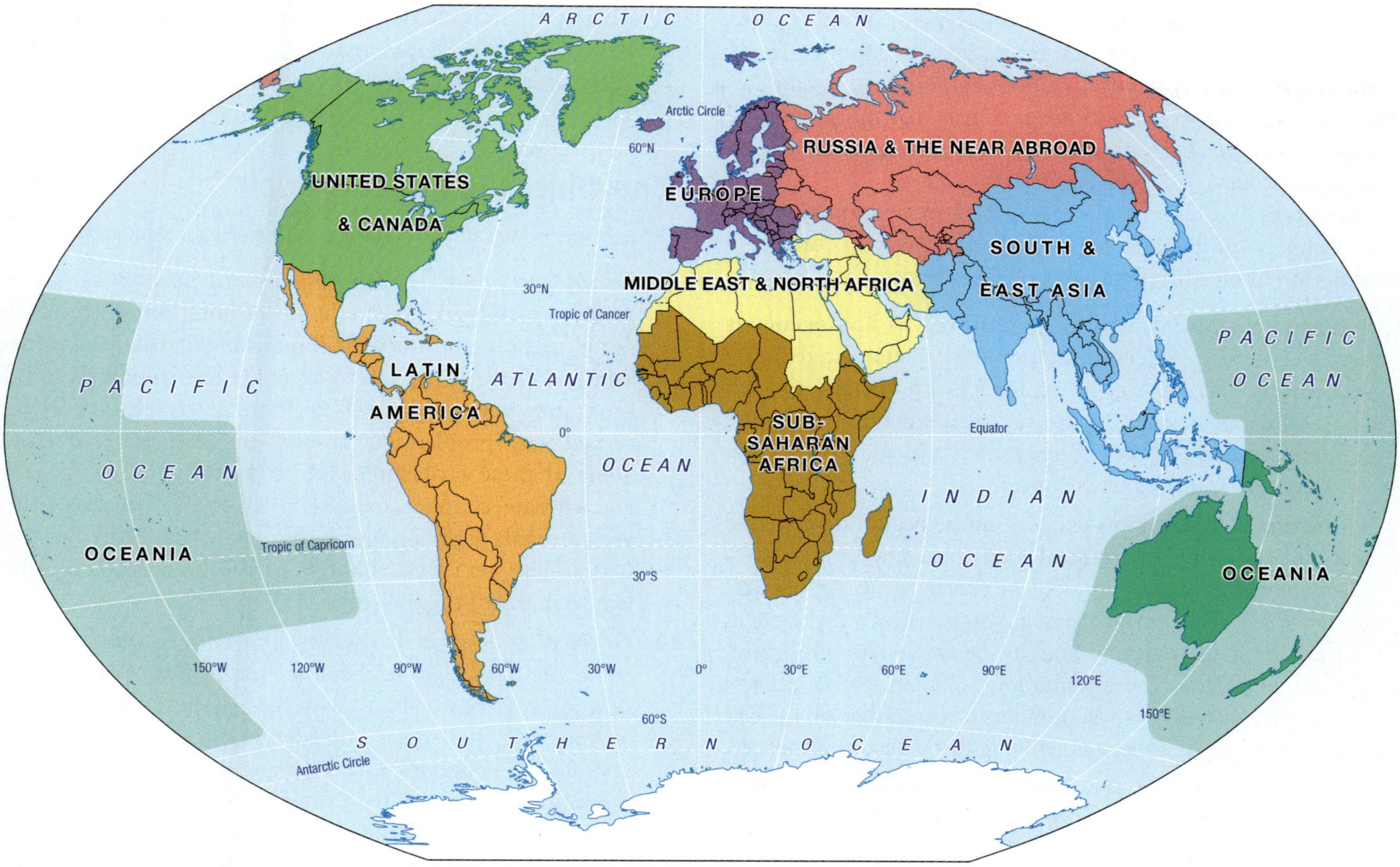

• **Figure 1.3** World regions as identified and used in this book.

Table 1.1 The Major World Regions: Basic Data

Region	Area (thousands; sq mi)	Area (thousands; sq km)	Estimated Population (millions)	Estimated Population Density (sq mi)	Estimated Population Density (sq km)	Annual Rate of Natural Increase (%)	Human Development Index	Urban Population (%)	GNI PPP Per Capita ($US)
Europe	1,959.3	5,072.0	532.3	272	105	0.0	0.873	72	31,070
Russia and the Near Abroad	8,533.2	22,100.8	282.0	33	13	0.2	0.726	64	12,300
Middle East and North Africa	5,416.1	14,027.8	503.4	93	36	1.8	0.650	62	9,460
Monsoon Asia	8,265.3	21,407.1	3,829.2	463	179	1.0	0.621	42	6,080
Oceania	3,306.8	8,564.5	37.3	11	4	1.1	0.822	67	27,440
Sub-Saharan Africa	8,655.2	22,417.2	846.5	98	38	2.5	0.423	32	2,080
Latin America	7,946.2	20,580.7	595.8	75	29	1.2	0.724	79	10,560
North America	8,403.8	21,765.8	346.3	41	16	0.5	0.909	79	44,800
World Regional Totals	**52,485.9**	**135,938.5**	**6,972.8**	**133**	**51**	**1.2**	**0.647**	**51**	**10,240**
United States (for comparison)	3,717.8	9,629.1	311.7	84	32	0.5	0.910	79	45,640

Sources: World Population Data Sheet, Population Reference Bureau, 2011; U.N. Human Development Report, United Nations, 2010; World Factbook, CIA, 2011.

ground." People create and draw boundaries around regions that share relatively similar characteristics. A region is simply a convenience and a generalization, helping us become acquainted with the world and preparing us for more detailed insights. This book recognizes eight world regions; others have more or less.

Three types of regions are recognized by geographers and used in this book. Each is helpful in its own way in conveying information about different parts of the world:

- A **formal region** (also called a **uniform** or **homogeneous region**) is one in which all the population shares a defining trait or set of traits. A good example is a political unit such as a county or a state, where the regional boundaries are defined on a map. Figure 4.1 on page 66 is a formal region map showing the countries of Europe.
- A **functional region** (also called a **nodal region**) is a spatial unit characterized by a central focus on some kind activity (often an economic activity). At the center of a functional region, the activity is most intense, whereas toward the edges of the region the defining activity becomes less important. A good example is the distribution area for a metropolitan newspaper, with the highest numbers of subscribers in the city and diminishing numbers at growing distances from the city.
- A **vernacular region** (or **perceptual region**) is a region that exists in the minds of many people. This region may play an important role in cultural identity, but does not necessarily have official or clear-cut borders. Good examples are the South, the Bible Belt, and the Rust Belt in the United States (•**Figure 1.4**). These regional terms have economic and cultural connotations, but 10 people might have 10 different definitions of the qualities and boundaries of these regions. Vernacular or perceptual regions, created by individuals and cultures, represent the regional identities that help us organize, simplify, and make sense of the world around us.

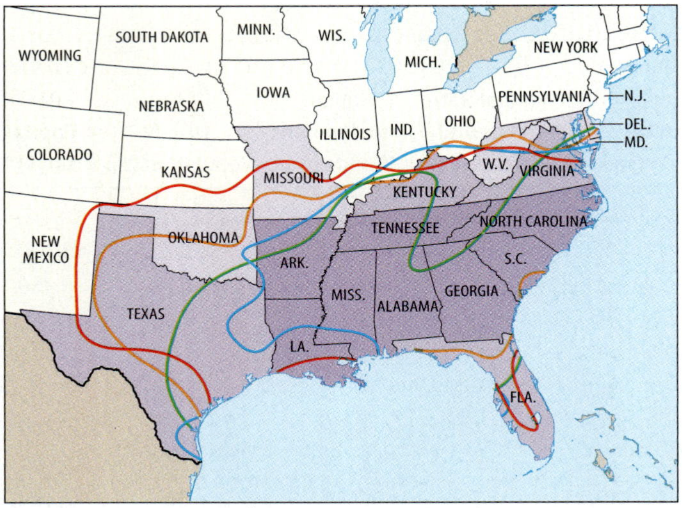

• **Figure 1.4** Definitions of a vernacular region, the American South. Purple shading represents three state-based delineations; colored lines delimit various religious, linguistic, and cultural "Souths." These are just a few of the many different interpretations of the region.

The Objectives of This Book

This book is written to help you achieve four objectives:

1. *To understand Earth's problems and potential solutions to these problems.* As the home of humankind, the Earth needs to be maintained. It has problems that need attention and that can be solved or made less threatening. Like geography in general, world regional geography is very concerned with problems in human–environment interaction. Some of these problems, such as overpopulation, poverty, and climate change, are global in scope, whereas others are national, regional, and local, or are manifested at these scales. The next two chapters of the book will introduce you to many of these issues.
2. *To develop a habit of synthesizing information to understand the world.* To understand the problems of Afghanistan (or any country) and why these issues may be relevant to your life, you must consider many factors: resources, population, economic development, ethnicity, history, and geopolitical interests, for example. Is that too much information for you to take in? Not if you have the right tool for filtering and synthesizing the information. World regional geography is that tool. Again, geography is all about context and connections. Using geography's holistic and integrative approach in a regional framework, you will look at relationships between places and people. You will be using information, techniques, and perspectives from both the natural sciences and the social sciences. You will become familiar with issues in political science, history, economics, anthropology, sociology, geology, atmospheric science, and other areas. In pulling these issues and perspectives together and finding the links among them, you will be doing geography.

   This approach will be rewarding for you in both tangible and intangible ways. Your overall university experience should become richer as you become better able to link the various courses you are taking. In the future, you will be more likely to interact with the world with new wisdom that can be rewarding both professionally and personally. More complete knowledge of the world—good geography—is also good business. In the competitive environment of the global economy, better understanding of cultures and environments throughout the world helps boost the "bottom line." You may be surprised how much your geographic knowledge will help you in your career, whatever it turns out to be.
3. *To understand current events.* This book is carefully written to set the stage of world events for you. With this book and your professor's guidance, you should become able to read or view news with a much better understanding of the issues underlying world events. Incidents like violence in

Joe Hobbs

• **Figure 1.5** Where is this place?[5] What clues on the landscape or in the man's appearance might tell you where you are?

the Middle East, earthquakes and tsunamis in the western Pacific, and disease epidemics are not random, unpredictable events. They are rooted in consistent, recognizable problems that have geographic dimensions. You should find it satisfying to feel so well informed about a problem that "suddenly" appears in the news.

4. *To develop the ability to interpret places and "read" landscapes.* As you saw in the exercise on Ground Zero, in
16 doing geography you are concerned both with **space** (the exact placement of locations on the face of the Earth) and with **place** (the imprecise but important physical and cultural contexts of a location). Place is much more subjective than space because, like a vernacular region, it often is defined by the meanings of a particular location. For example, your perceptions of New York City may be very different from those of your friend and may be shaped by personal experience in the "Big Apple" or by photographs or movies you have seen. In this book, there is much discussion of the "sense of place" that individuals and groups have about locations and regions. Perception of place can have a very strong influence on how we make decisions and interact with other people. Perception of place can even have a strong impact on world events. For example, in Chapter 6 on the Middle East, you will see how Jewish and Muslim perceptions of sacred places located within a few meters of each other in Jerusalem play crucial roles in conflict and peacemaking in the Middle East and beyond.

As an example of how you can define and identify place, study the photograph in •Figure 1.5, and try to name the place—or at least the region—where this photograph was taken. What clues in the physical and human geographies of this place help you locate it? As you use this book and progress through your course, you should be increasingly skilled in looking at images of a place with insight into many of the features that make up its identity. These include the climate, vegetation, and landforms of the physical environment and the language, religion, history, and livelihoods of the people who live in that environment. The skill of interpreting place will also serve you well when you travel. For this place, see note 5 on page 19.

1.2 The Language of Maps

How skilled are you at reading maps? What about imagining or drawing them? (See the Insights feature on page 8.) Did you ever wish you were better at these skills? This section should help you—and it will be useful as you navigate this book.

We turn now to the most important tool that geographers use to explore and explain relationships on our planet: the map. As geographers study people, places, and environments, they usually (but not always) collect and depict information that can be mapped. In other words, they are interested in the **spatial** context of things. As noted in the first essential element of geography, *spatial* means "pertaining to space."

A **map** is a representation of various phenomena over all or a part of the Earth's surface, usually rendered on a flat surface such as paper or a computer monitor. The science of making maps is called **cartography.** There are two basic types of maps: reference maps and thematic maps. **Reference maps** are concerned mainly with depicting the locations of various features, both natural and human-made, on the Earth's surface (road atlas maps are a good example, as are the opening maps for each regional chapter, such as Europe in Figure 4.1). **Thematic maps** show the spatial distribution of one or more attributes across

INSIGHTS Mental Maps

When someone asks you, "Could you draw me a map of how to get there?" you might quickly draw some lines, write down some street names, talk about some familiar landmarks, and apologize for how crude your map is. Your map would probably end up looking very different from that of another person asked the same question. Our understanding of location is not completely objective. Each of us has a personal sense of space and place and associations with them.

A **mental map,** like a vernacular region, is a collection of personal geographic information that each of us uses to spatially organize the images and facts we have about places, both local and distant. We constantly draw upon that geographic information to make our way through daily life, and are always revising and updating that information as we succeed or fail on our way. Sometimes we use that information to create actual maps. These maps are not accurate, precise, or scientific, but they portray useful information and tell us much about the individuals and cultures that create them (• **Figure 1.A, B**).

Joe Hobbs
A

Joe Hobbs
B

• **Figure 1.A, B** The man in Figure 1.A is Saleh Ali, a Bedouin tribesman from the Eastern Desert of Egypt. During his 75 years of living, he has never used a map, but has drawn on a wealth of personal experience and oral traditions—his mental maps—to find his way across one of the world's most rugged natural environments, the Sahara. On this day, he used his "mental maps" to take the author to a valley called Wadi Abu Haadh, which, he said, was "a place of many vipers." You can see him probing the gravel under an acacia tree—using his mental map of that habitat—looking for one of these poisonous snakes. In Figure 1.B he has found one of the creatures. "Don't step there," he warned me. Can you see the head of this deadly animal?[6] See note 6 on page 19.

a given area. Thematic maps can be divided into two categories: quantitative and qualitative. Quantitative thematic maps show the spatial distribution of numerical information (such as population density or income levels, as in **Figure 3.14 on page 55**), whereas qualitative thematic maps display nonnumeric data (such as the distribution of climates or languages, as in **Figure 2.4 on page 24**.

As maps are an essential tool in the study of world regional geography, it is important that you know how to read them. The main elements of the "language of maps" are *scale, coordinate systems, projections,* and *symbolization.*

Scale

A map is a reducer; it shrinks an area to the manageable size of a chart, piece of paper, or computer monitor. The amount of reduction appears on the map's **scale**, which shows the actual distance on Earth as represented by a given linear unit on the map. A common way to depict scale is with a fraction or ratio, such as 1:10,000 or 1:10,000,000. In the fraction, one linear unit on the map (for example, 1 inch or 1 centimeter) represents 10,000 or 10,000,000 such real-world units on the ground. A **large-scale map** is one with a relatively large representative fraction (for example, 1:10,000 or even 1:100) that portrays a relatively small area in more detail. A **small-scale map** has a relatively small representative fraction (such as 1:1,000,000 or 1:10,000,000) that portrays a relatively large area in more generalized terms. Compare the two maps in •**Figure 1.6**. **Figure 1.6a** is a small-scale map showing San Francisco and surrounding parts of the Bay Area. **Figure 1.6b** is a large-scale map that zooms in on part of San Francisco. Remember this inverse relationship: A small-scale map shows a large area, and a large-scale map shows a small area.

Coordinate Systems

Maps cannot convey the subjective meanings associated with place, but they are very effective in conveying information about space and location. In this book, you will be concerned with two kinds of **location**: relative location and absolute location.

Relative location defines a place in relationship to other places and is information you can derive from many maps. It is one of the most basic reference tools of everyday life; you might say you live south of the city, just west of the shopping mall, or next door to a good friend. Relative location will become part of your basic geographic knowledge and your critical thinking about geography. You might look at **Figure 10.1. on page 351** to see, for example, how tantalizingly close the landlocked country of Bolivia is to the Pacific Ocean, and wonder, "Why on earth is Bolivia landlocked?!" Understanding the implications of relative location will be quite useful for you in following world affairs; keep your eye on Bolivia's efforts to recover in international courts the sea access it lost to Chile through warfare in 1879.

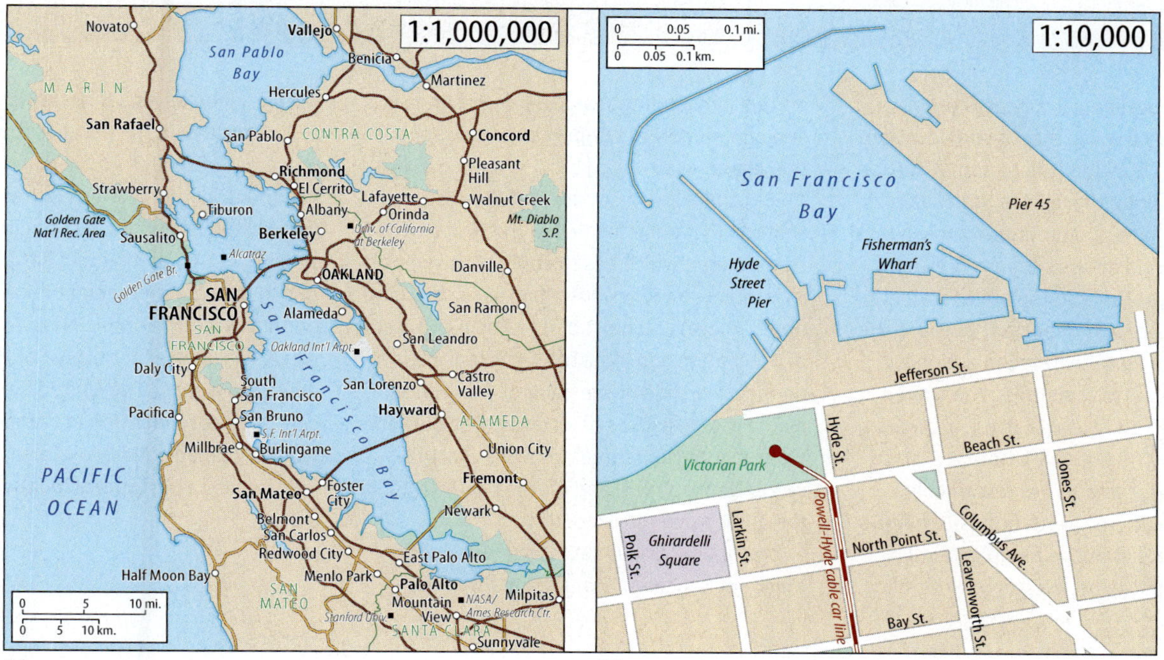

(a) (b)

• **Figure 1.6** **(a)** Small-scale and **(b)** large-scale maps of San Francisco and environs.

Absolute location, also known as **mathematical location,** is crucial for reference maps but is not always necessary for thematic maps. **Coordinate systems** are used to determine absolute location. These coordinate systems use grids consisting of horizontal and vertical lines covering the entire globe. The intersections of these lines create addresses in a global coordinate system, giving each location a specific, unique, and mathematical placement (as appears, for example, as a waypoint in the common GPS device).

The most common coordinate system uses **parallels** of **latitude** and **meridians** of **longitude.** The term *latitude* denotes position with respect to the equator and the poles (see •**Figure 1.7**). Latitude and longitude are measured in **degrees** (°), **minutes** (′), and **seconds** (″). Each degree of latitude, which is made up of 60 minutes, is about 69 miles (111 km) apart; these distances vary a little because Earth is a slightly flattened (oblate) sphere or ellipsoid. Each minute of latitude, which is made up of 60 seconds, is therefore roughly a mile apart. The **equator,** which circles the globe east and west midway between the poles, has a latitude of 0°. All other latitudinal lines are parallel to the equator and to each other, which explains why they are called *parallels*.

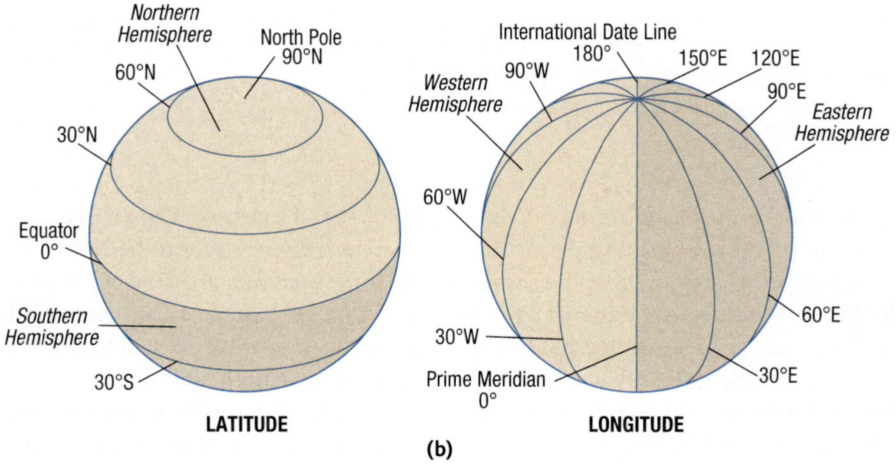

(a) (b)

• **Figure 1.7** **(a)** Earth's lines of latitude (parallels) in increments of 30 degrees, from the equator (0 degrees) to the North Pole (90 degrees north latitude). **(b)** Earth's lines of longitude (meridians) in increments of 30 degrees.

INSIGHTS Core Location and Peripheral Location

Among the geographical concepts used throughout this book is that of core location and peripheral location (the region of Europe is subdivided along these lines, for example).
65 Some locales have greater importance in local, regional, or world affairs because they have a central, or core, location relative to others. Other locales are less important because they are situated far from "where the action is." A comparison of two countries, the United Kingdom (U.K.) and New Zealand, provides a good example (• **Figure 1.C**). Both are island countries. Their climates are remarkably similar, although they are in opposite **hemispheres** and are about as far apart as two places on Earth can be. Westerly winds blow off the surrounding seas, bringing abundant rain and moderate temperatures throughout the year to both.

But there are important differences. The United Kingdom is located in the Northern Hemisphere, which has the bulk of the world's land (it is sometimes described as Earth's **land hemisphere**) and most of its principal centers of population and industry; New Zealand is on the other side of the equator, surrounded by the vast expanses of water in the Southern Hemisphere (sometimes known as Earth's **water hemisphere**) and off the beaten track of the globe's economic activity.

As Figure 1.C illustrates well, the United Kingdom is located near the center of the world's landmasses. Only a narrow channel separates it from the densely populated industrial areas of western continental Europe. Many major oceanic commercial routes converge on this western seaboard area of Europe. For centuries, the United Kingdom has played a major role in the economic and political development of northwestern Europe. New Zealand, meanwhile, has been a far outpost of that development and history. The United Kingdom has a central or **core location** in the modern framework of human activity on Earth, whereas New Zealand has a **peripheral location**.

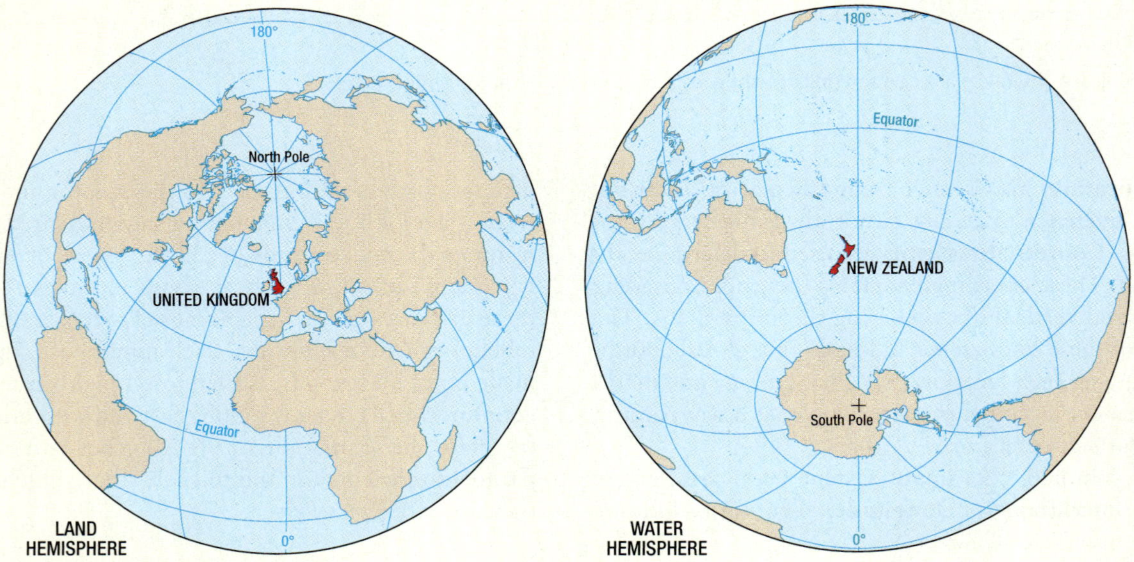

• **Figure 1.C** In the left map, note how the major landmasses are grouped around the margins of the Atlantic and Arctic Oceans. The British Isles and the northwestern coast of Europe lie in the center of the "land hemisphere," which constitutes 80 percent of the world's total land area and has about 90 percent of the world's population. In the map on the right, New Zealand lies near the center of the opposite hemisphere, or "water hemisphere," which has only 20 percent of the land and about 10 percent of the population.

Every point on a parallel has the same latitude (for example, places on the Equator in both South America and Africa are located at 0° latitude). Places north of the equator are in **north latitude.** Places south of the equator are in **south latitude.** The highest latitude a place can have is 90°N (the **North Pole**) or 90°S (the **South Pole**). Places located between the **Arctic Circle** at 65.56°N and the North Pole, and between the **Antarctic Circle** at 65.56°S and the South Pole, form the most commonly recognized boundaries of the **high latitudes**. Places located between the **Tropic of Cancer** and the **Tropic of Capricorn,** at 23.44°N and 23.44°S, respectively, are said to be in **low latitudes.** Places occupying an intermediate position with respect to the poles and the equator are said to be in the **middle latitudes.** Incidentally, there are no universally accepted definitions for the boundaries of the high, middle, and low latitudes. The northern half of the Earth between the equator and the North Pole is called the **Northern Hemisphere,** and the southern half between the equator and the South Pole is the **Southern Hemisphere.**

Meridians of longitude are straight lines connecting the poles (see Figure 1.7b). Every meridian runs due north–south. All the meridians converge at the poles and are farthest apart at the equator. Lines of longitude are not the same distance from one another across the globe, so their values vary. At the equator, the distance between lines of longitude is about 69 miles (111 km), whereas at the Arctic Circle it is only about 28 miles (45 km). The meridian most commonly used as a starting point is the one running through the Royal Astronomical Observatory in Greenwich (pronounced "Grenich"), England. Known as the Greenwich Meridian, or **prime meridian,** it has a longitude of 0°. Places east of the prime meridian are in **east longitude;** places west of it are in **west longitude.**

The meridian of 180°, exactly halfway around the world from the prime meridian, is the other dividing line between places east and west of Greenwich. All of the Earth's surface eastward from the prime meridian to 180° is in the **Eastern Hemisphere,** and all of the Earth's surface westward from the prime meridian to 180° is in the **Western Hemisphere.** This meridian of 180° has another purpose in addition to complementing the prime meridian. It serves as the **International Date Line,** where the beginning of one day and the end of another day meet. The Earth has 24 time zones, and there must be a line where the Earth's clock begins and ends. The line has a few zigzags in it for political and practical reasons (especially to fit a country or part of a country in a single time zone; see Figure 8.1, page 285). The date west of the line is one day ahead of the date east of the line. The person traveling west across the International Date Line gains a day (crossing from Monday to Tuesday, for example); and traveling east, loses a day (crossing from Monday to Sunday).

Latitude and Longitude — Try It

Here is a useful exercise to ensure that you understand how latitude, longitude, and absolute location work.

On the map in • **Figure 1.D**, the latitude of Madrid, Spain, is approximately 41 degrees north latitude, 4 degrees west longitude (41°N, 4°W).

What are the approximate latitude and longitude coordinates of Oslo, Norway? In which hemispheres (north/south; east/west) is Oslo located[7]?

Understanding absolute location is a simple and indispensible part of the language of maps.

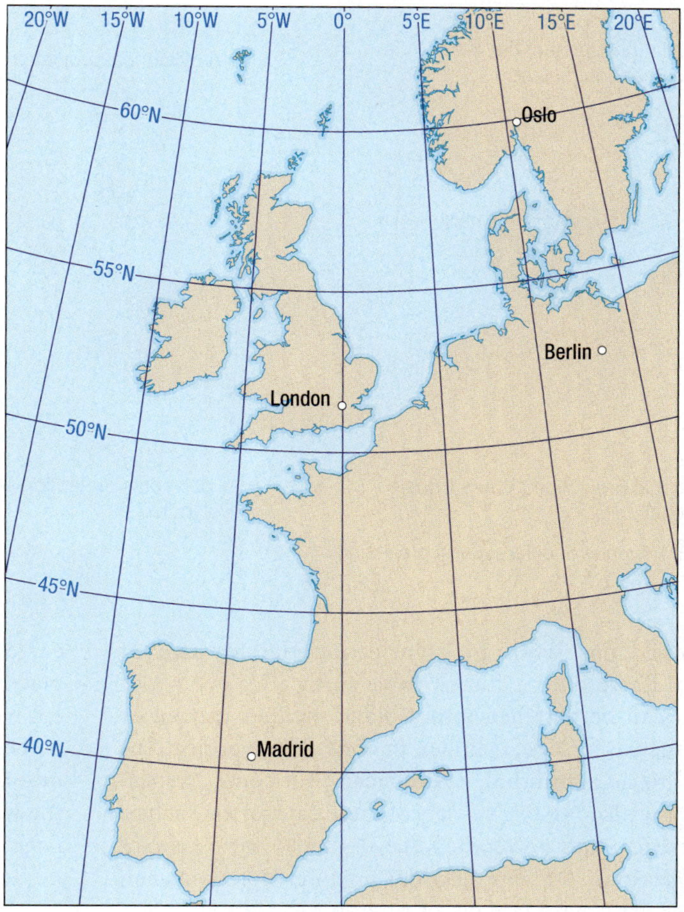

• **Figure 1.D** What are the approximate latitude and longitude coordinates of Oslo, Norway? The answer is in footnote 7 on page 19.

You now have the ability to create and interpret the absolute location of any spot on Earth. Try It—see the feature on this page.

Projections

Because the Earth is spherical, any map created on a flat surface will inevitably have some distortion. **Map projections** are mathematical attempts to minimize this distortion. There are thousands of projections, but there is no single "correct" or "perfect" projection for any particular map. All flat maps have varying amounts of distortion among the five basic properties of a globe: area, shape, scale, distance, and direction. On large-scale maps (such as a map of a city) the distortions are small enough that they may be disregarded for most purposes, but on maps of smaller scales (countries, continents, the entire world) distortion is an increasing problem as scale becomes smaller.

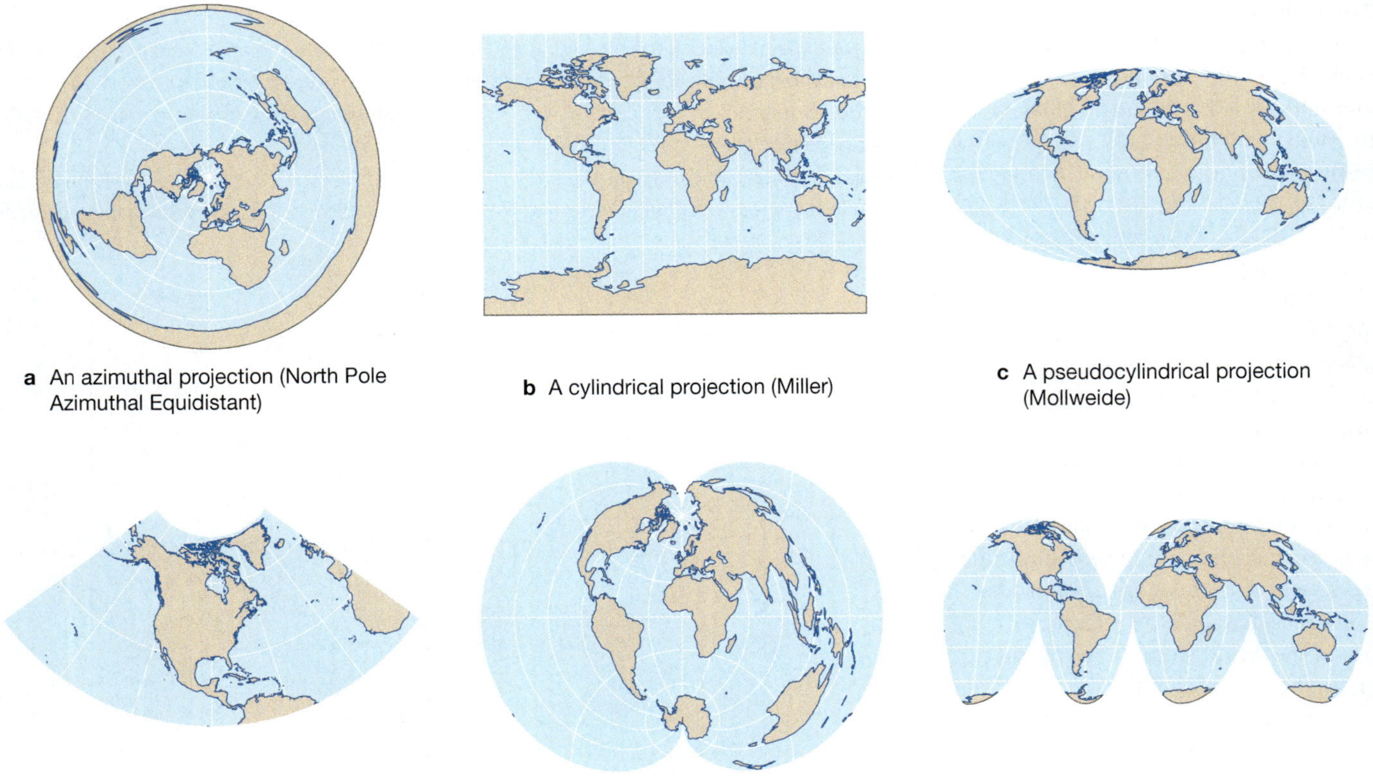

a An azimuthal projection (North Pole Azimuthal Equidistant)

b A cylindrical projection (Miller)

c A pseudocylindrical projection (Mollweide)

d A conic projection (Albers Equal-Area Conic)

e A polyconic projection (American Polyconic)

f An interrupted projection (Goode Homolosine)

• **Figure 1.8** Examples of different map projections.

Most projections work by transferring the features of the spherical Earth onto a "developable surface" (a geometric surface that can be flattened into a plane without tearing or stretching) such as a plane, cylinder, or cone. These projections are referred to as **azimuthal**, **cylindrical**, and **conic**, respectively (there are also notable subsets of these categories, such as pseudocylindrical and polyconic.) See •**Figure 1.8** for examples of these projections. A few projections are interrupted, instead of contiguous; the most famous of these is Goode's Homolosine, which has a "peeled orange" look.

Projections are also classified by which metric property they preserve (or distort the least). **Conformal projections**, which include the famous Mercator Projection discussed in the feature on page 13, preserves shapes well. **Equal-area projections**, as the name suggests, preserve area. But no equal-area projection can preserve shapes, and no conformal projection can preserve areas. **Equidistant projections** preserve distance from a specific point to all other points. Map projections that do not preserve any one metric, or try to distort all properties about equally for aesthetic purposes (making the map "look right"), are called **compromise projections**. The Winkel Tripel projection used for the world maps in this book is a compromise projection.

Another important property of a map is its **orientation**, the relationship between the direction on the map and the corresponding compass directions in reality. Almost all maps place north at the top, but there are sometimes reasons to orient a map differently, and it is possible to present a different perception of geographical space by changing a map's orientation. There are some radical orientations that literally turn the world upside down—see A Different Viewpoint feature on page 14.

Symbolization

Maps allow us to get information, to see patterns of distribution, and to compare these patterns with one another. But no map can be a complete record of a given area. In a process called *cartographic abstraction*, important details are chosen by the cartographer to convey the map's information, while less relevant details are often not shown. Understanding the abstraction process is an important skill to have when using maps: the user knows that no map can be "complete," and many details must be simplified or omitted to keep a map legible. Ironically, all maps must "lie" to some degree to inform their readers. There is a fine book about the use and abuse of maps entitled *How to Lie with Maps*.

The geographic features shown on maps are represented by symbols, such as lines, fills, shapes, colors, and type. For example, reference maps can show different types of roads by using lines of differing width, and make different countries easier to tell apart by using different fill colors for each.

Thematic maps often (but not always) use just one type of symbol to display their data, and they can be classified by which symbol they use. **Choropleth maps**, the most common type of thematic map in this book, display their data by filling in political units with differing colors. A good example of a choropleth map is, again, the map of Europe in Figure 4.1 on page 66. **Isarithmic maps** do not use political units, but instead use lines or bands of color to join points of equal value across the mapped area. A map showing high temperatures or a topographic map showing contour lines are both examples of isarithmic maps, as is the map of vernacular map of the South in Figure 1.4. **Graduated symbol maps** use simple symbols such as circles or bar graphs, scaled proportionally to the quantity of the attribute being mapped. **Dot maps** use dots to represent a stated amount of some phenomenon within a political unit (for example, if one dot equaled 1,000 people, 12 dots in an area would indicate 12,000 people). **Flow maps** use arrows of various widths to indicate the movement of people or goods from one area to another (see the map of people on the move in Figure 11.3 on page 394).

1.3 Geographic Technologies and Careers

To close this chapter, we turn to some of the most innovative tools and breakthroughs in geography and consider how you (or someone you know) might become part of the action in this growing field. This section should also open your eyes to some of the newest tools that will help you have an easier and yet deeper understanding of the world's regional geographies. The opportunities that may open to you by studying geography are abundant and exciting. "We are living in the era of the geographer," Stanford ecologist Hal Mooney wrote.[8]

The Geospatial Revolution

In recent decades, technological advances in the field of geography and related sciences have accelerated so quickly that they can truly be called revolutionary. These advances are not confined to the laboratory or the library. Like information on the Web, they have been democratized: you can use the power of geographic tools on your own computer, tablet, or smartphone.

A relatively new term relates to most of these technological advances: **geospatial**. This adjective means "pertaining to the geographic location and characteristics of natural or constructed features and boundaries on, above, or below the Earth's surface, especially referring to data that is geographic and spatial in nature."[9] Even with simple paper maps, geographers have always worked with geospatial information.

A closer LOOK

The Mercator Projection

Arguably the most well-known map projection is the one developed by Gerardus Mercator in 1569. Mercator (1512–1594) developed his cylindrical, conformal projection mainly for navigational use. The **Mercator Projection**, which you can see in • **Figure 1.E**, was designed to show lines of constant compass bearing (known as rhumb lines) as straight lines, which was very important to assist sailing vessels in charting their courses. This was a significant improvement over previous projections, and Mercator projections are still used for water navigation today. However, the projection's usefulness at sea makes it largely unsuitable for other purposes, including reference world maps. In order for the rhumb lines to be shown straight, the projection must continually increase the spacing between the parallels away from the Equator. This results in enormous distortions of size approaching the polar areas (in fact, the poles themselves cannot be shown on a Mercator map as they lie at infinity). On a Mercator map, Greenland and Africa appear similarly sized, whereas in reality Africa is about 14 times larger than Greenland!

Despite the objections of cartographers for many decades, the Mercator projection is still a common sight on world maps in school classrooms, TV newscasts, and online mapping services such as Google's. Its straight lines and convenient rectangular shape make it attractive for such uses. The Mercator Projection also makes it possible to show the United States or Europe, despite their northern locations, as being essentially at the "center of the world." This is done by removing most of Antarctica but keeping much of the Arctic in the frame. Indeed, the Mercator projection's exaggeration of mid-northern latitudes made it very popular in the West during the age of European colonialism. The 45
Mercator Projection was gradually accepted as being "authoritative," and some people are still slow to abandon it for less distorted projections. Most atlases and textbooks no longer use Mercator. But it is still prevalent enough in other media that it is important to recognize this projection's drawbacks and to think critically about how a map projection can influence our perceptions of the world.

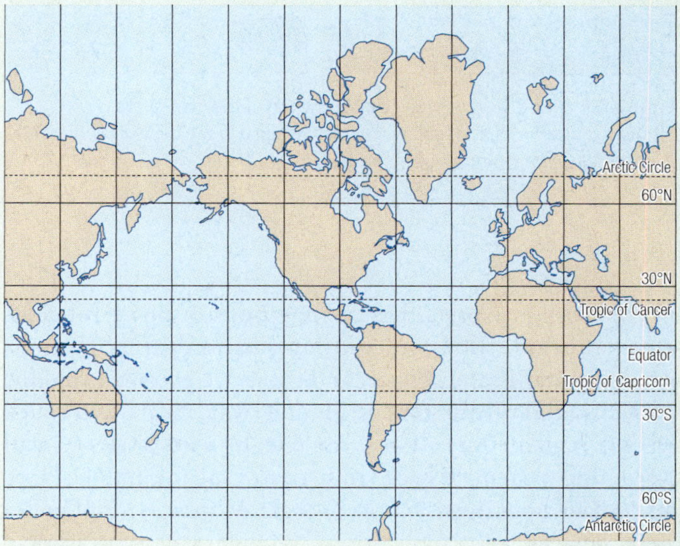

• **Figure 1.E** The Mercator Projection.

A DIFFERENT **viewpoint**

"What's Up: South!"

Now you know how world maps can look different, depending on what projection is used, and how they can be used to produce certain impressions (such as Europe or North America being at the "center of the world"). But imagine *not* being from North America or Europe and getting very tired of how your part of the world is depicted on world maps. Some Australians from "Down Under" have produced a map that challenges the basic premise that "Up Over" people have about the world: that north is at the top. These Aussie cartographers have produced a map called "What's Up: South!" with the world turned upside down (See •**Figure 1.F**). They argue that, like any ball, the Earth has no definitive top and bottom. The conventional choice to put north at the top is just that, a convention. The "South Up" map causes us to see the world in a different way. For one thing, our eyes see a world that is more watery. And we see that Australia and the other "southern" continents now dominate the world! Incidentally, there has been a long history of alternative map orientations. Classical Arab cartographers, producing the world's finest maps of their time, routinely placed south at the map's top (a good example is al-Idrisi's twelfth-century world map). For our purposes in this book, however, unless otherwise noted north is always at the top. This is true of most maps.

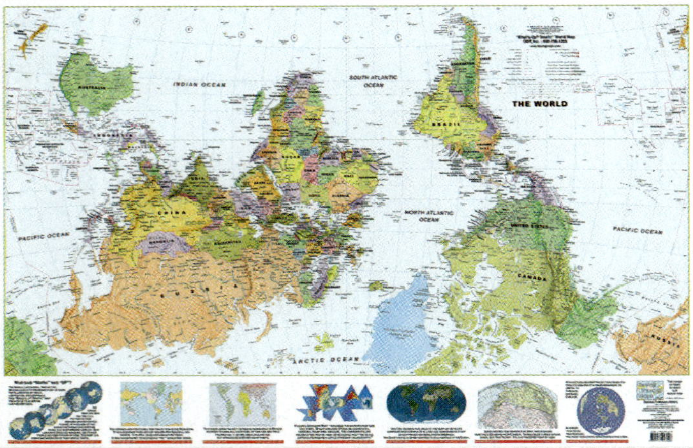

• **Figure 1.F** Who says north has to be at the top of the map? Here is an Australian projection that is literally meant to make you see the world differently. Source: Copyright © 2006, ODT Maps. All Right Reserved.

But the word geospatial has connotations of powerful computer hardware, software, and information-gathering tools including satellites. These technologies are evolving rapidly. The geospatial revolution is still underway, and it will touch your life in many ways. Have you seen or used a GPS (global positioning system) device, in a car or in the hand? Have you looked up directions on Google Maps or MapQuest? These devices and applications provide you with simple information, but underlying them are enormous and sophisticated technologies. Two of the most important geospatial technologies are geographic information systems (GIS) and remote sensing.

GIS—the familiar acronym for **geographic information systems**—represents the cutting edge of geography today. This is one of geography's most powerful tools, and is the field's leading area of growth and employment because of its value in addressing so many real-world problems. GIS is quite complex, but in essence it is a computer-based tool that allows people to create, view, manipulate, analyze, and store geospatial data. These data can be far more than just the latitude–longitude coordinates of a feature on the Earth's surface—a mouse-click on a road, for example, will bring up a box displaying the road's name; its length and width; when it was last repaved; its speed limit; whether it is owned by the city, county, or state; the amount of traffic it receives in a certain time frame; and so forth. These different types of information, called attributes, do not magically appear, but must be created by people who spend an enormous amount of time and energy to compile them from a variety of sources.

GIS data are created and displayed in "layers"—databases storing the locations and attributes of features belonging to a single theme (for example, with layers showing potential customers, the streets on which those customers live, their land ownership, the land's elevation, and land uses in their community, as in •**Figure 1.9**). A GIS can have dozens of layers displaying all kinds of geospatial data for a given area. GIS users can select and highlight certain attributes in various layers that relate to their interests or analysis, while turning off those attributes that are not needed. The ability to selectively

• **Figure 1.9** Geographic information systems (GIS) create and use layers of spatial data. GIS data, images, and models have an enormous range of applications.

INSIGHTS | GIS in Action

GIS makes it possible to see geographic patterns, problems, and connections easily and efficiently—a capability that Esri, the world's largest GIS software maker, calls "The Geographic Advantage."[10] One of the best ways to see what GIS can do is to visit Esri's Web site, http://www.gis.com, where you will see an amazing number of ways that GIS is being used all around you and all around the world. Here are a few examples of the geographic advantage:

- *Natural Resource Management.* In forestry, caring for existing and future trees ensures a steady supply of wood for the world's building needs. GIS provides tools to help determine where to cut today and where to seed tomorrow while minimizing negative impacts.
- *Business.* Every day, businesses deliver goods and services to clients all around a city. Each truck driver needs a route of how to most efficiently visit each client. GIS provides tools to create efficient routes that save time and money and reduce pollution.
- *Defense.* In the military, leaders need to understand terrain to make decisions about how and where to deploy their troops, equipment, and expertise. They need to know which areas to avoid and which are safe. GIS provides tools to help get personnel and materials to the place where they can best do their job.
- *Emergency Preparedness and Response.* During floods and hurricanes, emergency response teams save lives and property. GIS provides tools to help locate shelters, distribute food and medicine, and evacuate those in need.
- *Communications and Media.* In telecommunications, when phone service is out, it means part of the network may be disconnected. GIS provides tools to help find out what part of the network is affected and brings that information to the field so workers can get everyone talking again.
- *Planning.* Planners of all kinds—business analysts, city planners, environmental planners, and strategists from all organizations—use GIS to lay out a framework so that growth can occur in a managed way. As you will see in Chapter 7, GIS can even help us plan for a warmer world in which rising sea levels threaten to flood coastal cities and farms. 260

view certain types of geospatial data, as well as to add other data (such as from a handheld GPS device, from a digitized paper map, or from an image taken by an orbiting satellite), allows users to see spatial relationships with extraordinary ease and clarity. The knowledge they gain from understanding geographic relationships is extremely valuable for decision making and explains why GIS is known as a "critical thinking technology" (see Insights, "GIS in Action" on this page). GIS, in other words, is a high-tech example of what geography is all about.

Remote sensing, also known as *Earth observation*, is the science of acquiring information about the Earth's surface without being in direct contact with it. Most remote sensing data is obtained by aerial photography (cameras mounted on airplanes taking pictures of the ground) or by sensors on Earth-orbiting satellites. Remote sensing is not limited to cameras that capture visible light; much important information about processes and features on the surface or in the atmosphere is gleaned from satellite sensors that can "see" other parts of the electromagnetic spectrum such as infrared and microwave wavelengths. Radar (which measures the reflection of satellite-emitted radio waves bouncing off of ground features) and light detection and ranging ([LIDAR], which uses light in the same way) are also examples of remote sensing.

Remote sensing is an exceptionally good tool for helping geographers understand how people and natural processes modify the Earth. This book uses remote sensing to introduce you to many places, patterns, and problems. "A picture is worth a thousand words," the adage says, and a remote sensing image often goes far in describing a landscape. For example, •**Figure 1.G** (a remotely sensed image taken on April 4, 2010, by a satellite 425 miles above the Earth), is a fine depiction of the cultural landscape of Sendai, in northeast Japan. Just inland from the concrete wave breakers and scalloped shoreline, people have built homes between a greenbelt of trees and an irrigation canal. The canal is a straight and uniform feature—a good indication that it is the handiwork of people rather than of nature. Bridges and streets also stand out as built features. Continuing westward (left) across the image, you see many more homes, including some nestled against a patch of trees (the Japanese prize their forests, and go to great lengths to protect them). Finally, at the far left of the image the long brown rows of empty land are fields for wet rice, which would become green when planted and irrigated from the nearby canal. This satellite image provides an excellent snapshot of how the people of Japan modify nature to suit their needs and desires.

Illustrating dramatically how remote sensing allows us to appreciate landscape change, the next image (Figure 1.H) shows how nature reclaimed Sendai. On March 11, 2011, a magnitude 9.0 earthquake thrust a section of the Earth's crust upward from the seafloor about 80 miles from Sendai. The displaced water roared ashore as devastating

Try It Reading the Landscape with Remote Sensing

Remote sensing images have a way of bringing out the detective in all of us. Compare Figures G and H, examining and describing the details of this landscape's transformation. What changes has this natural disaster wrought on Sendai?

GeoEye/EyeQ

G

GeoEye/EyeQ

H

• **Figure 1.G, H** The shoreline of Sendai, Japan, before and after the gigantic earthquake and tsunami on March 11, 2011.

tsunami waves as high as 40 feet. The results are breathtaking: almost every feature on the landscape, both natural and cultural, has been destroyed or transformed. See the Try It feature on this page, which invites you to interpret these remote sensing images, and A Closer Look feature on
254 page 17.

Careers in Geography

This striking poster in •**Figure 1.10** was produced by the Association of American Geographers (AAG). The employment trends in favor of geography since 2000 have been remarkable. The U.S. Department of Labor has identified geospatial technology as one of the most important emerging and evolving fields in the technology industry. In 2010, this agency projected "much faster than average" growth—more than 20 percent—in jobs for geographers, geoscientists, cartographers, urban and regional planners, and other geographic professionals, through the year 2018.[11]

Much of this job growth will occur in professions in which people feel like they can have a positive impact. The Association of American Geographers had these observations of global trends that are contributing "to a renaissance of geography and its potential for making a difference in society and the world:"

> These include globalization at an increasing pace and scale, phenomena that compel greater understanding of the world, places, people, and natural systems that affect us as a planet and as global citizens and consumers. It includes a recent proliferation of geographic technologies, once fairly obscure and now pervasive in our daily lives, such as GPS in cell phones and cars, online mapping at your fingertips, cable news reports using spatial visualizations, and many more applications in modern business and government services that underlie operations, planning, and progress in all sectors everywhere we live and work. It also includes an academic trend toward greater interdisciplinarity, especially a renewed focus on big questions that matter but that require a breadth of knowledge and multiple fields to tackle. Geography's long-standing intellectual traditions in crossing those usual disciplinary boundaries are now better understood, increasingly seen as relevant and more widely respected in scholarly circles. These trends have produced unprecedented growth in the field.[12]

More and more students are graduating with geography degrees—bachelors, masters, and doctoral—and more of them are finding jobs. Here are just a few examples of starting jobs of recent graduates from the University of Texas-Austin Department of Geography. B.A. students became a Nature Conservancy preserve manager, a transportation planner, a natural science teacher, and a geospatial engineer. Masters students became a U.S. Geological Survey water specialist, the chief executive officer of an insurance company in the West Indies, and a vice president for market research in real

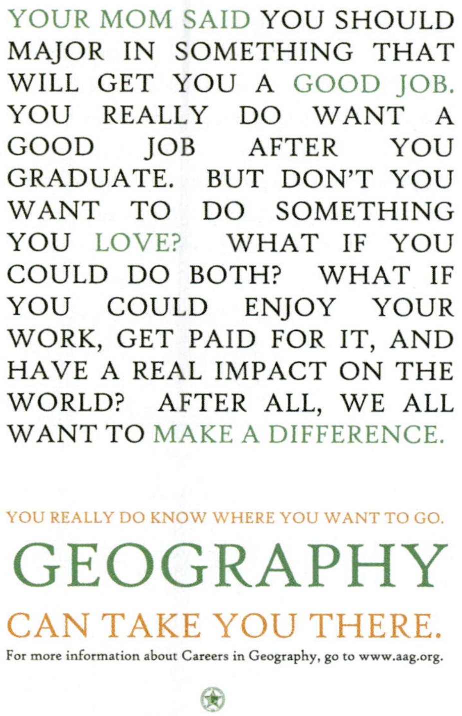

• **Figure 1.10** Geography is awesome.

Source: Dr. Patricia Solís, Association of American Geographers, copyright registered 2004. Reprint permission granted for educational and dissemination purposes only; please do not reprint, translate, or otherwise alter without express written permission by the author.

estate.[13] Many Ph.D. graduates became geography professors (ask your prof which of the 60 specialty groups of the AAG she or he belongs to), whereas others took executive positions in GIS departments and companies or became decision makers in foundations and research institutions all around the world. But even if you are not a geography major or you have not yet seen an advertisement proclaiming "Geographer Wanted," chances are that many professional opportunities will arise if you use geography's tools to grow knowledge of the world in which you live. This book will help you do just that.

A closer LOOK

Google Earth

I would like to recommend two tools that can greatly enhance your understanding and enjoyment of world regional geography. One is a globe. As we saw earlier, flat maps distort representations of the Earth, but a globe puts a representation of the true world in your hands. If you keep a globe handy, and use it as you work your way through the book, it will help you learn more about the Earth and remember what you learn.

The other tool is a virtual globe, map, and geographic information software program like Google Earth or NASA's World Wind. If you have not already done so (and if you have permission to do so), on the PC or Mac, navigate to http://www.earth.google.com (apps are also available for most smartphones and tablets). Download the free Google Earth application, open it, bookmark it, and use it often—*PLEASE!* With Google Earth you are in control of an easy-to-use but incredibly powerful geospatial toolkit. It is so powerful, in fact, that it enables you to become a rather effective spy: you can see not only your home or apartment, but top secret North Korean and Iranian nuclear facilities. A Russian intelligence official grumbled of Google Earth, "Terrorists don't need to reconnoiter their target. Now an American company is working for them!"[14]

By the way, when I photograph a bridge, power plant, or dam (as I do often, thinking I might use the photo in one of my classes, or for this book), I am sometimes detained and questioned by security guards. That is ironic, considering what Google Earth can do.

Google Earth's base map is composed of many thousands of remotely sensed images, from both aerial photography and orbiting satellites, put together as a giant mosaic. And although Google Earth is not a full-fledged GIS, its "Layers" sidebar works in a similar fashion: you can turn on and off certain features like country boundaries and road networks. With the "Weather" layer, you can get real-time views of storm systems and temperatures. How hot or cold is it in Arabia or Siberia right now? It's easy to find out. You might also be interested in the "Global Awareness" layer that lets you explore some of the world's environmental problems. And then there is the "Street View" layer for many locations, infamous for news stories like this: "A woman, checking out a female friend's house on Google Maps, was surprised to see her husband's Range Rover out front. A divorce is underway."[15]

summary

- A recent study showed that Americans generally have poor knowledge of world geography.
- There are six essential elements of the national geography standards: the world in spatial terms, places and regions, physical systems, human systems, environment and society, and the uses of geography.
- Geography means "description of the Earth" and is also defined as "the study of the Earth as the home of humankind."
- The four main objectives of the text are for readers to understand Earth's problems and potential solutions for these problems; develop a habit of synthesizing information to understand the world; understand current events; and develop the ability to interpret places and "read" landscapes.
- Maps are the geographers' most basic tools. The basic language of maps includes the concepts and terms of scale, coordinate systems, projection, and symbolization. Maps can depict spatial data in a variety of ways.

- Individuals and cultures generate their own unique "mental maps." Regions are in effect mental maps that help us make sense of a complex world. Modern geographic thought derives from a long legacy of interest in how people interact with the environment. The dominant approach has been to understand how people have changed the landscape or the face of the Earth.
- The discipline of geography may be divided into regional and systematic specialties, with the systematic fields having the most followers. Their concerns overlap many disciplines in the natural and social sciences. Geographers are employed in many private and public capacities. The strongest growth area with the most jobs is in geographic information systems (GIS). The "geospatial revolution" employs many geographic tools and touches many parts of our daily lives.

Key Terms + Concepts

Review Questions

1. What is geography? What are some of its characteristic approaches?
2. What are the six essential elements of geography, as defined by the National Geographic Society? What does each element indicate about geography's concern with space, place, or the environment? What are the five themes of geography? Are you comfortable applying them to your knowledge of places?
3. What does *spatial* mean, and how does geography's interest in space differentiate it from other disciplines?
4. What geographic features make the United Kingdom and New Zealand different?
5. What are the major terms and concepts associated with scale, coordinate systems, projections, and symbolization?
6. Why is a map made with the Mercator projection more suitable for navigation than a map made with other projections?
7. What is the difference between a dot map and a choropleth map?
8. What is a mental map?
9. What is GIS, and what typically makes it different from old-fashioned manual cartography? What are some applications of remote sensing? How can modern geospatial tools help us with longstanding concerns in geography?
10. What do geographers do for a living?

Notes

1. The abbreviation B.C.E. stands for "before the Common Era," which is a reference to the dating system invented by European Christians that sets the birth of Jesus Christ as year 1. In Christian cultures, dates before that year are expressed as B.C., meaning "before Christ," and later years are identified as A.D., which stands for *anno Domini* (Latin, "in the year of our Lord"). Religion-neutral dating systems such as the one used in this book employ B.C.E. (before the Common Era) and C.E. (Common Era), respectively, but the years are numbered the same in the two systems.
2. Quoted in Geoffrey J. Martin and Preston E. James, *All Possible Worlds: A History of Geographical Ideas* (New York: Wiley, 1993), p. 150.
3. Quoted in Andrew Goudie, *The Human Impact on the Natural Environment* (Oxford: Blackwell, 1986), p. 6.
4. The anthropologist Kathleen A. Dahl authored this definition of culture.
5. You are in Monument Valley, northeastern Arizona near the Utah border, in the southwestern region of the United States. Place clues include an arid landscape, sandstone buttes that you have probably seen in American Western movies, and a Native American of the Navajo group, wearing typical Navajo turquoise and silver jewelry.
6. The viper's triangular head is just left of the center of the photo. This snake has buried himself in the shallow sand and gravel, a "virtual land mine" ready to explode on its prey or foe: a mouse, a Bedouin, or a geographer.
7. Using this map, you should determine Oslo's location as about 60 degrees north, 11 degrees east (60°N, 11°E). Oslo is therefore in the Northern Hemisphere, the Eastern Hemisphere, and the land hemisphere.
8. Committee on Strategic Directions for the Geographical Sciences in the Next Decade, National Research Council, 2010, Understanding the Changing Planet: Strategic Directions for the Geographical Sciences, Washington, DC: The National Academies Press.
9. http://www.dictionary.com.
10. Esri, at http://www.gisday.com/cd2009/fliers/what_is_gis.pdf.
11. http://www.aag.org/cs/careers
12. http://www.aag.org/cs/jobs_and_careers/geography_can_take_you_there
13. Information courtesy of Bill Doolittle, University of Texas at Austin.
14. Lieutenant General Leonid Sazhin, Federal Security Service, quoted in Katie Hafner and Saritha Rai, "Google Offers a Bird's-Eye View, and Some Governments Tremble," *New York Times,* December 20, 2005, p. 1.
15. http://gawker.com/5191459/cheating-husband-said-caught-via-google-street-view.

Online Resources

CourseMate: Make the most of your study time by accessing everything you need to succeed in one place. Read your textbook, take notes, review flashcards, watch videos, complete activities, take practice quizzes, and more—online with CourseMate. Log in at **www.cengagebrain.com**.

"The acquisition of wisdom is above that of pearls. The topaz of Ethiopia cannot equal it, Nor can it be valued in pure gold. Where then does wisdom come from? And where is the place of understanding?"

—JOB 28

Joe Hobbs

Joe Hobbs

Russia's Arctic wilderness of Franz Josef Land, not far from the North Pole. The dark patches of ground are permafrost, or "permanently" frozen ground. But as the Arctic warms, permafrost is melting and adding greenhouse gases to the Earth's atmosphere. The Arctic is the "canary in the coal mine" of climate change.

2

Physical Processes and World Regions

Many issues in world regional geography relate to the interaction of people and the natural environment, and how these relationships have changed the face of the Earth. Physical processes are the main focus in this chapter, and human processes the focus in Chapter 3. But these processes are so intertwined that human forces receive much attention in this chapter, and natural forces much attention in the following chapter.

This discussion of physical geography introduces you to the four "spheres" that make up the Earth's habitable environment: the **lithosphere**, its outer "rind" of rock; the **hydrosphere**, made up of all the world's water features; the **atmosphere**, the layer of gases surrounding the Earth; and the **biosphere** (also known as the ecosphere), which is the global ecological system including all the relationships played out among the lithosphere, hydrosphere, and atmosphere. Earth's lithosphere is a work in progress, and you will see how its changing surfaces provide both opportunities and threats to people. We will consider the climate and vegetation types that play such large roles in human activities, and appreciate the rich diversity of wild plant and animal species. We will look briefly at the planet's often-overlooked oceans and the resources they hold. Finally, we will examine how the climate may be changing, what these changes may mean for life on Earth, and what people can do to prevent or lessen some of the most dangerous climatic changes.

chapter outline

chapter objectives

This chapter should enable you to

- Understand the tectonic forces behind some of the world's major landforms and natural hazards
- Recognize consistent global patterns in the distribution of vegetation types and climates
- Identify the natural areas most threatened by human activity and explain how natural habitat loss may endanger human welfare
- Appreciate the important roles of the world's oceans
- Describe the potential impacts of global climate change and international efforts to prevent them

2.1 Geologic Processes

The Earth's surface is in motion. Those of you living on the West Coast of the United States have probably had some personal experience of the unforgettable sensation of the ground moving beneath you in a mild "temblor" or even a large earthquake. These events reflect processes that are constantly
16, 254, 261 changing the face of the Earth. Here we explore just one set of geological processes—arguably the most important of them all because of its role in shaping global landforms—plate tectonics.

Plate Tectonics

About a century ago, a German geologist named Alfred Wegener came up with what seemed to many like an outlandish theory known as **continental drift.** Pointing to such things as the jigsaw puzzle–like geometry of Africa's west coast and South America's east coast, he proposed that the continents were once joined in a supercontinent (which he named Pangaea) but that they drifted apart over time. However, Wegener was not able to explain the forces behind these movements, and he was derided by a scientific community that seemed unwilling to accept anything other than a *terra firma* that was truly firm. Peers pronounced his conclusions "utter damned rot" and "mere geopoetry."[1] But later discoveries in deep-sea science established Wegener's basic proposition as factual, and today a good deal is known about how continental drift occurs.

The Earth's lithosphere, which varies from 50 to 125 miles in thickness, is made up of about a dozen giant and several smaller sections of rock called **plates.** These plates move in various directions in processes known collectively as **plate tectonics** (•Figure 2.1). On the ocean floors in places such as the Mid-Atlantic Ridge and the East Pacific Rise, new lithosphere is "born" as molten material rises from the Earth's mantle and cools into solid rock (•Figure 2.2). Plate tectonics are often explained by the useful analogy of a conveyor belt (the convection cell in Figure 2.2) in constant motion. On either side of the long, roughly continuous ridges, the two young plates move away from one another, carrying islands with them; this process is called **seafloor spreading.**

Seafloor spreading has few impacts on people, but when the Earth's plates collide, there is cause for great concern: such **tectonic forces** are among the planet's greatest natural hazards. The most dangerous tectonic forces are related to **seismic activity** (*seismic* refers to Earth vibrations, mainly earthquakes) that causes earthquakes and **tsunamis,** and the **volcanism** (movement of molten earth material, especially in volcanoes) often associated with plate movements. The globe's most active and deadly realm of tectonic activity is the so-called "Ring of Fire" on the rim of the Pacific Ocean (see Figure 2.1 and page 358).

Plates collide and converge in different ways and with different consequences; among the most notable consequences are the building of Earth's great mountain ranges. In some parts of the world—off the east coast of Japan, as we saw in Chapter 1, and 16, 254, 358, 396
off the coast of the U.S. Pacific Northwest, for example—one plate "dives" below another in a process known as **subduction** (see Figure 2.2). The descending lithosphere is melted again as it dives into the Earth's superheated mantle along a deep linear feature known as a **trench** (for example, the Mariana Trench off Japan). Subduction is another stage along the "conveyor belt" process that will eventually see this material recycled as newborn lithospheric crust.

This subduction process releases enormous amounts of energy. Periodically, the great stress of one plate pushing beneath another is released in the form of an earthquake. The world's largest recorded earthquakes—registering 9.5 (Chile, 1960), 9.2 (United States, 1964), 9.1 (Indonesia, 2004), and 9.0 (Japan, 2011), respectively, on the **Richter scale,** which measures the strength of the earthquake at its source—have struck along these subduction zones. This sudden displacement of a section of oceanic lithosphere is also what triggers 214, 358, 396
a tsunami, one of nature's most powerful and destructive processes. See pages 17, 254, and 288 for more insight into tsunamis.

In other places where they meet, lithospheric plates grind and slide along one another, as in California and Turkey. The processes of rock crowding together or pulling apart along these fracture lines is known as **faulting,** with movement along various kinds of **faults** causing earthquakes, the emergence of new landforms, and other consequences.

Volcanism generally takes place along and near **subduction zones** (see Figure 2.2) and also in the world's several dozen geologic hot spots where molten material has broken through the crust as a "plume" (as in Yellowstone National Park and in Hawaii—see page 288). Despite posing a host of natural hazards

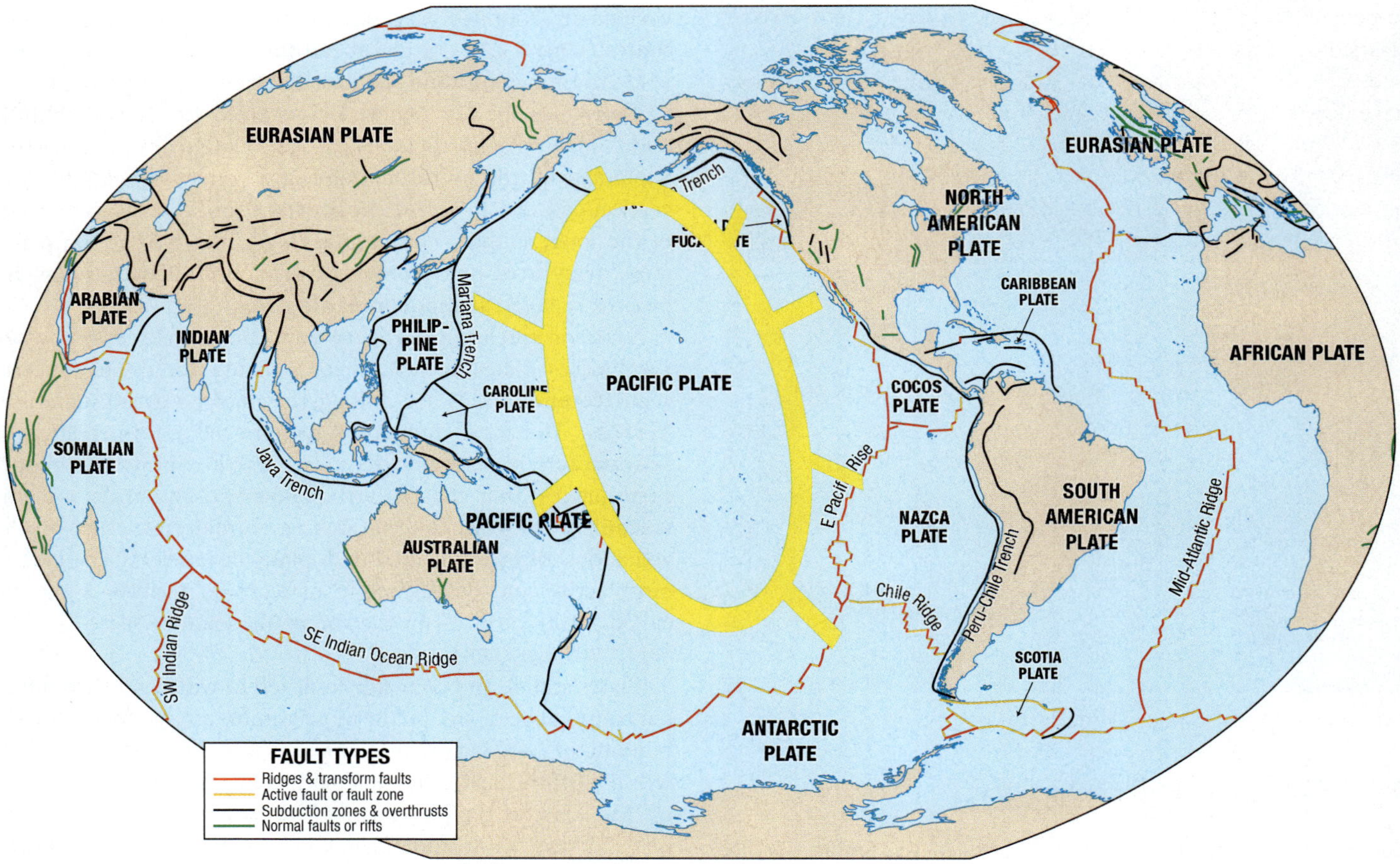

• **Figure 2.1** Major tectonic plates and their general direction of movement. Earthquakes, volcanoes, and other geologic events are concentrated where plates separate, collide, or slide past one another. Where they separate, rifting produces very low land elevations (well below sea level at the Dead Sea of Israel and Jordan, for example) or the emergence of new crust on the ocean floor (in the middle of the Atlantic Ocean, for example). The yellow ring points to the rough outline of the "Ring of Fire."

Source: Adapted from NASA, "Global Tectonic Activity Map of the Earth," DTAM-1, 2002.

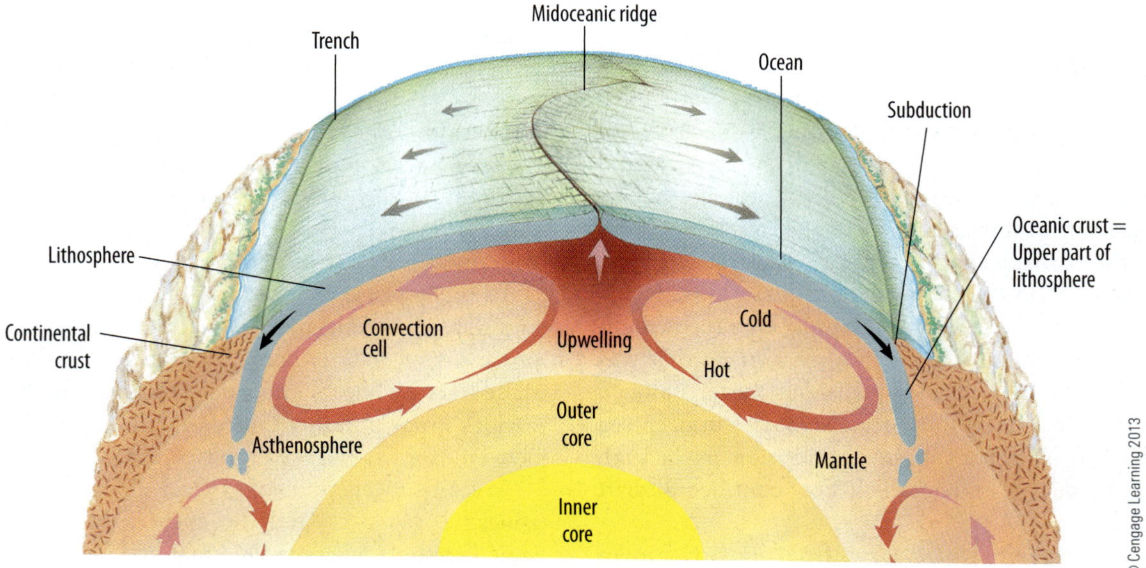

• **Figure 2.2** Generalized cross section of the Earth, showing its main concentric layers and the process by which its lithosphere is recycled.

to people living on their slopes or downwind—including pyroclastic flows (fast-moving currents of rock fragments, hot gases, and ash), lava flows, ashfalls, and other dangers—volcanoes have beneficial qualities. Volcanic rock usually breaks down to form fertile soils (as in Ethiopia and the Nile Valley), and the range of climate and vegetation types on their slopes creates op-
181, 311 portunities for growing crops and raising livestock.

We turn now to global patterns of climate and vegetation.

2.2 Patterns of Climate and Vegetation

"Everybody talks about the weather, but nobody does anything about it," Mark Twain reportedly said. We will see in this chapter that we have actually done a lot about the weather. First, we look at natural patterns in vegetation and climate that present different opportunities for people around the world. Once you recognize these patterns, you will be more prepared to explore all the world's regions.

As you experience a warm, dry, cloudless summer day or a cold, wet, overcast winter day, you are encountering the **weather**—the atmospheric conditions occurring at a given time and place. **Climate** is the average weather of a place over a long time period. Along with surface conditions such as elevation and soil type, climatic patterns have a strong correlation with patterns of natural vegetation and in turn with human opportunities and activities on the landscape.

Precipitation and temperature are the key variables in weather and climate. Warm air holds more moisture than cool air, and **precipitation**—rain, snow, sleet, and hail—results from processes that cool the air to release moisture. Water is, of course, essential for life on Earth. In this book, you will see many examples of how the precious resource of fresh water is increasingly imperiled and contested. Some geographers would tell you that 180, 250, 259, 265, 436
the map of the world's precipitation—•**Figure 2.3**—is the most important of all maps in understanding life on Earth.

Combinations of precipitation, temperature, latitude, and elevation produce a great variety of local climates. Geographers group these local climates into a number of major climate types, each of which occurs in more than one part of the world (•**Figure 2.4**) and is associated closely with other types of natural features, especially vegetation. Geographers recognize 10 to 20 major types of terrestrial ecosystems, called **biomes,** which are categorized by dominant types of natural vegetation (•**Figures 2.5** and **2.6**). When you are using maps of biomes, it is important to keep in mind that these depict vegetation that would dominate in the absence of human activity. For example, Figure 2.5 depicts a vast, unbroken area of temperate mixed forest in the eastern United States, but of course much of this region has been

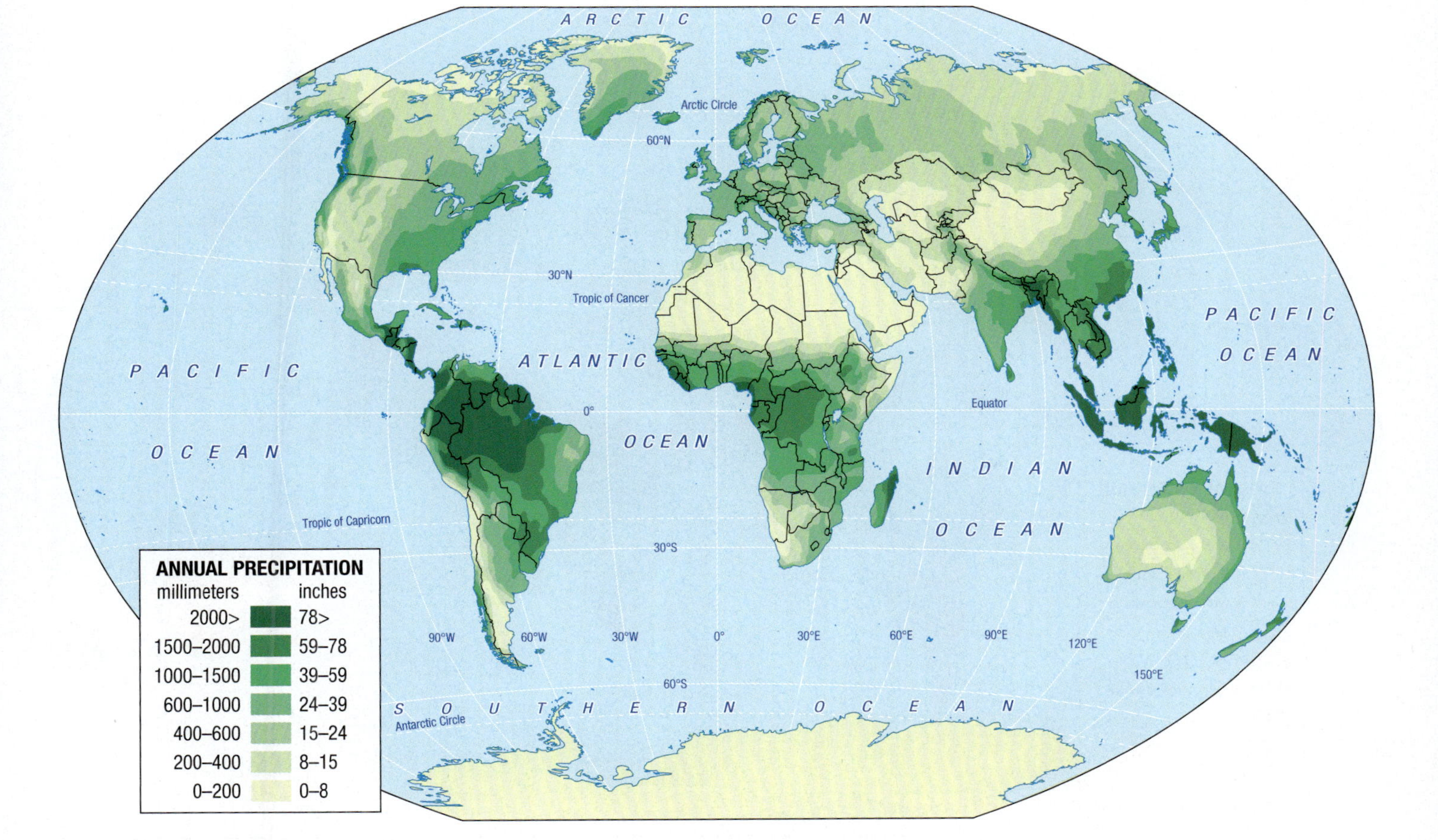

• **Figure 2.3** World precipitation map. Is this, as some authorities say, the most important of all world maps?
Source: Adapted with permission from UNEP/GRID and UEA/CRU, "Mean Annual Precipitation," GRID - Geneva Data Sets - Atmosphere, GNV174.

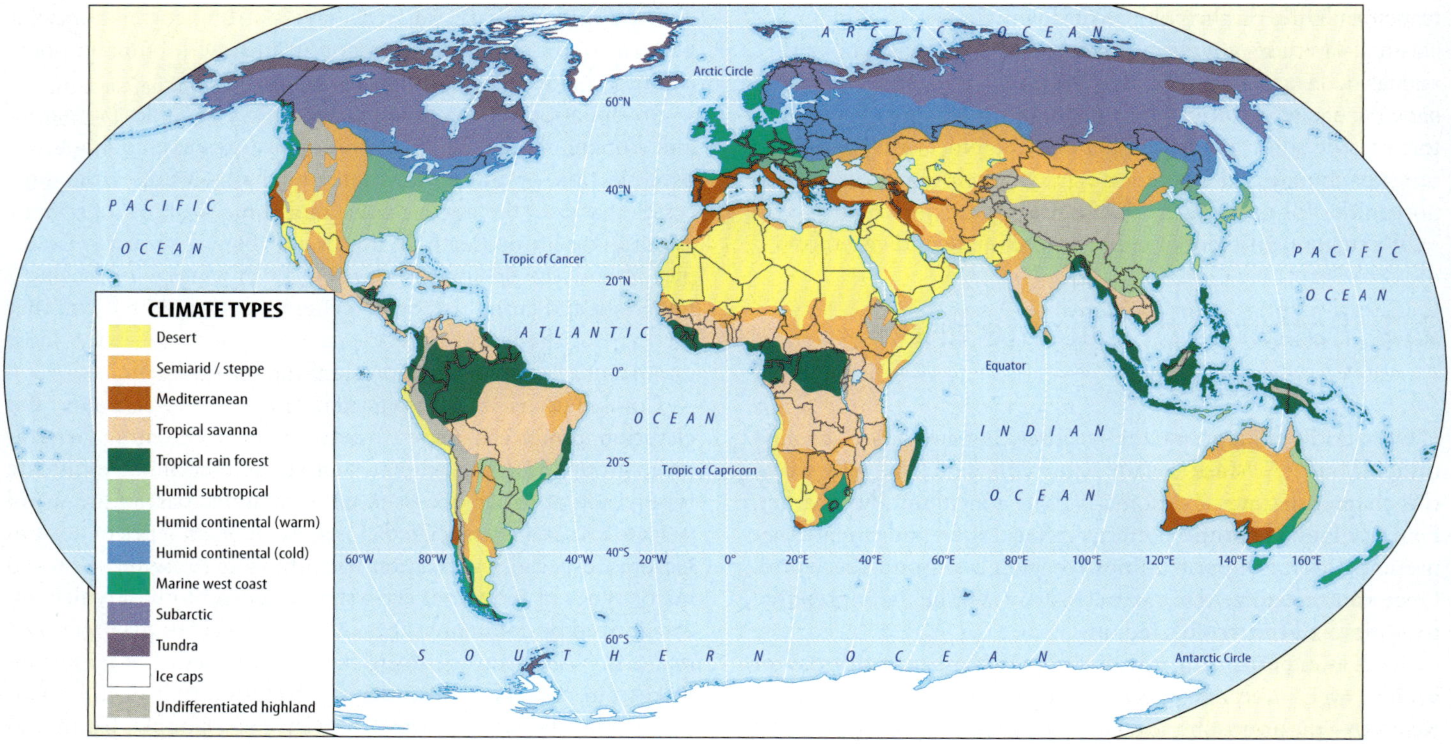

• Figure 2.4 World climates.

Source: Based on Rand McNally's Classroom Atlas, 2003.

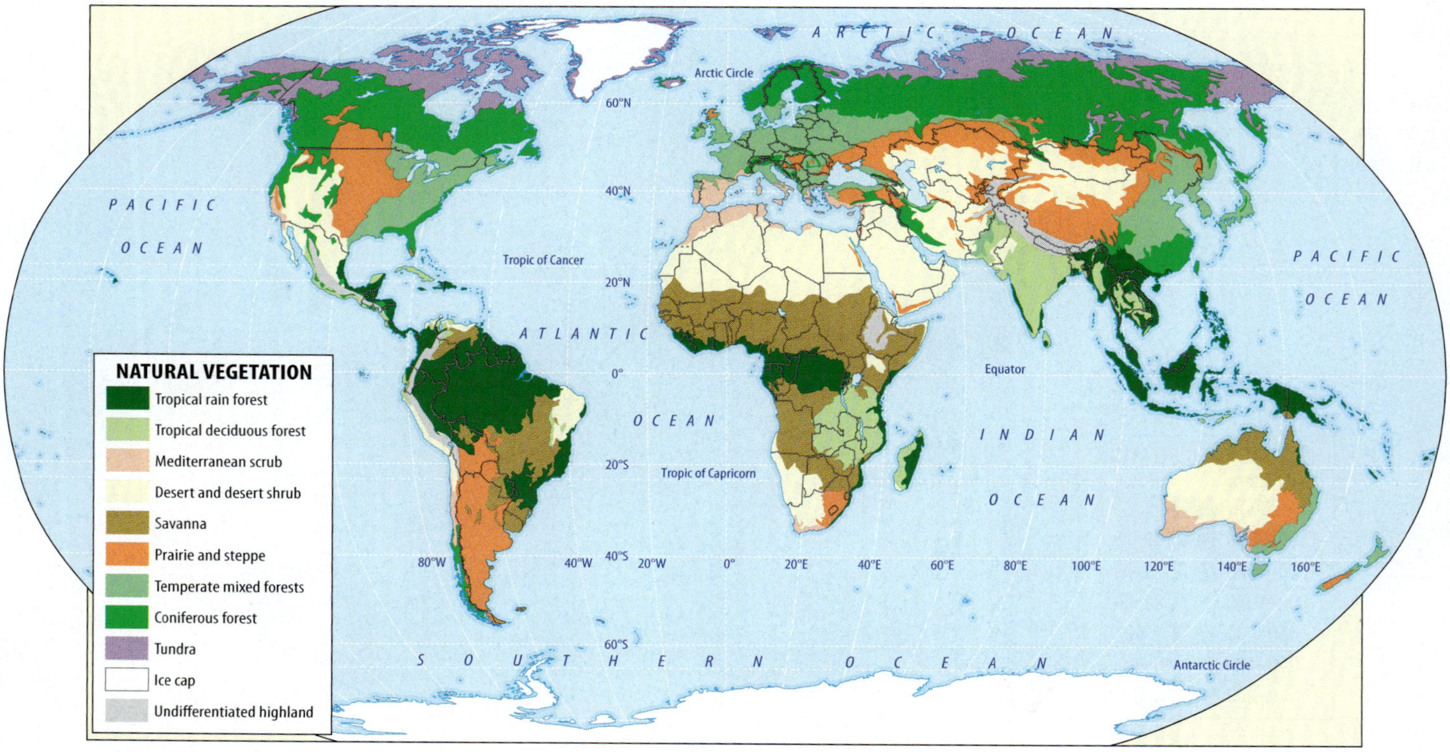

• Figure 2.5 World biomes (natural vegetation) map.

Source: Based on World Wildlife Fund ecoregions data, 1999.

transformed by agriculture and urbanization. The map of the "human footprint" on page 42 in the next chapter shows what people have done to change the face of the Earth.

Climate plays the main role in determining the distribution of biomes, but differing soils and landforms may promote different types of vegetation where the climatic pattern is essentially the same. Vegetation and climate types are sufficiently related that many climate types take their names from vegetation types—for example, the tropical rain forest climate and the tundra climate. You may easily see the geographic links between climate and vegetation by comparing the maps in Figures 2.4 and 2.5. The spatial distributions of climate and vegetation types do not overlap perfectly, but there is a high degree of correlation (see Try It on this page).

Your Global Environmental Counterparts — Try It

Examine the maps in Figures 2.4 and 2.5 (by the way, are these large-scale or small-scale maps?). In each map, find approximately where you live. What kind of climate region is this? What is the characteristic biome or natural vegetation there? Now find one or more other areas in the world that have the same climate and biome types—what places are most like where you live? Remember, these maps are greatly generalized, so you may need to be forgiving about the details.

In the **ice cap, tundra,** and **subarctic** climates, the dominant feature is a long, severely cold winter, making agriculture difficult or even impossible. Summers are short and cool. Perpetual ice in the form of glaciers may be found at very high elevations in the lower latitudes (even in equatorial regions) and in the polar realms. The ice cap biome (•Figure 2.6a) has no plant life, except in those very few spots where enough ice or snow melts in the summer to allow tundra vegetation to grow. Tundra vegetation (•Figure 2.6b) is made up of mosses, lichens, shrubs, dwarfed trees, and some grasses. Needleleaf evergreen **coniferous trees** can stand long periods when the ground is frozen, depriving them of moisture. **Coniferous forests** (often called **boreal forests** or *taiga,* their Russian name; •Figure 2.6c) thrive in areas of subarctic climate,
121 including large parts of Russia's Siberia region.

In **desert** and in **semiarid** or **steppe** climates, the dominant feature is aridity (extreme dryness) or semiaridity (mainly dry conditions, with sparse precipitation). Deserts and steppes occur in both low and middle latitudes. Agriculture in these areas usually requires irrigation. The Earth's largest dry region extends in a broad band across northern
Africa and southwestern and central Asia. The deserts of the 151, 160
middle and low latitudes are generally too dry for either trees or grasslands. They have **desert shrub** vegetation (•Figure 2.6d), often only in dry riverbeds, and in many places have no vegetation at all. Grasslands dominate in the moister steppe climate, which represents a transitional zone between very arid

Joe Hobbs

• **Figure 2.6** **(b)** Tundra, northern Norway.

Joe Hobbs

• **Figure 2.6** **(a)** Ice cap biome, Franz Josef Land, Russian Arctic region.

Antropov/Shutterstock.com

• **Figure 2.6** **(c)** Coniferous forest.

Joe Hobbs

• **Figure 2.6** **(d)** Desert shrub, southern Sinai Peninsula, Egypt.

Joe Hobbs

• **Figure 2.6** **(f)** Tropical rain forest.

deserts and more humid areas. The biome composed mainly of short grasses is also called the steppe, or **temperate grassland** (•**Figure 2.6e**). The temperate grassland region of the United States and Canada originally supported both tall-grass and short-grass
399 vegetation types, known in those countries as **prairies.**

The **tropical rain forest** and **tropical savanna** climates are rainy, low-latitude climates. The key difference between them is that the tropical savanna type has one or more prolonged dry seasons, whereas the tropical rain forest climate has a steadier rainfall regime. Heat and moisture dominate in the tropical rain forest biome (•**Figure 2.6f**), where broadleaf evergreen trees (which lose their leaves, but not seasonally or in tandem) dominate the vegetation. In dry tropical areas that have enough moisture to support trees, **tropical deciduous forest** replaces the rain forest. Here the broadleaf trees are not green throughout the year; they lose their leaves and are dormant during the dry season and then add foliage and resume their growth during the wet season. The tropical deciduous forest approaches the luxuriance of the tropical rain forest in wetter areas but thins out to low, sparse scrub and thorn forest (•**Figure 2.6g**) in drier areas. **Savanna** vegetation (•**Figure 2.6h**) has taller grasses than the steppe and usually includes some trees, if found in areas of greater overall rainfall and more pronounced wet and dry seasons. You have seen much savanna vegetation in documentaries
about the great animal herds of East Africa. 314

In the **marine west coast** climate, occupying the western sides of continents in the higher middle latitudes (the Pacific Northwest region of the United States, for example), warm ocean currents moderate the winter temperatures, and summers tend to be cool. Coniferous forest dominates in some cool, wet areas of marine west coast climate; a good example is the redwood and sequoia forest of northern California (•**Figure 2.6i**). These humid middle-latitude regions have mild to hot summers, and winters ranging from mild to cold, with several other types of climate interspersed.

The **Mediterranean** climate (named after its most prevalent area of distribution, the lands around the Mediterranean Sea) is typically found between a marine west coast climate and a

Joe Hobbs

• **Figure 2.6** **(e)** Steppe, eastern Turkey.

Joe Hobbs

• **Figure 2.6** **(g)** Scrub and thorn forest, northern Zimbabwe.

Joe Hobbs

• **Figure 2.6** **(h)** Savanna, southern Kenya.

76 lower-latitude steppe or desert climate. In the summer high-sun period, it lies under high atmospheric pressure and is typically rainless. In the winter low-sun period, it usually receives precipitation from low-pressure systems carried by westerly winds. **Mediterranean scrub forest** (•**Figure 2.6j**), known locally by such names as *maquis* and *chaparral,* characterizes Mediterranean climate areas. Many of these shrubs contain oils that are highly aromatic, repelling wild grazing animals but attracting human culinary interest: sage and thyme

Joe Hobbs

• **Figure 2.6** **(i)** Marine west coast forest type; giant sequoias in King's Canyon National Park, central California.

Joe Hobbs

• **Figure 2.6** **(j)** Mediterranean scrub forest (dark area in left foreground) east of Los Angeles. The mountains produce a "rain shadow" effect: the light area in the right and background is desert.

are good examples. Much of California has Mediterranean climate and vegetation.

The **humid subtropical** climate occupies the eastern portion of continents between approximately 20 and 40 degrees of latitude and is characterized by hot summers, mild to cool winters, and ample precipitation for agriculture. The **humid continental** climate lies poleward of the humid subtropical type; it has cold winters, warm to hot summers, and enough rainfall for agriculture, with the greater part of the precipitation falling in the summer. This climate type is often subdivided into warm and cold subtypes, indicating the greater severity of winter in the zones closer to the poles. In middle-latitude areas with these humid subtropical and humid continental climate types, a **temperate mixed forest** with mostly broadleaf but also coniferous trees (as in the U.S. Midwest and Northeast; •**Figure 2.6k**) is found. As cold winter temperatures freeze the water within reach of plant roots, broadleaf trees shed their colorful leaves and cease to grow, thus reducing water loss. Then they produce new foliage and grow vigorously during the hot, wet summer.

Joe Hobbs

• **Figure 2.6** **(k)** Temperate mixed forest, central Missouri, United States.

Joe Hobbs

• **Figure 2.6** **(l)** Undifferentiated highland vegetation, San Juan Mountains, Colorado, United States. Note the "timber line" above which trees do not grow. Here, biomes change with elevation, and so the vegetation types have to be characterized as "undifferentiated" on small scale maps.

Coniferous forests can thrive in some hot and moist locations where porous, sandy soil allows water to escape downward, giving conifers (which can withstand drier soil conditions) an advantage over broadleaf trees. Pine forests on the coastal plains of the southern United States are an example.

Undifferentiated highland climates have a range of conditions according to elevation and exposure to wind and sun. The vegetation in these climates (•**Figure 2.6l**) differs greatly, depending on elevation, degree and direction of slope, and other factors. These climates are "undifferentiated" in the context of world regional geography in that a small mountainous area may contain numerous biomes, and it would be impossible to map them on a small scale. The world's mountain regions have a complex array of natural conditions and opportunities for human use. With increasing elevation, temperatures fall by a predictable **lapse rate,** on average about 3.6°F (2.0°C) for each increase of 1,000 feet (305 m) in elevation. In climbing from sea level to the summit of a high mountain peak near the equator (for example, in western Ecuador), a person would experience many of the major climate and biome types found in a
356 sea-level walk from the equator to the North Pole!

2.3 Biodiversity

All of Earth's natural regions have remarkable life forms. But scientists recognize the exceptional importance of some biomes because of their **biological diversity** (or **biodiversity**)—the number of plant and animal species present and the variety of genetic materials these organisms contain.

The most diverse biome is the tropical rain forest. From a single tree in Peru's Amazon region, the entomologist Terry Erwin recovered about 10,000 insect species. From another tree several yards away, he counted another 10,000, many of which differed from those of the first tree. Alwyn Gentry of the Missouri Botanical Garden recorded 300 tree species in a 1-hectare (2.47-acre) plot of the Peruvian rain forest. Less than 50 years ago, scientists calculated that there were 4 million to 5 million species of plants and animals on Earth. Now, however, their estimates are much higher, in the range of 40 million to 80 million. This startling revision is based on research, still in its infancy, on species inhabiting the rain forest. Many of these are small insects. Incidentally, the number of animal species far exceeds that of plant species.

The Importance of Biodiversity

Such diversity is important in its own right, but it also has vital implications for nature's ongoing evolution and for people's lives on Earth. Humankind now relies on a handful of crops as staple foods. In our agricultural systems, the trend in recent decades has been to develop high-yield varieties of grains and to plant them as vast **monocultures** (single-crop plantings). This trend, which is the cornerstone of the so-called Green Revolution, is controversial. On the one hand, it puts more food on the
global table. But on the other, it may render agriculture more 249
vulnerable to pests and diseases and thus pose long-term risks of famine. In evolutionary terms, the Green Revolution has reduced the natural diversity of crop varieties that allows nature and farmer to turn to alternatives when adversity strikes. At the same time, while we remove tropical rain forests and other natural ecosystems to provide ourselves with timber, agriculture, or living space, we may be eliminating the foods, medicines, and raw materials of tomorrow, even before we have collected them and assigned them scientific names. "We are causing the death of birth," lamented biologist Norman Myers.[2]

Regions where human activities are rapidly depleting the rich 290,
variety of plant and animal life are known as **biodiversity hot** 344
spots, which scientists believe deserve immediate attention for study and conservation. Thirty-four priority regions identified by Conservation International are depicted in •**Figure 2.7**. By referring to the map of the Earth's biomes in **Figure 2.5**, you will see that most of these hot spots are in tropical rain forest areas. Many are islands that tend to have high biodiversity because species on them have evolved in isolation to fulfill special roles in these ecosystems and because human pressures on island eco-
systems are particularly intense. Efforts are under way in most 290,
of these hot spots to establish national parks and other protected 344
areas. Conservationists are also turning their attention increasingly to the state of the world's oceans, which play critical roles in the Earth's physical processes and sustain many people.

2.4 The World's Oceans

Life on Earth would be impossible without the roles and resources of the hydrosphere, which includes both oceans and freshwater sources such as lakes and rivers. This book deals often with issues of freshwater, which is appropriate considering the projection by the World Commission on Water that by 2025, half of the world's population will live under conditions of severe water stress.[3] In this section, the focus is on the world's oceans.

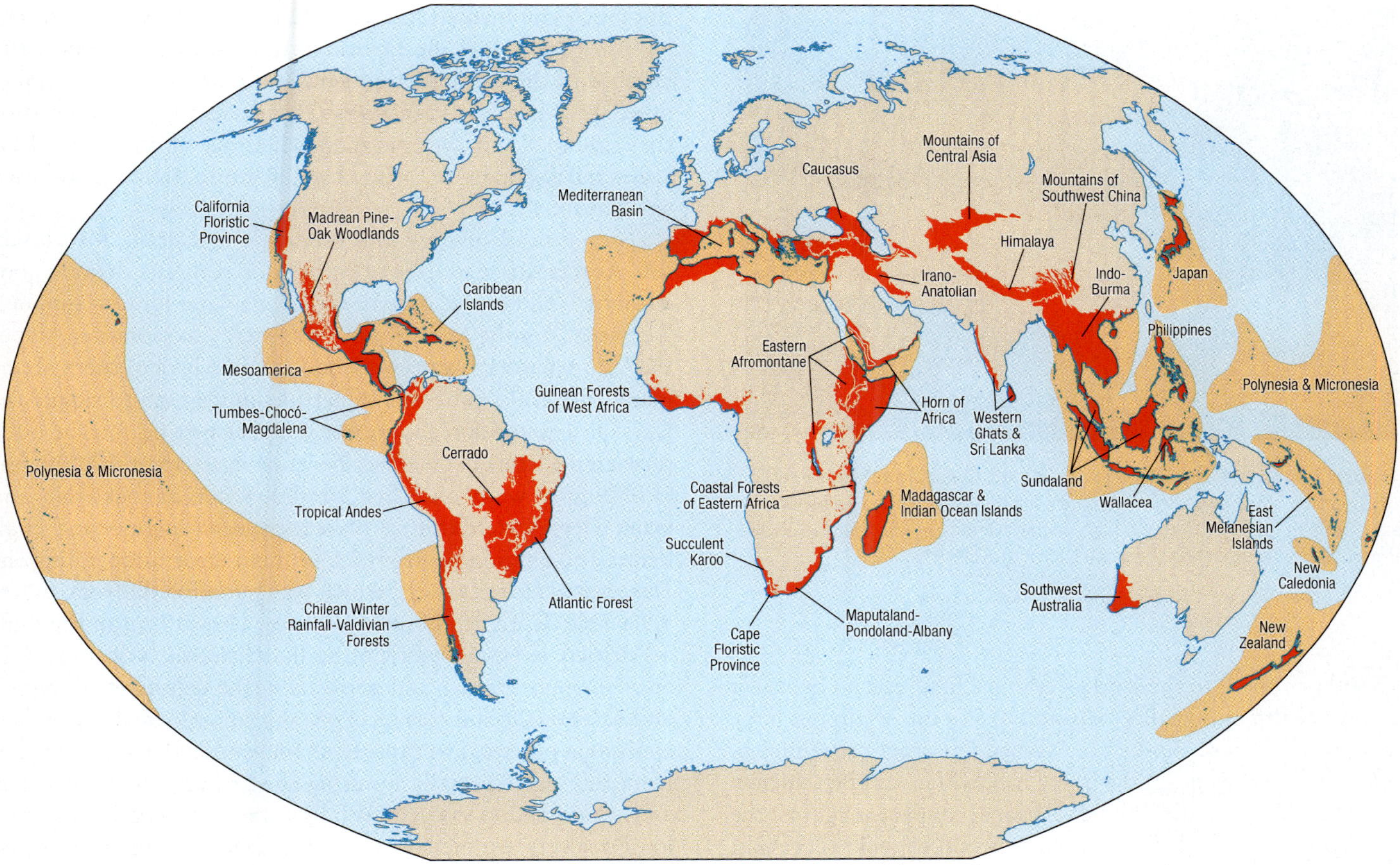

• **Figure 2.7** World biodiversity hot spots.

Source: Adapted from "Threatened Ecoregions" map, National Geographic Family Reference Atlas of the World, p. 40. Copyright © 2002 National Geographic Society. Source data provided by Conservation International.

Why Should We Care about Oceans?

I live in Missouri, where a day at the beach or a plate of fresh seafood is just a dream. Even those of you living close to the sea may have very little interaction with it. Why bother thinking about the world's oceans, much less worrying about them?

It's a watery world. About 70 percent of the world's surface is comprised of water. That is a huge proportion, and it is very important to consider how water shapes life on the planet. Oceans have the largest role in the **hydrologic cycle,** as the sun evaporates seawater into vapor that is then released as freshwater precipitation. Without the seas, the Earth's usable freshwater resources would be extremely limited.

The oceans feed us. About a billion people, or 15 percent of the world's population, rely mainly on fish for their protein (•**Figure 2.8**). Growing human populations, increased buying power among the middle classes in developing countries, and technological improvements in the fishing industry are putting these resources under unprecedented pressure.[4] Since 1980, there has been a spectacular 40 percent growth in worldwide demand for seafood. Some scientists have issued a stern warning on the future of the world's fisheries: Unless major steps are taken quickly, there will be a "global collapse" of all species currently fished by 2050.[5]

Defining "collapse" as a population less than 10 percent of its previous levels, these authorities noted that 30 percent of the fish species currently exploited have already collapsed. The declines are most extreme among stocks of large predatory fish like sharks, tuna, and swordfish. An estimated 75 million sharks are killed each year (mainly by Spanish and Indonesian fishermen) to provide shark fin soup (which can cost up

• **Figure 2.8** Seattle's Pike Street Market. Humanity's demand for fish and seafood is growing, but what about supplies? Stocks are falling, and in many species plummeting.

Joe Hobbs

• Figure 2.9 Shark fins at a Hong Kong market. There are several streets in this section of the city filled with shops selling shark fins and edible birds' nests. Both are specialty items in high demand among ethnic Chinese people throughout the world. Note the small Daoist prayer niche.

to $100 per bowl) to the world's ethnic Chinese, who believe it bestows health and virility (**•Figure 2.9**). An uncertain but large percentage of these sharks are "finned," a cruel procedure in which the fins are cut off the living fish, which are then thrown back into the sea to die. Public relations campaigns against the consumption of shark fin soup have begun to yield successes. Sales of shark fins in Hong Kong, the main center for the shark fin trade, have declined by about a third in recent years. (In one television ad, retired basketball star Yao Ming says, "Remember, when the buying stops, the killing can too.") Populations of the bluefin tuna—a magnificent, huge fish built for speed and endurance—are under pressure because their flesh is prized for *sushi* and *sashimi*. Not surprisingly, Japan has consistently blocked efforts to put the bluefin on a United Nations
100 (UN) list of protected species.

Some good news is that these grim realities and scenarios can be changed by action on several fronts, including reducing the number of unwanted fish caught in nets and cracking down on overfishing where it is recognized to be a problem. Governments will probably need to intervene. Legislation like the United Nations Convention on the Law of the Sea (see page 272) can encourage marine conservation and include mechanisms to enforce the law. Like the climate change treaties discussed later, it is easy for a country to sign an international agreement. But then this legislation needs to be ratified, and most importantly it needs to be enforced.

More good news is that **aquaculture** (the cultivation of aquatic organisms for food, including **fish farming**) has the potential to increasingly substitute for wild-caught fish stocks. Fish farms already supply half of the seafood consumed around the world. There has been particularly explosive growth in aquaculture in China (which produces 70 percent of the world's farmed fish). Some of this aquaculture is what may be described as "sustainable seafood," including the unlikely harvest of shrimp raised in well water deep in the Arizona desert.[6] But much commercial fish farming is controversial because of inputs of antibiotics, pesticides, fertilizers, and hormones, and because of the prospect that genetically modified salmon and other fish (dubbed "Frankenfish" by detractors) would escape their confines and alter the genetic heritage of wild fish. These issues mirror many of those raised in conventional land-based agriculture.

They provide energy and other raw materials for human use. As terrestrial sources of conventional energy dwindle, our fossil fuel–addicted economies create demand for new supplies. Deep seas are the final frontier for energy exploration. Under the U.S. territorial waters of the Gulf of Mexico, there are as many as 40 billion barrels of petroleum (enough to supply the U.S. oil demand for about 5 years). But getting to that oil is problematic: much of it lies beneath more than 10,000 feet (3,050 m) of water and then 5 miles (8 km) of rock, salt, and sand. Deep-water drilling presents special engineering challenges and environmental risks. Human error in operating the *Deepwater Horizon* well, which was pumping oil from an unbelievable depth of about 35,000 feet (10,000 m) in the Gulf
of Mexico, led to a historic oil spill and cleanup effort in 2010. 417
Similar opportunities and perils lie in the depths of the Arctic and Atlantic Oceans (see pages 427 and 382). Meanwhile, there is enormous potential to capture unconventional energy supplies from the sea, especially by using the power of surface winds and waves and of rising and falling tides to generate electricity. Prospects are increasing for the deep-sea mining of other minerals, including gold, silver, and the copper, cobalt, nickel, and "rare earth" minerals (many of which are used in high-tech devices) held within manganese nodules strewn across much of
the world's seafloor. 273

They play important roles in trade and commerce. The seafaring days are far from over. A remarkable 90 percent of global trade is seaborne. You know all those Chinese-made goods that are practically everywhere you look in the United States? About 98 percent of them travel across the Pacific on cargo ships.[7] There is an economic rationale for this. Airfreight can cost 20 times as much as sea freight. As we will see in many cases, protecting the world's sea-lanes from enemies has long been one of the world's most important geopolitical concerns. These concerns are greater now than at any time in history, particularly where oil is involved (see, for example, pages 177 and 240).

All of these features of the world's oceans help to explain a remarkable fact of modern life on Earth: Over half of the world's people live within 60 miles (100 km) of a coastline. Ten percent of the world's people live within 6 miles (10 km) of the coast. You can see these spatial characteristics in the map of the world's population on page 55.

2.5 Global Environmental Change

In the atmosphere—the layer of oxygen, nitrogen, and other gases extending from the Earth's surface to about 60 miles (100 km) above it—changes are occurring that will almost certainly have

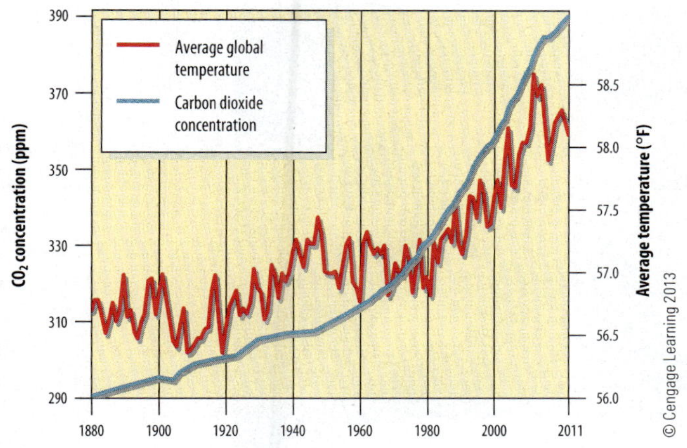

• **Figure 2.10** Industrialization, the burning of tropical forests and other factors have produced a steady increase in carbon dioxide emissions. Many scientists believe that these increased emissions explain the corresponding steady increase in the global mean temperature.

profound effects on natural systems and on human uses of the Earth. Many of these changes are attributable to human activities, and the final section of this chapter is a prelude to the following chapter on human processes affecting the planet.

Until about 2000, there was considerable uncertainty in the science of climate change. Most scientists at that time insisted that human activities were responsible for a documented warming of the Earth's surface (by 1.4° Fahrenheit, or about 0.8° Celsius, since the late nineteenth century), but a significant minority either insisted that warming was not occurring or, if it was, that a natural climatic cycle was responsible. Today, there is far less scientific uncertainty about the human role in climate change.

The roughly 2,500 atmospheric scientists from more than 130 countries who make up the United Nations–sponsored **Intergovernmental Panel on Climate Change (IPCC)** have concluded that global warming is "unequivocal." Human production of **greenhouse gases** such as **carbon dioxide (CO_2)** is, according to the IPCC, "very likely" responsible for this warming (•**Figure 2.10**). In the scientists' jargon, "very likely" means a more than 90 percent probability.[8]

The Greenhouse Effect

In 1827, French mathematician Jean-Baptiste Fourier established the concept of the **greenhouse effect,** noting that the Earth's atmosphere acts like the transparent glass cover of a greenhouse (•**Figure 2.11**). You can also think of the atmosphere as being like a car's windshield. Visible sunlight passes through the glass to strike the planet's surface. Oceans and land, like the floor of the greenhouse or the car's upholstery, reflect the incoming solar energy back as heat (invisible infrared radiation). Acting like the greenhouse glass or car windshield, the Earth's atmosphere traps some of that heat.

The greenhouse effect is not a bad thing. In fact, if not for naturally occurring greenhouse gases such as carbon dioxide

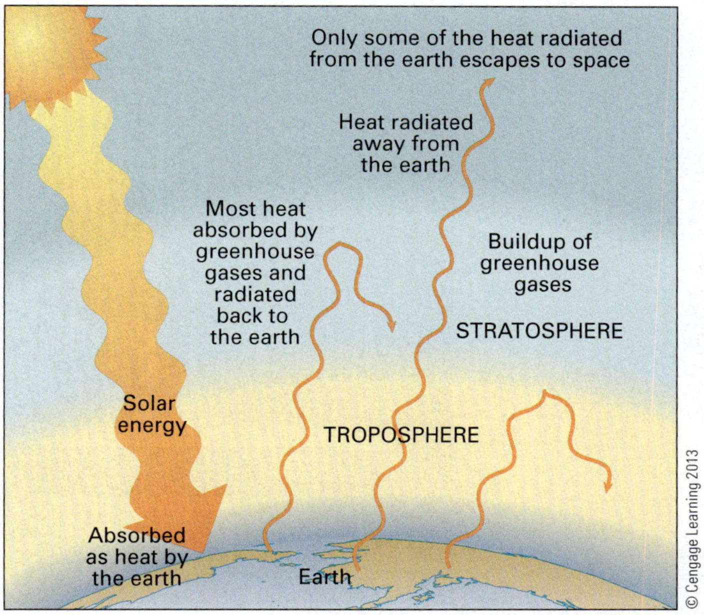

• **Figure 2.11** The greenhouse effect. Some of the solar energy radiated as heat (infrared radiation) from the earth's surface escapes into space, while greenhouse gases trap the rest. Naturally occurring greenhouse gases make the earth habitable, but carbon dioxide and other greenhouse gases emitted by human activities accentuate the greenhouse effect, making the planet unnaturally warmer.

and water vapor, the Earth would be too cold for most forms of life (with average temperatures well below freezing). What has so many people worried is what happens when human activities add newer and greater quantities of greenhouse gases into the atmosphere, trapping unnatural amounts of heat; these are the so-called "anthropogenic" causes of climate change. Carbon dioxide released into the atmosphere from the burning of coal, oil, and natural gas is the greatest source of concern, but **methane** (from rice paddies and the guts of ruminating animals like cattle), **nitrous oxide** (from the breakdown of nitrogen fertilizers), and **chlorofluorocarbons (CFCs),** used as coolants and refrigerants) are also greenhouse gases resulting from human activities. CFCs also destroy stratospheric ozone, a gas that has the important effect of preventing much of the sun's harmful ultraviolet radiation from reaching the Earth's surface. In the car-as-Earth metaphor, continued production of these greenhouse gases has the effect of rolling up the car windows on a sunny day, with the result of increased temperatures and physical overheating of the occupants.

The Effects of Global Warming

The trend toward a warmer world is remarkable. Although formal records of meteorological observations began only about a century ago, past climates left evidence in the form of marine fossils, corals, glacial ice, fossilized pollen, and annual growth rings in trees. These indirect sources, along with formal records dating back to 1861, indicate that the twentieth century was the warmest of the past six centuries and that the seven

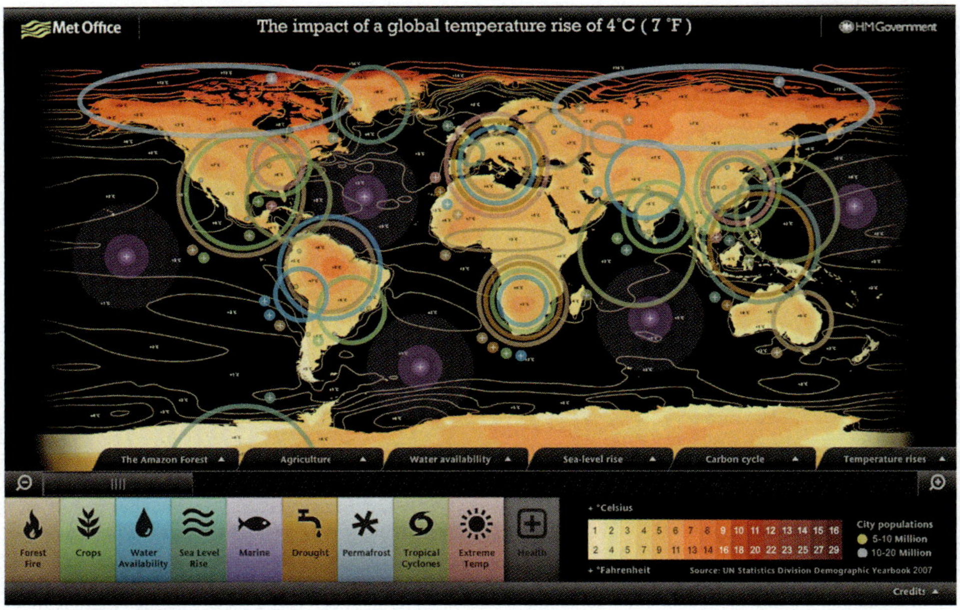

• Figure 2.12 Some of the world's climate changes anticipated with a global mean increase of 7°F (4°C). http://www.egov.vic.gov.au/government-initiatives/environment/climate-change-interactive-map.html

warmest years in the past 600 years were clustered recently; in descending order they were 2010, 1998, 2002, 2003, 2005, and 2007.

And what is the forecast? The IPCC uses a number of different "emission scenarios" that consider such variables as rates of population, economic and industrial growth, as well as efforts that might be made to curb greenhouse gas emissions. The "best estimate range" of the IPCC is that the mean global temperatures will warm from 3°F to 7°F (1.8°C to 4°C) by 2100. Briefly, these are the climate changes that are anticipated in this warming world (see **•Figure 2.12**).

A warmer climate overall, but . . . . Among our first questions in the study of the world regional geography of climate change is *where* the anticipated changes will occur and what effects these changes will have. Computer models suggest that there will not be a uniform temperature increase across the entire globe. Instead, the increases will vary spatially and seasonally. Some places will actually become cooler. Different computer models sometimes produce contradictory results about the timing and impacts of warming. For example, models predicting warmer summers generally forecast that crop productivity will decline, whereas models predicting warmer winters generally forecast an increase in crop yields.

More precipitation overall, but also more pronounced drought. Higher temperatures will mean increased evaporation from the world's oceans. This will result in more precipitation, but it will be distributed irregularly. The IPCC forecasts more precipitation in the higher latitudes and less precipitation, with intense and longer droughts, in the lower latitudes. What the IPCC calls "heavy precipitation events" will be more common. Even unusually large snowfall trends are expected, and the magnitude of hurricanes (whose strength increases with warmer ocean surface waters) will increase. 260, 437

Pronounced warming in the polar regions. Geographically, the impacts of climate change are expected to be greatest at the higher latitudes. Permafrost, the "permanently" frozen ground typical of the polar regions, has been melting at an alarming rate. Among other things, this has been forcing Arctic peoples to relocate from traditional settlements. 20 Sailing on a Russian nuclear-powered icebreaker, I saw open water at the North Pole in August 1996. The Russian Ice Master on board said that in 30 years of working on icebreakers in the Arctic, he had never seen anything like this: vast expanses of the Arctic Ocean completely ice-free or just dotted with sea ice and icebergs. Ever since that summer, there have been reports of unprecedented change at sea and on land in the Arctic. The IPCC report estimates that the average coverage of Arctic sea ice has shrunk about 3 percent per decade since 1978, with summertime ice decreasing 7 percent per decade. The trend is self-perpetuating: the darker ocean waters that replace the white ice cover absorb ever more solar radiation (think of cars here again: How hot would the surface of a dark blue car feel to your touch on a summer day compared with the surface of a white car?). That warming melts even more ice, creating more dark surface water and more warming (**•Figure 2.13**). In some parts of the world, this is seen as good news. A further retreat of polar ice would bode well for maritime shipping through the long-icebound Northwest Passage across Arctic Canada, and through the Northern Sea Route across the top of Russia, which in the past has been navigable only in summer and only with the aid of icebreakers. 148, 427

Joe Hobbs

• **Figure 2.13** A polar bear feeding on a seal in the Arctic Ocean. The bear's scientific name, *Ursus maritimus*—meaning "marine bear"—aptly suggests that this great predator spends much of its time at sea. But it must come up on the sea ice to feed, sleep, hibernate, and bear young. As sea ice retreats, there are fewer places for the bear to eat and rest, and there are increasing accounts of bears swimming to their death in open water. The sea ice seen here deflects the sun's rays, but the dark blue water absorbs them, contributing to warmer water and more ice melt. See page 32.

Shifting biomes, with species extinction and agricultural changes. There is general agreement that in a warmer world, the distribution of climatic conditions typical of biomes will shift poleward around the world, and upward in mountainous regions. There is particular concern about the Arctic and subarctic biomes of Canada, Scandinavia, and Russia, where studies suggest that climate change could alter 70 percent of these natural habitats' areal coverage, and cause the extinction of 20 percent of the habitats' species.[9]

Many animal species will be able to migrate immediately to keep pace with changing temperatures, but plants, being stationary, will not. Conservationists who have struggled to maintain islands of habitat as protected areas are particularly concerned about rapidly changing climatic conditions. Agriculture would also be affected as growing zones move poleward. And although the right temperature and rainfall combinations for productive agriculture might emerge in new areas, the right soil conditions would have to exist as well (the hard rock of the Canadian Shield will never make good farmland, for example).

Rising sea levels. There is also general agreement that if global temperatures rise, so will sea levels as glacial ice melts, and as seawater warms and occupies more volume (just as mercury in a thermometer does, causing the mercury to climb the glass tube). Sea levels rose 6 to 9 inches (15 to 23 cm) in the twentieth century. The IPCC forecasts a further rise of 7 to 24 inches (18 to 61 cm) rise by 2100, but other scientists call these figures far too conservative because they do not account for any melting of the vast Greenland and Antarctic land ice sheets. Projections of sea level rise based on the melting of these ice sheets are breathtaking.

Decision makers in coastal cities throughout the world, in island countries, and in nations with important urban or agricultural lowland areas adjacent to the sea are worried about the implications of rising sea levels. The low-lying coastal countries of the Netherlands and Bangladesh have for many years been outspoken advocates of reduced greenhouse gases. Their natural allies are the roughly 40 vulnerable island countries around the world who make up the advocacy group known as the Alliance of Small Island States. The president of alliance member country Maldives, an Indian Ocean country composed entirely of islands barely above sea level, pleaded "We are an endangered nation!" 250

Geopolitical instability. There is growing consensus that some of the poorest and most overcrowded nations will be hit hardest by climate change and that political and economic instability will result. The IPCC forecasts that rising sea levels will flood tens of millions of poor people out of their homes each year and that by 2080, possibly 200 million to 600 million people will face starvation because of failing crops. By then, one billion to three billion people in developing countries will face shortages of freshwater related to growing aridity. Epidemics of malaria and other killer diseases will be more widespread.

Such environmental crises will likely lead to large movements of "climate change refugees" and to conflicts over resources, upsetting the fragile balance of power between countries. In a published report, U.S. military thinkers concluded that climate change warming "presents significant national security challenges to the United States," adding, "Projected climate change will seriously exacerbate already marginal living standards in many Asian, African and Middle Eastern nations, causing widespread political instability and the likelihood of failed states. The chaos that results can be an incubator of civil strife, genocide and the growth of terrorism. The U.S. may be drawn more frequently into these situations."[10]

There is some scientific discussion of climate changes that cannot be avoided. From emissions already in the Earth's atmosphere, temperatures will continue to climb because of the oceans' "lag time": they warm up slowly and will take a long time to respond to greenhouse gases already in the atmosphere. Some scientists have begun speaking of a **tipping point** in climate change, a point at which feedback effects amplify temperature changes, producing some changes that are irreversible. The tipping point would bring a rapid acceleration in temperature and of related effects like more violent storms, dramatic crop losses, and spreading deserts.

What Can Be Done about Global Climate Change?

Fortunately, there is growing international political will to take the strong and sometimes costly measures that may be necessary to prevent even the direst scenarios. Scientists and policy-makers are advocating actions to limit future warming of the Earth's atmosphere to no more than 2°C (3.6°F)—well below the IPCC forecast of up to 4°C (7°F) by 2100.

Efforts to reduce the production of greenhouse gases focus on the world's wealthiest nations, but there is a good reason for the growing pressure on China to join the fight. By far the greatest producers of carbon dioxide emissions, China and the United States together account for roughly 40 percent of the entire global emissions output. As recently as 1994, China was predicted to catch up with the United States' CO_2 emissions levels by 2019. The fact that China achieved this distinction in 2007 instead is testimony to the dizzying pace of that country's development—a theme that echoes throughout this book.

There are essentially two approaches to confronting climate change: mitigation and adaptation. **Mitigation** measures aim to avoid the adverse impacts of climate change in the long term; these include steps like switching from coal to cleaner fossil fuels to produce electricity, replacing fossil fuels with other energy resources, reducing energy consumption, and removing greenhouse gases from the atmosphere by boosting photosynthesis (for example, by planting more trees, which absorb carbon dioxide in the photosynthetic process of producing plant tissue and oxygen). **Adaptation** measures are designed to cope with and reduce the unavoidable impacts of climate change in the short and medium terms; these include building sea walls and dikes to prevent flooding related to rising sea levels, relocating people from flood-prone areas to higher ground, and developing crop varieties that are more suited to expected changes in precipitation and temperature. Generally, mitigation is seen as a problem that the world's richer countries plus China should deal with, because they are the main producers of greenhouse gases. Adaptation is typically seen as a problem for the poorer countries, which are expected to suffer the most from climate change. Many of the world's largest population clusters are in low-lying river deltas in poor countries: Vietnam's Mekong, Bangladesh's Brahmaputra/Ganges, and Egypt's Nile, for example. These are the broad associations of adaptation and mitigation. In reality, most poor and wealthy countries alike will have to take measures to both adapt to and mitigate climate change.

Following are some of the main approaches to mitigating climate change, and thereby avoiding the kinds of almost-unthinkable adaptations that would be needed in vulnerable countries like Vietnam (see page 260).

Negotiate and implement international treaties to reduce greenhouse gas emissions. Most scientists and policymakers agree that the best way to confront global climate change is to implement international treaties to reduce emissions (see Geography of Energy, page 35, and •Figure 2.14). Countries have proven they can unite in effective action against greenhouse gases. Thanks to the **Montreal Protocol** and its amendments, signed by 37 countries in the late 1980s, the production of CFCs worldwide was reduced in phases to zero by 2010. As a result there should a marked reduction in the size of stratospheric ozone "holes" that have been observed seasonally over the southern and northern polar regions since about 1985. CFC molecules have very long life spans, but scientists predict that the ozone layer should recover to pre-1980 levels by 2050. This recovery will also depend on the successful phase-out of other ozone-destroying refrigerants known as hydrochlorofluorocarbons (HCFCs).

Cut emissions through market-based incentives. Compared with richer nations, the less developed countries (LDCs), other than China and India, release relatively little CO_2 into the atmosphere through industrial activities; most of their emissions come from deforestation. Because the less developed countries produce relatively little carbon dioxide, policymakers are coming up with innovative ways to keep them from doing so, and to encourage them to develop clean industrial technologies. One approach is the system of **cap-and-trade** or **emissions trading.** In these schemes, already being used by the United Nations, the European Union, and the unregulated "voluntary market," managers require the richer nations to achieve a net reduction in carbon emissions. It is very expensive for the already highly energy-efficient, richer country to cut its greenhouse gas emissions, but relatively inexpensive for the rich country to pay a poorer, energy-inefficient country to cut its emissions. These schemes therefore allow the richer countries either to make the cuts at home or to offset their emissions by buying credits from poorer countries that actually cut carbon emissions. The poorer country selling the credits is obliged to use the income to invest in energy-efficient and nonpolluting technologies. In some of the systems, the LDCs can also sell carbon credits for pollution they are not actually creating; with relatively little manufacturing, their emissions may be lower than what they are allowed to produce in the trading scheme.

China is making big profits from emissions trading, using European money to reduce emissions of HFC-23, a potent greenhouse gas produced in many Chinese refrigerator factories. Russia is another big advocate of emission-trading schemes like these. Its industries are producing well below their benchmark 1990 carbon dioxide emission levels (they declined 36% between 1990 and 2009), so the country has much to gain by selling its unused emission rights. One buyer is Denmark, which is paying to convert coal-fired electricity plants to cleaner-burning gas in Russia's Siberia region. In such a trading scheme, both countries apparently benefit, as does the global atmosphere. However, this practice is still young, and critics say it ultimately lets the big polluters in the industrialized countries avoid the hard work of cutting emissions further at home.

Increase carbon sequestration. **Carbon sequestration** refers mainly to the natural capture and long-term storage of carbon in forests, farmlands, or in the oceans, so that the build-up of carbon dioxide in the atmosphere will reduce or slow (there are also artificial means of carbon sequestration, for example, pumping carbon dioxide deep into underground repositories).[11] In the process of photosynthesis, plants absorb carbon dioxide. Collectively, the world's forests, farmlands, and seas (which contain carbon-absorbing phytoplankton) thus serve as giant **carbon sinks.** In international climate-change negotiations, forest- and farm-rich countries want to receive credit for having these natural buffers against climate change.

Climate change can also be mitigated by reducing deforestation, especially in the poorer countries. People in these countries rely heavily on trees and other vegetation for fuel. When burned, trees release the CO_2 that they had stored in the process of photosynthesis. At the same time, in being burned they cease to exist as organisms that, in the process of photosynthesis, remove CO_2 from the atmosphere. Deforestation for agricultural expansion, conversion to pastureland, infrastructure

GEOGRAPHY OF ENERGY | The Kyoto Protocol and Beyond

In 1997, at a world conference on climate change, 84 countries (of the 160 countries represented at the conference) signed the **Kyoto Protocol**, a landmark international treaty for climate change. The agreement required 38 more developed countries (MDCs, also known as the Annex I countries) to reduce carbon dioxide emissions to 5 percent or more *below their 1990 levels* by the year 2012. The United States pledged to cut its emissions to 7 percent below 1990 levels by that date. The European Union made a promise of 8 percent below 1990 levels, and Japan promised a 6 percent reduction below 1990 levels.

Meeting these pledges would require substantial legislative, economic, and behavioral changes in these more affluent countries. Transportation and other technologies that use fossil fuels would have to become more energy efficient, making these technologies at least temporarily more expensive. Advocates of the protocol argued that the initial sacrifices would soon be rewarded by a more efficient and competitive economy powered by cleaner and cheaper sources of energy derived from the sun. Opponents, however, felt that higher fuel prices would be too costly for the United States and other industrialized economies to bear. Their position made it difficult for some signatories of the Kyoto Protocol to ratify the treaty. But in 2005, after the European Union countries, Japan, and finally Russia ratified the Kyoto Protocol, it went into effect for the countries that ratified it.

The United States signed the Kyoto Protocol during the Clinton Administration. Even then, the political will for ratification did not exist. When President George W. Bush assumed office in 2001, he rejected the Kyoto Protocol outright. The Bush administration had two objections: the potentially high economic cost of implementing the treaty and the fact that China, along with all the world's less developed countries, was not required by the Kyoto Protocol to take any steps to reduce greenhouse gas emissions. China argued then, as it still does now, as the world's leading greenhouse gas producer, that it is a developing country whose growth should not be hampered by pressure from countries that became rich by burning fossil fuels. China's economy continues to grow by as much as 10 percent annually, and so do its carbon dioxide emissions, up 206 percent between 1990 and 2009. In the same period, U.S. emissions rose by 7 percent. European Union (EU) member country achievements have been remarkable: Britain's emissions were down 15 percent for that period, and Germany's down 21 percent.

CHRISTIAN ASLUND/AFP/Getty Images/Newscom

• **Figure 2.14** Protestors calling for action on negotiating the climate change treaties in Copenhagen in 2009.

The EU is particularly upset that the United States has not taken leadership on climate change. The 27 countries of the European Union see themselves as the industrialized, wealthy peers of the United States, with the same challenges in investing in leaner and greener manufacturing technologies. But although the Europeans kept "upping the ante"—proposing to cut emissions to at least 20 percent and as much as 40 percent *below* 1990 levels by 2020, and taking actions to keep up with original Kyoto Protocol reduction pledges—the United 98
States talked about cutting emissions. So did China. The American and Chinese political systems could not be more different, but each of these two leading contributors to greenhouse gas emissions has lacked the political will to confront climate change. Much-anticipated climate change talks in Copenhagen in 2009 failed to make progress, and it was not clear what kind of agreement if any would succeed the Kyoto Protocol when it expired in 2012.

development, logging, and other purposes accounts for about 20 percent of global greenhouse gas emissions (more than the global transportation sector and second only to the energy sector.[12] Since 2008, the United Nations has been developing a program called **Reducing Emissions from Deforestation and Forest Degradation (REDD).** This is an effort to create a financial value for the carbon stored in forests, offering incentives for developing countries to reduce carbon emissions from forested lands and invest in low-carbon development. REDD has other potential benefits, like alleviating poverty, enhancing the status of indigenous peoples (who often have their own systems of protecting environments), and protecting biodiversity. REDD is emerging as part of the carbon-trading scheme described above, allowing richer countries to purchase carbon credits from poorer countries. But in the long run, the richer countries may also be expected to donate funds to REDD in the interest of the planet's welfare. 253, 384

The next chapter discusses many ways in which the world's developed and less developed countries must confront problems like global climate change differently. Like this chapter, it will conclude with some specific ideas about how to deal with some of the most pressing issues of our time.

summary

- The Earth's three layers of habitable space are the hydrosphere, atmosphere, and lithosphere. The lithosphere is made up of separate plates that are in motion, a process known as plate tectonics. These movements result in mountain building, volcanic activity, earthquakes, and other consequences.
- Weather refers to atmospheric conditions prevailing at one time and place. Climate is a typical pattern recognizable in the weather of a region over a long period of time. Climatic patterns have a strong correlation with patterns of vegetation and in turn with human opportunities and activities in the environment.
- Geographers group local climates into major climate types, each of which occurs in more than one part of the world and is associated with other natural features, particularly vegetation. Geographers recognize 10 to 20 major types of ecosystems or biomes, which are categorized by the type of natural vegetation. Vegetation and climate types are so sufficiently related that many climate types are named for the vegetation types.
- Some biomes are particularly important because of their biological diversity—the number of plant and animal species and the variety of genetic materials these organisms contain. Regions where human activities are rapidly depleting a rich variety of plant and animal life are known as biodiversity hot spots, places scientists believe deserve immediate attention for study and conservation.
- Oceans cover about 71 percent of the Earth's surface. They play the key role in the hydrologic cycle, sustain large numbers of people through the protein in fish and seafood, and contain valuable mineral resources.
- Most of the sun's visible short-wave energy that reaches the Earth is absorbed, but some of it returns to the atmosphere in the form of infrared long-wave radiation, which generates heat and helps warm the atmosphere. This is the Earth's natural greenhouse effect.
- The scientific community is convinced with 90 percent or greater certainty that human activities, particularly the production of carbon dioxide and other greenhouse gases, are responsible for global warming. Computer-based climate change models use the scenario of a doubling of atmospheric carbon dioxide and indicate that the mean global temperature might warm up by an additional 11 degrees Fahrenheit (6.1 degrees Celsius) by the year 2100. The general scientific consensus is that with global warming, the distribution of climatic conditions typical of biomes will shift poleward and upward in elevation, sea levels will rise, and mean global precipitation will increase (but with drought intensified in some areas).
- There are two approaches to confronting climate change: Mitigation measures aim to avoid the adverse impacts of climate change in the long term, and are usually associated with more-developed countries (MDCs). Adaptation measures are designed to cope with and reduce the unavoidable impacts of climate change in the short and medium terms, and are usually associated with less-developed countries (LDCs).
- In the policy arena, the European Union countries tend to exert the strongest leadership in fighting climate change. The United States has often been slow or unwilling to take strong, specific actions. China has tried to be exempted from mandatory steps.
- The Kyoto Protocol was an international agreement requiring the industrialized countries that ratified it to make substantial cuts in their carbon dioxide emissions to reduce global warming. The United States dropped its support for the treaty. It went into effect for the ratifying countries after Russia ratified it in 2004 and expired in 2012. Efforts continued to reduce carbon dioxide emissions, especially in the MDCs and China, and to maximize ways to reduce deforestation, especially in the LDCs.

Key Terms + Concepts

adaptation (p. 34)
aquaculture (p. 30)
atmosphere (p. 20)
biodiversity (p. 28)
biodiversity hot spots (p. 28)
biological diversity (p. 28)
biomes (p. 23)
  boreal forest (p. 25)
  coniferous forest (p. 25)
  desert shrub (p. 25)
  Mediterranean scrub forest (p. 27)
  prairie (p. 26)
  savanna (p. 26)
  steppe (p. 25)
  taiga (p. 26)
  temperate grassland (p. 26)
  temperate mixed forest (p. 27)
  tropical deciduous forest (p. 26)
biosphere (p. 20)
cap-and-trade system (p. 34)
carbon sequestration (p. 34)
carbon sink (p. 34)
climate (p. 23)
  desert (p. 25)
  humid continental (p. 27)
  humid subtropical (p. 27)
  ice cap (p. 25)
  marine west coast (p. 26)
  Mediterranean (p. 26)
  semiarid (p. 25)
  subarctic (p. 25)
  tropical rain forest (p. 26)
  tropical savanna (p. 26)
  tundra (p. 25)
  undifferentiated highland (p. 28)
coniferous trees (p. 26)
continental drift (p. 21)
emission-trading system (p. 34)
fault (p. 21)
faulting (p. 21)
fish farming (p. 30)
greenhouse effect (p. 31)
greenhouse gases (p. 31)
  carbon dioxide (CO_2) (p. 31)
  chlorofluorocarbons (CFCs) (p. 31)
  methane (p. 31)
  nitrous oxide (p. 31)
hydrologic cycle (p. 29)
hydrosphere (p. 20)

Intergovernmental Panel on Climate Change (IPCC) (p. 31)
Kyoto Protocol (p. 35)
lapse rate (p. 28)
lithosphere (p. 20)
mitigation (p. 34)
monoculture (p. 28)
Montreal Protocol (p. 34)
plates (p. 21)
plate tectonics (p. 21)
precipitation (p. 23)
Reducing Emissions from Deforestation and Forest Degradation (REDD) (p. 35)
Richter scale (p. 21)
seafloor spreading (p. 21)
seismic activity (p. 21)
subduction (p. 21)
subduction zone (p. 21)
tectonic forces (p. 21)
tipping point (p. 33)
trench (p. 21)
tsunami (p. 21)
volcanism (p. 21)
weather (p. 23)

Review Questions

1. What are the three "spheres" of habitable life on Earth?
2. What is plate tectonics, and what are some of the main consequences of tectonic activity?
3. What is the difference between weather and climate? What are the main forces that produce precipitation and aridity?
4. What are the major climate types and their associated biomes? Where do they tend to occur on Earth?
5. Where are the biodiversity hot spots? In what kinds of locations and biomes do many of them occur?
6. What important roles do the world's oceans play, and what are their major resources? In what ways are these resources threatened?
7. What apparent long-term effects on the Earth's atmosphere can be attributed to modern technology? What predictions are there for future changes, and what steps are being taken or considered to avert these changes through mitigation and through adaptation?

Notes

1. Wolf Roder, "Three Near Misses: Are These Science or Pseudoscience?" *Cincinnati Skeptics Newsletter*, June 1996. Accessed November 14, 2006, from http://www.cincinnatiskeptics.org/newsletter/vol5/n5/misses.html.
2. Norman Myers, *Primary Source: Tropical Forests and Our Future* (New York: Norton, 1984), p. x.
3. World Bank, "On World Water Day, World Bank Calls for Investments in Water Infrastructure and Better Governance," press release, March 22, 2007.
4. Review the United Nations' *Annual Report of the State of the World's Fisheries and Aquaculture* at www.fao.org/sof/sofia/index_en.htm.
5. Boris Worm et al., "Impacts of Biodiversity Loss on Ocean Ecosystem Services," *Science*, November 3, 2006, pp. 787–790.
6. See the Monterey Bay Aquarium's guide to sustainable seafood, "Seafood Watch," at www.mbayaq.org/cr/seafoodwatch.asp.
7. Amy Roach Partridge, "Global Trucking Woes," *Global Logistics*, November 2006, www.inboundlogistics.com/articles/global/global1106.shtml. Accessed September 27, 2007.
8. Intergovernment Panel on Climate Change, "Climate Change 2007," www.ipcc.ch. Accessed July 4, 2007.
9. Sarah Lyall, "Global Warming Report Predicts Doom for Many Species," *New York Times*, September 1, 2000.
10. Brad Knickerbocker, "Could Warming Cause War?" *Christian Science Monitor*, April 19, 2007, p. 2.
11. Source on carbon sequestration: www.carbonventures.com/policy/article.php?list=The%20Kyoto%20Protocol&id=4467&link=Carbon%20Sequestration/
12. Source on deforestation and carbon: www.un-redd.org/AboutREDD/tabid/582/Default.aspx/

Online Resources

CourseMate: Make the most of your study time by accessing everything you need to succeed in one place. Read your textbook, take notes, review flashcards, watch videos, complete activities, take practice quizzes, and more—online with CourseMate. Log in at **www.cengagebrain.com.**

"Try to learn something about everything, and everything about something."

—THOMAS HUXLEY (1825–1895)

Joe Hobbs

Joe Hobbs

Above, Bedouin on the move in Egypt. Their environmental knowledge is great, and their environmental "footprint" small. Left, a young European visitor to an Indian camp near Iquitos, Peru. The impact of her culture on the landscape is much greater.

3 Human Processes and World Regions

This chapter continues your introduction to geography's basic vocabulary, focusing on how people have interacted with the environment to change the face of the Earth, with an emphasis on the human roles. Modern trends in human–environment interactions are seen as the products of revolutionary changes in the past: the arrival of agriculture and of industrialization. The chapter uses some of the concepts of economic geography to explain where rich and poor countries are located on the Earth's surface and account for some of these patterns of prosperity and poverty. It considers where and why populations are increasing and what the implications of that growth are. Finally, it shares ideas about how to solve some of the most important global problems of our time.

chapter outline

chapter objectives

This chapter should enable you to

- Gain a historical perspective on the capacity of human societies to transform environments and landscapes.
- Understand why some countries are rich and others poor and recognize the geographic distribution of wealth and poverty.
- Explain the simultaneous trends of falling population growth in the richer countries and rapid population growth elsewhere.
- Explore the principles of sustainable development.

Joe Hobbs

• Figure 3.1 Until the relatively recent past—just a few thousand years ago—people were exclusively hunters and gatherers. This rock art in Egypt's Sinai Peninsula was created over a long time span, as it depicts Neolithic-period hunting of ibex, later uses of camels and horses, and writing from the Nabatean period (first century C.E.).

3.1 Two Revolutions That Have Changed the Earth

The geographer's approach to understanding a current landscape—in almost all cases, a cultural landscape that has been fashioned by human activity—is often historical, involving study of its development from the prehuman or early human natural landscape. The current spatial patterns of our relationship with the Earth may be seen as products of two "revolutions": the Agricultural Revolution that began in the Middle East about 10,000 years ago and the Industrial Revolution that began in eighteenth-century Europe. Each of these revolutions transformed humanity's relationship with the natural environment. Each increased substantially our capacity to consume resources, modify landscapes, grow in number, and spread in distribution. Why should we consider these historic processes? The answer lies in the name the Geological Society of America gave to its annual meeting in 2011: "The past is the key to the future."

Hunting and Gathering

Until about 10,000 years ago, our ancestors lived by **hunting and gathering** (also known as **foraging**). We were quite good at it—it served us well for more than 100,000 years, until we began experimenting with the revolutionary technologies of agriculture.

Foraging was quite different from the farming and industrial ways of living that succeeded it. Joined in small bands of extended family members, hunters and gatherers were nomads with no villages, homes, or other fixed dwellings. They moved to take advantage of changing opportunities on the landscape. These foragers scouted large areas to locate foods such as seeds, tubers, foliage, fish, and game animals (•Figure 3.1). Moving their small group from place to place, they had a relatively limited impact on the natural environment, especially compared with the impacts left by agricultural and industrial societies.

Hunters and gatherers may have been the **"original affluent society."**[1] Many scholars have praised these preagricultural people for the apparent harmony they maintained with the natural world in both their economies and their spiritual beliefs. After short periods of work to collect the foods they needed, they enjoyed long stretches of leisure time. Studies of those few hunter-gatherer cultures that lingered into modern times, such as the San (Bushmen) of southern Africa and several Amerindian groups of South America, suggest that although their life expectancy was low, they suffered little from the debilitating social and psychological problems found in industrialized societies.

Hunters and gatherers did modify their landscapes. These foragers were not always at peace with one another or with the natural world. With upright posture, stereoscopic vision, opposable thumbs, an especially large brain, and no mating season, *Homo sapiens* (Latin for "wise man") became, after its emergence in southern Africa about 200,000 years ago, an **ecologically dominant species**—defined as one that competes more successfully than other organisms for nutrition and other essentials of life, or that exerts a greater influence than other species on the environment. Using fire to flush out or create new pastures for the game animals they hunted, preagricultural people shaped the face of the land on a vast scale relative to their small numbers. Many of the world's prairies, savannas, and steppes where grasses now prevail developed as hunters and gatherers repeatedly set fires.

These people also overhunted and in some cases eliminated animal species. The controversial **Pleistocene overkill hypothesis** states that rather than being at harmony with nature, hunters and gatherers of the late Pleistocene Epoch (a geological period from about 2 million to 10,000 years ago) hunted many species to extinction, including the elephant-like mastodon of North America.

Farming: Welcome to the Anthropocene

Despite these excesses, the environmental changes that hunters and gatherers could cause were limited. Humans' power to modify landscapes took a giant step with **domestication**, the controlled breeding and cultivation of plants and animals. Domestication brought about the **Agricultural Revolution**, also known as the **Neolithic Revolution** or the **Food-Producing Revolution**.

This was the dawn of a new age, the likes of which the world had never experienced. In conventional geologic time scales, the Agricultural Revolution that began about 10,000 years ago coincided with a great retreat of glacial ice sheets and the onset of the Holocene Epoch in which we now live (Holocene comes from Greek words meaning "entirely new"). But there is a fascinating scientific effort underway to rewrite our understanding of the age in which we live. The goal is to rename the Holocene as the **Anthropocene Epoch** or simply "the Anthropocene," meaning "the age of humans." Peoples' impacts on the atmosphere, lithosphere, and hydrosphere have been so monumental in the last 10,000 years—and especially within the last 250 years—that they rank alongside the great natural forces that defined past ages.

It all began with agriculture. Why people started to produce rather than continue to hunt and gather plant and animal foods—first in the Middle East and later in Asia, Europe, Africa, and the Americas—is uncertain. Although there are many ideas about this process, surprisingly little hard evidence exists to explain it confidently. Two theories are most often put forward. One is that the climate changed. Increasing drought and reduced plant cover may have forced people and wild plants and animals into smaller areas, where people began to tame wild herbivores and sow wild seeds to produce a more dependable food supply.

A more widely accepted theory is that their own growing populations in areas originally rich in wild foods compelled people to find new food sources, so they began sowing cereal grains and breeding animals. The latter process may have begun about 8000 B.C.E. in the Zagros Mountains of what is now Iran. The culture of domestication spread outward from there (in the geographic process known as **diffusion**), but also developed independently in several world regions.

Now that humans were producing as well as consuming foods, their landscape uses, cultures, social organizations, and other characteristics changed dramatically. Among other things, in choosing to breed plants and animals, people settled down. Gradually abandoning the nomadic life and **extensive land use** of hunting and gathering, they came to favor the **intensive land use** of agriculture and animal husbandry. They could sow and harvest crops in specific places year after year. With less need to move around, they began living in fixed dwellings, at least on a seasonal basis. These developed into villages, small settlements with fewer than 5,000 inhabitants.

• **Figure 3.2** The Tigris River in southeastern Turkey. The brown areas are rainless in the long, hot summers and are capable of producing only one crop a year through dry (unirrigated) farming. The green areas along the river are irrigated and can produce two or three crops a year. Irrigation was thus a revolutionary technology that greatly increased the number of people the land could support.

People in these settlements raised larger and more reliable stocks of food, making it possible to support their growing numbers. Through **dry farming**, which involved planting and harvesting according to the seasonal rainfall cycle, population densities could be 10 to 20 times higher than they were in the hunting and gathering mode. By about 4000 B.C.E., people along the Tigris, Euphrates, and Nile Rivers began **irrigation** of crops—bringing water to the land artificially by using levers, channels, and other technologies—an innovation that allowed them to grow crops year-round, independent of seasonal rainfall or river flooding (•Figure 3.2). Irrigation allowed even more people to make a living off the land; irrigated farming yields five to six times more food per unit area than dry farming. In ecological terms, the expanding food surpluses of the Agricultural Revolution raised the Earth's **carrying capacity**, the size of a species' population (in this case, humans) that an ecosystem can support. When domestication of plants and animals began about 10,000 years ago, there were probably fewer than 5 million people on Earth. Today, there are more than 7 billion.

Culture became more complex, and society became more stratified. The steep increase in food production freed more people from the actual work of producing food, and they undertook a wide range of activities unrelated to subsistence needs. Irrigation and the dependable food supplies it provided thus set the stage for the development of **civilization**, the complex culture of urban life characterized by the appearance of writing, economic specialization, social stratification, and high population concentrations. By 3500 B.C.E., for example, 50,000 people lived in the southern Mesopotamian city of Uruk, in what is now Iraq. Other **culture hearths**—regions where civilization followed the domestication of plants and animals—emerged between 8000 and 2500 B.C.E. in China, Southeast Asia, the Indus River Valley, Egypt, West Africa, Mesoamerica, and the Andes.

Human impacts on the natural environment increased. The agriculture-based urban way of life that spread from these culture hearths had larger and more lasting impacts on the natural environment than either hunting and gathering or early agriculture. Acting as agents of humankind, domesticated plants and animals proliferated at the expense of the wild species that people came to regard as pests and competitors. For example, *Bos primigenius*, a wild bovine that was the ancestor of most of the world's domesticated cattle, was hunted to extinction by 1627. Farmers who resented the animals' raids on their crops were probably among the most ardent hunters. Agriculture's permanent and site-specific nature magnified the human imprint on the land, while the pace and distribution of that impact increased with growing numbers of people.

The Industrial Revolution

The human capacity to transform natural landscapes took another giant leap with the **Industrial Revolution**, which began in Europe around 1750 C.E. This new pattern of human-land relations was based on breakthroughs in technology that several factors made possible (•**Figure 3.3**):

- Western Europe had the economic capital necessary for experimentation, innovation, and risk. Much of this money came from the lucrative trade in gold, undertaken initially
81 in the Spanish and Portuguese empires after 1400.
- Significant improvements in agricultural productivity took place in Europe by about 500. New tools such as the heavy plow and more intensive and sustainable use of farmland led to increased crop yields. Human populations grew correspondingly.
- Population growth itself was a factor. More people freed from work in the fields could devote their talents and labor to trying new things, including tinkering with technologies that would improve crop yields. As agricultural innovations spread and industrial productivity improved, more and more of the growing European population was freed from farming, and for the first time in history, a region emerged in which city dwellers outnumbered rural folk. The process of industrialization, having now spread to nearly all parts of the world, continues to promote urbanization today.

•**Figure 3.3** L. S. Lowry's *Industrial Landscape* has all the essential elements of industrializing Britain: cotton mills and factories, coal mines, and working families' cottages.

Most geographers see population growth today as a drain on resources, but the Industrial Revolution illustrates that given the right conditions, more people do create more resources. Innovations such as the steam engine tapped the vast energy of fossil fuels—initially coal and later oil and natural gas. Fossil fuels, which are the stored-up carbon products created by photosynthesis in ancient ecosystems, allowed the Earth's carrying capacity for humankind to be raised again, this time into the billions.

Industrialization, Colonization, and Environmental Change

As they began to deplete their local supplies of resources needed for industrial production, Europeans started to look for these materials abroad. As early as their **Age of Discovery**, also known as the **Age of Exploration**, which began in the fifteenth century, Europeans probed ecosystems across the globe to feed a growing appetite for innovation, economic growth, and political power. The process of European **colonization**—the extension of European countries' political and economic control over foreign areas—was thus linked directly to the Industrial Revolution. Mines and plantations from such faraway places as central Africa and India supplied the copper and cotton that fueled economic growth in colonizing countries such as Belgium and England.

No longer dependent on the foods and raw materials they could procure within their own political and ecosystem boundaries, Europeans of the Industrial Revolution had an impact on the natural environment that was far more extensive and permanent than that of any other people in history. The world today is heir to their legacy: we tend to think of countries and economies being "industrialized" or "industrializing," either possessing that enormous power or on the pathway to acquiring it. It is difficult to summarize the impacts of agriculture and industrialization during the Anthropocene; much of this book tells that story. But these are some good examples:[2]

- The vast majority of Earth's ecosystems reflect the presence of people.
- Today, fully 40 percent of the Earth's land-based photosynthetic output is dedicated to human uses, especially in agriculture and forestry.
- Since 1750, the total forested area on Earth has declined by more than 20 percent. Today, there are more trees on farms than in wild forests.
- Since 1750, total cropland has grown by nearly 500 percent, with more expansion in the period from 1950 to today than in the century from 1750 to 1850. Much of this expansion has been made possible by the use of fertilizers produced from fossil carbon.
- Since 1750 people have released quantities of fossil carbon that the planet took hundreds of millions of years to store away. This has given people a commanding role in the planet's carbon cycle.

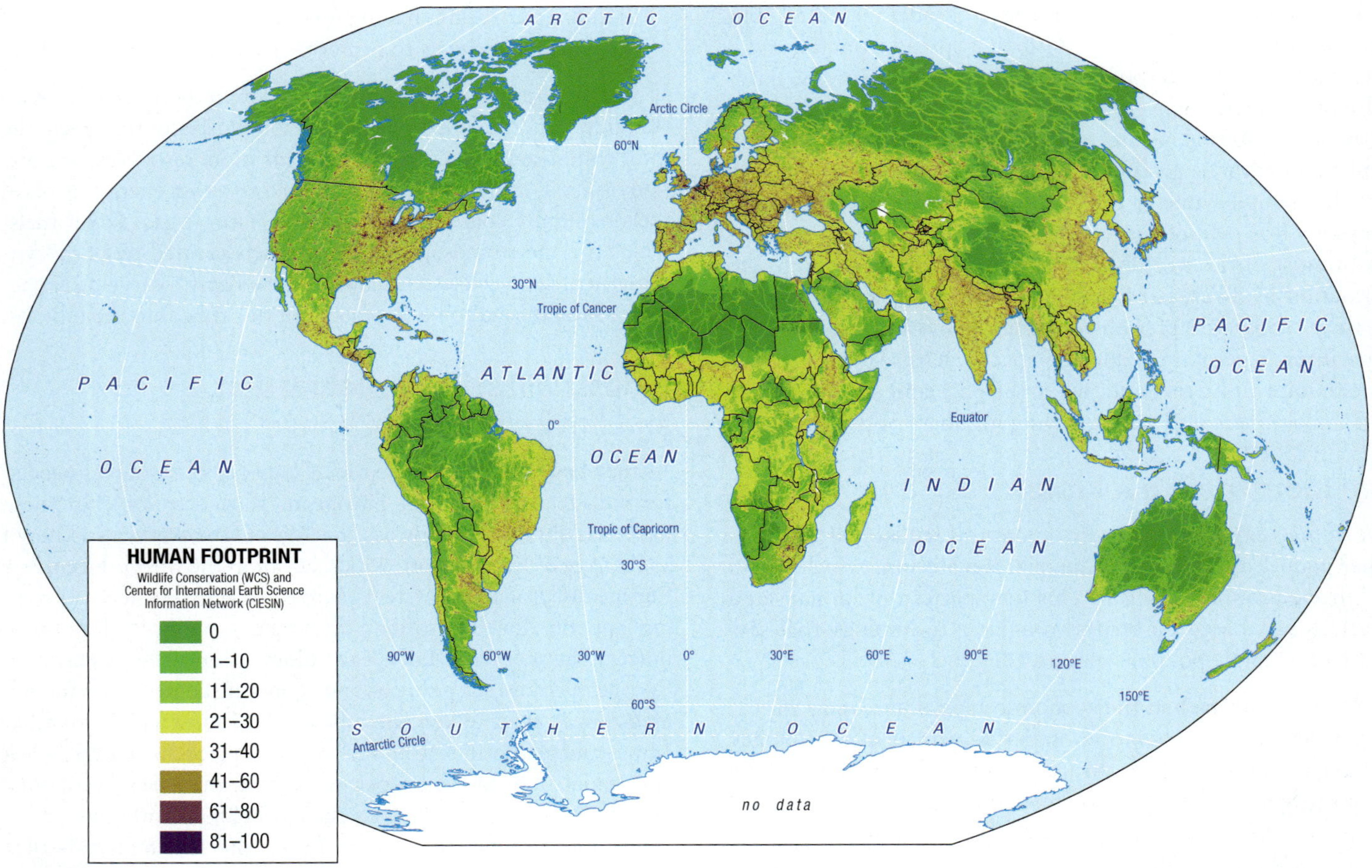

• **Figure 3.4** The biome map on page 24 depicts what the world's vegetation would look like without human activity. Here we see how strongly people have changed the natural environment. This isarithmic map of the human footprint is a quantitative analysis of human impacts on the Earth's biomes. A score of 1 indicates the least human influence in a given biome. A higher number means greater impact. However, because each biome has its own independent scale, a score of 1 in a tropical rainforest might reflect a different level of human activity than in a broadleaf forest.

Source: Wildlife Conservation Society (WCS) and Center for International Earth Science Information Network (CIESIN).

The global impacts of human activities are best summarized with the map of the "human footprint" depicted in •**Figure 3.4**. It shows uneven distributions of human impacts across the globe. These are the results of economic, population, and other processes we will now explore.

3.2 The Geography of Development

One of the most striking characteristics of human life on Earth is the large disparity between wealthy and poor people, both within and between countries. At a high level of generalization, the world's countries can be divided into "haves" and "have-nots" (•**Figure 3.5** and •**Table 3.1**). Writers refer to these distinctions variously as "developed" and "underdeveloped," "developed" and "developing," "more developed" and "less developed," "industrialized" and "industrializing" or "industrialized" and "nonindustrialized," and "North" and "South," based on the concentration of wealthier countries in the middle latitudes of the Northern Hemisphere and the abundance of poorer nations in the Southern Hemisphere. This text uses the terms **more developed countries (MDCs)** and **less developed countries (LDCs)**. This framework is an introductory tool and cannot account for the tremendous variations and continuous changes in economic and social welfare that characterize the world today. Some countries, including those known as the "Asian Tigers," are best described as **newly industrializing countries (NICs)** because they do not fit the MDC or LDC idealized types. The relevant regional chapters describe these cases. 233

Measuring Development

There is no single, universally acceptable standard for measuring wealth and poverty on the global scale. But you are likely to encounter these kinds of indexes and figures as you read about development.

Annual per capita gross domestic product. **Gross domestic product (GDP)** is the total output of goods and services that a country produces for home use in a year. Divided by the country's population, the resulting figure of **per capita GDP** is one

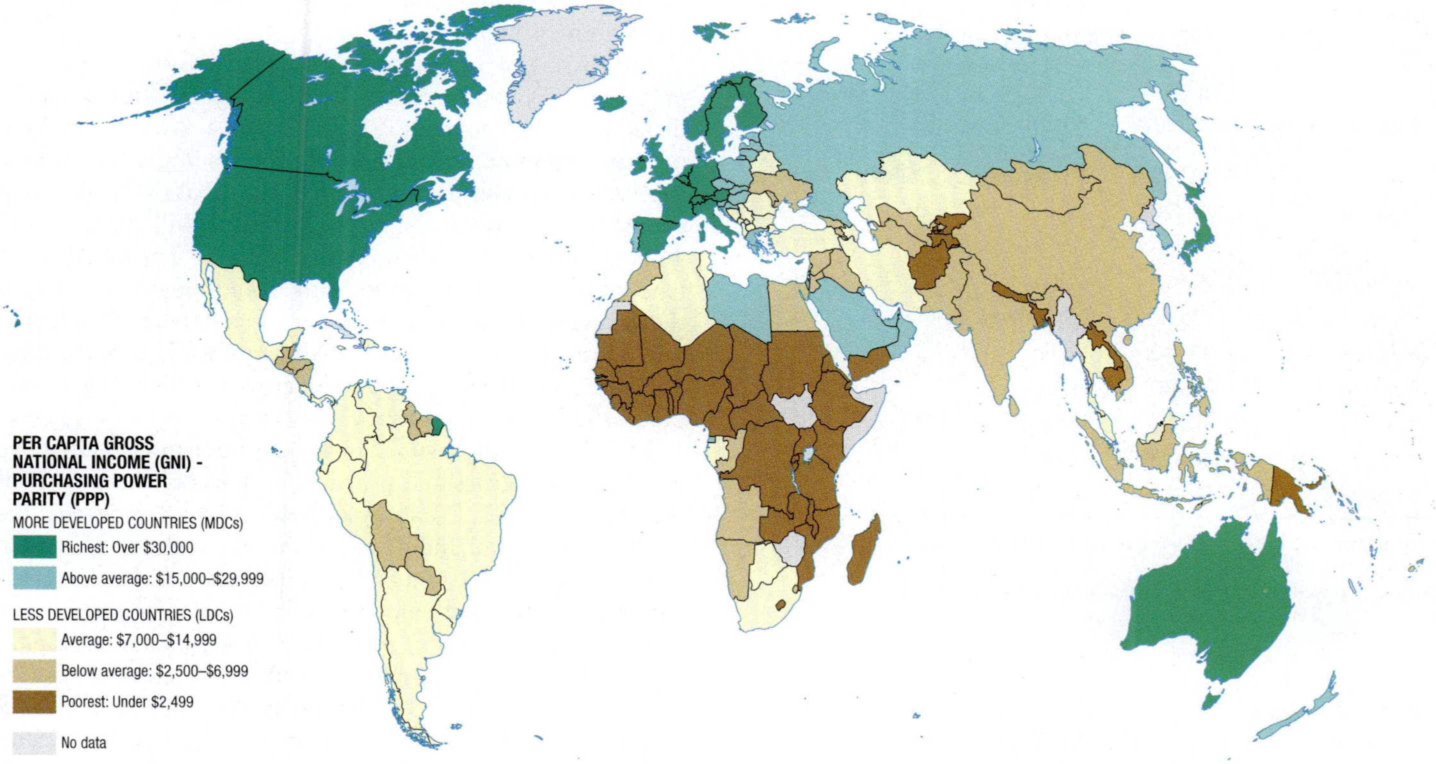

• **Figure 3.5** Wealth and poverty by country. Note the concentration of wealth in the middle latitudes of the Northern Hemisphere, and Africa's great poverty.

Source: Data from World Factbook, CIA, 2004.

Table 3.1 Characteristics of More Developed Countries (MDCs) and Less Developed Countries (LDCs)

Characteristic	MDC	LDC
Per capita GDP and income	High	Low
Percentage of population in the middle class	High	Low
Percentage of population involved in manufacturing	High	Low
Energy use	High	Low
Percentage of population living in cities	High	Low
Percentage of population living in rural areas	Low	High
Birth rate	Low	High
Death rate	Low	High[a]
Population growth rate	Low	High
Percentage of population under age 15	Low	High
Percentage of population that is literate	High	Low
Amount of leisure time available	High	Low
Life expectancy	High	Low

[a]Although death rates are high in the LDCs relative to the MDCs, in most countries they are quite low compared to what they once were in the LDCs.

of the most commonly used measures of economic well-being. A closely related measure is per capita **gross national product (GNP)**, which includes foreign output by domestically owned producers.

Annual per capita gross national income purchasing power parity. Although quite a mouthful, **per capita gross national income purchasing power parity (per capita GNI PPP)** is a useful figure in the study of world regional geography. **Gross national income (GNI)** includes gross domestic production plus income from abroad from sources such as rents, profits, and labor. **Purchasing power parity (PPP)** conversion factors consider differences in the relative prices of goods and services, providing a better overall way of comparing the real value of output between different countries' economies. GNI PPP is measured in current "international dollars," which indicate the amount of goods and services one could buy in the United States with a given amount of money. Definitions vary, but in this book, an MDC is a county with an annual per capita GNI PPP of $15,000 or more; all others are considered LDCs. By the way, economics is known sarcastically as "the dismal science," and you should probably not be too concerned with how the figures were calculated; instead, use them for simple comparisons.

The gulf between the world's richest and poorest countries is startling (•Table 3.2). The average per capita GNI PPP in the MDCs is nearly 6 times greater than in the LDCs. With a per capita GNI PPP of $59,950, Luxembourg is the world's richest country, whereas Liberia is the poorest with only $290. Forty-eight percent of the world's people live on less than $2 per day. These raw numbers suggest that economic productivity and income alone characterize **development**, which, according to a common definition, is a process

Table 3.2 Top and Bottom Countries in Per Capita GNI-PPP (U.S. Dollars, 2011)*

Top Countries	GNI-PPP Per Capita	Bottom Countries	GNI-PPP Per Capita
Luxembourg	59,590	Afghanistan	860
Norway	55,420	Togo	850
Singapore	49,780	Sierra Leone	790
Switzerland	47,100	Malawi	780
United States	45,640	Central African Rep.	750
Netherlands	39,740	Niger	680
Denmark	38,780	Eritrea	580
Australia	38,510	Burundi	390
Austria	38,410	Congo, Dem. Rep.	300
Sweden	38,050	Liberia	290

*This table excludes city-states, territories, colonies, and dependencies. Only countries with available data are listed.

Source: 2011 World Population Sheet, Population Reference Bureau.

of improvement in the material conditions of people through diffusion of knowledge and technology.

The Human Development Index. Definitions like that, and statistics like per capita GDP and per capita GNI PPP, reveal little about measures of well-being such as income distribution, gender equality, literacy, and life expectancy. Recognizing the shortcomings of strictly economic definitions, the United Nations Development Programme created the **Human Development Index (HDI)**, a scale that considers attributes of quality of life. This book uses HDI in the Basic Data tables of the chapter that introduce each region (for example, Table 4.1 on page 67). On the HDI scale, a measure of 1.0 means "perfect." According to this index, with a rating of 0.938, Norway is "the world's best place to live," although it ranks second in per capita GNI PPP. In Chapter 4, we will see why the Norwegians and other Scandinavians are living so well. Following Norway, in descending order, are Australia, New Zealand, the United States, and Ireland. In HDI terms, with a measure of just 0.140, Zimbabwe is the world's worst place to live; the Democratic Republic of the Congo, Niger, Burundi, and Mozambique fare only slightly better. Note that all five of these low-rated countries are in Africa, for reasons discussed in Chapter 9.

On the basis of per capita GNI PPP, 1.2 billion, or 18 percent, of the world's people, inhabit the MDCs. Most citizens of these countries, such as the United States, Canada, Japan, South Korea, Australia, New Zealand, and the nations of Europe, enjoy an affluent lifestyle with freedom from hunger. Employed in industries or services, most of the people live in cities rather than in rural areas. Disposable income, the money that people can spend on goods beyond their subsistence needs, is generally high. There is a large middle class. Population growth is low as a result of low birth rates and low death rates (these demographic terms are explained in Section 3.3 on population). Life expectancy is long, and the literacy rate is high.

Life for a very large number of the planet's other 82 percent, or about 5.7 billion people, is quite different. In the LDCs, including many countries in Latin America, Africa, and Asia, poverty and often hunger prevail. The leading occupation is subsistence agriculture, and the industrial base is small. Most people live in rural areas, but that statistic will change soon; 46 percent of the LDCs' population is urban. The middle class tends to be small but growing, with an enormous gulf between the vast majority of poor and a very small wealthy elite, who own most of the private landholdings. With high birth rates and falling death rates, population growth is high. Life expectancy is short, and the literacy rate is low.

LDCs are not predestined to be in poverty forever. Effective development policies and practices have changed the outlook for many countries. Doomsayers predicted for decades that the colossal country of India would one day collapse under the weight of overpopulation and poverty. As we will see in Chapter 7, India is instead emerging as a global economic power to be reckoned with, for example, by using its human capital to generate wealth from software and outsourced services (see Insights feature on page 235). A few decades ago, China was a desperately poor country, but today it has the world's second largest economy. Still, because of their large populations, both countries face many challenges—including big gaps between their haves and have-nots—and neither ranks as a more developed country. 232

With four-fifths of the world's people living in the poorer, less developed countries, it is important to understand the root causes of underdevelopment and to appreciate how wealth and poverty affect the global environment in very different but equally profound ways. With this knowledge, it becomes possible to envision solutions.

Why Are Some Countries Rich and Others Poor?

Many theories attempt to explain the disparities between MDCs and LDCs. It is important to recognize that there is no single, widely accepted explanation about development in general or in a particular region or country. Probably the best thing you can do with your critical thinking skills is to weigh the various explanations and see which one, or which combination of them, seems to fit the situation of a given country or region. Here are the main explanations you are likely to come across.

Embraced most strongly in the LDCs, **dependency theory** argues that the worldwide economic pattern established by the Industrial Revolution and by the forces of colonialism that accompanied it persists today. In his book *Ecological Imperialism*, the historian and geographer Alfred Crosby explains how dependency led to the rich–poor divide, depicting the two very different ways in which European powers used foreign lands during the Industrial Revolution.[3] In the pattern of **settler colonization**,

Europeans sought to create new Europes, or **neo-Europes**, in lands much like their own: temperate middle-latitude zones with moderate rainfall and rich soils where they could raise wheat and cattle. And so, between 1630 and 1930, more than 50 million Europeans emigrated from their homelands to create European-style settlements in what are now Canada, the United States, Argentina, Uruguay, Brazil, South Africa, Australia, and New Zealand. These lands were destined to become some of the world's wealthier regions and countries.

In contrast to their preference to settle familiar mid-latitude environments, Europeans viewed the world's tropical lands mainly as sources of raw materials and markets for their manufactured goods. The environment was too different from home to make settlement attractive. In establishing a pattern of **mercantile colonialism**, Crosby explains, Europeans were less inhabitants than conquering occupiers of the colonies, overseeing indigenous peoples and resettled slaves in the production of primary or unfinished products: sugar in the Caribbean; rubber in Latin America, West Africa, and Southeast Asia; and gold and copper in southern Africa, for example. Colonialism required huge migrations of people to extract the Earth's resources, including 30 million slaves and contracted workers from Africa, India, and China to work mines and plantations
294, 322, 365, 409 around the globe.

In the mercantile system, the colony provided raw materials to the ruling country in return for finished goods; this meant that people in India would purchase clothing made in England from the raw cotton they themselves had harvested. The relationship was most advantageous to the colonizer. England, for example, would not allow its colony India to purchase finished goods from any country but England. It prohibited India from producing any raw materials the empire already had in abundance, such as salt (India's Mohandas Gandhi defiantly violated this prohibition in his famous "March to the Sea"). Finished products are **value-added products**, meaning they are worth much more than the raw materials they are made from, so manufacturing in the ruling country concentrated wealth there while limiting industrial and economic development in the colony.

The colony was obliged to contribute to, but was prohibited from competing with, the economy of the ruling country—a relationship that dependency theorists insist continues today. Dependency theory asserts that to participate in the world economy, the former colonies, now independent countries, continue to depend on exports of raw materials to and purchases of finished goods from their former colonizers and other MDCs, and this disadvantageous position keeps them poor. Dependency theorists call this relationship **neocolonialism**. With independence, the former colonies needed revenue. To earn that money, they continued to produce the goods for which markets already existed—generally the same unprocessed primary products they supplied in colonial times. Dependency theorists argue that when the former colonies try to break their dependency by becoming exporters of manufactured goods, the MDCs impose trade barriers and quotas to block that development (see Insights, page 420). Dependency theory and neocolonialism also envision the world in terms of core and periphery—a concept of duality you have seen already in Chapter 1 and will encounter often in the book—with wealth flowing from the periphery of the world's poor countries to its core of wealthy countries.

Whether because of neocolonialism or a more complex array of variables, many developing countries continue to rely heavily on income from the export of a handful of raw materials. This makes them vulnerable to the whims of nature and the world economy. The economy of a country heavily dependent on rubber exports, for example, may suffer if an insect pest wipes out the crop or if a foreign laboratory develops a synthetic substitute. When demand for rubber rises, that country may actually harm itself trying to increase its market share by producing more rubber, because in doing so, it drives down the price. (Consider oil: members of the Organization of Petroleum Exporting Countries [OPEC] cartel of oil-producing countries drove oil prices to all-time lows in the mid-1980s when they overproduced oil in a bid to earn more revenue.) If the rubber-producing country withholds production to shore up rubber's price, it provides consuming countries with an incentive to look for substitutes and alternative sources. (Again, look at oil: after OPEC embargoed shipments of oil to the United States in the 1970s, the United States began developing domestic oil supplies and becoming more energy-efficient, thus temporarily reducing oil prices and OPEC's revenues.) The developing country is in a dependent and disadvantaged position—even if it produces something as vital as oil.

Many geographers view dependency theory and neocolonialism as simplistic and politically biased explanations of development. They consider a wider and more complex set of factors, including culture, location, and natural environment, to explain why some countries are wealthy and others are poor. These include the following factors:

Advantageous and disadvantageous location. Does a country's location play a role in how rich or poor it is? In some cases, yes. Location can influence a country's economic fortunes. For example, as discussed in Chapter 1, because it is situated close to a great mainland with which to trade, the island of Great Britain enjoys a core location favorable for economic development. Japan has a similar location relative to the Asian landmass. In contrast, landlocked nations such as Bolivia in South America and numerous nations in Africa have locations unfavorable for trade and economic development, and they have not overcome this disadvantage. But it is important to recognize that geographic location is never the sole decisive factor in development. Like Japan and Britain, Madagascar and Sri Lanka are island nations situated close to large mainlands, but for a variety of (mainly political) reasons discussed in the relevant chapters, neither has experienced prosperity.

Resource wealth or poverty. Having or lacking a diversity or abundance or **natural resources** (naturally occurring materials that people can use) plays a significant role in development.

INSIGHTS Globalization: "The World Is Flat"

One of the most remarkable international trends of recent decades has been **globalization**—the spread of free trade, free markets, investments, and ideas across borders and the political and cultural adjustments that accompany this diffusion. There has been much debate and sometimes-violent conflict over the pros and cons of globalization. For good or ill, globalization is a reality. *The World Is Flat*, journalist Thomas Friedman entitled his book on the subject. He observed, ironically, that in the search for India, Christopher Columbus proved the world was round; but in visiting India, Friedman himself found the world to be flat: the "playing field" of the international economy had been leveled: "It is now possible for more people than ever to collaborate and compete in real time with more people on more different kinds of work from more different corners of the planet and on a more equal footing than at any previous time in the history of the world—using computers, e-mail, fiber-optic networks, teleconferencing and dynamic new software."[4]

Supporters of globalization argue that the newly emerging global economy will bring increased prosperity to the entire world. Innovations in one country will be transferred instantly to another country, productivity will increase, and standards of living will improve. They propose that one obvious solution to the problem of the LDCs' inability to compete in the world economy, and therefore escape their dependency, is to reduce the trade barriers that MDCs have erected against them (some of these barriers are discussed on page 420 in Chapter 11). With free trade, free enterprise will prosper, pumping additional capital into national economies and raising incomes for all. Many of the inputs into globalization come from **multinational companies** (also called *transnational companies*, meaning companies with operations outside their home countries) as they increase their investments abroad. Most of the companies are based in the MDCs, but multinational corporations have also grown in LDCs such as China, India, and Mexico.

Opponents of globalization argue that the process will actually increase the gap between rich and poor countries, and even within countries; a selected few developing nations (or people within a country) will prosper from increased foreign investment and resulting industrialization, but the hoped-for global wealth will bypass other countries (or people within a country) altogether. And the increasing interdependence of the world economies will make all the players more vulnerable to economic and political instability. The multinational companies will recognize huge profits at the expense of poor wage laborers. In addition, environments will be harmed if environmental regulations are reduced to a lowest common denominator (for example, the high standards of air quality demanded by the U.S. Clean Air Act would be deemed noncompliant with World Trade Organization rules because they make it harder for countries with "dirtier" technologies to compete in the marketplace). Such concerns often lead to massive protests against "corporate-led globalization," especially at meetings of the World Trade Organization, the International Monetary Fund, and the World Bank. Many more confrontations like these can be anticipated. Typical protesters' demands are that working conditions be improved in the foreign "sweatshops," where textiles and other goods are produced at low cost for U.S. corporations, and that corporations like Starbucks should sell only "fair trade" coffee beans bought at a price giving peasant coffee growers a living wage rather than at the "exploitive" price typically paid. Corporations like Starbucks have often been quick to respond to the demands of the environmentally and socially conscious consumer.

It is often assumed that the rapid growth and spread of computer and wireless technologies are benefiting everyone, but in fact there is still a **digital divide** to overcome. The divide is between the handful of countries that are the leaders in **information technology (IT)** and the majority of nations that have little ability to create, purchase, or use new technologies. The fear in the technology-lagging countries is that the growth of e-commerce will concentrate the wealth generated by that commerce in the technology leadership countries. This would enable the leaders to make even further advances in technology and realize even bigger economic gains from a so-called **knowledge economy** based on innovation and services, while the technology laggards fail to catch up and simply become poorer (•**Figure 3.A**). There are many efforts underway to empower the world's poor through access to information technology; the "One Laptop per Child" program seeks to provide the world's children with rugged, low-cost, low-power, Internet-connected laptops (see http://.onelaptop.org). Rapid developments and falling prices in smartphone technology could also help level the playing fields of globalization. 331

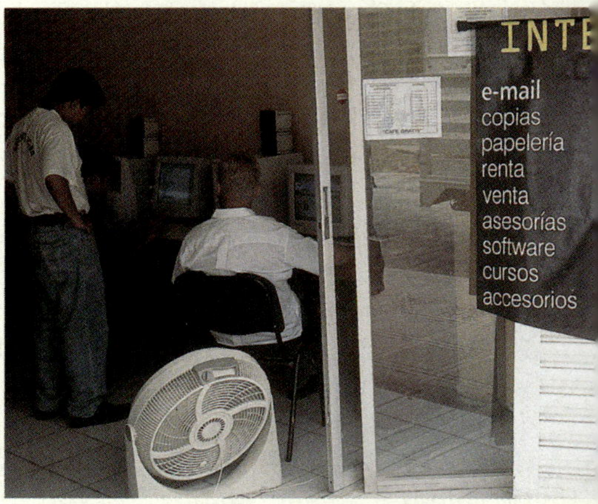

Joe Hobbs

• **Figure 3.A** An Internet café in Mérida, Yucatán, Mexico. Information technology—especially in the form of computers, the Internet, and cellular telephones—is spreading rapidly around the 196
world. Some people believe that there is now an unfolding Information Revolution comparable to the Agricultural and Industrial Revolutions in its capacity to change humanity's relationship with the earth.

Superabundance of an especially valuable resource (for example, oil in the Persian/Arabian Gulf countries) or a diversity of natural resources has helped some countries become more developed than others. The former Soviet Union and the United States achieved superpower status in the twentieth century in large part by using the enormous natural resources of both countries.

Cultural and historical factors. In some cases, human industriousness has helped compensate for resource limitations and helped to promote development. For example, Japan has a rather small territory with few natural resources (including almost no petroleum). Yet in the second half of the twentieth 270
century, it became an industrial powerhouse largely because the

Japanese people united in a common purpose to rebuild from wartime devastation, placing priorities on education, technical training, and seaborne trade from their advantageous island location. Conversely, cultural or political problems like corruption and ethnic factionalism can hinder development in a resource-rich nation, as in the mineral-wealthy Democratic Republic of
340 Congo.

Incidentally, many students have told me they enjoyed reading the geographer Jared Diamond's book *Guns, Germs, and Steel* because he writes so engagingly about the forces that propelled countries forward or held them back. Diamond's book has come to be known as "A Short History of Everybody for the Last 13,000 Years"!

Environmental Impacts of Underdevelopment

Geographers are very interested in the environmental impacts of relations between MDCs and LDCs, especially on the poorer countries. LDCs generally lack the financial resources needed to build roads, dams, energy grids, and other infrastructure critical to development. Then they turn to the World Bank, International Monetary Fund (IMF), and other institutions of the MDCs to borrow funds for these projects. Many borrowers are unable to pay even the interest on these loans, which is sometimes huge; debtor nations have been known to spend as much as 40 percent of annual revenues on interest payments to the lenders, or more than they spend on education and health combined.

This debt cycle can be very harmful to natural environments. When lender institutions threaten to cut off assistance, borrowing countries often try to raise cash quickly to avoid this prospect, generally by one or two methods, or both. One method is to dedicate more land to the production of **cash crops** (also known as **commercial crops**). These are items such as coffee, tea, sugar, coconuts, and bananas exported to the MDCs, where they are luxuries or staples (illegal drug crops are also cash crops, of course, but fall outside most statistics and many of the general associations made here). Governments or foreign corporations often buy out, force out, or otherwise displace subsistence food farmers in the search for new lands on which to grow these commercial crops. In this process, known as **marginalization**, poor subsistence farmers are pushed onto fragile, inferior, or marginal lands that cannot support crops for long and end up depleted by cultivation. In Brazil's Amazon Basin, for example, peasant migrants arrive from Atlantic coastal regions, where government and wealthy private landowners cultivate the best soils for sugarcane, soybeans, and other cash crops. The newcomers to Amazonia slash and burn the rain forest to grow rice and other crops that exhaust the soil's limited fertility in a few years. Then they move on to cultivate new lands, and in their wake come cattle ranchers, whose land use further degrades the
221, soil. This is a very common, important problem in the LDCs,
222, and you will see many examples of it in the following chapters.
384 The second way for the LDC debtor countries to raise cash quickly is to simply sell off their natural assets. Who is buying them? The wealthier countries, and increasingly and most
strikingly, China—an issue that this book covers often. LDCs 329,
often face difficult choices between using natural resources for 333,
immediate gain or for longer-term, more sustainable growth. 367
In most cases, they feel compelled to take short-term profits, by cash cropping and other strategies, a choice that can have devastating environmental consequences. Most LDCs have resource-based economies that rely not on industry, but on stocks of productive soils, forests, and fisheries. The countries' long-term economic health could benefit through protection and careful use of these assets. But to pay off international debts and meet other needs, the LDCs often draw down their ecological capital faster than nature can replace it. In ecosystem terms, they exceed **sustainable yield** (also known as the 61
natural replacement rate), the highest rate at which a **renewable resource** can be used without decreasing its potential for renewal. One figure that comes up often in the literature is that in African countries as a whole, 28 trees are cut down for every tree that is planted.

A country where 28 trees are cut for every one planted is approaching or is in a state of **ecological bankruptcy**, the exhaustion of environmental capital. This environmental poverty plays itself out in several damaging ways. First, there are negative effects on the health and well-being of the people living in such countries. People in the LDCs feel the impacts of environmental degradation more directly than people in the MDCs (•**Figure 3.6**). Many of the world's poor drink directly from untreated water supplies; an estimated 1 billion people lack access to clean water. They tend to cook their meals with fuelwood rather than with fossil fuels. They are more dependent on nature's abilities to replenish and cleanse itself, and so they suffer more when those abilities are diminished (see Insights, page 48).

In addition, many political and social crises can emerge from the ecological bankruptcy process. Revolutions, wars, and refugee migrations in developing nations often have underlying

Joe Hobbs

• **Figure 3.6** The Thu Bon River in Vietnam. In the LDCs, it is very common for people to rely on polluted water sources for drinking, bathing, cooking, and washing their clothes and eating utensils. Health effects can be severe; one of the most tragic is infant diarrhea, typically a waterborne illness that kills an estimated 2 million youngsters worldwide every year.

INSIGHTS Deforestation and The Fuelwood Crisis

Deforestation—the removal of tree cover—has many detrimental effects in the LDCs. You can see how these relationships play out in •**Figure 3.B**. As people remove trees to use as fuel, for construction, or to make room for crops, their existing crop fields lose protection against the erosive force of wind. Less water is available for crops because in the absence of tree roots to funnel water downward into the soil, it runs off quickly. Increased salinity (salt content) generally accompanies increased runoff, causing the quality of irrigation and drinking water downstream to decline. Eroded topsoil resulting from reduced plant cover can choke irrigation channels, reduce water delivery to crops, raise floodplain levels, and increase the chances that floods will destroy fields and settlements. As reservoirs fill with silt, hydroelectric generation, and therefore industrial production, is diminished. Upstream, where the problem began, fewer trees are available to use as fuel.

As they deplete nearby fuelwood supplies from trees and other vegetation, rural people make their own lives more difficult and have to change their activity patterns. Women and children, in almost every society the main fuelwood collectors, must walk farther, and then farther again, to gather fuel. Using animal dung and crop residues for fuel can solve one problem but create another: it deprives the soil of the fertilizers these poor people often use when farming. Reduced food output is the result. A family may eventually give up one cooked meal a day or put up with colder temperatures in their homes because fuel is lacking—steps that have a negative impact on the family's health. These impacts of fuelwood depletion are known collectively as the **fuelwood crisis**.

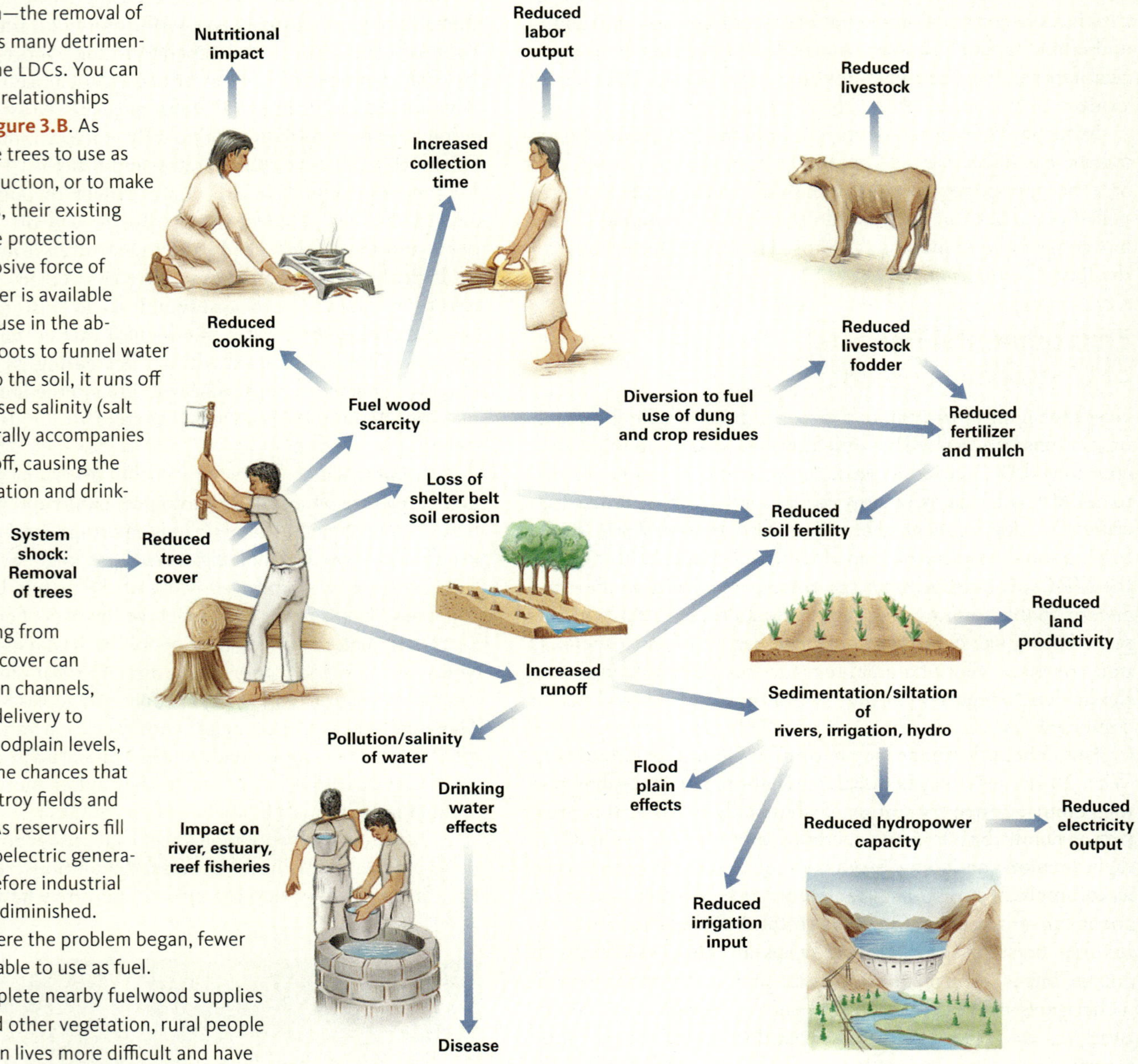

• **Figure 3.B** Impacts of deforestation in the LDCs.

The fuelwood crisis is a problem that governments and societies can confront successfully. A combination of new technologies, government policies, and sound economic growth can halt and even reverse deforestation. The biogas digester, a simple household device that can be used in warm climates, ferments animal manure for the production of methane cooking gas and thus spares fuelwood. Governments can protect forests while also allowing people to earn their livelihoods in them from **nontimber forest products (NTFPs)**, such as handicrafts made from bamboo (see www.ntfp.org). The process of economic development can itself lead to less pressure on forest resources. The world's wealthier countries have more forest now than they did in 1990, and forest cover has actually increased in about half of the world's 50 most forested countries since that time. A key challenge for the wealthier forested countries is to slow their demand for timber from the LDCs, and thus help break one of the cycles of deforestation, debt, and poverty.

environmental causes. Such problems, as related in the earlier discussion of climate change refugees in Chapter 2, are central to the national interests of the United States and other coun-
33 tries far beyond the affected nations. Deforestation in Mexico or Haiti, for example, is among the root causes why rural people from those countries try to move to the United States, legally or otherwise. When their home environments are so degraded that they simply cannot support farming as they once did, people will move.

3.3 The Geography of Population

Population may be the most critical issue in geography. It certainly is one of the most important issues for human life on Earth. The welfare of humanity and of the planet's other species and natural habitats is tied closely to two related issues: the number of people there are and the rates at which we consume resources. In some quarters, there are fears that the human **population explosion** the world has experienced since 1800 will lead to a precipitous crisis: a massive disease outbreak, famine, or some other kind of catastrophe triggered by too many people living too close together without sufficient food and other resources to sustain them. On the other hand, there is optimism that the human population growth rate is slowing, and our numbers are expected to stabilize—ideally at a population that can be sustained with minimal risk of famine or other suffering.

Meanwhile, large numbers of people move across and within political borders, usually by choice but in some cases by force. Migrants bring new cultures, ideas, and opportunities with them, but their reception is not always warm: tension and violence have characterized relations between majority and migrant minority groups in recent years in Europe and Australia, for example, and in the United States there is a heated debate about "immigration reform" directed at illegal migrants, especially from Latin America. Migration is one of the major themes discussed throughout this book.

Geography's interests in population are shared by many other fields, including biology, sociology, anthropology, and political science. The study of population is known as **demography**. The field of demography is most concerned with patterns of birth, death, marriage, and related issues in themselves, with less attention to issues of migration and population distributions. What most distinguishes population geography is its focus on spatial variations. This section of the book examines how many people have lived on Earth and how many we may be in the future, mindful of who these people are and why they have been so few or so many. It pays especially close attention to the fundamentally geographic issue of where they are. This discussion will be a very important reference for you as you work your way forward through the world regions. I hope you will find it interesting, and recommend that you explore the best Internet resource on world population (where you can also download a free world population data table): the Population Reference Bureau site at prb.org.

How Many People Have Ever Lived on Earth?

Around 100,000 years ago, our *Homo sapiens* ancestors came out of Africa across the land bridge of Suez and began to populate Eurasia. By around 10,000 years ago, about the time plants and animals began to be domesticated, there were probably about 5.3 million humans in the world—roughly the current number of residents of the Atlanta metropolitan area or the country of Finland. By 1 C.E., humans numbered between 250 and 300 million, about the population of the United States today. The first billion was reached around 1800. Then a staggering population explosion occurred in the wake of the Industrial Revolution. The second billion came in 1930, the fourth in 1975, the sixth in 1999, and the seventh in 2011 (•Figure 3.7). *Homo sapiens* is now by far the most populous large mammal on Earth and has succeeded where no other animal has in extending its range to the world's farthest corners. Using a 24-hour time period to represent the 200,000 years of our species' history, we reached our first billion within just the last three minutes. The people alive today represent about 6 percent of all the people who have ever lived on Earth!

Figure 3.7 illustrates very dramatically that right after about 1800, the human population surged. What happened after tens of thousands of years that made our numbers suddenly skyrocket? In addition to the factor, addressed earlier, of greater food surpluses that made it possible to feed more people, we answer that question by considering some of the basic vocabulary of population geography.

How Can We Measure Population Changes?

Excluding the issue of migration for now, two main variables determine population change in a given village, city, or country or the entire planet: birth rate and death rate. The **birth rate** is

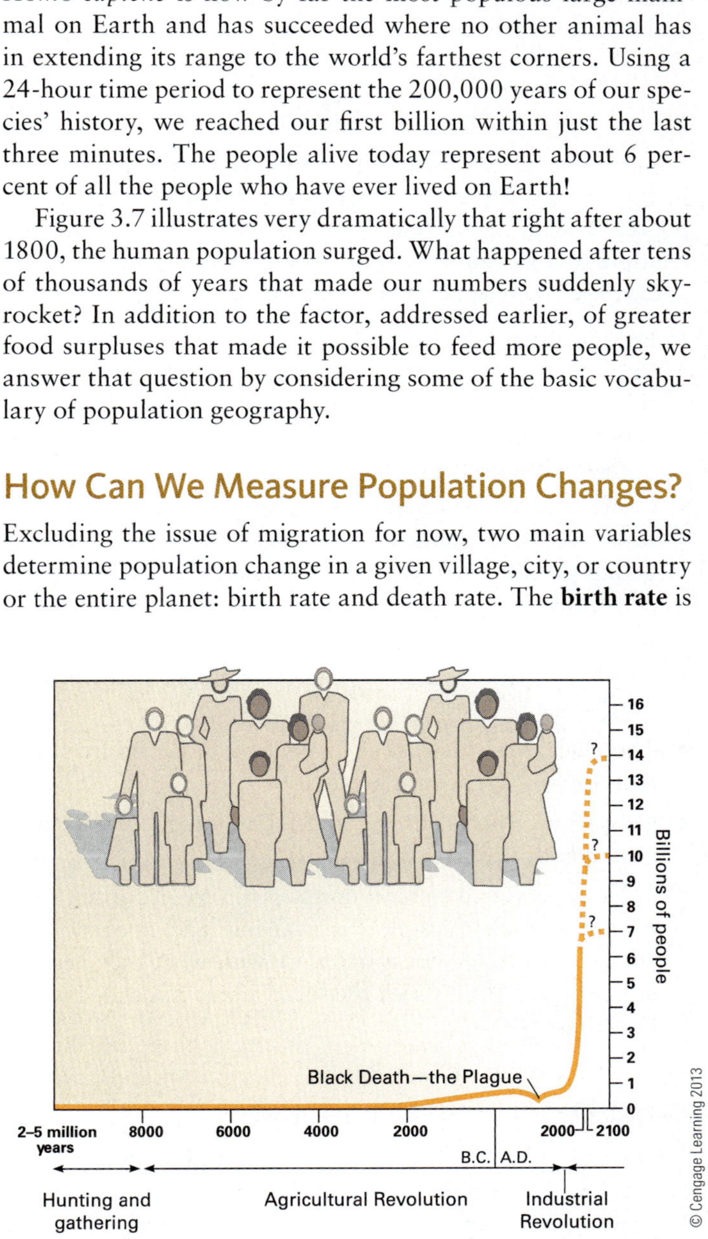

• **Figure 3.7** The human population has exploded since the Industrial Revolution, in a classic J-shaped curve of exponential growth.

the annual number of live births per 1,000 people in a population. The **death rate** is the annual number of deaths in that same sample population of 1,000. The **population change rate**—the figure that is often called the "population growth rate" but that may represent either growth or loss—is the birth rate minus the death rate in that population.

We can put these measures to work to appreciate the Earth's population now. Suppose there were a perfect sample of 1,000 people representing the world's population in 2011. Their birth rate was 20 per 1,000 and their death rate was 8 per 1,000. This means that by the end of that year, among the 1,000 people, 20 babies had been born and 8 people had died, resulting in a net growth of 12. That figure, 12 per 1,000, or 1.2 percent, represents the 2011 population change rate for the world.

We may now look at some of different combinations of birth rates, death rates, and population change rates and the various forces behind them.

What Determines Family Size?

Many factors affect birth rates, which tend to be much higher in the less developed countries of the world and among the poorer residents of more affluent countries. Some of the motivations and circumstances may seem unfamiliar at first, but you will see there are good reasons for them:

- Better-educated and wealthier people have fewer children. The parents of most of you reading this book probably considered the economic cost of raising you and sending you to college and made decisions about family size in part because of their education and yours.
- Conversely, less educated and poorer people generally want and have more children. One reason for this is economic: poorer parents are often convinced that "extra" children will help bring more income to the family by working in fields or factories and will help care for them in their old age.
- People in cities tend to have fewer children than those in rural areas.
- Those who marry earlier generally have more children (their reproductive life span is longer).
- Couples with access to and understanding of contraception generally have fewer children.
- Value systems and cultural norms play very important roles. Even where contraception is available and understood, a couple may decide not to interfere with what they perceive as God's will or may seek the social status associated with a larger family.

What Determines Death Rates?

Death rates are correlated mainly with health factors, particularly the level of nutrition and level of medical care available. Improvements in food production and distribution help reduce death rates. Better sanitation, better hygiene, and cleaner drinking water eliminate fatal diseases such as infant diarrhea, a common cause of infant mortality in the less developed countries. The availability of antibiotics, immunizations, insecticides, and other improvements in medical and public health technologies also correlate with death rates.

Human death rates overall have been on a steady trend of decline for decades. But death rates sometimes rise, of course, especially with the outbreak of epidemics such as HIV/AIDS. 316
Large proportions of the world's population have in fact been killed in disease epidemics and natural disasters. The "Black Death" (caused by bubonic plague) in Europe and Asia killed about 15 percent of the world's population between 1334 and 1349; at least 40 percent of Europe's population died. An influenza (flu) epidemic in 1918–19 killed as many as 100 million people. We cannot assume that such devastation will never happen again. Earlier this century, there were fears that the "bird flu" might kill as many as 150 million people if it mutated to a human strain.

Closely related to the measure of death rates is that of *life expectancy,* the number of years a person may expect to live in a given environment (typically defined as a country—see •**Figure 3.8**—and differentiated between women, who usually live longer, and men). As death rates fall, life expectancy increases, and the reverse is also true. In the United States in 2011, life expectancy for women was 80 years, and for men, 75 years. In Zimbabwe, a southern African country hit hard by the HIV/AIDS epidemic, a woman could expect to live 45 years and a man 46 years. But as we will see in **Chapter 9**, that is an improvement for Zimbabwe as it and other sub-Saharan African countries begin to make headway against the scourge of AIDS. 316

What Determines the Population Change Rate?

Throughout history, natural disasters, diseases, and wars have taken huge bites out of our numbers. Overall, however, with birth rates higher than death rates, the trend line has been one of growth—and since 1800, of spectacular growth. In 1968, the rate of population growth hit an all-time high of 2.0 percent. To appreciate how rapid that growth rate was, you can calculate the **doubling time**. Applying the often-used "Rule of 70," in which 70 is divided by the growth rate, the doubling time is the number of years required for the human population to double (assuming that the rate of growth would be unchanged over the entire period). In 1968, the human population was growing at a rate that would, if unchanged, have doubled in 35 years. That rate did slow down, however, illustrating the limited usefulness that doubling time has in projecting future growth. At the 2011 population change rate of 1.2 percent, our numbers would double in 58 years.

Doubling time may be only approximate, but it is a good tool for comparing among countries and also illustrates that human populations have the potential for exponential growth; that is, not an incremental or arithmetic increase from 1 to 2 to 3 to 4, and so on, but geometric growth from 2 to 4 to 8 to 16, for example. (Exponential growth in the context of the famous Malthusian scenario is discussed on **page 58**.) The annual rate of population change worldwide is depicted by country in •**Figure 3.9**.

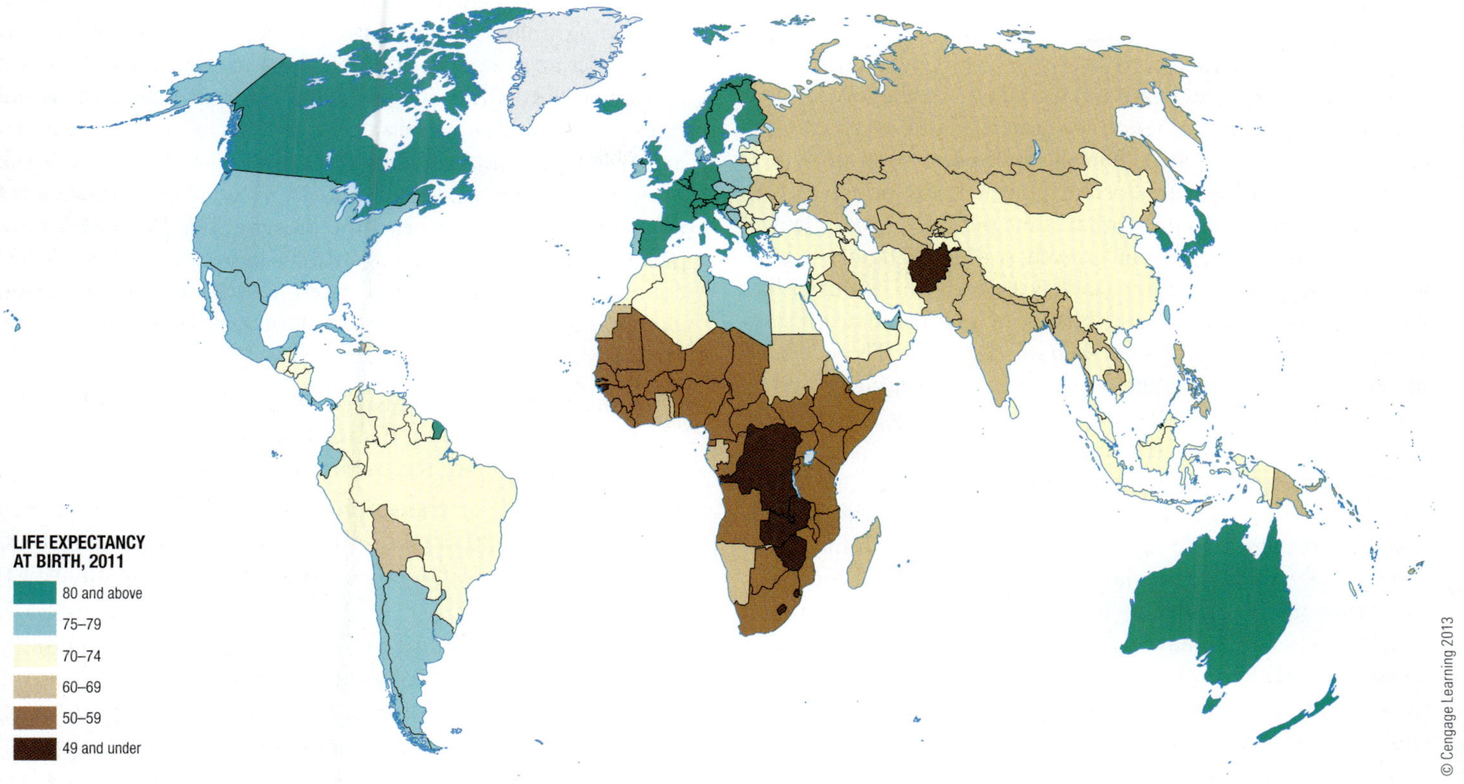

• **Figure 3.8** Life expectancy is closely tied to economic well-being; people live longer where they can afford the medicines and other amenities and technologies that prolong life.

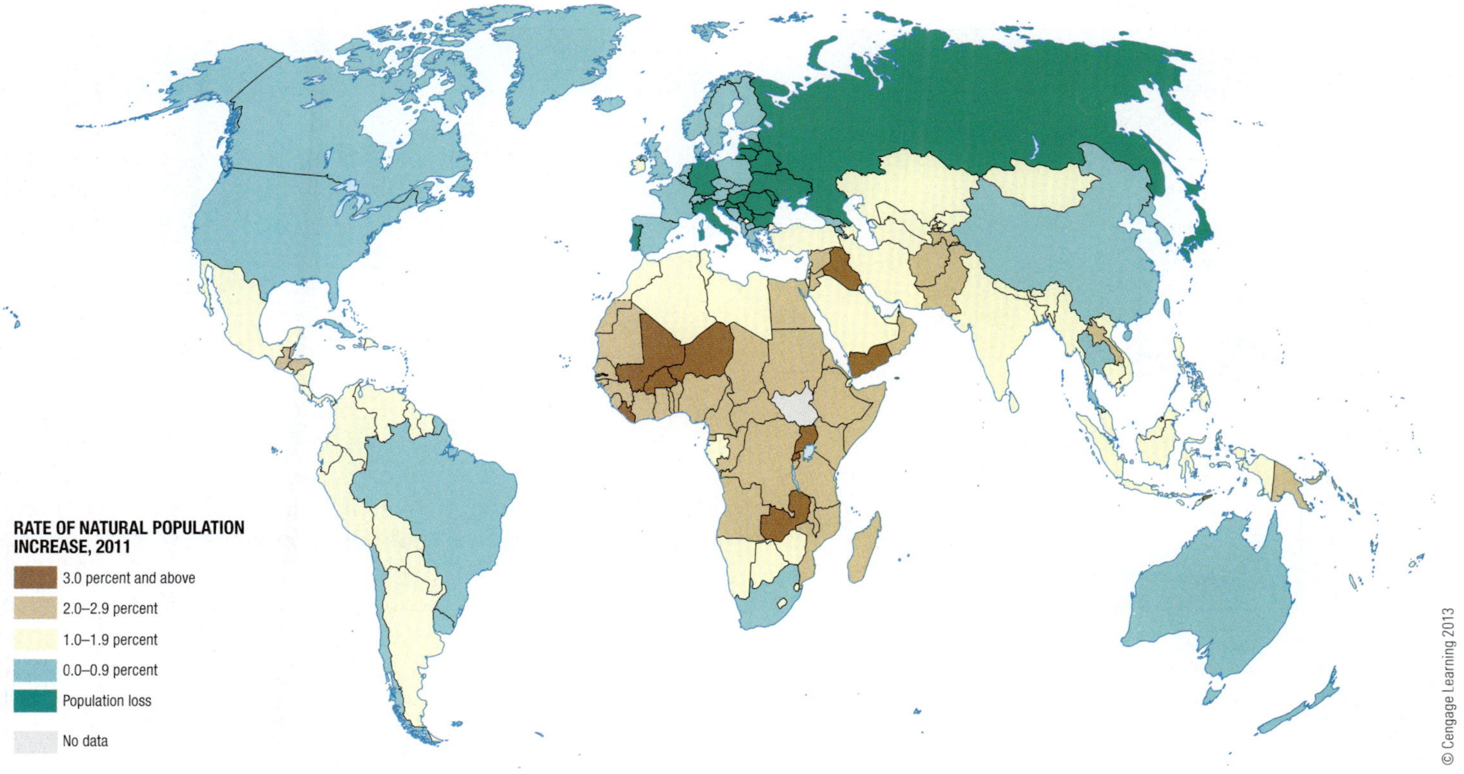

• **Figure 3.9** Population change rates are highest in the countries of Africa and other regions of the developing world and lowest in the more affluent countries.

Why Has the Human Population "Exploded?"

If both birth rates and death rates are high (as they were when our ancestors were hunters and gatherers and as recently as the dawn of the Industrial Revolution), population growth is minimal: many people are born to a given population in a given period, but many also die in that same period, so they "cancel each other out." Population growth is also negligible if both birth and death rates are low (as they are today in countries like Japan and Italy). But when the birth rate is high and death rates are low, population surges, as seen in •Figure 3.10. This scenario of high birth rates, plunging death rates, and surging growth is exactly what played out for our species beginning around 1800 in western Europe and after about 1950 in the LDCs.

It is vital to appreciate the fact that the explosive growth in world population since the beginning of the Industrial Revolution is the result not of a rise in birth rates, but of a dramatic decline in death rates, particularly in the less developed countries. The death rate has fallen as improvements in agricultural and medical technologies have diffused from the richer to the poor countries. Until recently, however, there were no strong incentives for people in the less developed countries to have fewer children. With birth rates remaining high and death rates falling quickly, the population has grown sharply; the LDCs are generally in or emerging from stage 2 of an important model that demographers call the **demographic transition** (Figure 3.10).

In its entirety, from stages 1 to 4, as seen in Figure 3.10, the demographic transition model depicts the change from high birth rates and high death rates to low birth rates and low death rates that accompanied economic growth in the more developed countries (for example, western European nations, Japan, and the United States). The first two are the same two stages that humankind as a whole experienced from our earliest days until the present. The latter two have generally been experienced only by people in the wealthier countries. Note how birth rates, death rates, population change rates, and economic development correspond in this model:

- In the **first**, or **preindustrial, stage** (from the earliest humans to about 1800 C.E.), birth rates and death rates were high, and population growth was negligible.
- In the **second**, or **transitional, stage**, birth rates remained high, but death rates dropped sharply after about 1800 due to medical and other innovations of the Industrial Revolution.
- In the **third**, or **industrial, stage**, beginning around 1875, birth rates began to fall as affluence spread.
- Finally, after about 1975, some of the industrialized countries entered the **fourth**, or **postindustrial stage**, with both low birth rates and low death rates and therefore (once again, as in stage 1), low population growth.

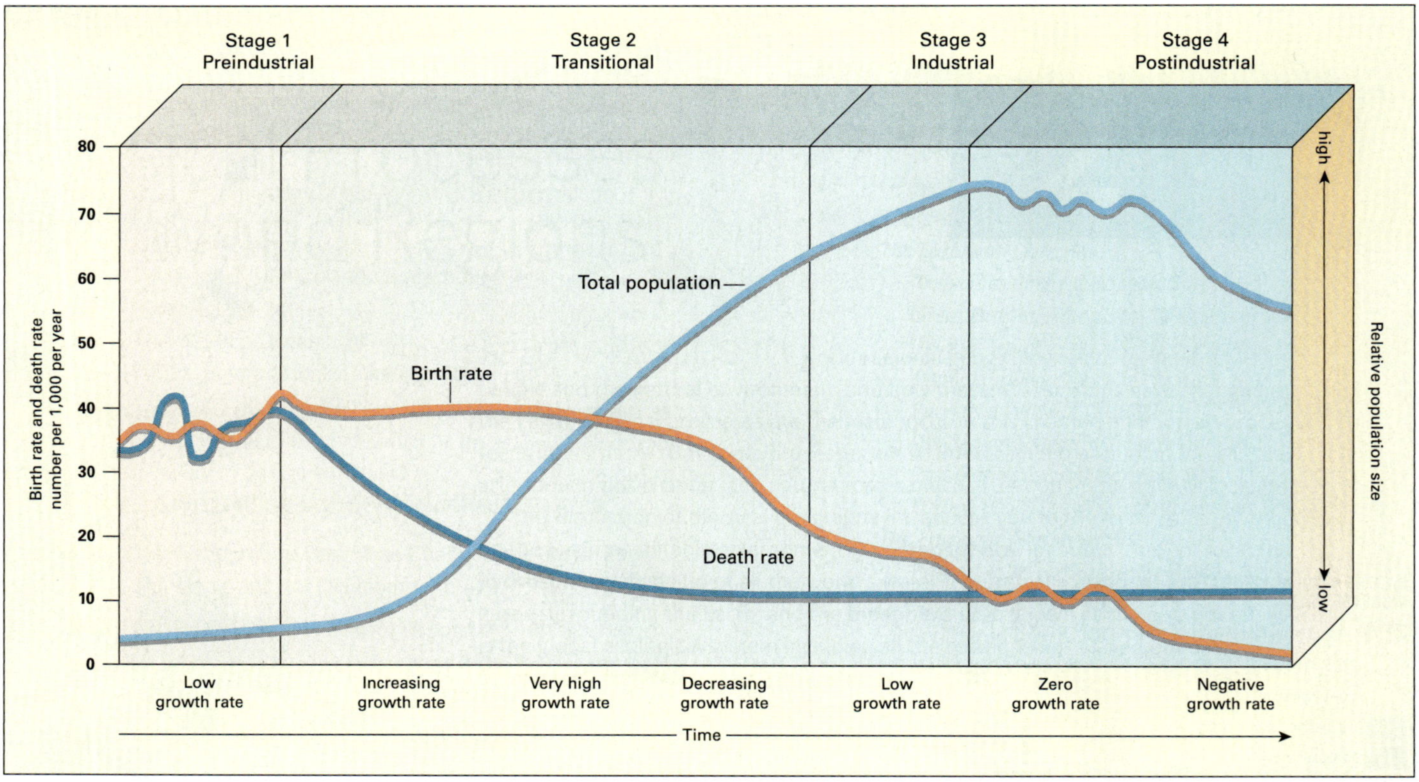

• **Figure 3.10** The demographic transition models population change in the world's wealthier countries. Note how the population surged in the wake of the Industrial Revolution as death rates fell while birth rates remained high, but then leveled out and began to decline as economic development advanced.

In this model, the United States, with population growth of about 0.5 percent per year, is in the early years of the postindustrial stage. Other industrialized, affluent countries have made their way well into that stage. Some wealthier countries, including Austria, Portugal, and Greece, have in recent years officially registered **zero population growth (ZPG)**, and many other European countries are close to ZPG. Because of the growing desire of women in some countries—including Japan and Germany—to pursue their own careers and postpone marriage, birth rates in those countries have fallen below death rates, meaning that these countries are actually losing population or even experiencing a **population implosion**. In other words, the fertility rate of Japan and some other postindustrial countries is below the **population replacement level**, the number of new births required to keep the population steady (generally calculated as 2.1 children per woman in the MDCs). Other countries—notably Russia and Hungary—that qualify as MDCs on the basis of per capita GNI PPP are experiencing population losses, unfortunately due as much to rising death
118 rates as to falling birth rates. Trends of this sort can be most easily appreciated by looking at a very useful and informative device, the age structure diagram.

The Age Structure Diagram

About 9 of every 10 babies born in the world today are in the poorer countries. Both current and projected rates of population growth are distributed quite unevenly between the poorer and richer nations. This phenomenon is apparent in the age structure diagrams typical of these countries. An **age structure diagram** (often called a **population pyramid**) classifies a population by gender and by five-year age increments (•**Figure 3.11**). One important index these profiles show is the percentage of a population under age 15. A country like Niger in Figure 3.11 is typically poor and faces the prospect of increasing poverty because so many new jobs, food, and other resources will have to be created to meet the demands of those children as they mature and have their own children. The bottom-heavy age structure diagram also suggests a continued surge in population as those children grow to enter their reproductive years. A large, youthful population in which competition for jobs, education, and land is intense is a social environment ripe for discord. A study by Population Action International found that 80 percent of the civil conflicts of recent decades took place in countries where at least 60 percent of the population was younger than 30.[5] Disaffection among huge populations of impoverished, unentitled young people was one of the main forces that brought down several authoritarian
governments in the "Arab Spring" of 2011. 192

The bottom-heavy, pyramid-shaped age structure diagram of Niger contrasts markedly with the more chimney-shaped structures of the United States and Germany in Figure 3.12. These wealthier countries have a much more even distribution of population through age groups, with a modest share under age 15. Such profiles suggest that, not considering migration, their population growth will be low in the near future.

Collectively, about 29 percent of the population of the poorer countries is under age 15, whereas the corresponding figure for

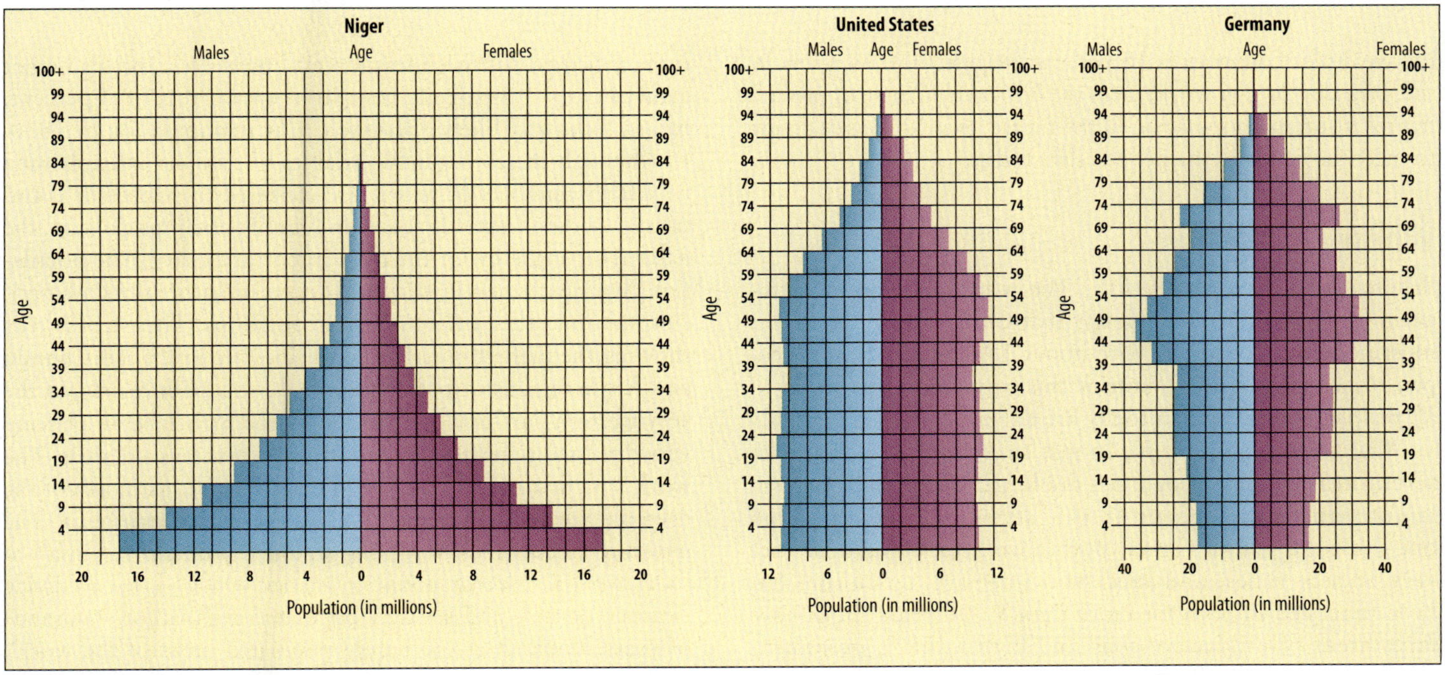

• **Figure 3.11** Population by age and gender in representative countries. The pyramid-shaped age structure diagram for Niger contrasts remarkably with those of the far more affluent United States and postindustrial Germany, with their chimney-like shapes. A poor country, Niger has a relatively high birth rate, with about 49 percent of the population under age 15. The U.S. population is growing slowly, while Germany and some other industrialized nations are losing populations.

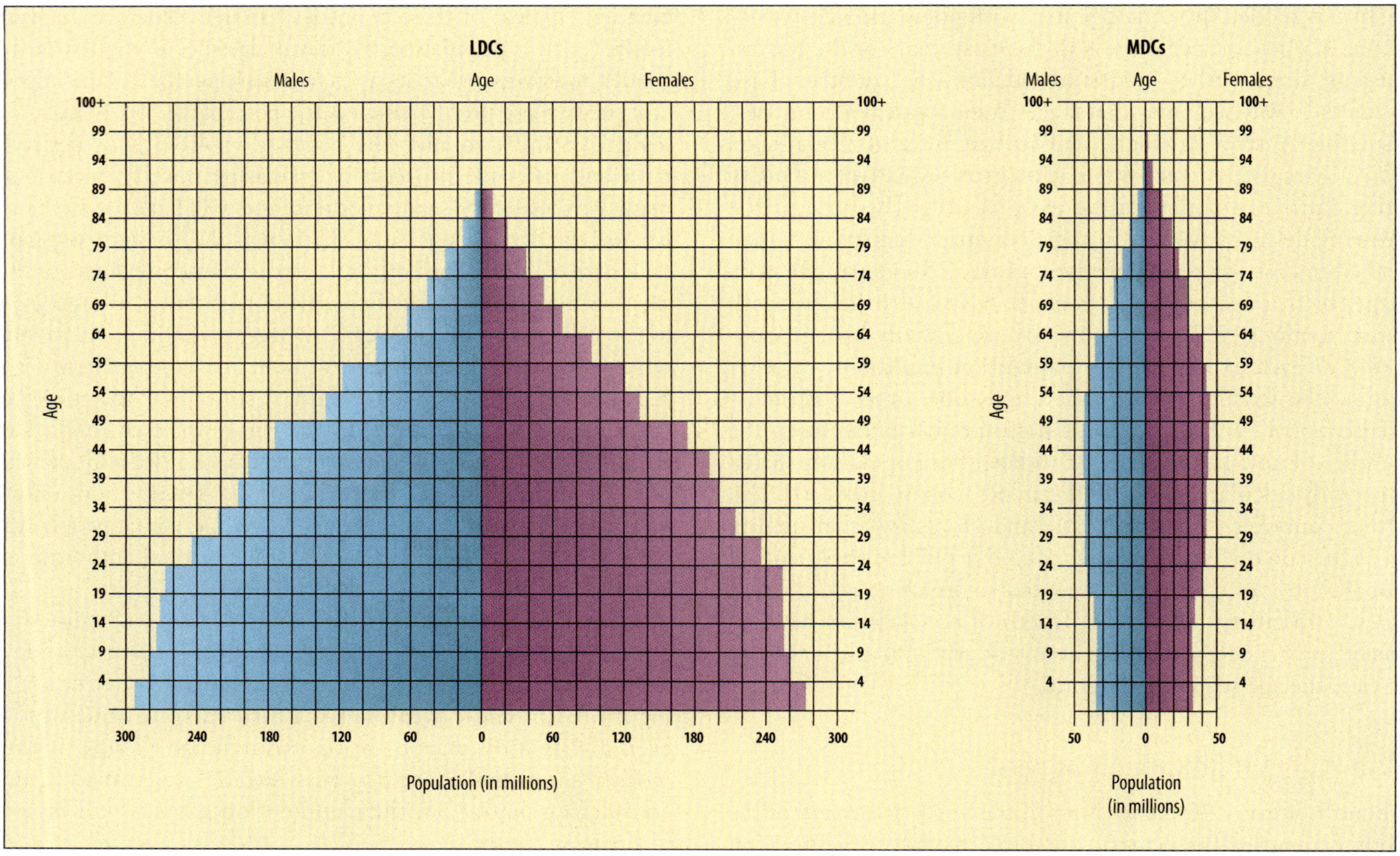

• **Figure 3.12** Population by age and gender worldwide, in terms of economic development. This diagram summarizes some of the most important facts about the human population: the lion's share lives in the poorer countries, and the lion's share of that population is young.

the wealthier countries is 16 percent (•**Figure 3.12**). As a whole, then, the developing world faces the critical challenge of providing for a surging population in the future, even while struggling to meet the needs of the people alive today.

Where Do We Live?

Where do all these people live? The world population cartogram in •**Figure 3.13** shows clearly that China and India are the most populous countries, with about 1.4 and 1.3 billion people, respectively. Think about it this way: about one-third of all people living on Earth today are either Chinese or Indian!

Why are so many people in just two countries? There are several reasons. Both countries are large. China is 20 percent larger than the contiguous 48 U.S. states, and India is about one-third the size of China. Both have been populated since very ancient times, and successful intensive agriculture has been practiced in both for more than 4,000 years. Both have large areas of productive soils and high rainfall, promoting successful farming. Both are developing countries in which birth rates remained high while death rates fell. Their population explosions prompted China's government to adopt an aggressive population control policy several decades ago, and India followed suit with less forceful measures. India's higher growth rate—1.5 percent, compared with China's 0.5 percent, means India will likely overtake China in population by 2030.

Although there are many variables to consider in explaining why large numbers of people are clustered in particular countries, including cultural factors and family planning policies, the natural setting is by far the most important factor. The population densities shown in •**Figure 3.14** correlate generally with agricultural and other environmental conditions. The deepest reds showing the highest population densities are in the more humid and fertile regions of both China and India. Conversely, in the very western part of China, including the high Tibetan Plateau, very dry conditions limit agriculture to a few favored areas. The world's highest mountains, the Himalayas, rise just north of the deep red area of high population density in northern India. The moisture-laden winds that bring so much productive rainfall to India cannot cross that mountain barrier, which has been a divide between densely and sparsely populated regions for thousands of years. Looking at the lightly populated areas of the world, you can recognize similar environmental factors at work: in the Sahara of northern Africa, the Arctic of northern Canada, and the Amazon Basin of South America, conditions are too dry, too cold, or too wet and infertile to support large numbers of people.

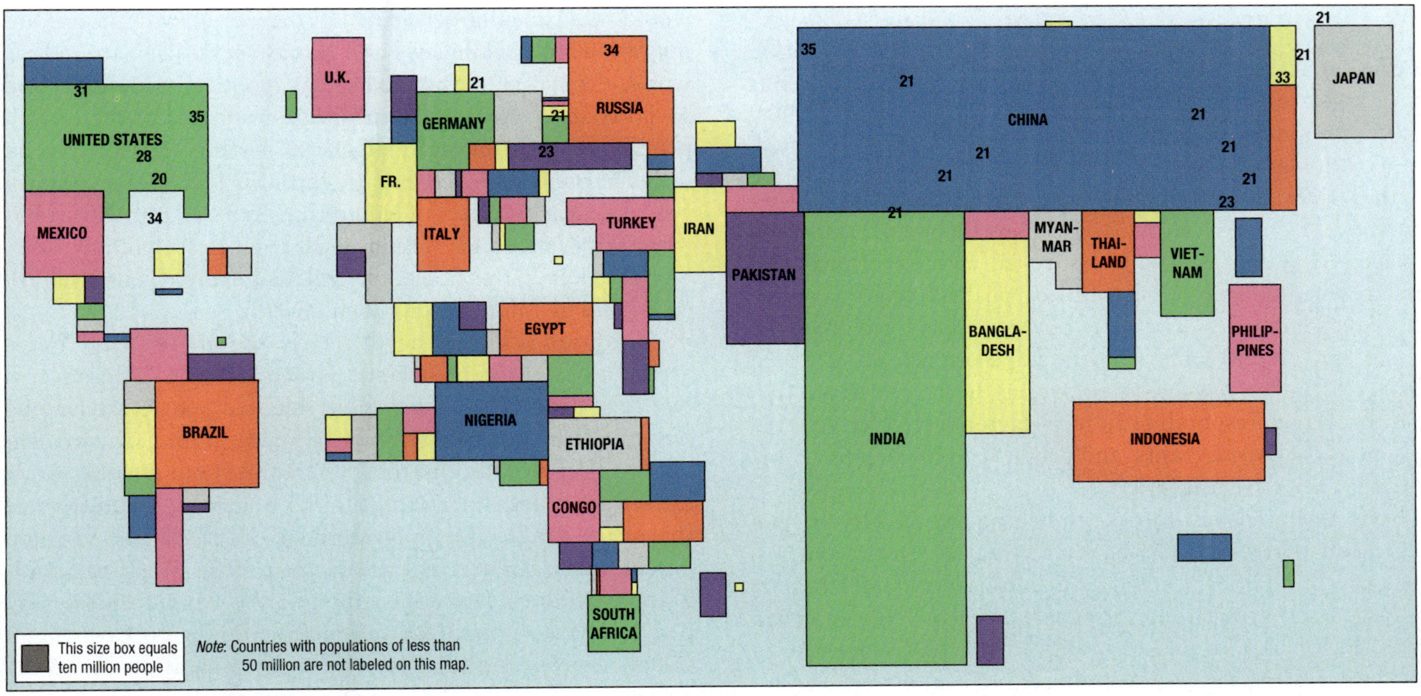

• Figure 3.13 The demographic heavyweights of China and India stand out in the world population cartogram. The United States and Indonesia, the world's third and fourth most populous countries, are prominent too.

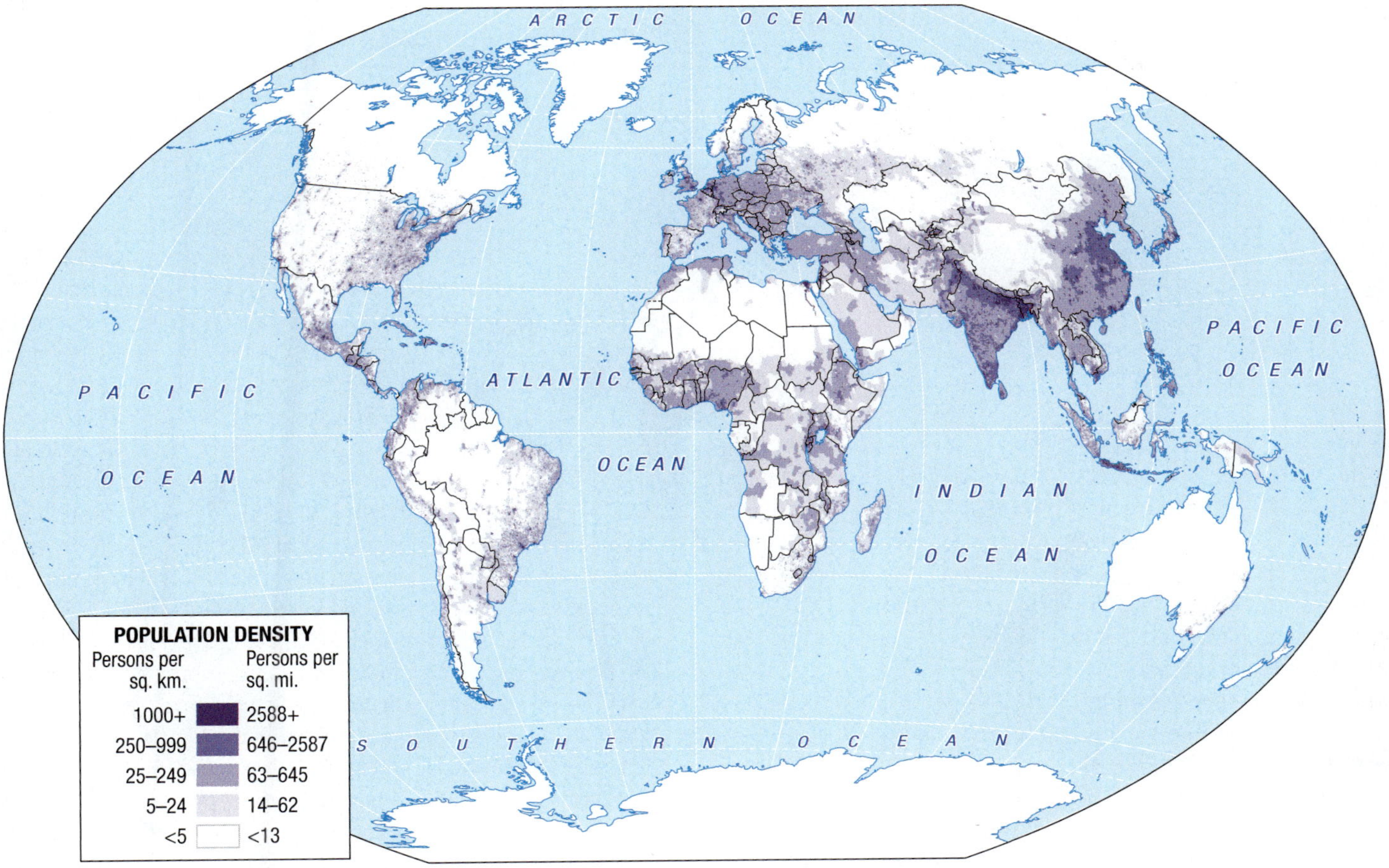

• Figure 3.14 This isarithmic map of population density shows the approximate distribution of people around the world.

Source: Based on data from Population Reference Bureau, 2004.

The Geography of Migration

Migration refers to the movement of people from one location to another in any setting, whether within a community, within a country, or between countries. Migration is one of the most dynamic and most problematic human processes on Earth (•**Figure 3.15**). A migrant is always both an **emigrant** (one who moves from a place) and an **immigrant** (one who moves to a place). Migration is usually associated with either **push factors**, as when hunger or lack of land "pushes" peasants out of rural areas into cities or warfare pushes people from one place to another, or **pull factors**, when, for example, an educated villager is "pulled" by a job opportunity in the city or another country. People responding to push factors are often referred to as **nonselective migrants**, and people reacting to pull factors are called **selective migrants**.

Both push and pull forces are behind the **rural-to-urban migration** pattern that is characteristic of most countries. The growth of cities, known as **urbanization**, is due in part to the natural growth rate of people already living in urban areas, but it is particularly strong in the LDCs because of this internal migration of both selective and nonselective
226 migrants.

Within and between countries today, there are considerable movements of **refugees**, the victims of such severe push factors as persecution, political repression, and war. They may be on the move either as illegal immigrants or as people granted **asylum**, or given permission to immigrate on the grounds that they would be harmed or persecuted in their country of origin. Immigration and asylum laws and quotas vary widely around the world, depending on host countries' political, economic, and social systems. Among the most disadvantaged of the world's peoples are the **internally displaced persons (IDPs)** who are dislodged and impoverished by strife in their home country but have little prospect of emigrating. Sub-Saharan Africa has more IDPs (about 12 million in 20 countries) than any other world region. In this book, you will read many accounts of others moved by forces beyond their control.

Migration is almost never clearly "good" or "bad" but is almost always a controversial mixed blessing. Migration is bad for the country whose most talented people are leaving but good for the country they go to; the Indian doctor benefits your community in the United States but may be sorely missed in India, for example. The doctor is illustrative of the **brain drain**, the emigration of educated and talented people from a place that needs them. The Mexican immigrant who does low-wage labor in the United States may be perceived as an **illegal alien** threatening to overwhelm social services and take jobs away from local people, or as a **guest worker** who performs services that few others want to do but that are critical to the country's economic well-being. Some host-country peoples are more welcoming of new cultures that enrich their ethnic mosaic (Americans are generally perceived as among the most accommodating cultures in the world), whereas others fear losing their ethnic
majorities and privileges. 71, 393

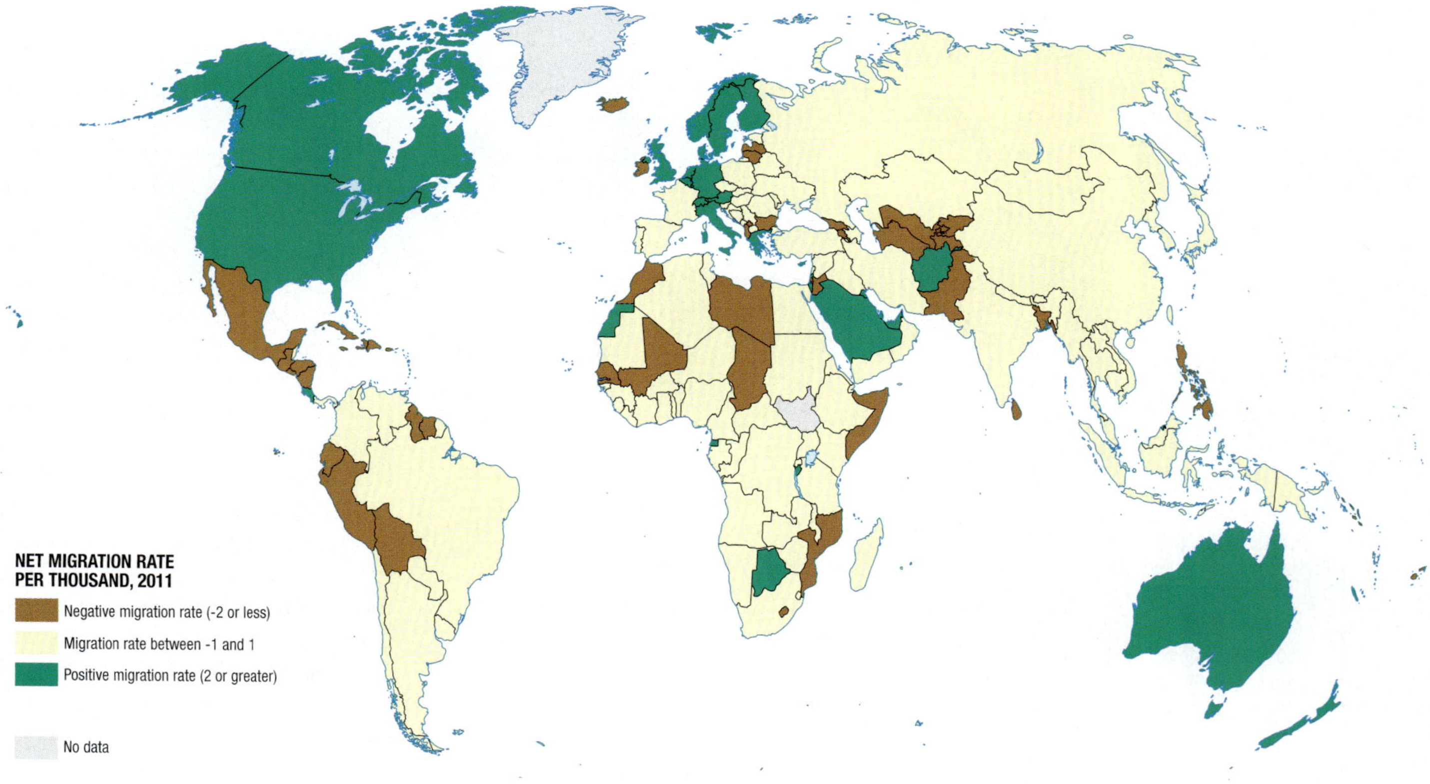

• **Figure 3.15** The global picture of people on the move. The major trends are of migrants in search of work in more affluent countries and of refugees driven by warfare or environmental adversity.

How Many People Will Live on Earth?

Population geographers are fairly confident in their calculations of how many people have lived on Earth at various times in the past. Projecting future numbers is another matter. There are many uncertainties. Will birth rates fall faster than expected in the developing world? Will death rates surge because of HIV/AIDS or some other epidemic? These are some of the wild cards in the population deck.

In asking how many people will live on Earth in the future, we are essentially asking, "Will the poorer countries of the world go through the demographic transition?" Will the current, relatively high birth rates in the less developed countries continue their present slow decline? In 1970, Kenya had a birth rate of 51 and Bangladesh had a birth rate of 45; by 2011, Kenya's birth rate had dropped to 37 and Bangladesh's to 22. Considering what might have been, given the country's huge population, China's decline may be the most impressive of all, from a rate of 38 in 1965 to 12 in 2011. The Chinese government's generally harsh **one-child policy**, which uses a combination of incentives and punishments, has been the main reason for the decline. But the demographic transition model also suggests to us that many increasingly affluent Chinese couples would probably have make the free
216 choice to limit their children, if given that opportunity. In Kenya and Bangladesh, growing literacy and the slow but steady economic progress of women have brought down the birth rate.

Family planning policies and levels of education and economic well-being have played various roles in the poorer countries, but have collectively combined to bring birth rates down since 1968 (•**Figure 3.16**). If the processes of increasing economic and social development continue in the LDCs as a whole, they should make steady progress through the demographic transition, and the Earth's population should cease to grow and should stabilize.

Joe Hobbs

• **Figure 3.16** National family planning programs can have a pronounced impact on birth rates if the actions go beyond slogans. This sign on the famous Tahrir Square in Cairo urges parents to have no more than two children "for the sake of a better life." Egypt's birth rate fell from 40 in 1970 to 25 in 2011.

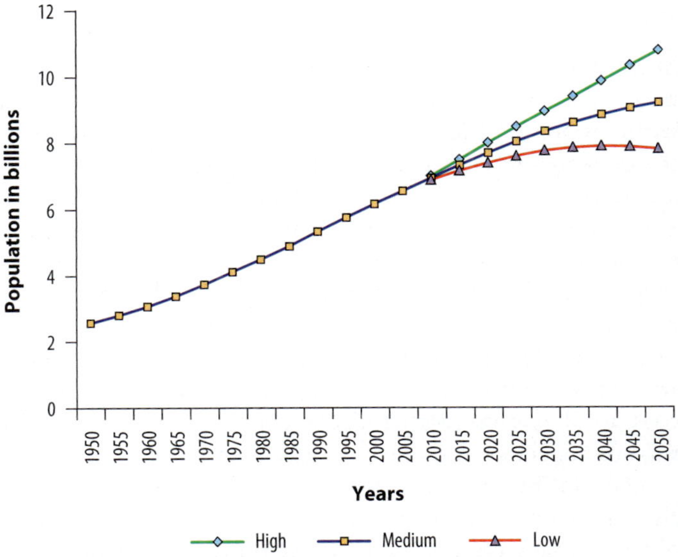

• **Figure 3.17** United Nations projections for population growth.

But will the poorer countries complete the transition successfully? An unsavory but possible scenario would see death rates rise dramatically (due to disease or famine, for example) in at least some of the countries, in effect pushing them back to stage 1 of the demographic transition, where both birth and death rates are high. Others could remain in stage 2, with high birth rates and falling death rates, long enough to bring unexpectedly high numbers of people into the world. With such different scenarios in mind, the United Nations prefers to use a widely respected model with three projections: high, moderate, and low growth (•**Figure 3.17**).

In view of declining birth rates worldwide, the United Nations has revised its projection for modest population growth downward. The agency predicts 9.3 billion by 2050 (half a billion fewer than it had estimated in earlier projections) and stabilization—the maximum number of people that will ever live on Earth at one time—at 10 billion in 2100.

This downward revision has prompted a rash of popular and academic articles proclaiming that "the population explosion is over" and even some essays arguing there would soon be too few people on Earth, particularly in the richer countries. Other experts have cautioned that it is too soon to declare the population bomb defused. "World population growth turned a little slower," a Population Institute report argues. "The difference, however, is comparable to a tidal wave surging toward one of our coastal cities. Whether the tidal wave is 80 or 100 feet high, the impact will be similar."[6] (To put that in some perspective, while population growth rates have been declining since 1968, the population base is so large that more people than ever before—about 83 million—are added to the world's population each year.) The United Nations has pledged to do its best to help stabilize the global population at a lower number than now projected. The organization's plan is to focus on enhancing the education and employment of women in the less developed countries as a means of bringing down birth rates.

The Malthusian Scenario

Even while the birth rate falls, the population increases. Already large and growing numbers pose fundamental questions: Can the Earth sustain 10 billion or more people? Will we exceed the planet's carrying capacity, and what will happen if we do? Early in the Industrial Revolution, an English clergyman named Thomas Malthus (1766–1834) postulated that human populations, growing geometrically or exponentially, would exceed food supplies, which grow only arithmetically or linearly. He predicted a catastrophic human die-off as a result of this irreconcilable equation (•Figure 3.18). He could not have foreseen that the exploitation of new lands and resources, including tapping the energy of fossil fuels, would permit food production to keep pace with or even outpace population growth for at least the next two centuries.

This **Malthusian scenario** of the lost race between food supplies and mouths to feed remains a source of constant and important debate today. On one side of the debate are optimists, the so-called **technocentrists** or **cornucopians**, who argue that human history provides insight into the future. Thanks to their technological ingenuity, people have always been able to conquer food shortages and other problems and therefore always will (•Figure 3.19). Julian Simon, a University of Maryland economist, argued that far from being a drain on resources, additional people create additional resources. Technocentrists therefore insist that people can raise the Earth's carrying capacity indefinitely and that the die-off that Malthus predicted will always be averted. Our more numerous descendants will instead enjoy more prosperity than we do.

In contrast, the **neo-Malthusians** (heirs of the reasoning of Thomas Malthus) argue that although we have been successful so far, we cannot increase the Earth's carrying capacity indefinitely. There is an upper limit beyond which growth cannot occur, with calculations ranging from 8 to 40 billion people. Neo-Malthusians insist that the poorer countries cannot remain indefinitely in the second stage of the demographic revolution. Either they must intentionally bring birth rates down further and make it successfully through the demographic transition, or they must unwillingly suffer nature's solution, a catastrophic increase in death rates. Thus by either the **birth rate solution** or the **death rate solution** to the problem, the neo-Malthusians argue, the LDCs must confront their population crisis (see pages 59–61).

Whereas technocentrists view the equation between people and resources passively, insisting that no corrective action is needed, neo-Malthusians tend to be activists who describe terrible scenarios of a death rate solution in order to motivate people to adopt the birth rate solution. The biologist Paul Ehrlich thinks that we are increasingly vulnerable to a Malthusian catastrophe, particularly as viruses diffuse around the globe with unprecedented speed. "The only big question that remains," Ehrlich wrote, "is whether civilization will end with the bang of an all-out nuclear war, or the whimper of famine, pestilence and ecological collapse."[7] Such dire warnings have earned many neo-Malthusians the reputation of being "gloom-and-doom pessimists." The neo-Malthusians insist that there are simply too many people already for the world to support, especially in the LDCs, where there are not enough resources to support them. But what does "too many people" mean? How many are too many, and based on what criteria?

What Is "Overpopulation"?

It is probably most useful to think about two distinct types of overpopulation, one characteristic of the poor countries and the other of the rich (•Figure 3.20). **People overpopulation** is an apparent problem in the poorer countries. The environmental problems characteristic of LDCs are intensified by the relatively high rates of human population growth in those countries. More people cut more trees, hunt more wildlife, and otherwise use more resources. Many persons, each using a small quantity of natural resources daily to sustain life, have a great collective impact on the environment and may add up to too many people for the local environment to support. Common consequences are malnutrition and even the famine emergencies in which richer countries are called on to provide relief.

Consumption overpopulation is characteristic of the MDCs. In the wealthier countries, there are fewer persons, but each uses a large quantity of natural resources from ecosystems around the world. Their collective impacts also degrade the environment, and even their smaller numbers may be "too

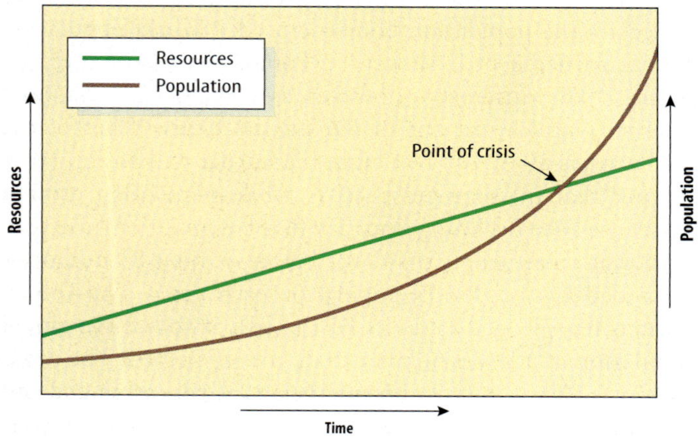

• **Figure 3.18** Malthus envisioned a race between people and resources, in which people lost.

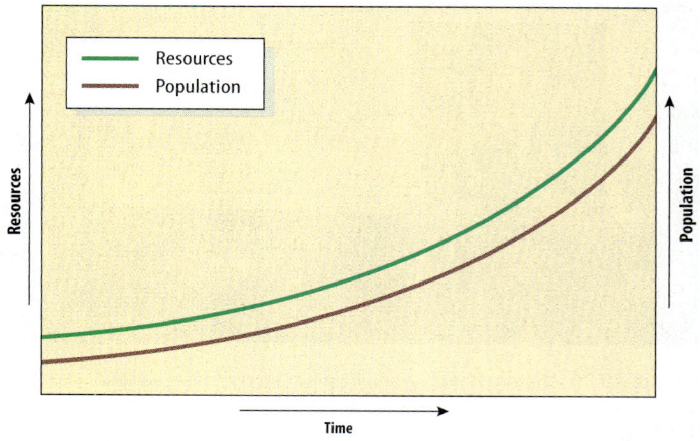

• **Figure 3.19** The technocentrists reason that production of food and other resources will always stay ahead of population growth.

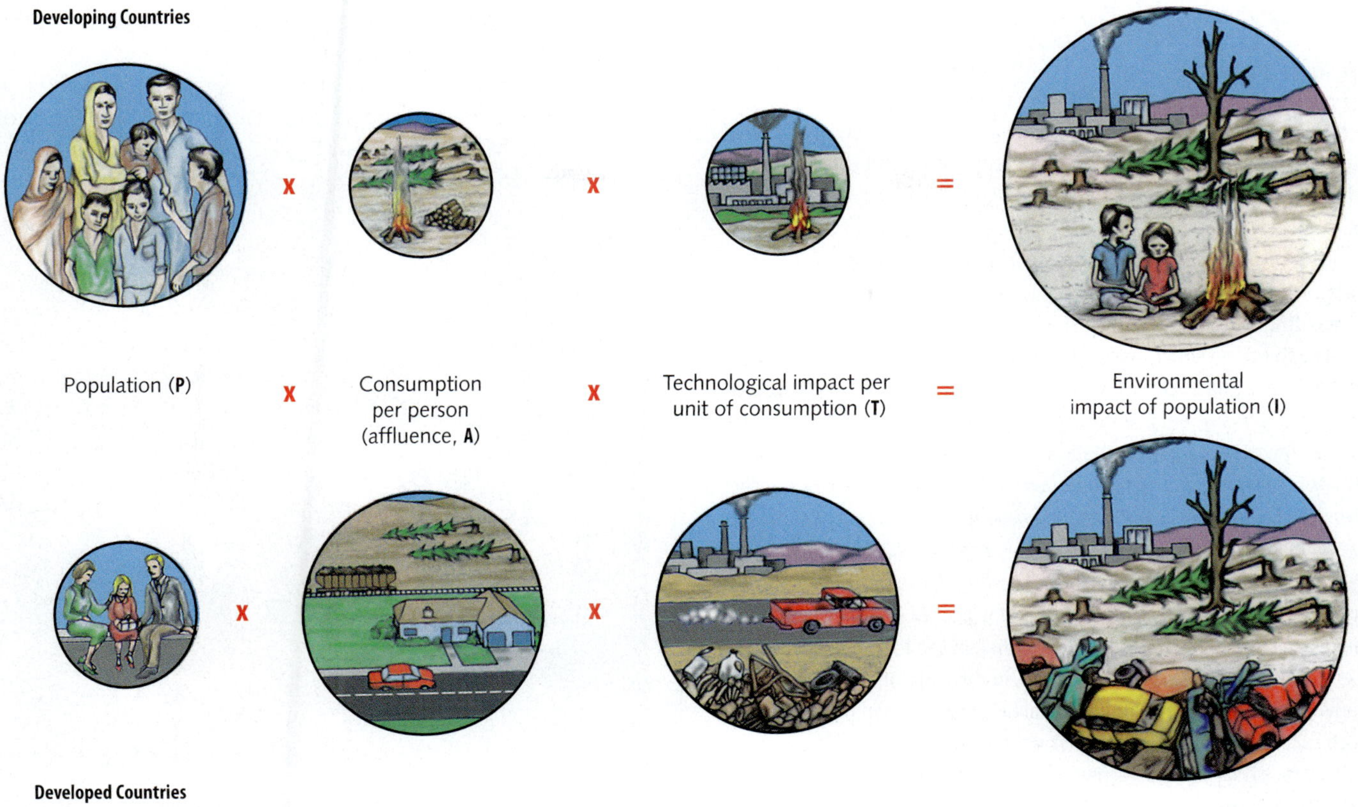

• **Figure 3.20** Two types of overpopulation, calculated according to this formula: *number of people* × *number of units of resources used per person* × *environmental degradation and pollution per unit of resource used* = *environmental impact*. Circle size shows the relative importance of each factor. People overpopulation is caused mostly by growing numbers of people and is typical of LDCs. Consumption overpopulation is caused mostly by growing affluence and is typical of MDCs.

Source: From MILLER, Environmental Science, 4E. © 1993 Cengage Learning.

many" at such unsustainably high levels of consumption, particularly with their habit of feeding so high on the food chain; much more energy is required to produce meat than to produce grains. With less than 5 percent of the world's population, the United States may be regarded as the world's leading overconsumer. According to a useful measure known as the **ecological footprint**—the amount of biologically productive land needed to sustain a person's consumption and absorb his or her wastes—for every acre needed to support an average Ethiopian, 69 acres are needed to support an average American. The disparity suggests that if Ethiopia is "underdeveloped," the United States is "overdeveloped" (see A Closer Look feature on page 60).

If the vast majority of the world's population were to consume resources at the rate that U.S. citizens do, the environmental results might be ruinous. Even if consumption levels in the LDCs do not rise substantially, the sheer increase in numbers of people in those countries suggests that degradation of the environment will accelerate in the coming decades.

3.4 An Action Plan for Global Problems

The cornucopian view is comforting: we have nothing to worry about. By keeping up the good, ingenious work as we have in the past, our futures will be secure. If, however, one diverges modestly from this premise or even accepts the most dire neo-Malthusian view (see Try It, page 61), it is appropriate to ask what can be done to prevent or solve some of the world's critical problems involving natural resources and human population numbers.

The Death Rate Solution and Lifeboat Ethics

Confronted with the Malthusian scenario, one option is the death rate solution, to "let nature take its course" and allow people imperiled by famine or other catastrophe to perish. The neo-Malthusian ecologist Garrett Hardin introduced **lifeboat ethics**—the question of whether or not the wealthy should rescue the "drowning" poor—in a distressing and challenging essay about the Malthusian scenario. "People turn to me," wrote Hardin, "and say, 'My children are starving. It's up to you to keep them alive.' And I say, 'The hell it is. I didn't have those children.'"[9] He described the world not as a single "spaceship *Earth*" or "global village" with a single carrying capacity, as many environmentalists do, but as a number of distinct "lifeboats," each occupied by the citizens of single countries and each having its own carrying capacity. Each rich nation is a lifeboat comfortably seating a few people. The world's poor are in lifeboats so overcrowded that many fall overboard. They swim to the rich lifeboats and beg to be brought aboard. What should the passengers of the rich lifeboat do? The choices pose an ethical dilemma for you to consider (•**Figure 3.21**).

A closer LOOK

Population and Energy

This book discusses energy a lot, and for good reasons: energy supply and demand issues impact our lives at many scales. But energy does not refer just to fossil fuels. Food energy, which is made available to ecosystems through the process of photosynthesis, and how that energy flows through ecosystems, are central to the question "How many people can the Earth support?"

Food chains (for example, in which a deer eats grass and a mountain lion eats a deer) are short, seldom consisting of more than four trophic (feeding) levels. The collective weight (known as **biomass**) and absolute number of organisms decline substantially at each successive trophic level (there are fewer leopards than antelopes). The reason is a fundamental rule of nature known as the **second law of thermodynamics** or the second energy law, which states that the amount of high-quality usable energy diminishes dramatically as the energy passes through an ecosystem. In living and dying, organisms use and lose the high-quality, concentrated energy that green plants produce and that is passed up the food chain. As an organism is consumed in any given link in the food chain, about 90 percent of that organism's energy is lost, in the form of heat and feces, to the environment. The amount of animal biomass that can be supported at each successive level thus declines geometrically, and little energy remains to support the top carnivores. So, for example, there are many more deer than there are mountain lions in a natural system.

The second law of thermodynamics has important consequences and implications for the human use of resources. The higher we feed on the food chain (the more meat we eat, for example), the more energy we use. The question of how many people the Earth's resources can support thus depends very much on what we eat. The affluent consumer of meat demands a huge expenditure of food energy because that energy flows from grain (producer) to livestock animal (primary consumer, with a 90 percent energy loss) and then from livestock animal to human consumer (secondary consumer, with another 90 percent energy loss). Each year, the average U.S. citizen eats 100 pounds of beef, 50 pounds of pork, and 45 pounds of poultry. By the time it is slaughtered, a cow has eaten about 10 pounds of grain per pound of its body weight; a pig, 5 pounds; and a chicken, 3 pounds. Thus in a year, a single American consumes the energy captured by two-thirds of a ton of grain (1,330 pounds) in meat products alone. By skipping the meat and eating just the corn that fattens these animals, many people could live on the energy required to sustain just one meat-eater. The point is not to feel guilty about eating meat but to realize that there is no answer to the question "How many people can the world support?" The question needs to be rephrased: "How many people can the world support based on people consuming ____?" (with the blank filled in with a specific number of calories per day or with a general lifestyle).

Development means a more affluent lifestyle, including eating more meat (feeding higher on the food chain). Some analysts fear the burden that the changing dietary habits accompanying development will pose to the planet's food energy supply. Lester Brown of the WorldWatch Institute wrote a provocative essay titled "Who Will Feed China?"[8] Keeping in mind the second law of thermodynamics, he argued that if China continues on its present course of growing affluence, with a move away from reliance on rice as a dietary staple to a diet rich in meats and beer, the entire planet will shudder. How could more than a billion and half people be sustained so high on the food chain?

Joe Hobbs

• **Figure 3.21** A dangerously overcrowded ferry in Tanzania. In Hardin's metaphor, what if this boat capsizes while a less crowded boat floats nearby?

Hardin set out the following scenario. There are 50 rich passengers in a boat with a capacity of 60. Around them are 100 poor swimmers who want to come aboard. The rich boaters have three choices. First, they could take in all the swimmers, capsizing the boat with "complete justice, complete catastrophe." Second, as they enjoy an unused excess capacity of 10, they could admit just 10 from the water. But which 10? And what about the margin of comfort that excess capacity allows them? Finally, the rich could prevent any of the doomed from coming aboard, ensuring their own safety, comfort, and survival—and ensuring the others would die.

Hardin's choice was the third: drowning. To preserve their own standard of living and ensure the planet's safety, the wealthy countries must cease to extend food and other aid to the poor and must close their doors to immigrants from poor countries. "Every life saved this year in a poor country diminishes the quality of life for subsequent generations," Hardin concluded. "For the foreseeable future, survival demands that we govern our actions by the ethics of a lifeboat."[10] Incidentally, Hardin jumped off his lifeboat, committing suicide in 2003.

The Birth Rate Solution and Sustainable Development

In recent decades, new concepts and tools for managing the Earth and its resources in an effective, long-term way have emerged. Known collectively as **sustainable development** (or **ecodevelopment**), these ideas and techniques consider what both MDCs and LDCs can do to avert the possible Malthusian dilemma and improve life on the planet. By promoting the birth rate solution and other concrete actions, sustainable development offers an activist agenda without the peril of doom depicted by the neo-Malthusians.

The World Conservation Union defines sustainable development as "improving the quality of human life while living within the carrying capacity of supporting ecosystems."[11] Sustainable development refutes what its proponents perceive as the current pattern of unsustainable development, whereby economic growth is based in large part on excessive resource use. Advocates of sustainable development point out that a country that depletes its resource base for short-term profits gained through deforestation increases its gross domestic product (GDP) and appears to be more "developed" than a country that protects its forests for a long-term harvest of sustainable yield. Deforestation appears to be beneficial to a country because it raises GDP through the production of pulp, paper, furniture, and charcoal. However, GDP growth does not measure the negative impacts of deforestation, such as erosion, flooding, siltation, and malnutrition. These consequences are known as **external costs**, or **externalities**, and they are not taken into account in the prices of goods and services. Advocates of sustainable development argue that these externalities should be added or "internalized" before a good or a service is marketed because the true high costs of producing these goods and services would be recognized. The lower true costs of less destructive practices would then be evident, providing stronger incentives for individuals, companies, and nations to invest in sustainable practices and technologies.

Sustainable development is a complex assortment of theories and activities, but its proponents call for eight essential changes in the way people perceive and use their environments:

1. People must change their worldviews and value systems, recognizing the finiteness of resources and reducing their expectations to a level more in keeping with the Earth's environmental capabilities. Proponents of sustainable development argue that this change in perspective is needed especially in the MDCs, where instead of trying to "keep up with the Joneses," people should try to enjoy life through more social rather than material pursuits.
2. People should recognize that development and environmental protection are compatible. Rather than viewing environmental conservation as a drain on economies, we should see it as the best guarantor of future economic well-being. This is especially important in the LDCs, with their resource-based economies.
3. People all over the world should consider the needs of future generations more than we do now. Much of the wealth we generate is in effect borrowed or stolen from our descendants. Our economic system values current environmental benefits and costs far more than future benefits and costs, and so we try to improve our standard of living today without regard to tomorrow.
4. Communities and countries should strive for self-reliance, particularly through the use of appropriate technologies. For example, remote villages could rely increasingly on solar power for electricity rather than be linked into national grids of coal-burning plants.
5. LDCs need to limit population growth as a means of avoiding the destructive impacts of people overpopulation. Advances in the status of women, improvements in education and social services, and effective family planning technologies can help limit population growth.
6. Governments need to practice land reform, particularly in the LDCs. Poverty is often not the result of too many people on too little total land area, but of a small, wealthy minority holding a disproportionately high share of quality land. To avoid the environmental and economic consequences of marginalization (as discussed on page 47), a more equitable distribution of land is needed.
7. Economic growth in the MDCs should be slowed to reduce the effects of consumption overpopulation. If economic growth, understood as the result of consumption of natural resources, continues at its present rate in excess of sustainable yield, the Earth's "environmental capital" will continue to diminish rapidly.
8. Wealth should be redistributed between the MDCs and LDCs. Because poverty is such a fundamental cause of environmental degradation, the spread of a reasonable level of prosperity and security to the LDCs is essential. Proponents argue that this does not mean that rich countries should give cash outright to poor countries. Instead, the lending institutions of MDCs can forgive some existing debts owed by LDCs or use such innovations as **debt-for-nature swaps**, in which a certain portion of debt is forgiven in return for the borrower's pledge to invest that amount in national parks or other conservation programs. Reducing or eliminating **trade barriers** that MDCs impose against products from LDCs would also help redistribute global wealth.

Some geographers and other scientists believe that sustainable development will bring about the "**Third Revolution**," a shift in human ways of interacting with the Earth so dramatic that it will be compared with the origins of agriculture and industry. The formidable changes called for in sustainable development are attracting increasing attention, perhaps because there are no other comprehensive strategies for dealing with some of the most critical issues of our time.

Weighing Lifeboat Ethics

Try It

Think about these options, with some concrete examples in mind. As an occupant of the rich lifeboat *United States*, for example, what would you do for the drowning refugees from lifeboat *Haiti* or lifeboat *Somalia*? Help bail out everyone in need? Assist just a few? Ignore them altogether? Why did you make this choice?

(This can be a difficult ethical question. I do this as an exercise in my class, with everyone making their choices anonymously on a piece of paper. I tally up the results of "save" and "drown," and we discuss as much as the students want to.)

summary

- A useful way to begin to appreciate the current spatial patterns of our relationship with the Earth is to view these patterns as products of two "revolutions" in the relatively recent past: the Agricultural or Neolithic (New Stone Age) Revolution and the Industrial Revolution. Each transformed humanity's relationship with the natural environment. After the Industrial Revolution, the human impact on Earth grew dramatically.
- The world is markedly divided between the haves and the have-nots, characterized at the largest scale by contrasts between more developed and less developed countries (MDCs and LDCs). Explanations for these disparities include dependency theory, cultural factors, geographic location, and natural resource base.
- LDC economies often rely on the export of a few raw materials or commercial crops and tend to draw down their natural "capital" quickly, with profound impacts on the environments.
- There is disagreement on whether globalization is beneficial or detrimental to poorer countries and to the poorer people who live in those countries.
- Population growth is measured by birth rates, death rates, and migration. Growth rates tend to be higher in the LDCs, due to relatively high but falling birth rates and much lower and falling death rates. The demographic transition, a model of what happened to populations in the MDCs, shows the wealthier countries passing from high birth rates and death rates to low birth rates and death rates.
- The explosive growth in world population since the beginning of the Industrial Revolution is the result not of a rise in birth rates, but of a dramatic decline in death rates, particularly in the LDCs. High rates of human population growth intensify environmental problems characteristic of the LDCs. More people use more resources, a phenomenon that suggests there is a problem of people overpopulation in the poorer countries. Consumption overpopulation is more characteristic of the MDCs, where fewer people use large quantities of natural resources from around the world.
- Human demands for more food and other photosynthetic products are growing, and it is questionable whether supplies can keep pace. The Malthusian scenario, which has not been realized so far, insists that a catastrophic die-off of people will occur when their numbers exceed food supplies.
- New concepts and tools for managing the Earth and its resources in an effective, long-term way have emerged in the past two decades. Known collectively as sustainable development or ecodevelopment, these ideas and techniques consider what humanity can do to avert the Malthusian dilemma and improve life on the planet.

Key Terms + Concepts

Age of Discovery (p. 41)
Age of Exploration (p. 41)
age structure diagram (p. 53)
Agricultural Revolution (p. 40)
Anthropocene Epoch (p. 40)
asylum (p. 56)
biomass (p. 60)
birth rate (p. 49)
birth rate solution (p. 58)
brain drain (p. 56)
carrying capacity (p. 40)
cash crops (p. 47)
civilization (p. 40)
colonization (p. 41)
commercial crops (p. 47)
consumption overpopulation (p. 58)
cornucopians (p. 58)
culture hearth (p. 40)
death rate (p. 50)
death rate solution (p. 58)
debt-for-nature swap (p. 61)
deforestation (p. 48)
demographic transition (p. 52)
  first (preindustrial) stage (p. 52)
  second (transitional) stage (p. 52)
  third (industrial) stage (p. 52)
  fourth (postindustrial) stage (p. 52)
demography (p. 49)
dependency theory (p. 44)
development (p. 43)
diffusion (p. 40)
digital divide (p. 46)
domestication (p. 40)
doubling time (p. 50)
dry farming (p. 40)
ecodevelopment (p. 60)
ecological bankruptcy (p. 47)
ecological footprint (p. 59)
ecologically dominant species (p. 39)
emigrant (p. 56)
extensive land use (p. 40)
external costs (p. 61)
externalities (p. 61)
food chain (p. 60)
Food-Producing Revolution (p. 40)
foraging (p. 39)
fuelwood crisis (p. 48)
globalization (p. 46)
gross domestic product (GDP) (p. 42)
gross national income (GNI) (p. 43)
gross national product (GNP) (p. 43)
guest worker (p. 56)
Human Development Index (HDI) (p. 44)
hunting and gathering (p. 39)
illegal alien (p. 56)
immigrant (p. 56)
Industrial Revolution (p. 41)
information technology (IT) (p. 46)
intensive land use (p. 40)
internally displaced persons (IDPs) (p. 56)
irrigation (p. 40)
knowledge economy (p. 46)
less developed countries (LDCs) (p. 42)
lifeboat ethics (p. 59)
Malthusian scenario (p. 58)
marginalization (p. 47)
mercantile colonialism (p. 45)
migration (p. 56)
more developed countries (MDCs) (p. 42)
multinational companies (p. 46)
natural replacement rate (p. 47)
natural resource (p. 45)
neocolonialism (p. 45)
neo-Europes (p. 45)

Neolithic Revolution (p. 40)
neo-Malthusians (p. 58)
newly industrializing countries (NICs) (p. 42)
nontimber forest products (NTFPs) (p. 48)
nonselective migrants (p. 56)
"original affluent society" (p. 39)
one-child policy (p. 57)
people overpopulation (p. 58)
per capita GDP (p. 42)
per capita gross national income (p. 43)
  purchasing power parity (p. 43)
  (per capita GNI PPP) (p. 43)
Pleistocene overkill hypothesis (p. 39)
population change rate (p. 50)
population explosion (p. 49)
population implosion (p. 53)
population pyramid (p. 53)
population replacement level (p. 53)
pull factors (p. 56)
purchasing power parity (PPP) (p. 43)
push factors (p. 56)
refugees (p. 56)
renewable resource (p. 47)
rural-to-urban migration (p. 56)
second law of thermodynamics (p. 60)
selective migrants (p. 56)
settler colonization (p. 44)
sustainable development (p. 60)
sustainable yield (p. 47)
technocentrists (p. 58)
"Third Revolution" (p. 61)
trade barriers (p. 61)
urbanization (p. 56)
value-added products (p. 45)
zero population growth (ZPG) (p. 53)

Review Questions

1. What were the Agricultural and Industrial Revolutions? In what ways did they initiate important changes in human–Earth relationships?
2. What are the typical differences between MDCs and LDCs?
3. According to dependency theory, what are the causes of disparities between MDCs and LDCs? What other factors may explain global wealth and poverty?
4. What is the fuelwood crisis? What are some of the effects of deforestation on human lives and economies in the LDCs?
5. What are the two types of overpopulation, and how do they differ?
6. What has been the main cause of the world's explosive population growth since the beginning of the Industrial Revolution?
7. What variables distinguish the four stages of the demographic transition, and what explains them?
8. Which world regions have the most and the fewest people, and what factors might account for these differences?
9. How do technocentrists and neo-Malthusians view the balance between people and resources?
10. What do Hardin's lifeboats represent—that is, what is his metaphor?
11. What are the goals and methods of sustainable development?

Notes

1. Marshall B. Sahlins, *Stone Age Economics* (Chicago: Aldine-Atherton, 1972).
2. Many of these points come from this special edition: "The Geology of the Planet: Welcome to the Anthropocene," *The Economist*, May 26, 2011.
3. Alfred W. Crosby, *Ecological Imperialism: The Biological Expansion of Europe, 900–1900* (New York: Cambridge University Press, 1986).
4. Friedman, Thomas. 2005. The World is Flat: A Brief History of the Twenty-First Century. New York: Picador.
5. Celia W. Dugger, "Very Young Populations Contribute to Strife, Study Concludes." *New York Times,* April 4, 2007, p. A6.
6. Cited in Steven A. Holmes, "Global Crisis in Population Still Serious, Group Warns." *New York Times,* December 31, 1997, p. A7.
7. Paul R. Ehrlich, "Populations of People and Other Living Things," in *Earth '88: Changing Geographic Perspectives,* ed. Harm De Blij (Washington, DC: National Geographic Society, 1988), p. 309.
8. Lester R. Brown, "Who Will Feed China?" *WorldWatch Magazine,* September–October 1994, p. 10.
9. Garrett Hardin, "The Tragedy of the Commons." *Science, 162,* 1968, pp. 1243–1248.
10. Garrett Hardin, "Lifeboat Ethics: The Case against Helping the Poor," *Psychology Today,* September 1974, p. 6; and "Living on a Lifeboat," *BioScience, 24,* 1974, p. 568.
11. World Wildlife Fund, *Sustainable Use of Natural Resources: Concepts, Issues and Criteria* (Gland, Switzerland: World Wildlife Fund, 1995), p. 5.

Online Resources

CourseMate: Make the most of your study time by accessing everything you need to succeed in one place. Read your textbook, take notes, review flashcards, watch videos, complete activities, take practice quizzes, and more—online with CourseMate. Log in at **www.cengagebrain.com.**

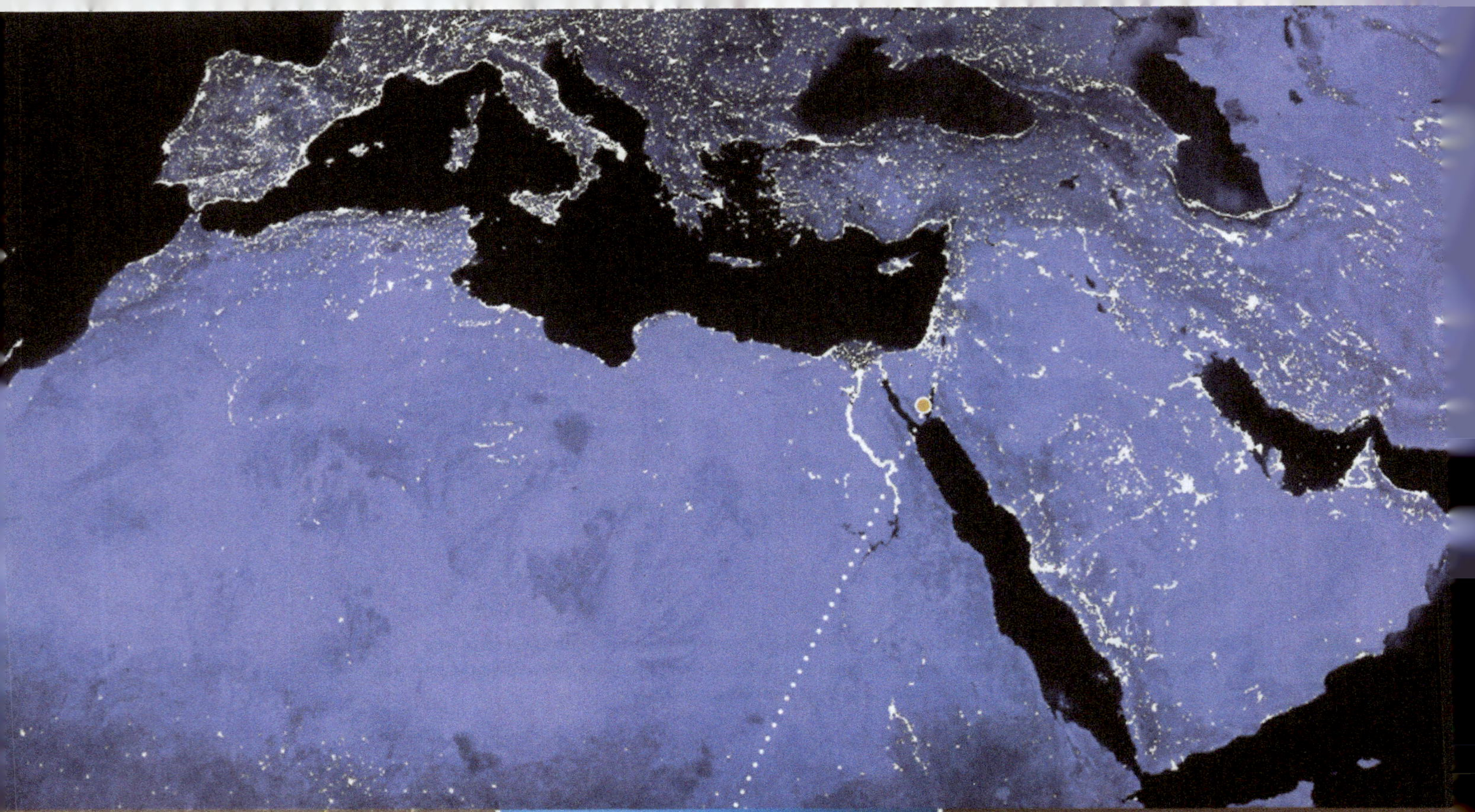

> *"Knowledge is light, and all who contribute bring light to our world."*
>
> —Bedouin Proverb

C. Mayhew & R. Simmon (NASA/GSFC), NOAA/ NGDC, DMSP Digital Archive

Joe Hobbs

Above: The Middle East and North Africa at night. Note especially the population concentrations along the great river valleys of the Nile, Tigris and Euphrates. Left: The Monastery of St. Katherine at the foot of Jebel Musa in Egypt's Sinai Peninsula. This is widely believed to be Mount Sinai, where God gave Moses the Ten Commandments.

6

The Middle East and North Africa

Welcome to the Middle East. What other part of the world outside your own is in the news every single day? What part of the world seems so threatening to your own, even if you feel dependent upon it just to drive across town? What part of the world seems to have so many problems for which no solutions seem to be in sight?

The good news is that there *is* good news coming out of the Middle East, and there always has been. This is the fountainhead of many faiths, civilizations, and sciences. Also good is that there *are* solutions to the region's seemingly intractable problems (and in fact, peace and stability in the Middle East will make your life better). This chapter will equip you with enough knowledge that you should feel capable of attempting to solve the Palestinian-Israeli conflict in this chapter's "Try It" feature.

chapter outline

chapter objectives

This chapter should enable you to

- Understand and explain the mostly beneficial relationships between villagers, pastoral nomads, and city dwellers in an environmentally challenging region.
- Know the basic beliefs and sacred places of Jews, Christians, and Muslims.
- Recognize the importance of petroleum to this region and the world economy.
- Identify the geographic chokepoints and oil pipelines that are among the world's most strategically important places and routes.
- Appreciate the problems of control over freshwater in this arid region.
- Know what al-Qa'ida and other Islamist terrorist groups are and what they want.
- Understand the major issues of the Arab-Israeli conflict and the obstacles to their resolution.
- Learn about the Arab Spring, how it evolved, and what can be expected in its wake.
- Balance the pros and cons of Egypt's Aswan High Dam and other large water engineering projects.
- Recognize how Islamist terrorism transformed American geopolitical perspectives and priorities.
- Trace the trajectory of Persian Gulf (Arabian Gulf) countries from underpopulated, wealthy welfare states to populous, vulnerable states overly dependent on a single commodity.
- See the U.S. wars on Iraq as the result of tragic miscalculations by Saddam Hussein, combined with strategic American interests in the region and, many analysts say, by American miscalculations.
- Recognize Turkey as an "in-between" country, a less developed Asian nation aspiring to be a European power.

The *Middle East* is a fitting designation for the places and the cultures of this vital world region: they really are in the middle. This is a physical crossroads, where the continents of Africa, Asia, and Europe meet, and the waters of the Mediterranean Sea and the Indian Ocean mingle. Its peoples—Arab, Jew, Persian, Turk, Kurd, Berber, and others—express in their cultures the coming together of diverse influences. They have bestowed on humankind a rich legacy that includes the ancient civilizations of Egypt and Mesopotamia and the world's three great monotheistic faiths of Judaism, Christianity, and Islam.

People tend to overlook contributions like these and instead associate the region with war and terrorism. These negative connotations are often accompanied by muddled understanding. This chapter makes the complex and misunderstood Middle East and North Africa more intelligible for you by illuminating their geographic context.

6.1 Area and Population

What is the Middle East, and where is it? The term itself applies to a Eurocentric vernacular region invented by the British, who 6
placed themselves in the figurative center of the world. They began to use the term before the outbreak of World War I, when the *Near East* referred to the territories of the Ottoman Empire in the eastern Mediterranean region, the *East* to India, and the *Far East* to China, Japan, and the western Pacific Rim. With *Middle East,* they designated as a separate region the countries around the Persian Gulf (known to Arabs as the Arabian Gulf and in this text often simply the Gulf). Gradually, the perceived boundaries of the region grew.

Sources today vary widely in their interpretation of which countries are in the Middle East. For some, the Middle East includes only the countries clustered around the Arabian Peninsula. For others, the Middle East spans a vast 6,000 miles (9,700 km) west to east, from Morocco in northwest Africa to Iran in central Asia, and a north–south distance of about 3,000 miles (4,800 km) from Turkey, on Europe's southeastern corner, to Sudan, which adjoins East Africa (**•Figure 6.1**). This larger area is the region covered in this chapter, where it is referred to as the Middle East *and* North Africa, and often by the acronym MENA, which is popular in many regional studies. North Africa is added to the Middle East because the North African peoples of Morocco, Algeria, and Tunisia generally do not consider themselves Middle Easterners; they are, rather, from what they call the *Maghreb,* meaning "western land."

Defined with these borders, the Middle East and North Africa include 21 countries, the Palestinian territories of the West Bank and Gaza Strip, and the disputed Western Sahara (**•Table 6.1**), occupying 5.6 million square miles (14.7 million sq km) and inhabited by about 503 million people in 2011. This area is about 1.8 times the size of the lower 48 United States and is generally situated at latitudes equivalent to those between Boston, Massachusetts, and San Jose, Costa Rica (see **•Figure 6.2**).

These half-billion people are not distributed evenly across the region but are concentrated in major clusters (**•Figure 6.3**). Three countries contain the lion's share of the region's population: Turkey, Iran, and Egypt, each with more than 70 million people. There is a very strong correlation between the distribution of population and the distribution of water resources. Where water is abundant in this generally arid region, so are people. Egypt has the Nile River, and parts of Iran and Turkey have bountiful rain and snow. Conversely, where rain seldom falls, as in the Sahara of North Africa and in the Arabian Peninsula, people are few.

There is a considerable range in the region's rates of population growth. Bahrain, the United Arab Emirates, Turkey, and Tunisia had the lowest rate of population growth, 1.2 percent in 2011. The highest is 3.1 percent in Yemen, followed by

• Figure 6.1 The Middle East and North Africa.

3.0 percent in Iraq and 2.8 percent in the Palestinian territories (consisting of the West Bank and Gaza Strip). These are some of the highest population growth rates in the world, on
313 par with many countries in Africa South of the Sahara. Yemen's
195 high population growth is related to its great poverty. In the Palestinian territories, rapid growth may be attributed in part to the Palestinians' poverty and probably also to the wishes of many Palestinians to have more children to counterbalance the demographic weight of their perceived Israeli foe. As we will see later in the chapter (on pages 187–189), there is an important demographic component to the Palestinian-Israeli conflict.

Between these extremes are countries with modest rates of population growth. Their governments have regarded most of these countries as too populous for their resources and economic bases, and have encouraged family planning. They have been successful in lowering birth rates, which bodes well for their development.

At the other end of the spectrum, some oil-rich but population-poor countries of the Gulf nations have been encouraging citizens to give birth to more citizens. If there were more resident nationals, there would be less need to import foreign laborers and technicians, and these countries would be more self-sufficient in their development. Some oil-rich countries of the Gulf region have far more foreigners than citizens living in them; about 80 percent of the working-age population of the United Arab Emirates, for example, is nonnative, mainly from South Asian countries. Many of these workers complain of dreams broken by poor living conditions and low wages, and in the worst cases, of physical and emotional abuse shared
by victims of human trafficking. 145

The Middle East and North Africa as a whole have a medium rate of population growth, suggesting that the region is moving out of the second and into the third phase of the demographic transition. Death rates have been on the decline for decades, and recently birth rates have begun falling. The result is a declining rate of population growth. The average annual rate of population change for the 21 countries, the Palestinian territories, and Western Sahara was 1.8 percent in 2011. This is higher than the world average of 1.2 percent but is a significant decline from the high of 3.0 in 1980.

These numbers seem promising for progress: as we saw in Chapter 3, a lowering of population in the LDCs is a fundamental requirement for their sustainable development. On closer examination, however, we can see a problem. The number of people in the lower age groups is so high that even with the declining rate of growth, very large numbers of people are young. This is the **youth bulge** present in all but a handful of the Middle East and North African countries. Sixty percent of MENA's people are less than 25 years old. The number of youth (people ages 15–24) in the region is estimated to grow until peaking at 100 million in 2035.

If there were jobs for such large numbers of people, there would be little reason for concern. Unfortunately, this region has the highest unemployment rate of all the world's regions (officially, 10% overall, with unemployment among young people over 25%, but the actual numbers may be considerably

Table 6.1 Middle East and North Africa: Basic Data

Political Unit	Area (thousands; sq mi)	Area (thousands; sq km)	Estimated Population (millions)	Estimated Population Density (sq mi)	Estimated Population Density (sq km)	Annual Rate of Natural Increase (%)	Human Development Index	Urban Population (%)	Per Capita GNI PPP ($US)
Middle East	**2,378.9**	**6,161.4**	**298.5**	**125**	**48**	**1.8**	**0.680**	**70**	**12,060**
Bahrain	0.3	0.8	1.3	4,333	1,673	1.3	0.806	100	33,690
Iran	630.6	1,633.3	77.9	124	48	1.3	0.707	70	11,470
Iraq	169.2	438.2	32.7	193	75	3.0	0.573	67	3,330
Israel	8.1	21.0	7.9	975	377	1.7	0.888	92	27,010
Jordan	34.4	89.1	6.6	192	74	2.6	0.698	83	5,730
Kuwait	6.9	17.9	2.8	406	157	1.6	0.760	98	29,200
Lebanon	4.0	10.4	4.3	1,075	415	1.5	0.739	87	13,400
Oman	82.0	212.4	3.0	37	14	2.6	0.705	73	24,530
Palestinian Territories	2.4	6.2	4.2	1,750	676	2.8	0.641	83	N/A
Qatar	4.2	10.9	1.7	405	156	1.0	0.831	100	N/A
Saudi Arabia	830.0	2,149.7	27.9	34	13	1.8	0.770	81	24,020
Syria	71.5	185.2	22.5	315	122	2.2	0.632	54	4,620
Turkey	299.2	774.9	74.0	247	95	1.2	0.699	76	13,500
United Arab Emirates	32.3	83.7	7.9	245	94	1.2	0.846	83	23,990
Yemen	203.8	527.8	23.8	117	45	3.1	0.462	29	2,330
North Africa	**3,037.2**	**7,866.4**	**204.9**	**67**	**26**	**1.8**	**0.606**	**51**	**5,680**
Algeria	919.6	2,381.8	36.0	39	15	1.5	0.698	67	8,110
Egypt	386.7	1,001.6	82.6	214	82	2.0	0.644	43	5,680
Libya	679.4	1,759.6	6.4	9	4	1.8	0.760	78	16,400
Morocco	172.4	446.5	32.3	187	72	1.3	0.582	56	4,400
Sudan	718.7	1,861.5	36.4	51	20	2.4	0.408	41	1,990
Tunisia	63.2	163.7	10.7	169	65	1.2	0.698	68	7,810
Western Sahara	97.2	251.7	0.5	5	2	2.3	N/A	82	N/A
Summary Total	**5,416.1**	**14,027.8**	**503.4**	**93**	**36**	**1.8**	**0.650**	**62**	**9,460**

Sources: World Population Data Sheet, Population Reference Bureau, 2011; Human Development Report, United Nations, 2011; World Factbook, CIA, 2011.

higher). The conditions of high population growth and lack of economic opportunities have long made this region ripe for unrest. As we will see, in the spring of 2011 the Middle East and North Africa exploded in a wave of popular revolutions. This was the "Arab Spring" that is presented on pages 192–197 of this chapter.

Many developing countries have economies largely dependent on subsistence agriculture, with low percentages of urban inhabitants. Perhaps surprisingly, however, the Middle East and North Africa have more urbanites than country folk. The average urban population among the 23 countries and territories is 62 percent. The most prosperous countries are also the most urban. Essentially city-states, Bahrain, Qatar, and Kuwait are 100 percent urban. The other oil-wealthy Gulf countries have urban populations over 70 percent. Consistent with its profile as a Western-style industrialized country without oil resources, Israel is 92 percent urban. At the other end of the spectrum, desperately poor Yemen has an urban population of 29 percent.

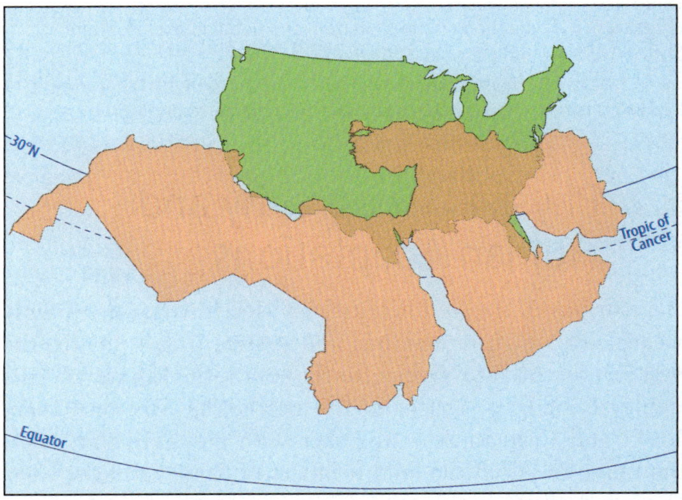

• **Figure 6.2** The Middle East and North Africa compared in latitude and area with the conterminous United States.

Source: Based on John Haywood, ed., Atlas of World History (Barnes and Noble Books, 1997).

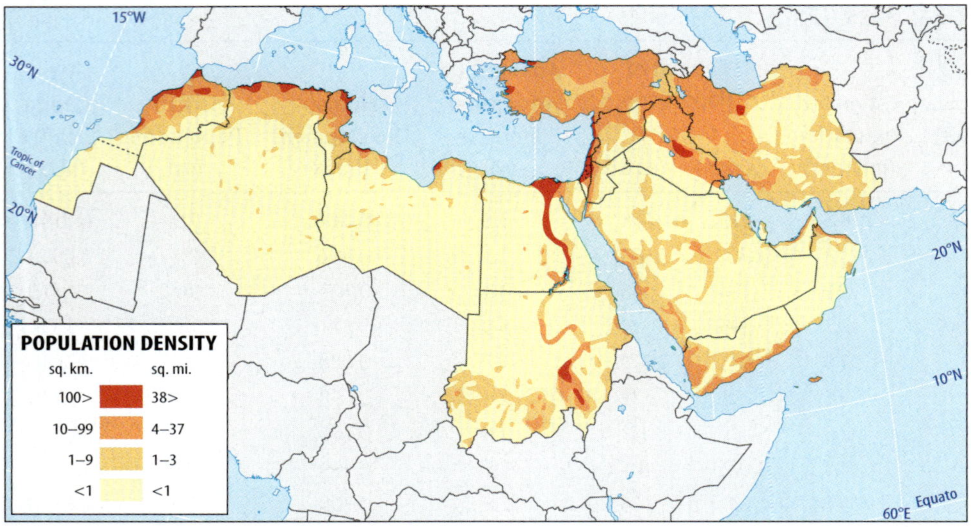

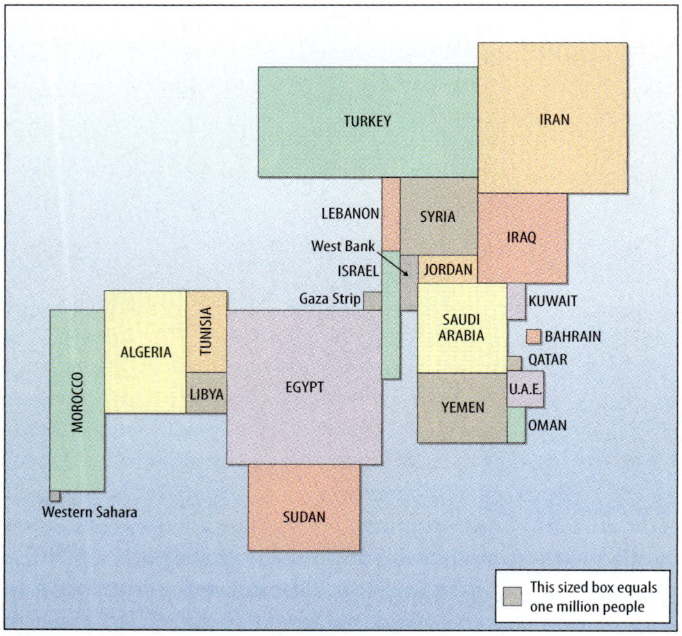

• Figure 6.3 Population distribution (top) and population cartogram (bottom) of the Middle East and North Africa.

Source: The Oxford School Atlas edited by Patrick Wiegand (OUP, 1997), copyright © Oxford University Press. Used by permission of Oxford University Press.

6.2 Physical Geography and Human Adaptations

The margins of the Middle East and North Africa are mainly oceans, seas, high mountains, and deserts (Figure 6.1). To the west lies the Atlantic Ocean; to the south, the Sahara and the highlands of East Africa; to the north, the Mediterranean, Black, and Caspian Seas, together with mountains and deserts lining the southern land frontiers of Russia and the Near Abroad; and to the east, the mountains on the Iran-Afghanistan frontier and the Baluchistan Desert straddling Iran and Pakistan. The land is composed mainly of arid plains and plateaus, together with large areas of rugged mountains and isolated "seas" of sand and "islands" or ribbons of fertility known as oases. Despite the environmental challenges, this region has given rise to some of the world's oldest and most influential ways of living.

A Region of Stark Geographic Contrasts

Aridity dominates the Middle East and North Africa (•Figure 6.4). At least three-fourths of the region has average yearly precipitation of less than 10 inches (25 cm), an amount too small for most types of dry farming (unirrigated agriculture). Sometimes, however, localized cloudbursts release moisture that allows plants, animals, and small populations of people—the Bedouin, Tuareg, and other pastoral nomads—to live in the desert. Even the vast Sahara, the world's largest desert, supports a surprising diversity and abundance of life. Plants, animals, and even people have developed strategies of **drought avoidance** and **drought endurance**. Migrating to avoid drought is a coping strategy that pastoral nomads and many animals use. Other organisms, such as trees, must endure drought by such adaptations as having very deep roots and small leaves. Populations of people, plants, and animals are all but nonexistent in the region's vast **sand seas**, including the Great Sand Sea of western Egypt and the Empty Quarter of the Arabian Peninsula (•Figure 6.5).

The region's climates have the comparatively large daily and seasonal ranges of temperature characteristic of dry lands. Desert nights can be surprisingly cool. Most days and nights are cloudless, so the heat absorbed on the desert surface during the day is lost by radiational cooling to the heights of the atmosphere at night. Summers in the lowlands are very hot almost everywhere. The hottest shade temperature ever recorded on earth, 136°F (58°C), occurred in Libya in September 1922. Many places regularly experience daily maximum temperatures over 100°F (38°C) for weeks at a time.

Human settlements located near the sand seas often experience the unpleasant combination of high temperatures and

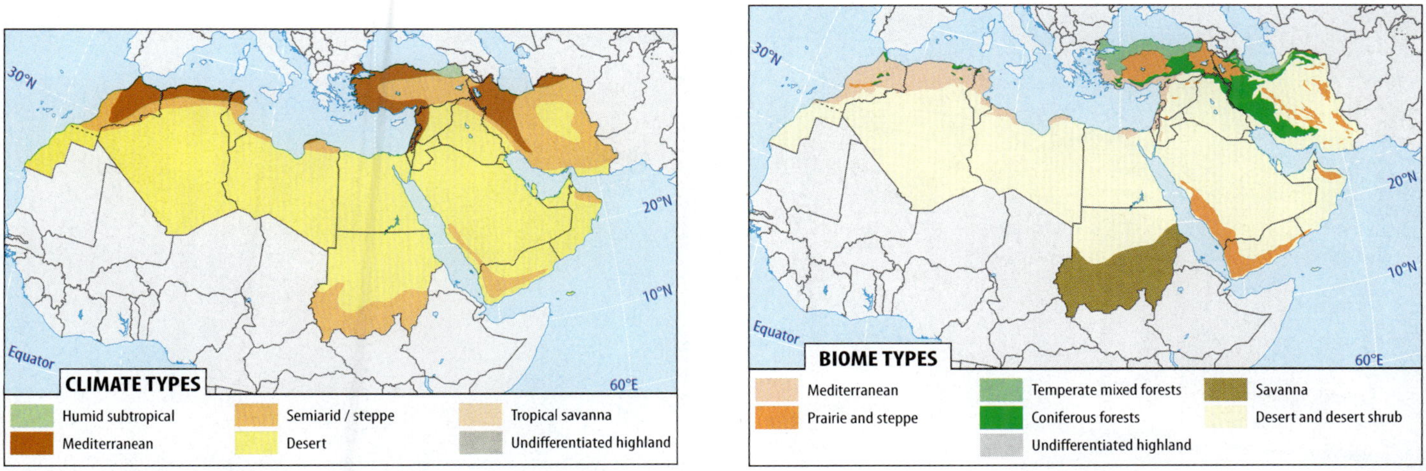

• **Figure 6.4** Climates (left) and biomes (right) of the Middle East and North Africa.
Sources: (left) Based on Rand McNally's Classroom Atlas, 2003. (right) Based on World Wildlife Fund ecoregions data, 1999.

hot, sand-laden winds, creating the sandstorms known locally by such names as *simuum* ("poison") *sirocco,* and *haboob* (the latter, ushered in by an ominous, towering wall of sand, is also familiar to the people of southern Arizona). Only in mountainous sections and in some places near the sea do higher elevations or sea breezes temper the intense midsummer heat. The population of Alexandria, on the Mediterranean Sea coast, explodes in summer as Egyptians flee from Cairo and other hot inland locations. In Saudi Arabia, the government relocates from Riyadh to the highland summer capital of Taif to escape the lowland furnace.

Lower winter temperatures bring relief from the summer heat, and the more favored places receive enough precipitation for dry farming of winter wheat, barley, and other cool season crops. Land uses across the region are depicted in •**Figure 6.6**. In general, winters are cool to mild. But very cold winters and snowfalls occur in the high interior basins and plateaus of Iran, Afghanistan, and Turkey (there are even glaciers in Iran and Turkey!). These locales generally have a steppe climate. Only in the southernmost reaches of the region, notably in southern Sudan, do temperatures remain consistently high throughout the year. A savanna climate and biome prevail there.

Most areas bordering the Mediterranean Sea have 15 to 40 inches (38 to 102 cm) of precipitation a year, falling almost exclusively in winter, whereas the summer is dry and warm—a typical Mediterranean climate pattern. Throughout history, 26 people without access to perennial streams have stored this moisture to make it available later for growing crops that require the higher temperatures of the summer months. The Nabateans, for example, who were contemporaries of the Romans in what is now Jordan, had a sophisticated network of limestone cisterns and irrigation channels (•**Figure 6.7**).

Along the southern and northern margins of the region, there is enough rainfall to support dry (nonirrigated) farming during the summer. There are some places that defy the usual perception of the region as bone-dry. The Black Sea side of Turkey's Pontic (Kuzey) Mountains, for example, is lush and moist

Joe Hobbs

• **Figure 6.5** Seas of sand cover large areas in Saudi Arabia, and parts of the Sahara. This is the edge of the Empty Quarter (al-Rub' al-Khali) in the United Arab Emirates.

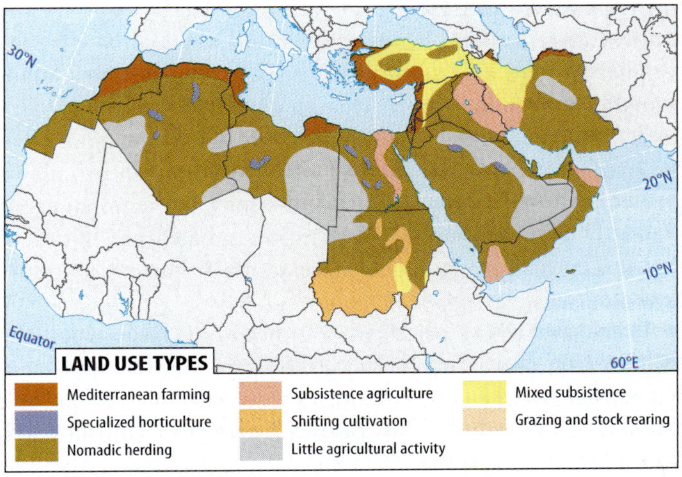

• **Figure 6.6** Land use in the Middle East and North Africa.
Source: The Oxford School Atlas edited by Patrick Wiegand (OUP, 1997), copyright © Oxford University Press. Used by permission of Oxford University Press.

Joe Hobbs

• **Figure 6.7** The Treasury, a temple carved in red sandstone, probably in the first century C.E., by the Nabateans at their capital of Petra in southern Jordan. The English poet Dean Burgen called Petra "a rose red city, half as old as time." The Nabateans built sophisticated networks for water storage and distribution. For scale, note the man at the lower right.

Joe Hobbs

• **Figure 6.8** Rainfall is heavy on the Black Sea side of Turkey's Pontic (also known as Pontus or Kuzey) Mountains. Note that roofs are pitched to shed precipitation; in the region's drier areas, most roofs are flat. The crop growing here is corn.

in the summer, and tea grows well there (•**Figure 6.8**). In the southwestern Arabian Peninsula, a monsoonal climate brings summer rainfall and autumn harvests to Yemen and Oman, probably accounting for the Roman name for the area: *Arabia Felix* ("Happy Arabia"). Mountainous areas in the region, like the river valleys and the margins of the Mediterranean, play a vital role in supporting human populations and national economies. Due to orographic or elevation-induced precipitation, the mountains receive much more rainfall than surrounding lowland areas.

There are three principal mountainous regions of the Middle East and North Africa (see Figure 6.1). In northwestern Africa between the Mediterranean Sea and the Sahara, the Atlas Mountains of Morocco, Algeria, and Tunisia reach over 13,000 feet (3,965 m) in elevation. Mountains also rise on both sides of the Red Sea, with peaks up to 12,336 feet (3,760 m) in Yemen. These mountains rose with the tectonic processes that are pulling the African and Arabian plates apart at the northern end of the Great Rift Valley. The hinge of this crustal movement is the Bekaa Valley of Lebanon, where the widening fault line follows the Jordan River Valley southward to the Dead Sea (•**Figure 6.9**). This valley is the deepest depression on the earth's land surface, lying about 600 feet (183 m) below sea level at Lake Kinneret (also known as the Sea of Galilee or Lake Tiberius) and nearly 1,300 feet (400 m) below sea level at the shore of the Dead Sea, the lowest point of land on the planet. The rift then continues southward to the very deep Gulf of Aqaba and Red Sea, before turning inland into Africa at Djibouti and Ethiopia, and continuing all the way down the Great Rift Valley of Africa (see page 317). By the way, should you visit the Dead Sea one day, just to try to sink. It's impossible—as this lake has no outlet, the waters are so salty at 32 percent that you will bob like a cork.

In tectonic processes, one consequence of rifting in one place is the consequent collision of the earth's crustal plates elsewhere. There are several such collision zones in the Middle East and North Africa, particularly in Turkey and Iran, where mountain

Image Science and Analysis Laboratory, NASA-JSC.

• Figure 6.9 It looks peaceful from space, but the land around the Jordan Rift Valley is one of the most contested areas on Earth. The Jordan River runs into and out of Lake Kinneret (the Sea of Galilee) near the top of the image, emptying into the Dead Sea near the bottom of the image. If political boundaries were drawn on the image, they would include parts of Israel, Lebanon, Syria and Jordan, and all of the West Bank, Gaza Strip, and Golan Heights.

building has resulted. These are seismically active zones—meaning earthquakes occur there—and rarely does a year go by without a devastating quake rocking Turkey or Iran.

A large area of mountains, including the region's highest peaks, stretches across Turkey and Iran. The loftiest mountain ranges in Turkey are the Pontic, Taurus (**•Figure 6.10**), and Anti-Taurus, and in Iran, the Elburz and Zagros Mountains. These chains radiate outward from the rugged Armenian Knot in the tangled border country where Turkey, Iran, and the countries of the Caucasus meet. In Turkey, the glacier-fringed summit of the extinct volcano Mt. Ararat soars to 16,804 feet (5122 m)

Joe Hobbs

• Figure 6.10 The Taurus Mountains of southeastern Turkey.

over the frontier between Turkey and Armenia. Armenians seethe in anger and resentment that this mountain, so sacred
to them, fell into Turkish hands. Some people believe that 149
Mt. Ararat is where Noah's Ark lies (and even go in search of it), for in Genesis it is written that the vessel came to rest "in the mountains of Ararat." (Incidentally, like so many things, the accounts of Noah are shared by all three Abrahamic traditions of Judaism, Christianity, and Islam.)

Extensive forests existed in early historical times in the Middle East and North Africa, particularly in these mountainous areas, but overcutting and overgrazing have almost eliminated them. Since the dawn of civilization in this area, around 3000 B.C.E., people have cut timber for construction and fuel faster than nature could replace it. Lebanon, described in ancient times as "an oasis of green with running creeks" and "a vast forest whose branches hide the sky," illustrates the result. If not for human impacts, forests would cover large parts of Lebanon today. On many mountainsides, the dominant tree would be the cedar, the tree that is the centerpiece of Lebanon's flag. This hardwood was highly valued throughout the Mediterranean Basin. The Phoenician civilization, with its base in Lebanon, grew wealthy by trading cedarwood. Ancient Egyptians were among those who prized the cedar. The solar boat of King Cheops, dating to the third millennium BCE, would carry this pharaoh's soul in the afterlife, and it had to be built of the best wood: cedar of Lebanon. King Tutankhamen's funerary shrines and Solomon's Temple in Jerusalem were built
of cedar of Lebanon (**•Figure 6.11**). Only a few isolated groves 170
of cedar remain in Lebanon. Some of these are on the grounds of monasteries and survive because they grow in these sacred places. One grove is called "The Cedars of God."

The "primordial" or potential vegetation of Morocco's Atlas Mountains can be easily reconstructed by looking at the vegetation in the untouched sacred space surrounding saints' tombs.[1] This is a fine example of the importance of

Joe Hobbs

• Figure 6.11 Ancient Egyptians believed that the solar boat of King Cheops of Egypt (c. 2500 B.C.E.), builder of the Great Pyramid, would carry the pharaoh's spirit through the firmament. It had to be made of the best wood—cedar from Lebanon. The solar boat was interred next to the pyramid and now stands in a specially constructed museum.

environmental perception in geography. The principle of environmental perception is that people, societies, and cultures perceive and use the environment in different ways, depending on their age, gender, socioeconomic standing, language, religion, nationality, and other variables. Often these perceptions have an important role in shaping the cultural landscape. In this case, Islamic beliefs about the sacred space surrounding saints' tombs have protected the plants and animals living there, offering us a window into the environmental history of the Atlas Mountains.

Villager, Pastoral Nomad, Urbanite: The Ecological Trilogy

In the 1960s, the American geographer Paul English developed a useful model for understanding relationships between the three ancient ways of life that, although highly modified, still characterizes the Middle East and North Africa today: villager, pastoral nomad, and urbanite. Each of these modes of living is rooted in a particular physical environment. Villagers are the subsistence farmers of rural areas where dry farming or irrigation is possible; pastoral nomads are the desert peoples who migrate through arid lands with their livestock, following patterns of rainfall and vegetation; and urbanites are the inhabitants of the large towns and cities, generally located near bountiful water sources but sometimes placed for particular trade, religious, or other reasons. Describing these ways of life as components of the **Middle Eastern ecological trilogy**, English explained how each of them has a characteristic, usually mutually beneficial, pattern of interaction with the other two (•**Figure 6.12**).

The peasant farmers of Middle Eastern and North African villages (the **villagers**) represent the cornerstone of the trilogy. They grow the staple food crops such as wheat and barley that feed both the city dweller and the pastoral nomad of the desert. Neither urbanite nor nomad could live without them. The village also provides the city, often unwillingly, with tax revenue, soldiers, and workers. And before the mid-twentieth century,

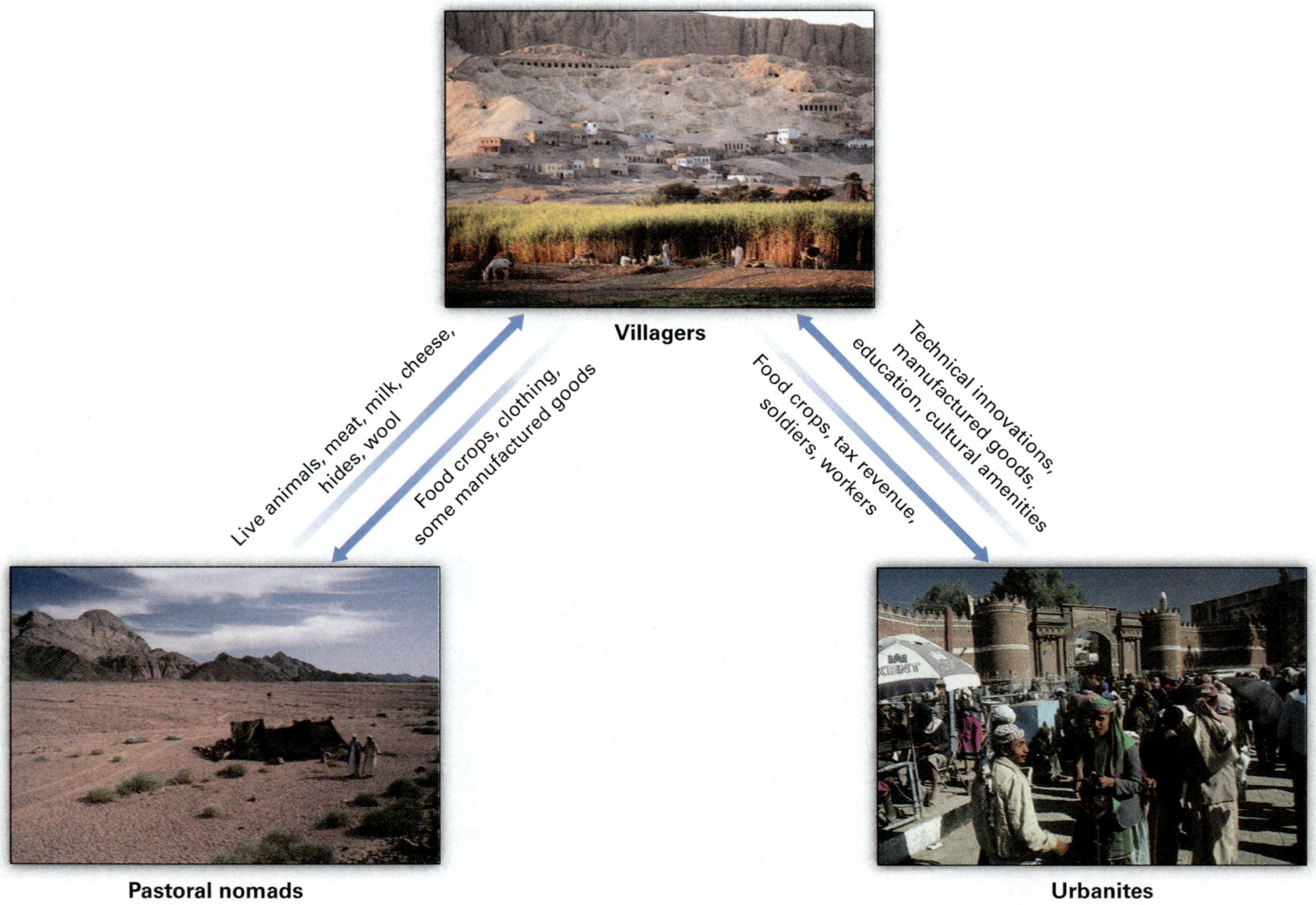

Joe Hobbs

• **Figure 6.12** People, environments, and interactions of the ecological trilogy (here, villagers harvesting sugarcane in upper Egypt, Bedouin at camp in Egypt's Eastern Desert, and shoppers at the main gate to the historic city of San'a, Yemen). This relationship is generally symbiotic, although historically both urbanites and pastoral nomads preyed on the villagers, who are the trilogy's cornerstone.

villages provided **pastoral nomads** with plunder as the desert dwellers raided their settlements and caravan supply lines. Generally, however, the exchange is mutually beneficial.

The nomads provide villagers with livestock products, including live animals, meat, milk, cheese, hides, and wool, and with desert herbs and medicines. Educated and progressive **urbanites** provide technological innovations, manufactured goods, religious and secular education and training, and cultural amenities (today, including films and music).

There is little direct interaction between urbanites and pastoral nomads, although some manufactured goods such as clothing travel from city to desert, and some desert folk medicines pass from desert to city. Historically, the exchange was sometimes violent, as urban-based governments have sought to control the movements and military capabilities of the elusive and sometimes hostile nomads. Pastoral nomads once plundered rich caravans plying the major overland trade routes of the Middle East and North Africa. Governments did not tolerate such activities and often cracked down hard on the nomads they were able to catch.

Later, in the 1970s, English wrote an article marking the "passing of the ecological trilogy." He noted that cities were encroaching on villages, villagers were migrating into cities and giving some neighborhoods a rural aspect, and pastoral nomads were settling down—thus the trilogy no longer existed. In reality, although the makeup and interactions of its parts have changed, the trilogy model is still valid and useful as an introduction to the major ways of life in the Middle East and North Africa. It is especially significant that a given man or woman in the region strongly identifies as a villager, a pastoral nomad, or an urbanite. This perception of self has an important bearing on how these people of very different backgrounds interact, even when they live in close proximity. Urban officials may work in rural village areas, but they remain at heart and in their perspectives city people and usually live apart from farmers. Extended families of pastoral nomads may settle down and become farmers, but they continue to identify themselves by affiliation with the nomadic tribe. Many continue to harvest desert resources on a seasonal basis and retain marriage and other ties with desert-dwelling relatives.

Let's take a closer look at each of the components of the ecological trilogy.

The Village Way of Life

Agricultural villagers historically represented by far the majority populations in the Middle East and North Africa; only in recent decades have urbanites begun to outnumber them. In the region's typically dry environments, the villages are located near reliable water sources with cultivable lands nearby. They are usually made up of closely related family groups, with many fields owned by an absentee landlord. Villagers typically live in closely spaced homes made of mud brick or concrete blocks. Production and consumption focus on a staple grain such as wheat, barley, or rice. As land for growing fodder is often in short supply, villagers keep only a small number of sheep and goats and rely in part on nomads for pastoral produce. Residents of a given village usually share common ties of kinship, religion, ritual, and custom, and the changing demands of agricultural seasons regulate their patterns of activity.

These patterns of village life have been increasingly exposed to outside influences since the mid-eighteenth century. European colonialism brought significant economic changes, including the introduction of cash crops and modern facilities to ship them. Improved and expanded irrigation, financed initially with capital from the West, brought more land under cultivation. Recent agents of change have been the countries' own government-supported doctors, teachers, and land reform officers. Modern technologies such as motor vehicles, gasoline-powered water pumps, radio, television, and—most recently—the Internet and cell phones have modified old patterns of living. The educated and ambitious, as well as the unskilled and desperately poor—motivated, respectively, by pull and push factors (see Chapter 3)—have been drawn to urban areas. Improved roads and communications have in turn carried urban influences to villages, prompting villagers to become more integrated into the national society.

The Pastoral Nomadic Way of Life

Pastoral nomadism emerged as an offshoot of the village agricultural way of life not long after plants and animals were first domesticated in the Middle East (about 7000 B.C.E.). Rainfall and the wild fodder it brings forth, although scattered, are sufficient resources to support small groups of people who migrate with their sheep, goats, and camels (and in some locales, cattle) to take advantage of this changing resource base (see Perspectives from the Field, next page). In addition to selling or trading livestock to obtain food, tea, sugar, clothing, and other essentials from settled communities, pastoral nomads hunt; gather; work for wages; and where possible, grow crops. Many now work in the tourist industry, as Westerners hungry for insight into traditional cultures seek them out. Their multifaceted livelihood has been described as a strategy of **risk minimization** based on the exploitation of multiple resources so that some will support them if others fail.

Although renowned in Middle Eastern legends, pastoral nomads have been described as "more glamorous than numerous." It is impossible to obtain adequate census figures on the number living in the deserts of the Middle East and North Africa, but they are very small relative to the numbers of villagers and urbanites. Recent decades have witnessed the rapid and progressive settling down, or **sedentarization**, of the nomads—a process attributed to a variety of causes. In some cases, prolonged drought virtually eliminated the resource base on which the nomads depended. Traditionally, they were able to migrate far enough to find new pastures, but modern national boundaries now inhibit such movements. Some have returned with the rains to their desert homelands, but others have chosen to remain as farmers or wage laborers in villages and towns. In the Arabian Peninsula, the prosperity

PERSPECTIVES FROM THE FIELD

Way-Finding in the Desert

The research for my doctoral dissertation in geography at the University of Texas–Austin was an 18-month journey with the Khushmaan Ma'aza, a clan of Bedouin nomads living in the northern part of Egypt's Eastern Desert. I studied their perceptions, knowledge, and uses of natural resources, and tried to understand how their worldviews and kinship patterns apparently helped them devise ways of protecting these scarce resources. I have worked with these people from the early 1980s to the present; I am now doing a multiyear study of how they carefully protect acacia trees, which offer them many resources when drought prevents the growth of annual plants. I have always been astonished by how different they are from me in perceiving the world around us, and how much better they are in their ability to "read" the ground and recall landscape details. In the following passage from my book *Bedouin Life in the Egyptian Wilderness,* I marvel at their way-finding abilities.

The process by which the Khushmaan nomads have developed roots in their landscape, fashioning subjective "place" from anonymous "space," and the means by which they orient themselves and use places on a daily basis deserve special attention: these are essential parts of the Bedouins' identity and profoundly influence how they use resources and affect the desert ecosystem.

The Bedouin have little need or knowledge of maps. On the other hand, when shown a map or aerial image of countryside they know, the nomads accurately orient and interpret it, naming the mountains and drainages depicted. Saalih Ali (**•Figure 6.A**) especially enjoyed my star chart. He would tell me what time the Pleiades would rise, for example, and ask me what time the chart indicated. The two were almost always in agreement.

Many desert travelers have been astonished by the nomads' navigational and tracking ability, calling it a "sixth sense." The Khushmaan are exceptional way-finders and topographical interpreters able, for instance, to tell from tracks whether a camel was carrying baggage or a man; whether gazelle tracks were made by a male or female; which way a car was traveling and what make it was; which man left a set of footprints, even if he wore sandals; and how old the tracks are. Bedouins are proud of their geographical skills, which, they believe, distinguish them from settled people. Saalih told me, "Your people don't need to know the country but we do, to know exactly where things are in order to live." Pointing to his head, he said, "My map is here."

The difference in the way-finding abilities of nomads and settled persons may be due to the greater survival value of these skills for nomads and to the more complex meanings they attribute to locations. Bedouin places are rich in ideological and practical significance. The nomads interpret and interact with these meanings on a regular basis and create new places in their lifetimes. Theirs is an experience of belonging and becoming with the landscape, whereas settled people are more likely to inherit and accommodate themselves to a given set of places.

The nomads' homeland is so vast, and the margin of survivability in it so narrow, that topographic knowledge must be encyclopedic. Places either are the resources that allow human life in the desert or are the signposts that lead to resources. A Khushmaan man pinpointed the role of places in desert survival: "Places have names so that people do not get lost. They can learn where water and other things are by using place names." Tragedy may result if the nomad has insufficient or incorrect information about places: many deaths by thirst are attributed to faulty directions for finding water.*

Joe Hobbs

• Figure 6.A Saalih Ali, a Bedouin of the Khushmaan Ma'aza, with the author in 2010, 30 years after they began fieldwork together.

Source: From BEDOUIN LIFE IN THE EGYPTIAN WILDERNESS by Joseph J. Hobbs, Copyright © 1989. Courtesy of the University of Texas Press.

and technological changes prompted by oil revenues made rapid inroads into the material culture—and then the livelihood preferences—of the desert people; many preferred the comforts of settled life. Some governments, notably those of Israel and prerevolutionary Iran, were frustrated by their inability to count, tax, conscript, and control a sizable migrant population and therefore forced the nomads to settle.

Pastoral nomads of the Middle East and North Africa identify themselves primarily by their tribe, not by their nationality. The major ethnic groups from which these tribes draw are the Arabic-speaking Bedouin of the Arabian Peninsula and adjacent lands, the Berber and Tuareg of North Africa, the Kababish and Bisharin of Sudan, the Yörük and Kurds of Turkey, and the Qashai and Bakhtiari of Iran. Members of a tribe claim common descent from a single male ancestor who lived countless generations ago; their kinship organization is thus a patrilineal descent system. It is also a segmentary kinship system, so called because there are smaller subsections of the tribe, known as clans and lineages, that are functionally important in daily life. Members of the most closely related families comprising the lineage, for example, share livestock, wells, trees, and other resources. Both the larger clans, made up of numerous lineages, and the tribes possess territories. Members of a clan or tribe typically allow members of another clan or tribe to use the resources within its territory on the basis of **usufruct**, or nondestructive mutual use. That arrangement has often proved to be lifesaving in an environment where water often is in dangerously low supply.

Although some detractors have depicted pastoral nomads as the "fathers" rather than "sons" of the desert, blaming them for wanton destruction of game animals and vegetation, there are numerous examples of pastoral nomadic groups that have developed indigenous and very effective systems of resource conservation. Most of these practices depend on the kinship groups of family, lineage, clan, and tribe to assume responsibility for protecting plants and animals.

The Urban Way of Life

The city was the final component to emerge in the ecological trilogy, beginning in about 4000 B.C.E. in Mesopotamia (modern Iraq) and 3000 B.C.E. in Egypt. Though they resembled villages in many ways, the early cities were distinguished by their larger populations (more than 5,000 people), the use of written languages, and the presence of monumental temples and other ceremonial centers. The early Mesopotamian city and, after the seventh century C.E., the classic Islamic city, called the **medina**, had several structural elements in common (•Figure 6.13). The medina had a high surrounding wall built for defensive purposes. The congregational mosque and often an attached administrative and educational complex dominated the city center. Although Islam is often characterized as a faith of the desert, religious life has always been focused in, and diffused from, the cities. The importance of the city's congregational mosque in religious and everyday life is often emphasized by its large size and outstanding artistic execution.

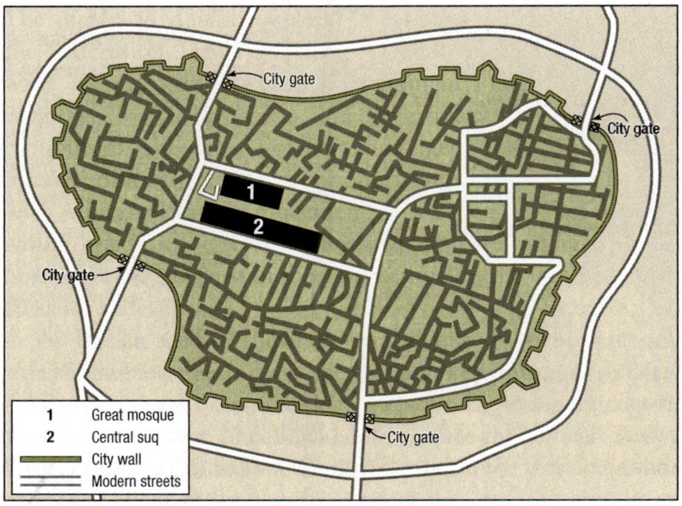

• **Figure 6.13** An idealized model of the classic medina, or Muslim Middle Eastern city.

A large commercial zone, known as a *bazaar* in Persian and a *suq* in Arabic and recognizable as the ancestor of the modern shopping mall, typically adjoined the ceremonial and administrative heart of the city (•Figure 6.14). Merchants and craftspeople selling various commodities occupied separate spaces within this complex, and visitors to an old medina today can still find sections devoted exclusively to the sale of spices, carpets, gold, silver, traditional medicines, or other particular goods. Smaller clusters of shops and workshops were located at the city gates.

Residential areas were differentiated as quarters, not by income group but by ethnicity; the medina of Jerusalem, for example, still has distinct Jewish, Arab, Armenian, and non-Armenian Christian quarters (see Figure 6.B). Homes tended to face inward toward a quiet central courtyard, buffering the occupants from the noise and bustle of the street. The narrow, winding streets of the medina were intended for foot traffic and small animal-drawn carts, not for large motor vehicles, a fact that accounts for the traffic jams in some old Middle Eastern and North African cities today and for the wholesale destruction of the medina in others.

• **Figure 6.14** A typical Middle Eastern *suq or bazaar* (market). This one is in Cairo, Egypt.

Table 6.2 Middle East and North Africa: Metropolitan Populations Data

1	Cairo, Egypt	16.4
2	Istanbul, Turkey	14.3
3	Tehran, Iran	13.2
4	Baghdad, Iraq	11.8
5	Khartoum, Sudan	10.1
6	Algiers, Algeria	6.7
7	Riyadh, Saudi Arabia	4.9
8	Alexandria, Egypt	4.6
9	Basra, Iraq	4.4
10	Ankara, Turkey	3.8
11	Aleppo, Syria	3.5
12	Tel Aviv, Israel	3.4
13	Casablanca, Morocco	3.3
14	Jiddah, Saudi Arabia	3.2
15	Isfahan, Iran	3.1
16	Izmir, Turkey	2.9
17	Mosul, Iraq	2.9
18	Damascus, Syria	2.8
19	Kuwait City, Kuwait	2.6
20	Mashhad, Iran	2.5

Population in millions.

The medinas that survive today are gently decaying vestiges of a forgotten urban pattern. Periods of European colonialism and subsequent nationalism changed the face and orientation of the city. During the colonial age, resident Europeans preferred to live in more spacious settings at the outer edges of the city, and later, the national elite followed this pattern. In recent times, independent governments have adopted Western building styles, with broad traffic arteries cutting through the old quarters, and large central squares near government buildings. This opening up of the cityscape has spread commercial activity along the wide avenues, diluting the prime importance of the central bazaar as the focus of trade.

Rural-to-urban migration and the city's own internal growth contribute to a rapid rate of urbanization that puts enormous pressure on services in the region's poorer countries. Governments often build high-rise public housing to accommodate the growing population, contributing to a cycle in which the urban poor move into the new dwellings, only to leave their old quarters as a vacuum to draw in still more rural migrants. In Cairo, millions of former villagers now live in the "City of the Dead," an extraordinary urban landscape composed of multistory dwellings erected above graves—a last resort for the poor who have no other place to go. The overwhelmingly largest city, or primate city, so characteristic of Middle Eastern and North African capitals,

• Figure 6.15 In Dubai, one of the city states of the United Arab Emirates, wealth created initially on oil revenues has made it possible for some to build almost anything, including islands in whimsical shapes. The islands, which are available for sale, have begun to erode and are sinking.

thus grows at the expense of the smaller original city. Two 95
of the primate cities of MENA, Cairo and Istanbul, are also **megacities**, defined as having more than 10 million inhabitants. The region's major metropolitan areas and their populations are listed in **•Table 6.2**.

Much of the rural-to-urban migration and subsequent urban gridlock and squalor could probably be avoided if governments invested more in the development of villages and smaller cities. The oil-rich countries with relatively small populations generally enjoy an urban standard of living equaling that of affluent Western countries. Modern industrial cities such as Saudi Arabia's Jubail and others founded on oil wealth were built virtually overnight, providing fascinating contrast to the region's colorful, complex ancient cities. With the recent resurgence of oil revenue into Saudi Arabia and the other Gulf countries, some of the cities boast the most elegant, sophisticated, and expensive amenities and architecture to be found anywhere in the world. There are no constraints in the over-the-top Persian Gulf emirate of Dubai. There is an indoor ski "resort" in a mall, and developers have even created islands in the shapes of the world's continents and of palm trees (**•Figure 6.15**).

6.3 Cultural and Historical Geographies

Peoples of the Middle East and North Africa have made many fundamental contributions to humanity. During the Agricultural Revolution, many of the plants and animals on which the world's agriculture is based today were first domesticated in the Middle East between 5,000 and 10,000 years ago. The list includes 40 wheat, barley, sheep, goats, cattle, and pigs, whose wild ancestors were processed, manipulated, and bred until their physical makeup and behavior changed to suit human needs. The interaction between people and the wild plants and animals they eventually domesticated took place mainly in the well-watered **Fertile Crescent**, the arc of land stretching from Israel to western Iran.

By about 6,000 years ago, people sought higher yields by irrigating crops in the rich soils of the Tigris, Euphrates, and Nile River Valleys. Their efforts produced the enormous crop surpluses that allowed civilization—a cultural complex based on an urban way of life—to emerge in Mesopotamia (literally, "the land between the rivers," the Tigris and Euphrates) and in Egypt. Accomplishments in science, technology, art, architecture, language, mathematics, and other areas diffused outward from these centers of civilization. Egypt and Mesopotamia are thus among the world's great culture hearths.

Ethnicities and Languages

The Middle East and North Africa are sometimes mistakenly referred to as the "Arab World"; in fact, the region has huge populations of non-Arabs. It is true that a majority of the region's inhabitants are Arabs. An **Arab** is best defined as a person of Semitic Arab ethnicity whose ancestral language is *Arabic*, a **Semitic language** spoken by about 250 million, or 49.7 percent, of the region's people. Originally, the Arabs were inhabitants of the Arabian Peninsula, but conquests after their majority conversion to Islam took them, their language, and their Islamic culture as far west as Morocco and Spain.

The region is also the homeland of the **Jews**, whose definition today is complex. Originally, Jews were both a distinct ethnic 170
and linguistic group of the Middle East who practiced the religion of Judaism. It is possible (although difficult) for non-Jews to convert to Judaism (the convert is known as a proselyte, or "immigrant"), so a strictly ethnic definition does not apply today. Many Jews do not practice their religion but still consider themselves ethnically or culturally Jewish. Whatever debate exists about what defines a Jew, Jewish identity is strong and resilient.

It is important to recognize that although political circumstances made them enemies in the twentieth century, Jews and Arabs lived in peace for centuries, and they share many cultural traits. Both recognize Abraham as their patriarch. Arabic is a Semitic language in the same **Afro-Asiatic language family** as *Hebrew*, which is spoken by most of the 5.7 million Jewish inhabitants of Israel (see the language map, •**Figure 6.16**).

There are other very large populations of non-Semitic ethnic groups and languages in the region. The greatest are the 60 million **Turks** of Turkey, who speak *Turkish*, a member of the Altaic

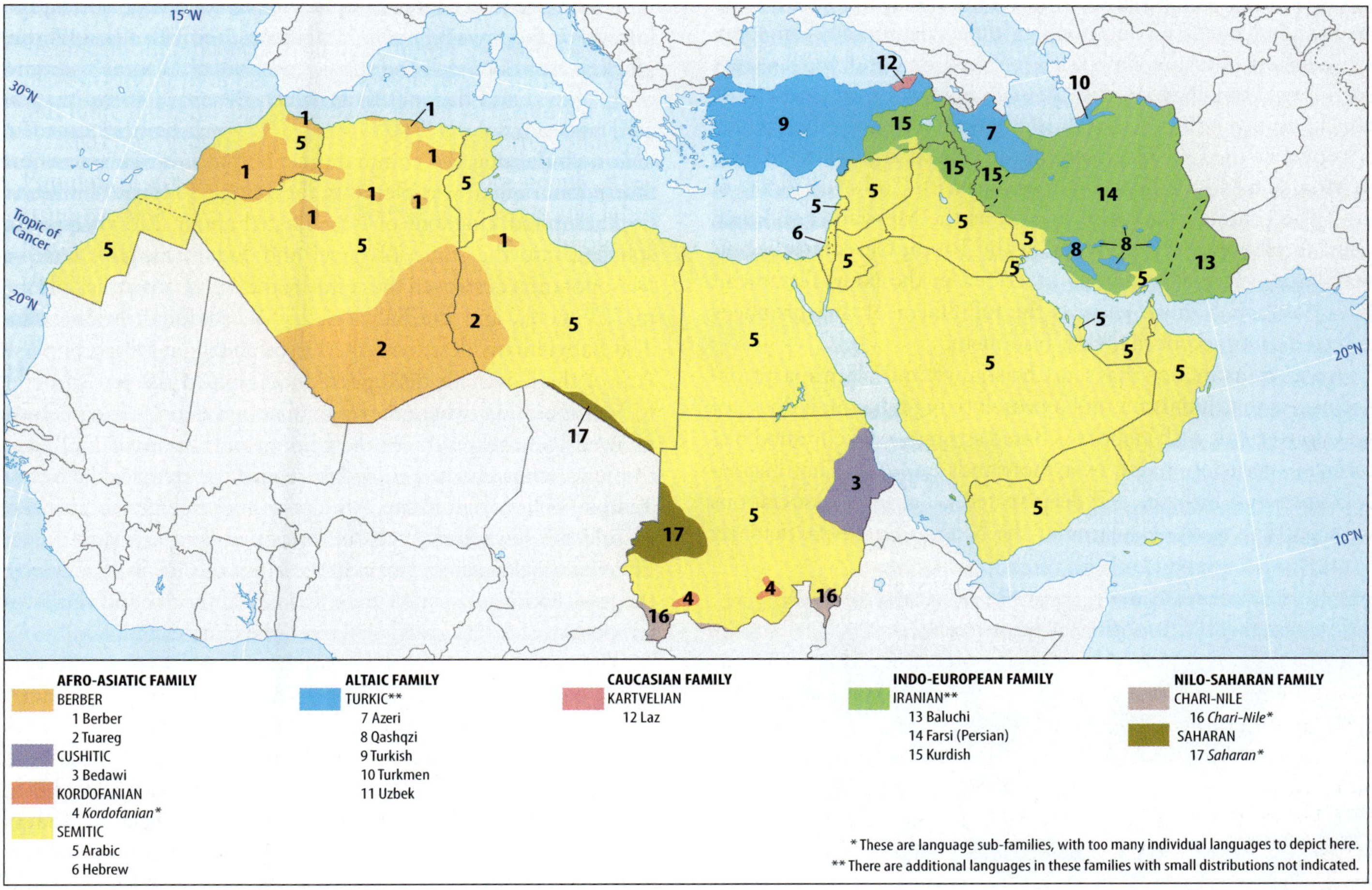

• **Figure 6.16** Languages of the Middle East and North Africa.

Source: Based on Bernard Comrie, et al., The Atlas of Languages, Revised Edition (New York: Facts on File, 2003).

language family, and the 41 million Persians of Iran, most of who speak **Persian** (*Farsi*) in the Indo-European language family. Both Persian and Arabic are written in Arabic script and so appear related, but they are not. Turkish was also written in Arabic script until early in the twentieth century, but since then it has been written in a Latin script. About 30 million **Kurds**—a people living in Turkey, Iraq, Iran, and Syria—speak *Kurdish*, which is also an Indo-European language. Many people in North Africa speak **Berber** and **Tuareg** (in the Afro-Asiatic language family).

The Abrahamic Faiths

Out of the Middle East came the closely related monotheistic faiths of Judaism, Christianity, and Islam. The region's human and political geographies are closely tied to these religions and their holy places (•Figure 6.17). All three are Abrahamic traditions; they regard Abraham as their patriarch, and they share many fundamental beliefs. The first to emerge was the faith of Judaism.

The Promised Land of the Jews

Judaism, the first significant monotheistic faith as far as we know, is today practiced by about 14 million people worldwide, mostly in Israel, Europe, and North America. September 2011 marked the start of the year 5772 in the Jewish calendar, but unlike its kindred faiths—Christianity and Islam—Judaism does not have an acknowledged starting point in time. Also unlike Christianity, Judaism does not have a fixed creed or doctrine. Jews are encouraged to behave in this life in compliance with God's laws, which, according to the Torah (the Jewish holy scripture, which is known as the Old Testament in the Christian Bible) God gave to Moses on Mount Sinai as a covenant with God's "chosen people."
156 The coming of a savior known as the Messiah ("Anointed One" in Hebrew) is prophesied in the Torah; Christians believe that Jesus was that savior, as described in the New Testament; Jews do not recognize Jesus as the fulfillment of that prophecy and so do not accept the New Testament.

Another distinction that sets Christianity and Islam apart from Judaism is that Judaism is not a proselytizing religion; it does not seek converts. Jewish identity is based strongly on a common historical experience shared over thousands of years. That historical experience has included deep-seated geographic associations with particular sacred places in the Middle East—particularly with places in Jerusalem, capital of ancient Judah (Judea), the province from which Jews take their name. Tragically, the Jewish history also has included unparalleled persecution.

• **Figure 6.17** Religions of the Middle East and North Africa.

Depending on one's perspective, the Jewish connection with the geographic region known as Palestine, the area now composed of Israel, the West Bank, and the Gaza Strip (see Figure 6.33), is most significant on a time scale of either about 4,000 years or about 100 years. According to the Bible, around 2000 B.C.E., God commanded Abraham and his kinspeople, known as **Hebrews** (later as Jews), to leave their home in what is now southern Iraq and settle in Canaan. God told Abraham that this land of Canaan—geographic Palestine—would belong to the Hebrews after a long period of persecution. The Bible says that the Hebrews did settle in Canaan until famine struck that land. At the command of Abraham's grandson Jacob, the Hebrews—known then as **Israelites**—relocated to Egypt, where grain was plentiful. That began the long sojourn of the Israelites in Egypt, which, according to the Bible, ended in about 1200 B.C.E. when Moses led them out in the journey known as the **Exodus**.

According to Jewish history, the prophecy of Abraham was first fulfilled when the Israelites settled once again in Canaan, their **Promised Land**. The Jewish King Saul unified the 12 tribes that descended from Jacob into the first united Kingdom of Israel in about 1020 B.C.E. In about 950 B.C.E. in Jerusalem—the capital of the kingdom enlarged by Saul's successor, David—King Solomon built Judaism's **First Temple**. He located it atop a great rock known to the Jews as *Even HaShetiyah*, the **Foundation Stone**, plucked from beneath the throne of God to become the center of the world and the core from which the entire world was created (see Geography of Sacred Space, next page). The Ark of the Covenant, containing the commandments that God gave to Moses atop Mount Sinai, was placed in the Temple's Holy of Holies.

The united kingdom of Israel lasted about 200 years before splitting into the states of Israel and Judah. Empires based in Mesopotamia destroyed these states: the Assyrians attacked Israel in 721 B.C.E., and the Babylonians sacked Judah in 586 B.C.E. The Babylonians destroyed the First Temple (at which point the Ark of the Covenant disappeared) and exiled the Jewish people to Mesopotamia, where they remained until conquering Persians allowed them to return to their homeland. In about 520 B.C.E., the Jews who returned to Judah rebuilt the temple (the **Second Temple**) on its original site. A succession of foreign empires came to rule the Jews and Arabs of Palestine: Persian, Macedonian, Ptolemaic, Seleucid, and around the time of Jesus, Roman. Herod, the Jewish king who ruled under Roman authority and was a contemporary of Jesus, greatly enlarged the temple complex.

The Jews of Palestine revolted against Roman rule three times between 64 and 135 C.E. The first revolt broke out as the profoundly monotheistic Jews refused to acknowledge the Roman emperor as a god. The Romans quashed these rebellions in a series of famous sieges, including those of Masada and Jerusalem. The Romans destroyed the Second Temple, and a third has never been built. All that remains of the Second Temple complex is a portion of the surrounding wall built by Herod. Today, this **Western Wall**, known to non-Jews as the Wailing Wall, is the most sacred site in the world accessible to Jews (Figures 6.B and •6.18). Some religious traditions

GEOGRAPHY OF SACRED SPACE | Jerusalem

The old city of Jerusalem is filled with sacred places and is one of the world's premier pilgrimage destinations. Over thousands of years, Jerusalem has been coveted and conquered by people of many different cultures and faiths. In the process, a place held sacred by one group has often come to be held sacred by a second and even a third.

The most important thing to notice on this map is that the location of the First and Second Jewish Temples is identical to that of the Muslims' Dome of the Rock. Judaism, Islam, and Christianity are very close in their origins and many of their basic precepts, but their faiths are not conjoined or syncretized in practice, as, for example, the Maya and Catholic faiths have been in Mesoamerica (see page 365), or Buddhism and other traditions have been in East Asia (see page 228). In each faith, there has been a long tradition of quarreling over ownership of sacred space. Nowhere in the world does contested sacred space have more potential to ignite violence and even war than in Jerusalem (which means, in Hebrew, "City of Peace"). The later section on Regional Issues and Landscapes offers more insight into the role of Jerusalem's holy places in the region's turbulent history (see page 187). •**Figure 6.B** depicts the major places sacred to Jews, Christians, and Muslims, and also the city's ethnic quarters; the walled old city is a classic medina (see page 167).

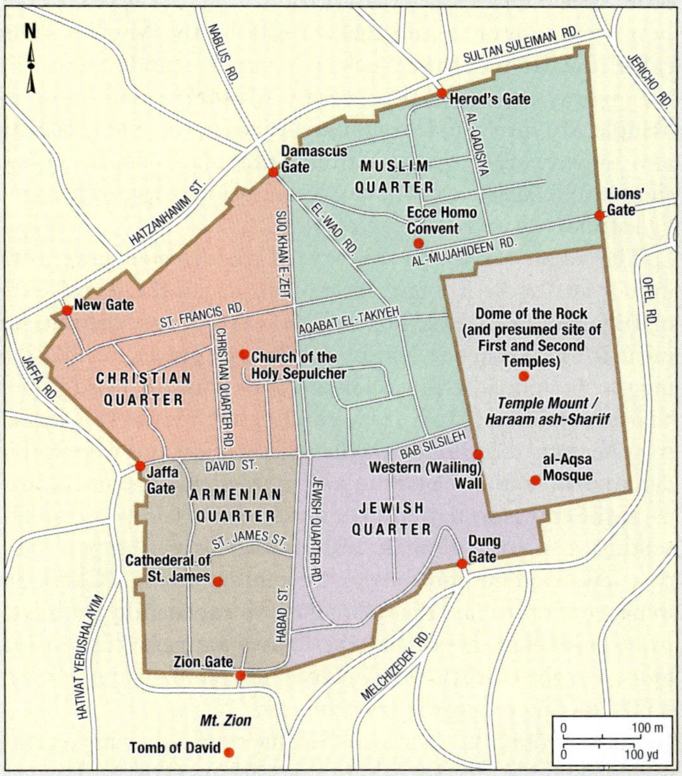

• **Figure 6.B** Sacred sites and ethnic quarters of the old city of Jerusalem.

• **Figure 6.18** Jerusalem's Western Wall, Temple Mount, and Noble Sanctuary. In this view are some of Judaism's and Islam's holiest places. At left, below the golden dome, is the Western Wall, which is all that remains of the structure that surrounded the Jews' Second Temple. The dome is the Muslims' Dome of the Rock. At far right, with the black dome, is the al-Aqsa Mosque, another very holy place in Islam.

prohibit Jews from ascending the **Temple Mount** above, the area where the temple actually stood, because it is too sacred. After the temple's destruction, that site was occupied by a Roman temple and then in 691 by the Muslim shrine called the **Dome of the Rock**, which still stands today (also in Figures 6.B and 6.18). The mostly Muslim Arabs know the Temple Mount as ***al-Haraam ash-Shariif***, the **Noble Sanctuary**. Supercharged with meaning, this place has in modern times often been the spark of conflagration between Jews and Palestinian Arabs.

The victorious Romans scattered the defeated Jews to the far corners of the Roman world. Thus began the Jewish exile, or **Diaspora**. In their exile, the Jews never forgot their attachment to the Promised Land. One Jewish prayer, recited during the weeklong Passover holiday, ends with the words "Next year in Jerusalem!" In Europe and Russia, Jews were subjected to systematic discrimination and persecution and were forbidden to own land or engage in a number of professions. Known as **anti-Semitism**, hatred of Jews developed deep roots. This sentiment in part grew out of the long-simmering perception that Jews were responsible for the murder of Jesus Christ and in part out of the fact that Christians were generally prohibited from practicing moneylending. Jews were permitted to loan money to the Christians, who accepted the loans but then resented having to pay the Jews interest.

In the 1930s, anti-Semitism became state policy in Germany under the Nazis, led by Adolf Hitler. Many German Jews fled to the United States, and others emigrated to Palestine in support of the **Zionist movement**, which, since the late 1890s, had aimed at establishing a Jewish homeland in Palestine, with **Zion** (Jerusalem) as its capital. Most Jews were not as fortunate as the Zionist emigrants. Within the boundaries of the Nazi empire that conquered most of continental Europe during World War II, Hitler's regime implemented its "Final Solution" to the "Jewish problem" by shipping to prison camps and eventually murdering all the Jews they could round up. The Nazi Germans and their allies killed an estimated 6 million Jews, along with millions of other minorities they deemed
81 "inferior," including Roma (Gypsies) and homosexuals. It was this Holocaust that prompted the victorious allies of World War II, from their powerful position in the newly formed United Nations, to create a permanent homeland for the Jewish people in Palestine. A seemingly endless cycle of violence, discussed beginning on page 185, ensued.

Christianity: Death and Resurrection in Jerusalem

Nearly a thousand years after Solomon established the Jews' First Temple, a new but closely related monotheistic faith emerged in Palestine. This was Christianity, named for Jesus Christ (*Christ* is Greek for "Anointed One," the equivalent of the Hebrew word for Messiah). Jesus, a Jew, was born near Jerusalem in Bethlehem, probably around 4 B.C.E. Tradition relates that when he was about 30 years old, Jesus began spreading the word that he was the Messiah, the deliverer of humankind long prophesied in Jewish doctrine. A small group of disciples accepted that he was the Messiah, the Son of God and a living manifestation of God Himself. They followed him for several years as he preached his message. He taught that the only path to God was through Him, and that faith in Christ as God's Son and as the redeemer of humanity's sins was the key to salvation.

Christianity is built upon a Jewish foundation: the common scriptures of Judaism and Christianity are known as the Hebrew Bible. Theologically, Christians would argue that Jews have the correct foundation but did not accept God's complete message, which continues beyond the Hebrew Bible to the new covenant, or the New Testament.

Jesus Christ's own teachings denied the validity of some Jewish doctrines, and around 29 C.E., a growing chorus of Jewish protesters called for his death. Palestine was then under Roman rule, and Roman administrators in Jerusalem placated the mob by ordering that he be put on trial. He was found guilty of being a claimant to Jewish kingship, and Roman soldiers put him to death by the particularly degrading and painful method of crucifixion. The cornerstone of Christian faith is that Jesus Christ was resurrected from the dead on the third day after his crucifixion and later ascended into heaven. For Christians, this was fulfillment of the scriptures: Jesus had come to redeem humanity's sins through His own death on the cross and through His resurrection. Christians believe that He continues to intercede with his Father on their behalf and that He will come again on Judgment Day, at the end of time.

After a period of relative tolerance, the Romans began actively persecuting Christians. Nevertheless, Christian ranks and influence grew. The turning point in Christianity's career came after 324 C.E., when the Roman Emperor Constantine embraced Christianity and established it as the official religion of the empire. The Christian Byzantine civilization that developed in the **"New Rome"** that Constantine established—Constantinople, now Istanbul, Turkey—created fine monuments at places associated with the life and death of Jesus Christ. These include the place long acknowledged as the center of the Christian world, Jerusalem's **Church of the Holy Sepulcher** (•Figures 6.19 and 6.B). This extraordinary, sprawling building, now administered by numerous separate Christian sects, contains the locations where tradition says Jesus Christ was crucified and buried. Incidentally, several of these Christian denominations have a long history of squabbling over turf within the church, and the key to this place is in the hands of a trusted, neutral party: a Muslim family.

Christianity has seldom been the majority religion in the land where it was born; only until Islam arrived in Palestine in 638 C.E. was the region primarily Christian. From then until the twentieth century, most of Palestine's inhabitants were Muslim, and since the mid-twentieth century, Muslims and Jews have been the major groups. Between the eleventh and fourteenth centuries, European Christians dispatched military expeditions to recapture Jerusalem and the rest of the Holy Land from the Muslims. These bloody campaigns, known as the **Crusades**, resulted in a series of short-lived Christian administrations in the region.

Joe Hobbs

• **Figure 6.19** Jerusalem's Church of the Holy Sepulcher, containing the locations where many Christians believe Jesus Christ was crucified and buried.

There are significant minority populations of Christians throughout the Middle East, including members of distinct sects such as the **Copts** of Egypt and the **Maronites** of Lebanon. Their share of the population has generally been declining, both because of emigration and lower birth rates than those of the majority Muslims. Christians have been persecuted in some countries, which is one of the push factors in their emi-
56 gration. Nevertheless, the Middle East remains the cradle of their faith for Christians the world over, and Jerusalem and nearby Bethlehem are the world's most important Christian pilgrimage sites. Whatever your faith, you would surely find Jerusalem to be one of the most compelling places on Earth. I hope you will have an opportunity to travel there.

The Message of Islam

Islam is by far the dominant religion in the Middle East and North Africa; only Israel. within its pre-1967 borders, has a non-Muslim majority. Because of Islam's powerful influence not merely as a set of religious practices but as a way of life, an understanding of the religious tenets, culture, and diffusion of Islam is vital for appreciating the region's cultural geography.

Islam is a monotheistic faith built on the foundations of the region's earliest monotheistic faith, Judaism, and its offspring, Christianity. Indeed, Muslims (people who practice Islam) call Jews and Christians **People of the Book**, and their faith obliges them to be tolerant of these special peoples. Muslims believe that their prophet, Muhammad, was the very last in a series of prophets who brought the Word of God to humankind. Thus they perceive the Bible as incomplete but not entirely wrong—Jews and Christians merely missed receiving the entire message (just as Christians would insist that Jews missed the entire message). Muslims do not accept the Christian concept of the divine Trinity (God manifested in the form of the Father, his son Jesus, and the Holy Spirit) and regard Jesus as a prophet rather than as God.

Muhammad was born in 570 C.E. to a poor family in the western Arabian (now Saudi Arabian) city of Mecca. Located on an important north–south caravan route linking the frankincense-producing area of southern Arabia (now Yemen and Oman) with markets in Palestine (now Israel) and Syria, Mecca was a prosperous city at the time. It was also a pilgrimage destination because more than 300 deities were venerated in a shrine there called the **Ka'aba** (the Cube; •**Figure 6.20**). Muhammad married into a wealthy family and worked in the caravan trade. Muslim tradition holds that when he was about 40 years old, Muhammad was meditating in a cave outside Mecca when the Angel Gabriel appeared to him and ordered him to repeat the words of God that the angel would recite to him. Over the next 22 years, the prophet

Nabeel Turner/Getty Images

• **Figure 6.20** The black-shrouded cubical shrine known as the Ka'aba (just right of center) in Mecca's Great Mosque is the object toward which all Muslims face when they pray and is the centerpiece of the pilgrimage to Mecca required of all able Muslims. Note the carpet of humanity spread across this image: this is the height of the *hajj* (pilgrimage), when millions of the devout converge on the holy places of Mecca and Medina.

INSIGHTS Sunni and Shiite Muslims

A schism occurred very early in the development of Islam, and it persists today. The split developed because the Prophet Muhammad had named no successor to take his place as the leader (caliph) of all Muslims. Some of his followers argued that the person with the strongest leadership skills and greatest piety was best qualified to assume this role. These followers became known as **Sunni**, or orthodox, Muslims. Others argued that only direct descendants of Muhammad, specifically through descent from his cousin and son-in-law Ali, could qualify as successors. They became known as **Shia**, or **Shiite**, Muslims.

The military forces of the two camps engaged in battle south of Baghdad at Karbala in 680 C.E., and in the encounter Sunni troops caught and brutally murdered Hussein, a son of Ali (Karbala is thus sacred to Shiites, as is nearby al-Najaf, where Ali is buried). The rift thereafter was deep and permanent. The martyrdom of Hussein became an important symbol for Shiites, many of whom still today regard themselves as oppressed peoples struggling against cruel tyrants, including some Sunni Muslims.

Today, only three of the region's countries, Iran, Iraq, and Bahrain, have Shiite majority populations. Significant minority populations of Shiites are in Syria, Lebanon, Yemen, and the Arab states of the Persian (Arabian) Gulf (see Figure 6.17).

related these words of God (whom Muslims call **Allah**) to scribes who wrote them down as the Qur'an (or Koran), the holy book of Islam.

During this time, Muhammad began preaching the new message, "There is no god but God (Allah)," which the polytheistic people of Mecca viewed as heresy. As much of their income depended on pilgrimage traffic to the Ka'aba, they also viewed Muhammad and his small band of followers as an economic threat. They forced the Muslims to flee from Mecca and take refuge in Yathrib (modern Medina), where a largely Jewish population had invited them to settle. There were subsequent skirmishes between the Meccans and Muslims, but in 630, the Muslims prevailed and peacefully occupied Mecca. The Muslims destroyed the idols enshrined in the Ka'aba, which became a pilgrimage center for their one God.

The Ka'aba is Islam's holiest place, and Mecca and Medina are its holiest cities. Jerusalem is also sacred to Muslims. Muslim tradition relates that on his **Night Journey**, the Prophet Muhammad ascended briefly into heaven from the great rock now beneath the Dome of the Rock (the same rock Jews regard as the Foundation Stone). Nearby on the Temple Mount (*al-Haraam ash-Shariif*) is **al-Aqsa Mosque**, a sacred congregational site. The proximity, even the duplication, of holy places between Islam and Judaism came to be the most difficult issue in peace negotiations between Palestinians and Israelis, as explained on page 190.

After Muhammad's death in 632, Arabian armies carried the new faith far and quickly. The two decaying empires that then prevailed in the Middle East and North Africa—the Byzantine or Eastern Roman Empire, based in Constantinople (Istanbul), and the Sassanian Empire, based in Persia (Iran) and adjacent Mesopotamia (Iraq)—put up only limited military resistance to the Muslim armies before capitulating. Local inhabitants generally welcomed the new faith, in part because administrators of the previous empires had not treated them well, whereas the Muslims promised tolerance. Soon the Syrian city of Damascus became the center of a Muslim empire. Baghdad assumed this role in 750 C.E.

Arab science and civilization flourished in the Baghdad immortalized in the legends of *The Thousand and One Nights*. Important accomplishments and discoveries were made in mathematics, astronomy, and geography. Scholars translated the Greek and Roman classics, and if not for their efforts, many of these works would never have survived to become part of the modern European legacy. It was an age of exploration, when Arab merchants and voyagers visited China and the remote lands of southern Africa. Many important discoveries by the Arab geographers were recorded in Arabic, a language unfamiliar to contemporary Europeans, and had to be rediscovered centuries later by the Portuguese and Spaniards. Arab merchants carried their faith on the spice routes to the East. One result, surprising to many today, is that the world's most populous Muslim country is not in the Middle East and North Africa, and its people are not Arabs; it is Indonesia, 5,000 thousand miles (8,000 km) east of Arabia.

Whether in Arabia or Indonesia, whether they be Sunni Muslims or Shiite Muslims (see Insights, above), all believers are united in support of the five fundamental precepts, or **Pillars of Islam**. The first of these is the **profession of faith**: "There is no god but God, and Muhammad is His Messenger." This expression is often on the lips of the devout Muslim, both in prayer and as a prelude to everyday activities.

The second pillar is **prayer**, required five times daily at prescribed intervals. Two of these prayers mark dawn and sunset. Business comes to a halt as the faithful prostrate themselves before God. Muslims may pray anywhere, but wherever they are, they must turn toward Mecca. There also is a congregational prayer at noon on Friday, the Muslim Sabbath.

The third pillar is **almsgiving**. In earlier times, Muslims were required to give a fixed proportion of their income as charity, similar to the concept of the tithe in the Christian church. Today, the donations are voluntary. Even Muslims of very modest means give what they can to those in need.

The fourth pillar is **fasting** during Ramadan, the ninth month of the Muslim lunar calendar. Muslims are required to abstain from food, liquids, smoking, and sexual activity from dawn to sunset throughout Ramadan. The lunar month of Ramadan falls earlier each year in the solar calendar and thus periodically occurs in summer. In the torrid Middle East and North Africa, that timing imposes special hardships on the faithful, who, even if they are performing manual labor, must resist the urge to drink water during the long, hot days.

The final pillar is the *hajj*, or **pilgrimage to Mecca**, Islam's holiest city. Every Muslim who is physically and financially able is required to make the journey once in his or her lifetime. A lesser pilgrimage may be performed at any time, but the prescribed season is in the 12th month of the Muslim calendar. Those days witness one of earth's greatest annual human migrations as about 3 million Muslims from all over the world converge on Mecca. Hosting these throngs is an obligation the government of Saudi Arabia fulfills proudly and at considerable expense. However, there has been some trepidation in recent years because of the security threat foreign visitors may pose to the host country and because accidents such as stampedes and tent city fires have cost numerous lives. Many pilgrims also visit the nearby city of Medina, where Muhammad is buried. Most Muslims regard the *hajj* as one of the most significant events of their lifetimes. All are required to wear simple seamless garments, and for a few days, the barriers separating groups by income, ethnicity, and nationality are broken. Pilgrims return home with the new stature and title of *hajji* but also with humility and renewed devotion.

All Muslims share the Five Pillars and other tenets, but they vary widely in other cultural practices related to their faith, depending on the country they live in; whether they are from the desert, village, or city; and their education and income. The governments and associated clerical authorities in Saudi Arabia and Iran insist on strict application of **Islamic law** (known as ***sharia***) to civil life; in effect, there is no separation between church and state. The Qur'an does not state that women are required to wear veils, but it does urge them to be modest, and it portrays their roles as different from those of men. Clerics in Saudi Arabia insist that women wear floor-length, long-sleeved black robes and black veils in public, that they travel accompanied by a male member of their families, and that they not drive cars. In Egypt, by contrast, Muslim women are free to appear in public unveiled if they choose. However, in most Muslim countries, conservative ideas about the role of women are still very strong: they should be modest, retiring, good mothers, and keepers of the home, although they can work. The Qur'an portrays women as equal to men in the sight of God, and in principle, Islamic teachings guarantee the right of women to hold and inherit property.

Most Muslim women argue that what others often see as "backward" cultural practices are in fact progressive. For example, they say, their modest dress compels men to evaluate them on the basis of their character and performance, not their attractiveness. They argue that segregation of the sexes in the classroom makes it easier for both women and men to develop their confidence and skills (•**Figure 6.21**). A married woman retains her maiden name. These apparent advantages can be weighed against the drawbacks that make women generally subordinate to men in public affairs.

Joe Hobbs

• **Figure 6.21** Education of women in the Muslim countries has increased dramatically in recent decades, but strong traditions keep most women out of the workforce. This is a geographic information systems (GIS) class in the Department of Geography, United Arab Emirates University.

6.4 Economic Geography

Overall, the Middle East and North Africa is a poor region; per capita GNI PPP for the 23 countries and territories averages only $9,460 (**see Table 6.1**). This may seem surprising in view of the "rich Arab" stereotype. Only the oil-endowed states around the Persian Gulf (Arabian Gulf) deserve the reputation for wealth, and by most measures other than per capita wealth, only non-Arab Israel is truly a more developed country (MDC). Israel's prosperity comes from its innovation in computer and other high-technology industries, the processing and sale of diamonds, large amounts of foreign (mostly U.S.) aid, and investment and assistance by Jews and Jewish organizations around the world.

Here we will examine the role of oil in the region's economies, and the lagging economic indicators and human conditions that should have served as the writing on the wall for the "Arab Spring" (**see page 192**).

"The Prize"

Vital to the industrialized countries as a source of fuels, lubricants, and chemical raw materials, petroleum is one of the world's most important natural resources. Daniel Yergin, recognized as the world's preeminent authority on oil, wrote the definitive book on this essential commodity. His title sums up what oil has come to represent in economic and geopolitical terms: *The Prize*.

© Jamal Nasrallah/epa/Corbis

• **Figure 6.22** Petroleum facilities in the interior wilderness of Arabia's Empty Quarter (al-Rub' al-Khali). Most of Saudi Arabia's production is from vast, shallow reservoirs of oil that are much closer to marine terminals on the Persian Gulf.

A crucial feature of world geography is the concentration of about two-thirds of the world's proven conventional petroleum reserves in a few countries that ring the Gulf. In the second half of the twentieth century, this geographical endowment transformed a handful of inconsequential countries and shaikhdoms into some of the most critical locations in the global economic system (•**Figure 6.22**).

By coincidence, the countries rich in oil tend to have relatively small populations, whereas the most populous nations have few oil reserves; Iran is an exception. All but about 1 percent of the Gulf oil region's proven reserves of crude oil are located in Saudi Arabia, Iraq, Kuwait, Iran, and the United Arab Emirates (UAE), with smaller reserves in Oman, Bahrain, and Qatar. Saudi Arabia, by far the world leader in reserves, has about 18 percent of the proven crude oil reserves on the globe. Venezuela is in second place, with about 14 percent of the world's proven oil; Canada is third, with 12 percent (not in the form of easily extractable crude oil, but as tar sands; see page 417); the next four are all along the Gulf: Iran (9%), Iraq (8%), Kuwait (7%), and the United Arab Emirates (6%). Libya also has significant oil reserves. The world's largest oil consumer, the United States, has 2 percent. China, the second largest consumer by a narrow margin, has 2 percent.

In the Gulf area, the great thickness of the region's oil-bearing strata and high reservoir pressures have made it possible to secure an immense amount of oil from a small number of wells. The productivity of each well makes each barrel inexpensive to extract, and it is simple to increase or reduce production quickly in response to world market conditions. Gulf oil is thus cheap to produce unless expenditures to maintain huge military forces and infrastructure to defend it are factored in (in economic terms, unless the "external costs" of oil are internalized). Some analysts—including a former U.S. Navy secretary—prefer to calculate the "real price" of oil with military expenditures calculated. If one accepts the premise that maintaining a costly military presence in the region is part of the United States' long-term strategy to maintain the flow of Gulf oil, that oil is no longer cheap by any measure.

Production, export, and profits of Middle Eastern and North African oil were once firmly in the hands of foreign companies. That situation changed after 1960, when most of the Gulf countries and other exporting nations formed the **Organization of Petroleum Exporting Countries (OPEC)**, with the aim of taking joint action to demand higher profits from oil. It changed again after 1972, when the oil-producing countries began to nationalize the foreign oil companies. OPEC was relatively obscure until the Arab-Israeli war of 1973, after which the organization began a series of dramatic price increases. In 1980, the organization's price reached $37 (U.S.) per barrel, compared with $2 a barrel in early 1973. Adjusted for inflation, that $37 would be the equivalent of about $100 per barrel in 2011, so the price hikes were truly titanic.

These events had enormous repercussions for the world economy. Immense wealth was transferred from the more developed countries to the OPEC countries to pay for indispensable oil supplies. The skyrocketing cost of gasoline and other oil products helped cause serious inflation in the United States and many other countries and contributed to the 1973 **energy crisis** in the United States. Desperately poor, less developed countries (LDCs) found that high oil prices not only hindered the development of their industries and transportation but also reduced food production because of high prices for fertilizer made from oil and natural gas. In the Gulf countries, the oil bonanza produced a wave of spending on military hardware, showy buildings, luxuries for the elite, and ambitious development projects of many kinds. Per capita benefits to citizens were greatest in the Arabian Peninsula, where small populations and immense inflows of oil money made possible the abolition of taxes, the establishment of comprehensive social programs, and heavily subsidized amenities such as low-cost housing and utilities, including freshwater distilled from the salty Gulf in desalination plants.

Then, as the 1980s began, the era of continually expanding OPEC oil production, sales, and profits seemed to come to an end. After 1973, the high price of oil stimulated oil development in countries outside OPEC. Oil conservation measures such as a shift to more fuel-efficient vehicles and factories were introduced. Substitution of cheaper fuels for oil increased. Coal replaced oil in many electricity-generating stations. Meanwhile, the world entered a period of economic recession due in part to high oil prices. Decreased business activity reduced the demand for oil. Profits of the world oil industry (and taxes paid to governments) were severely cut, large numbers of refineries had to close, and much of the world tanker fleet was idled.

Oil prices have risen and fallen, with predictable consequences for producers and consumers. But the immense oil and gas reserves still in the ground guarantee that the Gulf region will continue to have major long-term influence in world affairs and will remain prosperous as long as these finite resources are in demand—especially in the established industrialized countries and in the booming economies of China and India.

There are, of course, other resources and industries in the region's economic geography—**remittances** (earned income sent home by guest workers) in the oil-rich countries; revenues from ship traffic through the Suez Canal; and exports of cotton, rice, and other commercial crops, for example—but oil dominates the region's economy and is central to the global economy.

This depiction of MENA's oil wealth, which is concentrated in a few countries, does not consider the economic poverty so characteristic of the region's other countries and the great majority of its peoples. The worsening conditions in many of the Arab countries—the poverty, crowding, unemployment, and the growing gap between the great mass of the poor and the increasingly wealthy rich—were the economic ingredients of the Arab Spring described later.

6.5 Geopolitical Issues

The Middle East and North Africa have long been vital in world affairs, with their resources and geographical assets such as the Suez Canal coveted by outside interests. From very early times, overland caravan routes, including the great Silk Road, crossed the Middle East and North Africa with highly prized commodities traded between Europe and Asia. The security of these routes was vital. Geopolitical concerns are now focused on narrow waterways, access to oil, access to freshwater, and terrorism.

Chokepoints

One of the striking characteristics of the geography of the Middle East and North Africa is how many seas border and penetrate the region. In many cases, these bodies of water are connected to one another though narrow straits and other passageways. In geopolitical terms, such constrictions are known as **chokepoints**—strategic narrow passageways on land or sea that may be easily closed off by force or even the threat of force (•Figure 6.23). Chokepoints must be unimpeded if world commerce is to carry on normally. Keeping them open is therefore usually one of the top priorities of regional and external governments. Likewise, closing them is a priority to a combatant nation or a terrorist entity seeking to gain a strategic advantage. Many notable events in military history and the formation of foreign policy in the Middle East focus on these strategic places.

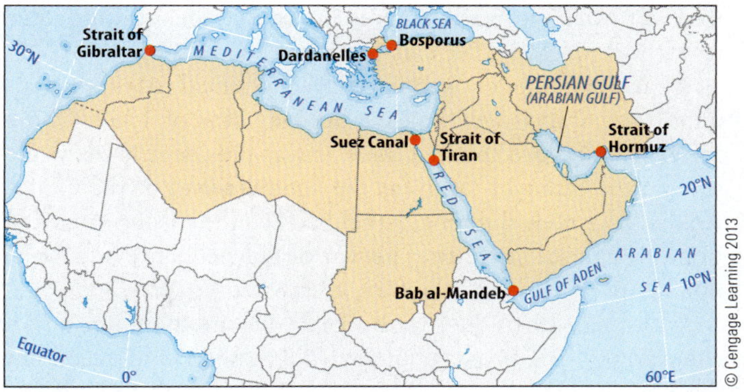

• **Figure 6.23** Chokepoints in the Middle East and North Africa.

One of the world's most important chokepoints is the Suez Canal, which opened in 1869. Slicing 107 miles (172 km) through the narrow Isthmus of Suez, the British- and French-owned Suez Canal linked the Mediterranean Sea with the Indian Ocean, saving cargo, military, and passenger ships a journey of many thousands of miles around the southern tip of Africa (•Figure 6.24). Keeping the Suez Canal in friendly hands was one of Britain's major military concerns during World War II. That objective led to a hard-fought and eventually successful British and Allied campaign against Nazi Germany in North Africa. Egyptian President Gamal Abdel Nasser's nationalization of the canal in 1956 led immediately to a British, French, and Israeli invasion of Egypt and a conflict known as the **Suez Crisis** and the **Arab-Israeli War of 1956**. Egypt effectively won

• **Figure 6.24** The strategically vital Suez Canal zone saw bitter fighting in the Middle East wars of 1956, 1967, and 1973.

that war when international pressure caused the invading forces to withdraw, leaving the canal in Egypt's hands.

Nearby, another chokepoint played a critical role in the most important Arab-Israeli war, the Six-Day War of 1967. One of the events that precipitated the war was President Nasser's closure of the Strait of Tiran (at the southern end of the Gulf of Aqaba) to Israeli shipping. Israel had won the right of navigation through the strait after its war of independence and would not accept its closure; therefore, it attacked Egypt.

On either side of the Arabian Peninsula are two more critical chokepoints. One is exceptionally vital to the world's economy: the Strait of Hormuz, connecting the Persian-Arabian Gulf with the Gulf of Oman and the Arabian Sea. Much of the world's oil supply passes through here in the holds of giant supertanker ships. Closure of the Strait of Hormuz would have devastating effects on the world's industrial and financial systems. Iran's plans to station Chinese-made missiles on the Strait of Hormuz and thus threaten international oil shipments led to a new level of U.S. involvement late in the Iran-Iraq War of 1980–1988. That war also prompted a flurry of new oil pipeline construction designed to bypass the Strait of Hormuz chokepoint (see Geography of Energy, next page). One of these pipelines, the Petroline route running east–west across the Arabian Peninsula, was built to bypass another important chokepoint, the Bab el-Mandeb, which connects the Red Sea with the Gulf of Aden and the Indian Ocean. Iran threatened to close the Strait of Hormuz when Western powers imposed sanctions against Iran for ramping up its nuclear program in 2012.

Turkey controls two more chokepoints that together are known as the Turkish Straits. The northernmost strait is the Bosporus, which cleaves the city of Istanbul into western (European) and eastern (Asian) sides, and the southern strait is called the Dardanelles. Their security has long been critical to the successful passage of goods between Europe and Asia and even more so to the successful passage of vessels between Russia (and the Soviet Union) and the rest of the world. As discussed on page 126, throughout the twentieth century one of the Soviet Union's constant strategic priorities was the right of navigation through the Turkish Straits.

Finally, the Strait of Gibraltar, connecting the Mediterranean Sea with the Atlantic Ocean, is also a chokepoint. Have a look at the map of Figure 6.23, and consider how geographic good luck gave the lands of the Mediterranean Sea access to the rest of the world through Gibraltar. Britain's insistence on maintaining control over its enclave of Gibraltar, decades after the decolonization of most of the world, is an excellent indicator of how critical this chokepoint is.

Access to Oil

The Suez Canal and other chokepoints, the cotton of Egypt's Nile Delta, and the strategic location of the region were important during colonial times and have remained so ever since. But oil has been and will remain (as long as fossil fuels drive the world's economies) what keeps the rest of the world interested economically in the Middle East and North Africa. The region's oil flows to many countries, but most of it is marketed in Western Europe, China, and Japan. The United States also imports large amounts of Gulf oil but has a smaller relative dependence on this source than Japan and Europe do; about 18 percent of its imported oil came from the Gulf in 2011. However, the Gulf region is very important to the United States because of the great dependence of close American allies on Gulf oil and because of the importance of the oil as a future reserve. American companies are also heavily involved in oil operations and oil-financed development in the Gulf countries. Gulf "petrodollars" are spent, banked, and invested in the United States, contributing significantly to the U.S. economy. Maintaining a secure supply of Gulf oil has therefore been one of the long-standing pillars of U.S. policy in the Middle East.

The United States has long had a precarious relationship with the key players in the Middle Eastern arena. The U.S. has pledged unwavering support for Israel, while at the same time courting Israel's oil-rich Arab foes, notably Saudi Arabia. That Arab kingdom and its neighbors around the Persian/Arabian Gulf possess more than 60 percent of the world's proven oil reserves. Thus they are vital to the long-term economic security of the Western industrial powers, China, India, and Japan. The United States and its Western allies made it clear they would not tolerate any disruption of access to this supply when Iraqi troops, under orders from Iraqi President Saddam Hussein, occupied Kuwait on August 2, 1990. U.S. President George H.W. Bush drew a "line in the sand," proclaiming, "We cannot permit a resource so vital to be dominated by one so ruthless—and we won't." In what came to be known as the **Gulf War**, the United States and a coalition of Western and Arab allies mounted a massive array of military might that ousted the Iraqi invaders within months and secured the vital oil supplies for Western markets. When the United States invaded and occupied Iraq in 2003, ostensibly to extinguish Iraq's ability to develop and deploy **weapons of mass destruction (WMD)**, including chemical, biological, and nuclear weapons, many critics insisted that this was just another example of America's determination to control Middle Eastern oil.

The Gulf War was not the first time the United States expressed its willingness to use force to maintain access to Middle Eastern oil. In 1979, in the wake of the Islamic revolution in Iran, the Soviet Union invaded neighboring Afghanistan. U.S. 242
military analysts feared that the Soviets might use Afghanistan as a launching pad to invade oil-rich Iran. The United States deemed this prospect unacceptable, and President Jimmy Carter issued the foreign policy statement that came to be known as the **Carter Doctrine**: "An attempt by any outside force to gain control of the Persian Gulf region will be regarded as an assault on the vital interests of the United States of America, and such an assault will be repelled by any means necessary, including military force." "Vital interests" meant oil, and defending them with military force meant that the United States was willing to go to war with the Soviet Union, presumably nuclear war.

Middle Eastern wars had already become proxy wars for the superpowers, with oil always looming as the prize. In the 1967 and 1973 Arab-Israeli wars, for example, Soviet-backed Syrian forces fought U.S.-backed Israeli troops. American support of Israel in this war prompted Arab members of OPEC to impose an **oil embargo**, refusing to sell their oil to the United States, thus precipitating the nation's first energy crisis. During 415, 440

GEOGRAPHY OF ENERGY | Pipelines and Chokepoints in the Middle East

The oil-exporting countries of the Middle East, often with the financial support of the United States and other leading oil-consuming countries, have invested enormous resources to ensure the safe passage of oil to world markets. The Middle East's oil region is not landlocked like the oil-rich "stans" of central Asia, but some of the same kinds of geopolitical issues are involved in oil export
140 planning. These concerns often point toward pipelines as the best means of routing oil shipments (**•Figure 6.C**). Pipelines are attractive because they shorten the time and expense involved in seaborne transport, and they bypass chokepoints. However, they are vulnerable to disruption. Weapons as small as grenades can disrupt supplies. A major challenge is how to route a pipeline so that it will not cross through a potential enemy's territory, and another is how to maintain friendships with the countries the oil crosses.

The following paragraphs will not be interesting at all if you don't use the map in **Figure 6.C.** Please use that map, and you will get a feel for the compelling intersection of economics, military thinking, and the consequences of war in pipeline geography.

One of the first major pipelines in the region was the 1,100-mile (1,760-km) Trans-Arabian Pipeline (Tapline), leading from the oil-producing eastern province of Saudi Arabia to a terminal on the Mediterranean coast of Lebanon. It has the advantage of bypassing three chokepoints (Strait of Hormuz, Bab el-Mandeb, and Suez Canal). However, at the time it was completed (1950), it could not have been anticipated that it would cross two of the worst impending conflict zones in the Middle East: the Golan Heights of Syria (which fell to Israel in the 1967 war) and southern Lebanon (a major theater of the Lebanese civil war of the 1970s and the subsequent Israeli–Lebanese Shiite struggles). Tapline was knocked out early by its unfortunate geography.

Another Middle East conflict, the 1967 Six-Day War between Israel and its neighbors, led to the closure of the Suez Canal (for about a decade) and thus to the construction of two new pipelines to bypass Suez: one across southern Israel from the Gulf of Aqaba to the Mediterranean Sea and another across Egypt from the Gulf of Suez to the Mediterranean (thus its acronym, SUMED).

No Middle Eastern country has been more dependent on pipelines than Iraq, and none has so systematically experienced the liabilities of pipelines. The Iran-Iraq war of 1980–1988 had major impacts on pipeline geography. Saudi Arabia supported Iraq in the war and so tried to ensure it could get oil to market without being threatened by Iran; this meant bypassing the Strait of Hormuz. That led to the construction of Petroline (opened in 1981), the east–west pipeline across the Arabian Peninsula. Iraq built a pipeline to ship some of its oil south to Petroline, thereby bypassing the Strait of Hormuz as well. Beginning in 1961, Iraq was also able to ship oil through pipelines almost due west though Syria to the Mediterranean Sea. But when it attacked Iran in 1980, Iraq lost that route because Syria was Iran's ally.

Iraq quickly responded to that setback by building a new pipeline leading almost due north to bypass Syrian territory and then making a sharp 90-degree turn westward through southern Turkey to the Mediterranean Sea. But in 1990, Iraq attacked Kuwait, threatened Saudi Arabia, and prompted a U.S.-led counterattack. Saudi Arabia responded by closing the southern link with Petroline. Turkey, a NATO ally of the United States, reacted by closing the northern link. The United Nations responded by issuing strict controls on Iraqi oil exports. In sum, by his military adventurism, Saddam Hussein did almost everything imaginable to deprive his country of oil exports.

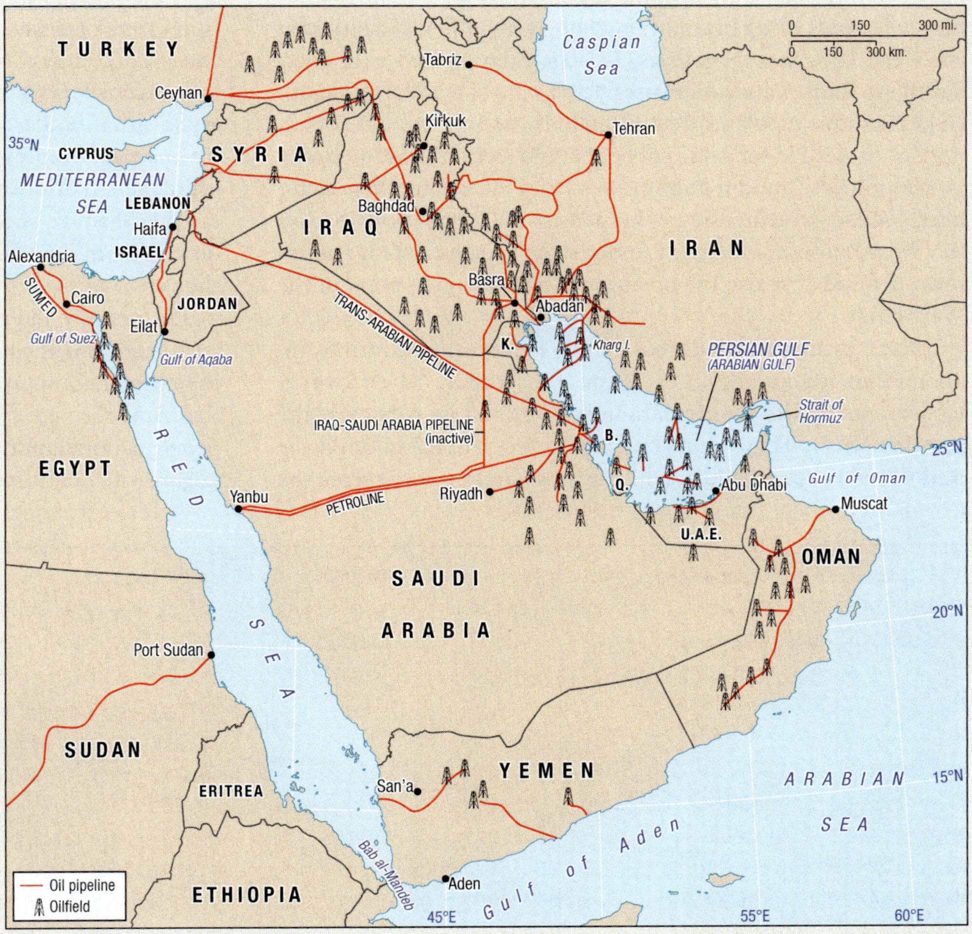

• Figure 6.C Principal oilfields and pipelines in the heart of the Middle East. Vulnerable chokepoints and volatile political relations have led to the construction and often indirect routing of many pipelines.
Source: Based on National Geographic's Atlas of the Middle East (National Geographic, 2003).

the 1973 war, the United States put its forces on an advanced state of readiness to take on the Soviets in a nuclear exchange if necessary. All of these events illustrate that the Middle East and North Africa form, as Eastern Europe did, a shatter belt—a large, strategically located region composed of conflicting
104 states caught between the conflicting interests of great powers.

Access to Freshwater

"The next war in the Middle East will be over water."

—Jordan's late King Hussein

Some of the most serious geopolitical issues in the Middle East and North Africa relate to **hydropolitics**, or political leverage and control over water. In this arid region, where most water is available either from rivers or from underground aquifers that cross national boundaries, control over water is an especially difficult and potentially explosive issue. An estimated 90 percent of the usable freshwater in the Middle East crosses one or more international borders.

Water is one of the most problematic issues in the Palestinian-
185 Israeli conflict. Freshwater aquifers underneath the West Bank supply about 40 percent of Israel's water. Palestinians point to Israel's control over West Bank water as one of the most troubling elements of its occupation. The average Jewish settler on the West Bank uses 74 gallons (278 l) per day, whereas the average West Bank Palestinian uses 19 (72 l). (The World Health Organization calculates that 13 gallons, or 50 liters per person per day, is needed for minimal health and sanitation standards.) Israeli policies prohibit Palestinians from increasing their water usage. Many Israeli policymakers insist that water resources in the West Bank must remain under strict Israeli control and on these grounds oppose the creation of a Palestinian state in the West Bank.

More promisingly, Jordan and Israel are working on agreements to share waters from the Jordan River (which forms a portion of their common border) and its tributary, the Yarmuk River. They are discussing a joint venture to build the **Red–Dead Peace Conduit**, which would send seawater from the Gulf of Aqaba to the Dead Sea via a network of canals and pipelines (see Figure 6.33). This would replenish the Dead Sea, which has retreated by about 3 feet (c. 1 m) per year for the last 25 years. The gravity flow of seawater to the Dead Sea would run generators to produce electricity, some of which would be used to desalinate the water. The two nations, and potentially the Palestinian Authority, could share this water and power.

Water is a critical issue blocking a peace treaty between Israel and Syria. If Syria were to recover all of the Golan Heights (which it lost to Israel in the 1967 war), it would have shorefront on Lake Kinneret and therefore, presumably, rights to use its water. That prospect is unacceptable to Israel. This body of water, also known as Lake Tiberias and the Sea of Galilee, is Israel's principal supply of freshwater, feeding the National Water Carrier system that transports water south to the Negev Desert (see Figure 6.33). In occupying the Golan Heights, Israel also controls some of the northern bank of the Yarmuk River on the border with Jordan. For many years, Israel has stated that it would never allow Syria and Jordan to construct their proposed Unity Dam, intended to store waters of the Yarmuk River to be shared between those countries. Israel thus implied it would bomb the dam rather than allow it to deprive Israel of water from the Jordan River.

A useful way to think about the geography of hydropolitics is in terms of **upstream** and **downstream countries**. Simply because water flows downhill, an upstream country is usually able to maximize its water use at the expense of a downstream country (a situation described as a **zero-sum game**, where any gain by one party represents an equivalent loss to the other). However, the situation between Israel and the countries upstream on the Yarmuk shows that this is not always true. Although downstream, Israel is far more powerful militarily and can use the threat of force to wrest more water out of the system.

Historically, Egypt has also defied the norm of upstream countries exercising their power over downstream countries. Egypt is the ultimate downstream country, at the mouth of a great river that runs through five countries and sustains about 170 million people (a population that is expected to double in about 20 years; •**Figure 6.25**). However, it has long been the strongest country in

• **Figure 6.25** Recent and proposed water developments in the Nile Basin. Note that this map is oriented such that north is at the left rather than at the top.

REGIONAL PERSPECTIVE | The Aswan High Dam

Egypt's enormous Aswan High Dam (•**Figure 6.25**; and see location in Figure 6.25) is located on a stretch of the Nile known as the First Cataract (the Nile's many **cataracts** are areas where the valley narrows and rapids form in the river). Its reservoir, Lake Nasser, stretches for more than 300 miles (480 km) and reaches into northern Sudan. Like all giant hydrological projects, the Aswan High Dam has generated benefits and liabilities, and its construction was controversial. On the plus side:

- By storing water in years of abundant rain for use in years of low rainfall, the High Dam provides a constant water supply to the Egyptian Nile Valley's 11,000 square miles (28,500 sq km) of irrigated fields, thus extending the cultivation period.
- Year-round irrigation has significantly boosted crop production in Egypt.
- The dam made possible the "reclamation" of previously uncultivated land on the desert fringe adjoining the western Nile Delta; there is enough surplus water to irrigate lands beyond the river's flood plain.
- Because water levels can be adjusted, there are fewer threats to ships navigating the Nile.
- The dam provides hydroelectricity to Egypt's industries and citizens.
- Excess waters stored in Lake Nasser help Egypt to avoid the damaging impacts of drought.
- The dam can hold back overabundant water, helping Egypt to avoid the damaging impacts of flooding.

The Aswan High Dam has also had drawbacks:

- The dam has caused the water table to rise, making it harder for irrigated soil to drain properly. When farmers use too much water, standing water evaporates and leaves a deposit of mineral salts—a problem known as **salinization**—that causes soil to lose its productivity.
- Canal waters, which are now high throughout the year, are ideal breeding grounds for the snail that hosts the parasite that causes schistosomiasis (bilharzia), a debilitating disease affecting a large proportion of Egypt's rural population.
- Mediterranean sardine populations, now deprived of the rich silt that nurtured their food supply of phytoplankton, have plummeted off the Nile Delta. The lucrative sardine fishing industry has declined by more than 90 percent.
- The dam prevents rich volcanic silt washed down the Nile from Ethiopia from reaching the floodplain below the dam. Without this "free fertilizer," Egyptian farmers must buy expensive artificial fertilizers.
- Because silt is no longer replenishing the soils of the Nile Delta, there is coastal erosion along the Mediterranean seafront. This is allowing seawater to destroy productive farmland.
- The Nubian people who lived in the area now inundated by the reservoir behind the dam had to be relocated from their homeland to an alien environment.
- Many of the archaeological treasures of ancient Egypt were drowned by the reservoir (but others were relocated).
- If the dam burst because of an earthquake or a bombing, the lives of nearly 70 million people downstream would be at risk. One analyst wrote:

  > If a bomb hit the top of the Dam, the water pouring out uncontrolled would make the Dam burst in its complete breadth within a few hours. The masses of water sweeping across the whole of Egypt (they would be at least the discharge of two years within hours or days) would cause utter devastation from Aswan to Damietta, destroy all cultivated land, and most probably about 99 per cent of the total population would lose their lives.[2]

Whenever a dam is built, there are questions about whether its advantages outweigh its drawbacks. When great dams are built in the LDCs—like the Aswan High Dam in Egypt or the Three Gorges Dam in China—such questions tend to be secondary to the symbolic importance of the dam. These massive engineering projects announce that a developing country is developed enough to master its environment and marshal the extraordinary engineering and other resources needed to build the dam. The dam announces that a country has "come of age."

---

the Nile Basin and has threatened to use its greater force if it does not get the water it wants. In 1926, when the British ruled Egypt and many other colonies in Africa, 10 countries located on the Nile and its tributaries upstream of Egypt were compelled to sign the **Nile Water Agreement**. This guaranteed Egyptian access to 75 percent of the river's flow, even though barely a drop of the Nile's waters actually originates in Egypt. Sudan was given access to 25 percent of the water, and Ethiopia was given none. The treaty forbids any projects that might threaten the volume of water reaching Egypt, prohibits use of Lake Victoria's water without Egypt's permission, and gives Egypt the right to inspect the entire length of the Nile to ensure compliance.

In recent years, however, one country after another has defied the treaty, calling it an outmoded legacy of colonialism. Kenya and Tanzania have plans to build pipelines to carry Lake Victoria waters to thirsty towns and villages inland. Uganda is building its controversial Bujagali Dam on the Nile, mainly for hydroelectricity production. With Chinese assistance, Ethiopia is building the huge Tekaze Dam, for hydropower and irrigation, on a tributary of the Blue Nile. Sudan is building the Merowe (Hamdab) and Kajbar Dams on its northern stretch of the Nile. South Korea, China, and India have purchased huge swaths of land in Sudan and Ethiopia to grow wheat and rice through irrigation. This will result in a massive new drawdown of Nile 329 waters. Predictably, Egypt has had a bellicose response to these developments. Egypt has even called Kenya's stated intention to withdraw from the Nile Water Agreement an "act of war."

Meanwhile, Egypt's demands on Nile waters are increasing. Following on the ambitious Aswan High Dam (see Regional Perspective), Egypt recently built the multibillion-dollar

Compliments of NASA/Earth Sciences and Image Analysis Laboratory at Johnson Space Center

• **Figure 6.26** Egypt's Aswan High Dam, on the First Cataract of the Nile River.

Toshka Canal, which transports water from Lake Nasser over a distance of 100 miles (160 km) to the Kharga Oasis of the Western Desert (see Figure 6.25). Proponents of the canal insist that it will result in the cultivation of nearly 2,350 square miles (c. 6,100 sq km) of "new" land and provide a living for hundreds of thousands of people. Critics argue that it is a waste of money and that salinization and evaporation will take a huge toll on the cultivated land and the country's water supply.

The familiar specter of population growth will play a big role in the hydropolitics of the Nile. The combined populations of Egypt, Sudan, and Ethiopia, at about 210 million now, are predicted to grow to 270 million by 2025 and 360 million by 2050. Every person's direct needs for water plus production of food for him or her means growing demands for Nile waters and the potential for conflict. The solution? Lester Brown of Worldwatch proposes three steps: Reduce population growth, grow crops that are less demanding of water, and go back to the negotiating table for a new Nile Water Agreement.[3]

Turkey—the source of four-fifths of Syria's water and two-thirds of Iraq's—exercises its upstream advantage on the Tigris and Euphrates rivers (•**Figure 6.27**). Turkey's position has long been that water in Turkey belongs to Turkey, just as oil in Saudi Arabia belongs to Saudi Arabia. Not surprisingly, downstream Syria and Iraq reject this position and are distraught by the diminished flow and quality of water resulting from Turkey's
205 comprehensive **Southeast Anatolia Project**. When completed, the project is expected to reduce Syria's share of the Euphrates waters by 40 percent and Iraq's by 60 percent. Also increasing the likelihood of serious future tension is a history of strained relations among Turkey, Syria, and Iraq, accompanied by the fact that no commonly accepted body-of-water law governs the allocation of water in such international situations.

Turkish leaders have said they will never use water as a political weapon, but Turkey has already wielded water to its advantage. In 1987, for example, Turkey increased the Euphrates flow into Syria in exchange for a Syrian pledge to stop support of Kurdish rebels inside Turkey. Turkey now says it wants to use water to promote peace in the Middle East by shipping it in converted supertankers for sale to such thirsty (and therefore potentially combative) countries as Israel, Jordan, Saudi Arabia, Libya, and Algeria.

Terrorism

During the two terms of U.S. President George W. Bush, almost all of the geopolitical issues related to this region—oil, economic

Joe Hobbs

• **Figure 6.27** The Tigris and Euphrates Rivers rise in Turkey, giving this non-Arab country control over a resources vital to the lives of millions of Arabs in downstream Syria and Iraq. This waterfall is on a tributary of the Tigris in far eastern Turkey.

development, trade, aid, the Arab-Israeli conflict, and more—were subsumed beneath the broader rubric of the president's declared **War on Terror**. Even the 2003 invasion of Iraq was justified as a response to that war. The open-ended, unconventional War on Terror would destroy the enemies who attacked the United States on 9/11 and use preemptive strikes if needed to protect the U.S. homeland against future attacks. President Obama much more frequently employed preemptive attacks on suspected terrorist targets, while avoiding the term *War on Terror*. One of his favorite tactics was the use of unmanned "predator drones" that fly quietly overhead and unleash deadly force on ground targets. Their use is controversial because they often
197, 238 result in civilian casualties.

Since 9/11, Americans have become accustomed to heightened security measures to provide protection against **terrorism**, which by the U.S. State Department definition is "premeditated, politically motivated violence perpetrated against noncombatant targets by subnational groups or clandestine agents, usually intended to influence an audience."[4] The terrorists pursued by the United States are almost without exception Islamist militants—best known as **Islamists**—and so it is useful to understand the nature of Islamic "fundamentalism" and radicalism. Not all Islamists are militant or terrorist, but they all reject what they view as the materialism and moral corruption of Western countries and the political and military support these countries lend to Israel. Both Sunni and Shiite Muslims have advanced a wide range of Islamist movements, notably in Iran, Lebanon, Egypt, Sudan, Algeria and the Palestinian Territories.

Although nominally religious, the more radical of these movements have political and cultural aims, particularly the destabilization or removal of U.S. and Israeli interests in the region and abroad. In 1993, followers of the radical Egyptian cleric Sheikh Umar Abdel-Rahman bombed New York City's World Trade Center as a protest against American support of Israel and Egypt's pro-Western government. In an attempt to destabilize and replace Egypt's government, which it viewed as an illegitimate regime too supportive of the United States, another Egyptian Islamist group attacked and killed foreign tourists in Egypt in the 1990s. In the 1980s, members of the pro-Iranian **Hizbullah**, or Party of God, in Lebanon kidnapped foreign civilians to use as bargaining chips for the release of comrades jailed in other Middle Eastern countries.

Within Israel and the autonomous Palestinian territories of the West Bank and Gaza Strip, Palestinian members of **Hamas** (an Arabic acronym for the Islamic Resistance Movement) and another organization called the al-Aqsa Martyrs Brigade carried out terrorist attacks on Israeli civilians and soldiers in a largely successful effort to derail implementation of the peace agreements reached in the 1990s between the Israeli govern-
189 ment and the Palestine Liberation Organization (PLO).

The Iranian Threat

In Western capitals, concern about "state-sponsored terrorism" has long focused on Iran. Iran has extended both open and clandestine assistance to a variety of Islamist terrorist groups, including Hizbullah and Hamas. There is great concern about Iran's nuclear weapons potential because such weapons might find their way to terrorist groups or be delivered by Iran itself
on its own missiles against Israel or another target. Iran in- 141
sists that its nuclear program is aimed only at electricity generation and medical research, and denies that it is developing weapons. American and Israeli intelligence agencies, however, believe Iran is on course to create nuclear weapons some time before 2016. The United States, Israel, and the United Nations' **International Atomic Energy Agency (IAEA)** are particularly watchful of Iran's acquisition of centrifuges, which are used to enrich uranium. The enrichment process is necessary for the production of fuel for civilian use in nuclear power production, which Iran is permitted to do, but can also be expanded for the production of nuclear weapons.

The United States was weighing three difficult options in response to Iran's nuclear weapons threat. As U.S. Defense Secretary Robert Gates noted, "There are no good options on Iran."[5] First, it could engage in diplomatic dialogue with Iran, as it had done with North Korea, offering a package of incentives and security guarantees in exchange for Iran's pledge to halt weapons development. The United States and Iran do share some common interests, notably stability in Iraq; Iran does not want to deal with strong new Sunni or Kurdish states there, or with the tide of refugees that more Iraqi conflict would send its way. Second, the United States could impose even harsher sanctions than it has already against Iran, especially to cut off investment and finance (the United Nations, European Union, and several countries also have levied sanctions against Iran).

Finally, with or without coordination with Israel, United States could launch military strikes on Iran's suspected nuclear weapons development sites, beginning at facilities northeast of the Shiite holy city of Qum, at nearby Natanz and Arak, and near the Gulf city of Bushehr (**•Figure 6.28**). Most analysts

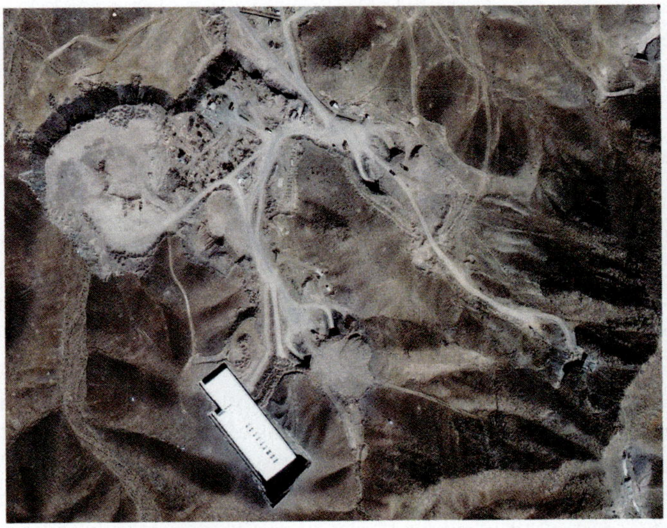

AP Photo/GeoEye Satellite Image, IHS Jane's Analysis

• Figure 6.28 Iran's suspected gas centrifuge uranium enrichment facility northeast of the holy city of Qum, in the rugged Zagros Mountains of northwest Iran. Just above and just below the center of the image you can see tracks leading into tunnel entrances. The critical equipment is presumably accessed from these tunnels, and may be so deep that "bunker buster" bombs cannot damage it.

believed that the military option would have far-ranging and dangerous consequences. Iran's nuclear program would be set back a few years, but the country would portray itself as a victim of U.S. aggression, galvanizing support across the Muslim world (even among Iran's traditional enemies, the Sunni Arabs), perhaps inciting its Hizbullah allies in Lebanon and Hamas allies in the Palestinian territories to attack Israel, encouraging al-Qa'ida–style terrorism against the West, and discouraging Iranians (particularly the younger ones) who favor peaceful relations with the United States. A military strike would probably also lead to a sharp reduction in oil exports from the Gulf, perhaps triggering a global economic crisis. Russia and China would meanwhile have an opportunity to enhance their geopolitical interests by offering diplomatic support and perhaps more to Iran throughout such a crisis. Both countries have considerable leverage in Iran.

Israeli officials have said publicly that Israel (which has its own nuclear weapons, a fact it never acknowledges officially) regards the Iranian bomb as an "existential threat" and will prevent its development, presumably by a single surprise air strike like the one Israel carried out against Iraq's nuclear facility in 1981. It is more likely that Israel rather than the United States would strike Iran. Again, Israel would delay the development of an Iranian bomb, but the regional and global consequences would be enormous. The United States has repeatedly warned Israel against attacking Iran because of the probable consequences discussed above.

The United States and Israel may have found a way to circumvent both diplomatic and military means to blunt Iran's nuclear weapons program: sabotage. In 2010, a computer worm (probably planted by Israeli and U.S. agents) called Stuxnet apparently did significant damage to centrifuges at Iran's nuclear facilities. The most sophisticated cyberweapon ever developed, the Stuxnet malware took over the control systems at Natanz and Bushehr, sending the centrifuges spinning wildly out of control while also issuing a sound that made control room staff think that systems were operating normally. By some intelligence estimates, the Stuxnet attacks may have set Iran's nuclear weapons program back by two years.

Al-Qa'ida

In 1998, the world began to hear about Osama bin Laden, a former Saudi businessman living in exile in Afghanistan, whose **al-Qa'ida** organization bombed U.S. embassies in Kenya and Tanzania as part of a struggle, or ***jihad***, against American imperialism and immorality. (In Islam, though, *jihad* has a very wide range of interpretations, from expressing one's commitment to a personal conviction, to an all-out war against nonbelievers.) Bin Laden's organization was also responsible for the 2000 bombing of the American naval destroyer *U.S.S. Cole* in Yemen's harbor of Aden. However shocking those assaults were, they pale in comparison to al-Qa'ida's attacks against targets in the United States on September 11, 2001. In the most ferocious terrorist actions ever undertaken to that date, members of al-Qa'ida "cells" (small groups of agents) in the United States hijacked four civilian jetliners and succeeded in piloting two of them into the twin towers of New York City's World Trade Center and one into the Pentagon, home of the U.S. Department of Defense, outside Washington, D.C. Passengers on a fourth plane succeeded in aborting the terrorists' likely mission of destroying the White House or damaging the U.S. Capitol, forcing the plane to crash in rural Pennsylvania. More than 3,000 people, mainly civilians, perished. Bin Laden and his followers cheered the carnage as justifiable combat against an infidel nation whose military troops occupied the holy land of Arabia, where Mecca and Medina are located.

As the United States counterattacked, ousting al-Qa'ida from its bases in Afghanistan and cracking down hard on its leadership around the world, al-Qa'ida evolved into a far more geographically diffuse organization. It carried out or supported attacks in Indonesia, Morocco, Algeria, Tunisia, Turkey, Spain, Uzbekistan, Egypt, Saudi Arabia, Jordan, the United Kingdom, Yemen, and Pakistan. A succession of al-Qa'ida videotapes and audiotapes promised an unrelenting and costly continuation of *jihad* against the United States and its allies. Even with Bin Laden dead, it will be many years before Americans and Europeans might be able to emerge from the shadow of this threat (•Figure 6.29). Al-Qa'ida clearly will not hesitate to use the most devastating weapons, even against large numbers of civilians; this deadly goal is a tactical priority. There is also growing consensus that al-Qa'ida is no longer a solitary severe threat to U.S. interests; instead, the *jihadist* cause has spread to smaller groups and to "lone wolves" that are also keen to take on the West. There is little evidence that al-Qa'ida's message has any broad appeal, and it had virtually no role in the tumultuous Arab Spring (see page 196). But the danger remains.

Using GIS and a variety of other geographic tools, many geographers—along with U.S. intelligence agencies, of course—are at work interpreting the geographic dimensions of terrorism. One of the most critical questions for geographers is where future al-Qa'ida attacks are likely to take place. Rohan Gunaratna, probably the greatest authority on al-Qa'ida, estimates that al-Qa'ida maintains a reserve of at least 100 targets worldwide.[6] Where are most of these? Suleiman Abu Ghaith, one of bin Laden's top aides, offered a clue, along with a geographic answer to the question of what al-Qa'ida wants: "Let the United States know that with God's permission, the battle will continue to be waged on its territory until it leaves our lands."[7]

In 9/11 and other atrocities, a tiny minority of Muslims carried out terrorist actions that the vast majority of Muslims condemned. Islamic scholars and religious authorities point out that murder of civilians is prohibited in Islamic law and that the attacks have no legitimate religious grounds. Mainstream Islamist movements are not military or terrorist organizations and have distinguished themselves through public service to the needy.

The following section offers more insight into the peoples and nations of this vital region. It begins with a continuation of the geopolitical theme, examining one of the world's most problematic, persistent, and influential conflicts, the one between Israel and its Arab neighbors.

REUTERS/Sean Adair/Landov

(a)

AP Photo/Virginia Mayo

(b)

• **Figure 6.29** Bookends to Osama bin Laden's engagement with the United States. In the 9/11/2001 attack on the World Trade Center **(a)**, note United Flight 175 about to strike the South Tower. Belgian newspapers trumpet bin Laden's death on May 3, 2011 **(b)**.

6.6 Regional Issues and Landscapes

Israel and Palestine

The Arab-Israeli Conflict

The Arab-Israeli conflict persists as one of the world's most intractable disputes. It has not been resolved, in part because the central issues are closely tied to such life-giving resources as land and water, and to deeply held and unyielding religious beliefs. When you understand these issues, you should be proud and prepared. Among other things, you will have a strong foundation from which to follow the news about this conflict for years to come.

The Arab-Israeli conflict is above all a conflict over who owns the land—sometimes very small pieces of land—and is therefore of extreme interest in the study of geography. It is also a conflict that has repercussions far beyond the boundaries of the small countries and territories involved. As long as it simmers or boils, there are other countries and entities that will use the unresolved Arab-Israeli conflict to advance their interests at the expense of others; both al-Qa'ida and Iran, for example, derive much beneficial propaganda value from it. A United Nations–sponsored group called the Alliance of Civilizations argues that the Palestinian-Israeli conflict is *the largest force behind global tensions*. Let's look at this in different terms: when it is resolved, the world will be a much more safer place.

A geographic understanding of this conflict requires familiarity with the events leading up to the creation of the state of Israel and with the wars that have followed, particularly as they have rearranged the boundaries of nations and territories.

The modern state of Israel was carved from lands whose fate had been undetermined since the end of World War I. The Ottoman Empire, based in what is now Turkey, had ruled Palestine (roughly the area now made up of Israel and the Palestinian territories) and surrounding lands in the eastern Mediterranean since the sixteenth century. After the British and French defeated the Ottoman Turks in World War I and destroyed their empire, they divided the region between themselves (•**Figure 6.30**). The British received the "mandate" (authority to establish a government) for Palestine, Transjordan (modern Jordan), and Mesopotamia (Iraq), and the French received the mandate for Syria (now Syria and Lebanon).

During World War I, British administrators of Palestine had made conflicting promises to Jews and Arabs. They implied that they would create an independent Arab state in Palestine and yet at the same time vowed to promote Jewish immigration to

•Figure 6.30 The Middle East and North Africa in 1920. The victorious allies of World War I carved up the Middle East among themselves. Growing difficulties of administration would drive them from the region within a few decades.

Source: Based on John Haywood, ed., Atlas of World History (Barnes and Noble Books, 1997).

Palestine with an eye to the eventual establishment of a Jewish state there. The **Palestinians**—Arabs who historically formed the largest majority of the region's inhabitants—did not welcome the ensuing Jewish immigration and rioted against both the migrants and the British administration. Militant Jews attacked British interests in Palestine, hoping to precipitate a British withdrawal.

Placing themselves in a no-win position with these conflicting promises and under increasing pressure from both Jews and Arabs, in 1947 the British decided to withdraw from Palestine and leave the young United Nations with the task of determining the region's future. The United Nations responded in 1947 with the **two-state solution** to the problem of Palestine. It established an Arab state (which would have been called Palestine) and a Jewish state (Israel).

The two-state plan was deeply flawed. The states' territories were long, narrow, and fragmented, giving each side a sense of vulnerability and insecurity (Figure 6.29). When Israel declared itself into existence in May 1948, the armies of the neighboring Arab countries of Transjordan, Egypt, Iraq, Syria, and Lebanon mobilized. In what Israelis call "the War of Independence" and Palestinians call "the Catastrophe" (*al-Nakba*), the smaller but better-organized and more highly motivated Israeli army defeated the Arab armies, and Israel acquired what have come to be known as its **pre-1967 borders** (the 1949 Armistice Agreement borders shown in **•Figure 6.31**). The boundary separating Israel from the West Bank would later come to be known as the "Green Line" (seen in Figure 6.36).

Prior to and during the fighting of this **1948–1949 Arab-Israeli war**, approximately 800,000 Palestinian Arabs chose or were forced to flee from the new officially Jewish state of Israel to neighboring Arab countries (**•Figure 6.32**). The United Nations established refugee camps for these displaced peoples in Jordan, Egypt, Lebanon, and Syria. Little was done to resettle them in permanent homes, and both the local Arab governments and the refugees themselves continued to insist on the right of the refugees to return to Israel and the restoration of their property there. This Palestinian **right of return** is one of the central problems of the modern peace process.

Other countries assumed control of those parts of the proposed Palestine that Israel did not absorb. Egypt occupied the Gaza Strip, a piece of land on the Mediterranean shore adjacent to Egypt's Sinai Peninsula that was inhabited mostly by Palestinian Arabs. Transjordan, renaming itself simply Jordan, occupied the predominantly Arab hilly region of central Palestine on the west side of the Jordan River, known as the West Bank, Fig. 6.3
and the entire old city of Jerusalem, including the Western Wall and Temple Mount, Judaism's holiest sites. The Palestinian Arab state envisioned in the UN partition plan was thus stillborn.

The **Six-Day War** of 1967 fundamentally rearranged the region's political landscape in Israel's favor, setting the stage for subsequent struggles and the peace process. This conflict was precipitated in part when Egypt, by positioning arms at the Strait of Tiran chokepoint, closed the Gulf of Aqaba to Israeli shipping. Egypt's President Nasser and his Arab allies took sev- 177
eral other belligerent but nonviolent steps toward a war they

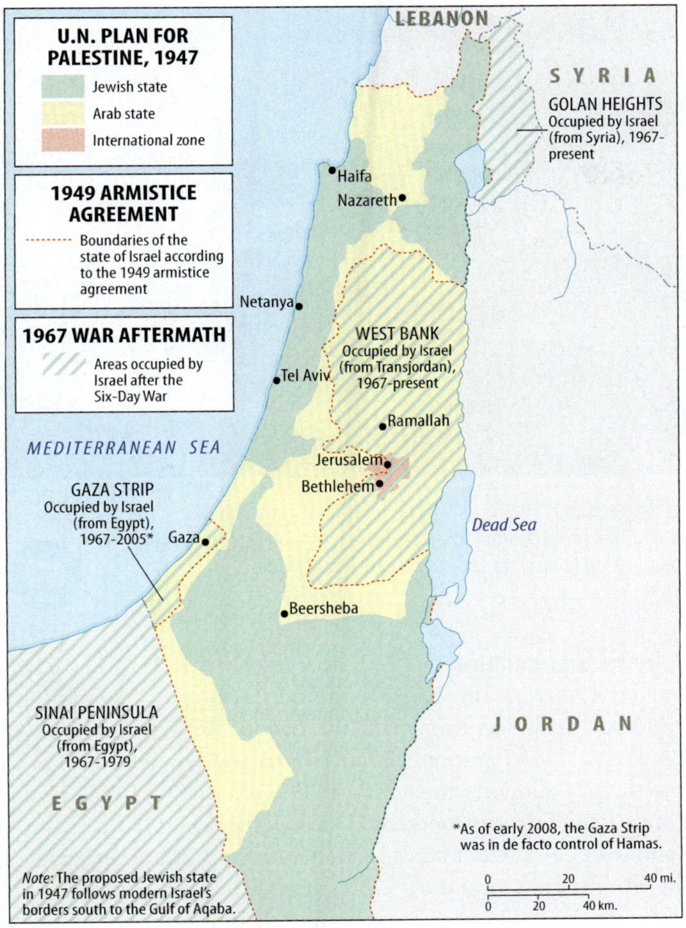

• **Figure 6.31** The 1947 UN partition plan for Palestine and Israel's original (pre-1967) borders. The 1948–1949 war, which began as soon as Britain withdrew from Palestine and Israel proclaimed its existence, aborted the UN plan and created a tense new political dynamic in the region.

Source: The Middle East and North Africa: A Political Geography by Drysdale & Blake (1985) Fig.9.7 The 1947 UN Partition Plan for Palestine and Israel's Original (Pre-1967) Borders p.280. By permission of Oxford University Press, Inc.

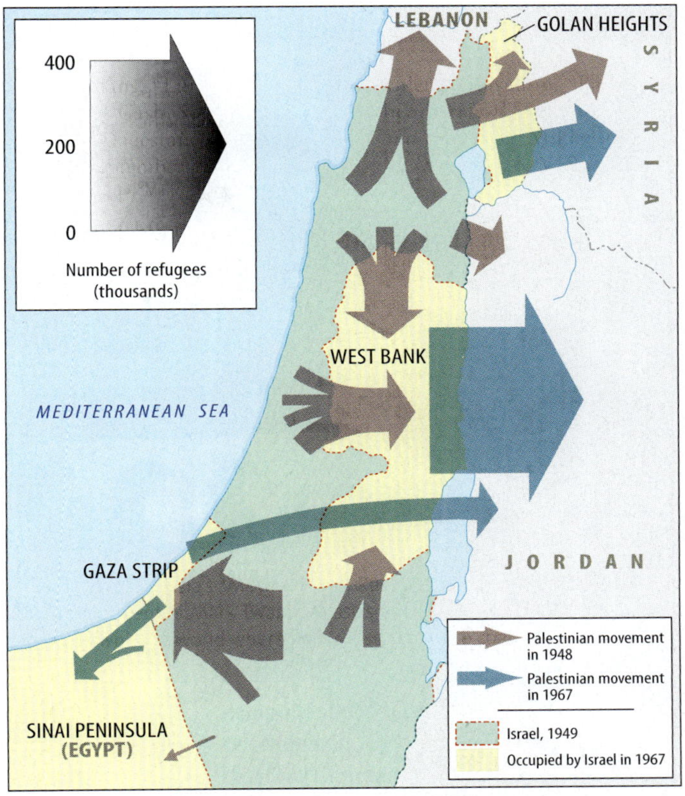

• **Figure 6.32** Palestinian refugee movements in 1948 and 1967. Many who fled in the first conflict relocated in the second.

Source: The Middle East and North Africa: A Political Geography, by A. Drysdale and G. Blake. 1985 Oxford University Press. Based on The Wandering Arabs, Geographical Magazine XLV (3) 1972 page 180.

were ill prepared to fight. Israel elected to make a preemptive strike on its Arab neighbors, virtually destroying the Egyptian and Syrian air forces on the ground. Israel gave Jordan's King Hussein an opportunity to stay out of the conflict. However, Jordan went to war and quickly lost the entire West Bank and the historic and sacred Old City of Jerusalem. The entire nation of Israel was transfixed by the news that Jewish soldiers were praying at the Western Wall. The Israeli army (Israeli Defense Forces, or IDF) also seized the Gaza Strip, Egypt's Sinai Peninsula (shutting down the Suez Canal), and the strategic Golan Heights section of Syria overlooking Israel's Galilee region. These three pieces of land would henceforth be known to the world as the **Occupied Territories**. Israel had tripled its territory in six days of fighting (•**Figure 6.34**).

The Palestinian refugee situation became more complex as a result of the war. Israel took over the areas where most of the camps were located. Many people in the camps again took flight, and Palestinians fleeing villages and towns in the newly Occupied Territories joined them as refugees (**see Figure 6.32**). Some later returned to their homes, but an estimated 116,000 either did not attempt to return or were denied permission by Israel authorities to do so.

The **1973 Arab-Israeli War** was a multifront Arab attempt to reverse the humiliating losses of 1967. On October 6, 1973, the Jewish holy day of Yom Kippur, Egypt and Syria launched a surprise attack on Israel, with the hope of rearranging the stalemated political map of the Middle East. Egypt's army initially showed surprising strength, penetrating deep into the Sinai Peninsula and overturning Israel's image of invulnerability. Israeli troops soon reversed the tide and surrounded Egypt's army, but in the ensuing disengagement talks Egypt won back the eastern side of the Suez Canal and by 1975 was able to reopen it to commercial traffic. In 1979, Israel agreed to return Sinai to Egypt under the U.S.-sponsored **Camp David Accords**, and Egypt recognized Israel's rights as a sovereign state. The Gaza Strip, West Bank, and Golan Heights continued to be held by Israel. Different Israeli administrations regarded them as lands that rightfully belong to Jews or as cards to be played in the regional game of peacemaking.

Arabs and Jews: The Demographic Dimension

In addition to issues of land, water, politics, and ideology, the Palestinian-Israeli conflict is about sheer numbers of people. Each side has wanted to maximize its numbers to the disadvantage of the other. To realize the Zionist dream of establishing a Jewish state, Jews began immigrating to Palestine around the start of the twentieth century. Jews made up 11 percent of

© Cengage Learning 2013

• **Figure 6.33** Israel and Arab territories as of 2012. Israel occupied the Sinai Peninsula in the 1967 war but returned it to Egypt following the Camp David Accords of 1979. Israel withdrew settlers and soldiers from the Gaza Strip in 2005 but kept them in the West Bank and the Golan Heights.

JODI COBB/National Geographic Stock

• **Figure 6.34** The Jewish settlement of Maaleh Adumim in the occupied West Bank has a commanding position overlooking the strategic West Bank road linking Jerusalem with the Dead Sea rift valley.

Palestine's population in 1922, 16 percent in 1931, and 31 percent in 1946, on the eve of Israel's creation. In keeping with national legislation known as the **Law of Return**, the state of Israel has always granted citizenship to any Jew who wishes to live there. Following the 1948–1949 war, waves of new immigrants from Europe (**Ashkenazi Jews**) joined the Middle Eastern **Sephardic Jews** who had inhabited Palestine and other parts of the Middle East (and until 1492, Spain) since early times.

In the 1990s, more waves of Jewish immigration followed the dissolution of the Soviet Union and a change of government in Ethiopia, where an ancient Jewish group called **Falashas** had lived in isolation for thousands of years. In 2011, Jews made up 76 percent of the population of 8 million people living within Israel's pre-1967 borders. About 1.5 million Palestinian and non-Palestinian Arabs, or 19 percent of the country's population, are Israeli citizens and residents; they are known as "Israeli Arabs." Palestinian Arabs without Israeli citizenship, along with a variety of other ethnic groups, comprise the balance of the population within Israel's pre-1967 borders.

Another 4 million people live in the West Bank, Gaza Strip, and Golan Heights—areas occupied by Israel in the wars of 1967 and 1973. About 300,000 of those are Jewish settlers, and most of the rest are Palestinian Arabs (mainly in the West Bank and Gaza Strip), **Druze** (members of a small offshoot of Shiite Islam, living mainly in the Golan Heights), and other Arabs. Another 200,000 Jewish settlers live in East Jerusalem, which is also part of the territories Israel conquered in 1967. There are approximately 6 million more Palestinians living outside Israel and the Occupied Territories. Jordan has the largest number (3 million), followed by Syria and Lebanon (with about 500,000 each). The remainder are in other Arab countries and in numerous nations around the world, comprising a sizable diaspora of their own that Palestinians call the ***ghurba***. In this respect, the Palestinians are in much the same situation as the Jews prior to Israel's creation.

After 1967, and especially after the election of its conservative **Likud Party** to power in 1977, Israel moved to strengthen its grip on the Occupied Territories through security measures and the government-sponsored establishment of new **Jewish settlements**. Most of these settlements were, and still are, spread through the West Bank (see Figure 6.36). Much smaller numbers of Jews settled in the troubled Golan Heights, where about 19,000 remained in 2012. These are generally not frontier farming settlements but communities inhabited by relatively prosperous middle-class Jews who commute to jobs in Israel. They were not built as temporary encampments but as permanent fixtures meant to create what have been described as **"facts on the ground"**—an Israeli presence so entrenched that its withdrawal would be almost inconceivable (Figure 6.34).

Some of the motivation for the settlements has been ideological, as the West Bank is composed of the biblical lands of **Judaea and Samaria**, which many devout Jews regard as part of Israel's historical homeland. The Jews believe strongly that God Himself declared that his "chosen people" should inhabit these lands forever. Put yourself in their shoes and imagine being told, as they were, that your occupation of this land was illegal, or even that you must move away as part of a peace deal.

The proliferation of Jewish settlements worsened relations between Israel and its Arab neighbors and even drove a wedge into the Jewish population of Israel itself. Many Israelis strongly oppose further settlement in the Occupied Territories and annexation of these territories to the Jewish state because if the territories were annexed, Israel would become a country whose population was only about 60 percent Jewish. With an Arab minority that has a far higher birth rate than that of the Jews, Jews could eventually become the minority population in their own country. The Israeli advocacy group Peace Now leaked official Israeli government documents showing that about 40 percent of the land on which Jewish settlements in the West Bank were built is privately owned by Palestinians. There is also international pressure against the settlements because they violate Geneva Conventions on activities allowed in territories captured in war. The International Court of Justice in the Hague has declared them illegal. **United Nations Resolutions 242 and 338** (passed after the wars of 1967 and 1973, respectively) called on Israel to withdraw completely from the Occupied Territories. The United States routinely calls the Jewish settlements "an obstacle to peace" but is loath to push Israel hard about them.

The fate of the Jewish settlements in the Occupied Territories is one of the key issues in the resolution of the Palestinian-Israeli conflict. It is a difficult problem to solve, but it is not insolvable.

The settlements and other aspects of Israeli occupation have fostered widespread and long-lasting Palestinian resistance against Israel. In 1987, Palestinians instigated a popular uprising in the Occupied Territories known as the ***Intifada*** (Arabic for "shaking"). Initially, they relied on rocks and bottles to engage Israeli troops, who answered with rubber and metal bullets. International media coverage of a Palestinian "David" fighting an Israeli "Goliath" did much to damage Israel's image abroad, and many Israelis came to question the justice and importance to security of Israel's continued occupation. Popular opinion and, by 1992, a new Israeli government led by the liberal **Labor Party** began to consider the previously unthinkable: giving the Palestinians at least some control over land and internal affairs in the Occupied Territories.

Land for Peace

They loom large in the headlines and in world affairs, but a closer look reveals that Israel and the Occupied Territories are almost unfathomably small. Within its legally recognized, pre-1967 borders, Israel's area is only 8,019 square miles (20,770 sq km), about the size of New Jersey or Slovenia (•Figure 6.35). Israelis often cite the small size of their country, nestled within the vastness of the Arab Middle East and North Africa, to highlight their vulnerability on the world stage. When and if Palestinians acquire their country, Palestine will be even smaller. The question "Whose lands are these?" has always been at the heart of the conflict between Israel and its neighbors.

President Anwar Sadat of Egypt and Prime Minister Menachem Begin of Israel set the precedent in 1979: Arabs could make peace with Israel on the formula of "**land for peace**," with Israel swapping Arab lands it occupied in 1967 in exchange for peace with its neighbors. The United States, Russia, and Norway initiated negotiations in the 1990s that made such a prospect appear possible. The process began in earnest in 1993, when Israel recognized the legitimacy of and began to negotiate with the **Palestine Liberation Organization (PLO)**, long recognized by the Palestinians as their legitimate governing body. In return, the PLO, under the leadership of Yasser Arafat, recognized Israel's right to exist and renounced its long-standing use of military and terrorist force against Israel.

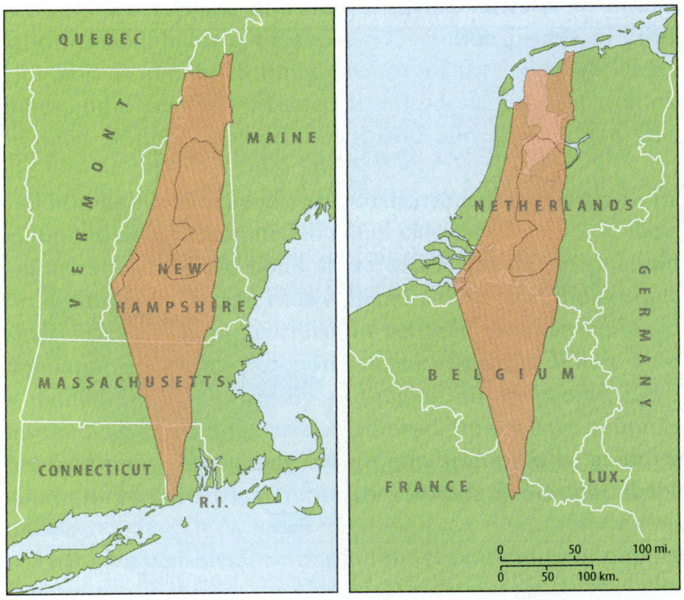

• **Figure 6.35** Israel and the Occupied Territories compared in size with New England and the Benelux countries.

U.S. President Bill Clinton orchestrated a historic handshake between Arafat and the Israeli prime minister, Yitzhak Rabin, and the leaders signed an agreement known as the **Gaza-Jericho Accord** (in its implementation, it came to be known as the **Oslo I Accord**, after the Norwegian capital where it was negotiated). It was designed to pave a pathway to peace by Israel's granting of limited autonomy, or self-rule, in the Gaza Strip and in the West Bank town of Jericho. Autonomy meant that the Palestinians were responsible for their own affairs in matters of education, culture, health, taxation, and tourism. Israel also pulled its troops out of these areas. Palestinians were allowed to form their own government, known as the **Palestinian Authority (PA)**, and they elected Yasser Arafat as its first president.

The accord established a five-year timetable for the resolution of much more difficult matters, the so-called **final status issues**. These included the political status of Jewish and Muslim holy places in Jerusalem and of the city itself, the possible return of Palestinian refugees (the "right of return"), the future of Jewish settlements in the Occupied Territories, and Palestinian statehood (independence). To this day, the final status issues are the ones that prevent the Palestinian-Israeli conflict from being settled. You will be asked to grapple with them as if you were a peacemaker, so please be acquainted with them.

In the wake of a deal known as the Wye Agreement, or **Oslo II Accord**, three successive Israeli prime ministers (Yitzhak Rabin, Benjamin Netanyahu, and Ehud Barak) promised to transfer more West Bank lands from Israeli to Palestinian control, and Arafat promised increased Palestinian efforts to crack down on Palestinian terrorists and so guarantee the security of Israelis in the Palestinian territories and in Israel. If finally implemented (it had not been as of 2012; see **•Figure 6.33**), this arrangement would have brought the total of West Bank lands under complete Palestinian control to 17 percent, leaving 57 percent completely in Israeli hands and 26 percent under joint control (**•Figure 6.36**).

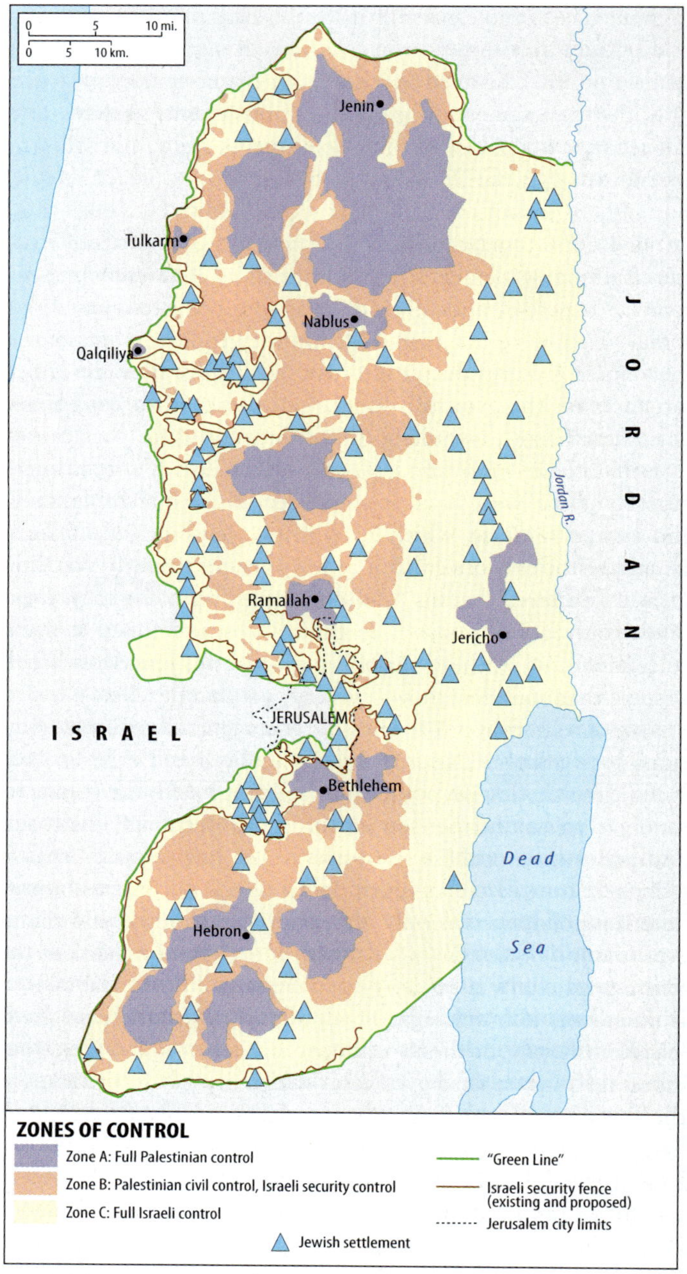

• Figure 6.36 The West Bank. Israeli and Palestinian areas of control were delimited in the peace process of the 1990s, but because of recurrent violence, almost all areas are effectively controlled by Israel. Also depicted are Jewish settlements, the completed and planned portions of the Israeli-built security fence, and the Green Line delimiting the internationally recognized border between sovereign Israel and the occupied West Bank.

Sources: Zone control data from Foundation for Middle East Peace, 2002; security wall route from IDF maps.

On the Brink of Peace During his final year in office in 2000, U.S. President Clinton sought to solidify his legacy as peacemaker by brokering a historic final settlement between Israelis and Palestinians. PLO Chairman Arafat and Israeli Prime Minister Barak huddled with Clinton and his advisers in the presidential retreat at Camp David, a site chosen because of its historical significance in Middle East peacemaking. Over weeks of tough negotiations, mostly over the "final status" issues, the two sides came close to a deal. It would have included the following:

- The creation of an independent Palestinian country in the Gaza Strip and the West Bank.
- A "land swap:" the Palestinian state would include 95 percent of the West Bank. The other 5 percent of the West Bank would be clusters of Jewish settlements that would remain under Israeli control. Palestine would receive 5 percent of Israel's land in exchange.

However, the talks broke down over two thorny issues:

- The status of Palestinian refugees abroad; Israel was unwilling to permit the "right of return" of the large numbers demanded by the Palestinians.
- The deal breaker, Jerusalem, which both Israelis and Palestinians proclaim as their capital. Israel agreed to have the Old City of Jerusalem divided, with Israel assuming sovereignty over the Jewish and Armenian quarters and Palestinians assuming sovereignty over the Muslim and Christian

quarters (see Figure 6.B). These issues were not as problematic as those involving the places held holy by each side. President Clinton put an interesting compromise on the table: neither side would have sovereignty over the Jewish Temple Mount/Muslim Noble Sanctuary (*al-Haraam ash-Shariif*). Instead, these would belong to God. This proposal, which came to be known as Divine Sovereignty, was rejected by both sides. Israel agreed to allow the Palestinians "soft sovereignty" over the Temple Mount/Noble Sanctuary, meaning that Palestinians could administer the area and fly a flag over it, but Israel would technically "own" it. Israel saw this point as essential because the First and Second Temples had stood on this spot. The Palestinians, however, insisted on full sovereignty over the Temple Mount and walked out on the negotiations.

Losing Ground Again Tragically, within weeks, this historic opportunity for peace evaporated and was replaced by a state of war. On September 28, 2000, the right-wing Israeli opposition leader (and subsequently prime minister) Ariel Sharon walked up to the Temple Mount under heavily armed escort. He meant to make the point that this sacred precinct belonged to Israel. Many observers claim he also meant to disrupt the peace process. Whether or not it was his intention, a disruption in the peace process resulted. Beginning that day and for the next few years, Palestinians took to the streets in the **al-Aqsa *Intifada*** to challenge Israeli forces with rocks, bottles, and guns. Hundreds of Palestinians and Israelis died.

Following the collapse of the Camp David talks and an escalation of suicide bombings of Israeli civilians, Israel chose to effectively suspend the peace process. Prime Minister Sharon branded Arafat an untrustworthy partner for peace negotiations, isolating him physically in the West Bank town of Ramallah and systematically dismantling the Palestinian Authority he headed. Israel also persuaded the United States to stop negotiating with Arafat, who died a forlorn figure in 2004. Israel and the United States insisted on the creation of the new post of Palestinian prime minister, hoping it would be filled by someone they could negotiate with. The United States led an effort joined by the other members of the so-called Quartet—the European Union, the United Nations, and Russia—to establish what they called a **"road map for peace"** between the Israelis and the Palestinian Authority. It was based largely on the 2000 Camp David formula and, if successful, would presumably have led to an independent Palestinian state consisting of the West Bank and the Gaza Strip.

Prime Minister Sharon, while resuming dialogue with the Palestinian Authority, decided unilaterally to withdraw Israeli settlers and troops from the Gaza Strip in August 2005 and to build an immense **security fence** separating the West Bank from Israel (the fence is depicted in Figure 6.36). Israel's stated goal in constructing the new barrier was to prevent Palestinian suicide bombers from reaching Israel. Noting that the security fence did not follow the internationally recognized **Green Line** of Israel's pre-1967 border but instead penetrated significant portions of the West Bank to pick up Jewish settlements there, Palestinians condemned it as a "land grab." The International Court of Justice concluded that the wall violated customary international law and several conventions on human rights to which Israel is a signatory.

Work on the barrier continued after Sharon lapsed into a coma and was succeeded as prime minister by a Likud Party colleague, Ehud Olmert, in 2006. Olmert vowed to continue Sharon's consolidation of Israeli territory, holding on to three major blocs of Jewish settlements in the West Bank and to "united Jerusalem." On the Palestinian side, a political earthquake occurred that year with the surprising majority vote for parliamentary candidates from the Islamist group Hamas, giving Hamas control of the Palestinian Authority. Palestinians had grown disillusioned with the showy wealth of many elected Palestinian Authority officials and by the PA's failure to deliver an independent country or any other tangible benefits. Hamas, meanwhile, won hearts and minds through its network of social services, often providing what the Palestinian Authority failed to, including education, medicine, and food rations.

Hamas's stunning victory in the parliamentary elections added another layer of uncertainty and discord to the region's political landscape. In the subsequent power-sharing agreement, Palestinian president Mahmoud Abbas of the PLO retained his post, but a Hamas leader, Ismail Haniyeh, took office as the Palestinian prime minister presiding over the Hamas-dominated parliament. Because the Hamas charter does not recognize Israel and in fact calls for its destruction, Israel and the Quartet suspended financial aid and discontinued dialogue with the Palestinian Authority. These sanctions would be lifted on three conditions: the Hamas-led PA must recognize Israel's right to exist, renounce violence, and abide by previous agreements reached between Israel and the Palestinians.

As the Palestinian economy deteriorated even further, clashes erupted between Hamas and PLO factions led by Abbas's Fatah Party. Hamas emerged victorious in fighting to control the Gaza Strip in 2007. Expelled from the territory, Abbas withdrew his recognition of Haniyeh as prime minister, and the Palestinian Authority government was dissolved. Abbas's Fatah Party consolidated its hold in the West Bank. In the winter of 2008–09, there was yet another conflict, the **Gaza War**, during which Israel sought to destroy as much of the weaponry and manpower of Hamas as possible.

In the wake of that conflict, U.S. President Barack Obama vowed to put the weight of his office behind the resolution of the Palestinian-Israeli conflict. The American leader put unprecedented political pressure on Israel, then led by conservative Likud Prime Minister Benjamin Netanyahu, to stop the construction of Jewish settlements in East Jerusalem and the West Bank. Netanyahu refused, and peace talks failed to get off the ground. Obama put the conflict on the back burner for nearly two years, then called for talks to restart—again insisting that Israel halt the settlements. Netanyahu again refused. With the U.S. political campaign for the 2012 presidential election moving into full swing, Obama again moved away from peacemaking in the volatile Middle East.

Try It Solve the Palestinian Arab Conflict

You know the problems now. Are you ready to propose a solution to the Palestinian-Israeli conflict? Try It.

Because many of these issues are sensitive, you may wish to remain anonymous if you are sharing your solution with anyone but yourself.

Many but not all of the issues you may wish to consider are the "final status" issues described earlier. You can start with those.

Palestinian statehood. Should the Palestinians have independence? Where and under what conditions? What concessions might be made in creating the country's boundaries? Hamas controls the Gaza Strip and calls for the destruction of Israel. Should Hamas participate in the creation of this state, and if so, under what conditions?

Palestinian refugees. Many Palestinians took flight in the 1948 and 1967 wars. Should they and their descendants have a "right of return" to an independent Palestine?

Jewish settlements. Many Jewish settlements have been built within the Green Line or pre-1967 borders of the West Bank. What should be done with them?

Jerusalem. Should the historic Old City of Jerusalem be divided between Israelis and Palestinians? How? Don't forget to use **Figure 6.B**. Who should control the sacred sites of the Western Wall and the Temple Mount/Noble Sanctuary (*al-Haraam al-Sharif*)?

A frustrated Mahmoud Abbas then took a risky but calculated political step. Putting aside the Americans, the Quartet, and all the conventions used since the peace process began, he went to the UN General Assembly in September 2011, appealing for the recognition of an independent Palestinian country inside the "pre-1967" (the Green Line) borders of East Jerusalem, the West Bank, and the Gaza Strip. The UN General Assembly is the same political body that, in November 1947, fulfilled the Zionist dream by portioning Palestine into
172 two countries, one Jewish and one Arab.

You have been very patient in sinking your teeth into the complexities of the Arab-Israeli conflict. I hope you will share my view that there are no shortcuts when it comes to understanding this critical problem. I hope you will read or view the news about this part of the world with much appreciation and interest. It is my hope that you will come to your own conclusions about how to solve the conflict. Are you ready to try? **See the Try It feature on this page.**

The Arab Spring

On December 17, 2010, in the Tunisian city of Sidi Bouzid, a 26-year-old vegetable vendor named Muhammad Bouazizi was shaken down for a bribe by a city inspector. Muhammad had paid the $7 to the inspectors many times before, but this time he protested. The inspector and her companions took apples from his rolling cart and confiscated his scale. The young man went to the local governor's office to lodge a complaint but was denied entry. Later in the day, he returned to the governor's office, shouting from the street, "How do you expect me to earn a living?" He poured paint thinner over his body and lit a match. Flames consumed his body.

This was the beginning of the Arab Spring.

Stirring from Stagnation

News of Bouazizi's self-immolation spread quickly throughout Tunisia and the Arab World by way of YouTube, Facebook, Twitter, and al-Jazeera television. This was an unfamiliar event, quite unlike suicide bombings or anything known even in the worst of times.

The young man's death struck a nerve among much of Tunisia's largely young and disaffected population. Tunisian resentment against the rule of the autocratic President Zine al Abidine Ben Ali had been growing throughout his more than two decades in power. Ben Ali's family was estimated to have directly or indirectly controlled half of the country's economy.

Protestors poured into the streets of Tunisian cities, bearing signs reading "We are all Muhammad Bouazizi!" and calling for Ben Ali's resignation. Demonstrators soon numbered in the hundreds of thousands, and Ben Ali began to make political concessions. It was too late. Within a month of the start of what became known as the **Jasmine Revolution**, Ben Ali's own generals turned against him. Ben Ali and his family fled to Saudi Arabia. The first domino had fallen.

Protests soon erupted in a dozen other countries across the Middle East and North Africa. The experiences in each of these countries were diverse, and it would take an entire book to describe them! Instead, I will discuss briefly some of the unique issues in the countries affected, focus mainly on the most important country affected—Egypt—and identify some patterns underlying the dissent.

The Pharaoh Falls

The next domino was the Middle East's demographic and political heavyweight, Egypt. There are several identifiable factors leading to this most important revolution in the region in half a century.

The first are two problems you know about: the general
trend of "people overpopulation," and MENA's own youth 58
bulge that is one of its consequences. Over the past decade, the 158
number of people of working age in Egypt, Jordan, Lebanon, Morocco, Syria, and Tunisia have grown by 2.7 percent, faster than any other world region except sub-Saharan Africa.

Egypt has MENA's biggest youth bulge: 60 percent of Egypt's 83 million people are under 25. In Arabic the word *shabaab* means "young people" and represents that youth bulge. It sometimes has the connotation of "bad boys." Bad or not, these boys have palpable frustrations. Without treading too much into psychosocial analysis, it is clear that the most basic desires and needs of young men that include courtship and marriage are extremely difficult to fulfill. They are expected to have steady work and to provide a home and other amenities to their brides, and this is hard given the economic challenges they face.

Unemployment and underemployment were the second set of ingredients in Egypt's Arab Spring. Egypt's annual rate of population growth is 2 percent. At this rate, Egypt's population will double in 35 years. Will Egypt double the number of jobs for its people in 35 years? That prospect is remote. In recent decades, South Korea, China, and other successful developing countries nurtured the growth of private sector employment, while Egypt and other MENA countries generally stayed with a Soviet style model of state-controlled enterprises that was unresponsive to people's needs and wants, and failed to grow
130 productive jobs. Outside the agricultural sector, 70 percent of working Egyptians are employed by the government. Private sector jobs are scarce.

Throughout MENA, the number of college graduates far exceeds positions to hire them. Egypt cranks out huge numbers of college graduates, but the quality of their education is questionable. All Egyptian college graduates are guaranteed jobs in the public sector. They are stuffed into already bloated bureaucracies where 10 people do the work that could be done by 1. Egypt and many other MENA countries suffer in this way from **underemployment** (the underutilization of labor). Their wages cannot possibly sustain a family, and most people "moonlight" with second or even third jobs. Under such circumstances, the visionary Steve Jobses of MENA are unlikely to ever be heard of, unless they flee with their brains to the West or the East.

Third, Egyptians suffered from government repression and a lack of individual freedoms. I heard the same complaints in Egypt in 1977 and on the eve of the Arab Spring. Egyptians had no control over their lives. They could not express dissatisfaction with government policies they were unhappy with, which were many. There was no real democratic political system. Every few years, Egyptians endured the sham of elections, in which an amazing 99 percent of Egyptians always chose President Hosni Mubarak's party to lead them. They had no freedom of the press. They lived under emergency law, which allowed authorities to arrest people without charges, detain prisoners indefinitely, and limit freedoms of expression and assembly. Their ranks were penetrated by the countless plainclothes secret police, the dreaded *mukhabarat* whose duty was to look for any sign of sedition. Egyptian security forces had torture down to a fine art (one of the reasons why the United States sent al-Qa'ida suspects to Egypt under the Bush and Obama programs of "extraordinary rendition").

The fourth factor in Egypt's uprising was the yawning gap between the haves and the have-nots, another problem of LDCs described in the book's introduction (see page 47). There was a veneer of prosperity. Egypt had a gleaming culture of conspicuous consumption for the relative few who could afford to consume. The gap between rich and poor grew widely in Mubarak's Egypt, and 20 percent of Egyptians were classified as poor at the time of the Arab Spring. For the wealthy and the well-connected, getting things done was easy: you just had to pay bribes and use your favors. To know someone with influence always helped. Corruption, known as *wasta*, was just the way it was. And if you didn't know anyone or could not afford to pay bribes, you could feel just like Muhammad Bouazizi did.

The results of unleashed *shaabab* power can be dramatic, as I have seen. In January 1977, Egypt's government under the soon-to-be-assassinated president Anwar Sadat removed many of the subsidies that made it possible for Egypt's poor to buy bread, cooking oil, and many other basic needs. Egyptians were furious, and the *shabaab* took to the streets in Egypt's infamous "food riots." In the country's main cities, young men broke businesses' windows, overturned and burned vehicles, and attacked government buildings. A large, loud mob assembled in Cairo's Tahrir Square and marched on a nearby train station, determined to raze it. A curious college student, I was on their periphery. Riot police with batons and shields advanced on us and began firing tear gas. Many of the demonstrators picked up the fuming canisters and threw them back at the riot police, who did not have gas masks and were themselves in as much discomfort as we. I picked up a canister that had burned out, and it read "Made in USA. For outdoor use only."

The police began firing rubber bullets, which are painful and can be fatal. The crowd continued to advance, pushing back the police. The police finally escalated to live ammunition, and when demonstrators began falling dead and wounded, the crowd evaporated. That was the end of it. Authorities imposed a two-day curfew prohibiting mass assemblies—defined as more than six persons—and answered the angry public's demands by restoring most of the subsidies. The food riots were over, but real hatred against Egyptian authorities continued to seethe, especially among the country's huge population of poor people. That hatred persisted for decades, as did Egypt's miserable conditions. Conditions just before the spring of 2011 didn't seem like anything special to most observers.

I go into such detail to underscore why I, as well as so many Middle East analysts, was taken off guard by Egypt's revolution in the Arab Spring. I expected only the *shabaab* to take to the streets in Egypt—and then more to let off steam rather than articulate meaningful dissent. I expected them to melt away when the riot police that you saw on television began using living ammunition. I expected the Mubarak regime to hold on and for Egypt's disaffected people to soldier on. I was wrong on all counts.

One of the opposition parties trying to organize in the last months of Mubarak's rule called itself Kifaya, meaning "Enough." Enough of all of this. The people of Egypt—not just the *shabaab*, but the older generations (including the *shabaab* I had seen rioting in 1977); girls and women; Muslims and Christians; poor, middle class, and even the well-to-do who wanted democracy and freedom of expression—were inspired by Tunisia's Jasmine Revolution to challenge the government—a government led by an 80-year-old man who died his hair jet black and who was grooming his son to succeed him.

Egypt's revolution played out all over the country, but its epicenter was in Cairo's Tahrir ("Liberation") Square

KHALED DESOUKI/AFP/Getty Images

• **Figure 6.37** Egyptian masses protesting against the government of Hosni Mubarak in Cairo's Tahrir Square, February 2011.

(•**Figure 6.37**). Egypt's despised secret police used deadly force even when there was no need. Do you remember the scenes of armed men mounted on camels and horses charging against the demonstrators in Tahrir Square (which became known as the "Battle of the Camels")? Rather then disperse, over a period of 18 days the protestors held their ground in Tahrir and Egypt's other public squares and outside government buildings. Without arms, the demonstrators took casualties of dead and wounded. In the end, their people power prevailed. The army, an institution that Egyptians long revered, took up their cause. Mubarak offered political concessions to Egyptians, but it was too late. As the army assumed power, he fled from Cairo, and was soon imprisoned and put on trial.

The Libyan Domino

A worse fate would befall Libya's dictator, Muammar Qaddafi. Libya was not like Egypt: with a population of just 6.4 million, Libya did not have an unmanageable youth bulge, nor was it propped up by American aid. With its considerable oil reserves, more than 3 percent of the world total, Libya should have been a prosperous country. For more than 40 years, it was instead the toy of Muammar Qaddafi, an eccentric man (what other world leader would call for Switzerland to be abolished?). Colonel Qaddafi ran the country as if it were his fiefdom, favoring and distributing money to those who were loyal to him. Tribal and ethnic politics were decisive. Qaddafi loyalists drew mainly from his own tribe on the central coast and from other tribes he favored in the center and west. King Idris I, whom Qaddafi overthrew in 1969, had favored tribes in the eastern region known as Cyrenaica, especially around the city of Benghazi. Those tribespeople never forgave Qaddafi for ousting their leader, and Qaddafi made sure they never earned any graces from him. Qaddafi also treated the ethnic Berber tribes of Libya's western mountains (whose livelihood but not ethnicity is
165 much like that of the Bedouin) as second-class citizens.

The people of Cyrenaica along with the Berbers of the west were inspired by the tide of the Arab Spring to rise up against Qaddafi, and a brief civil war emerged between them and Qaddafi loyalists. The opposition to Qaddafi was advantaged by the commitment of NATO air power, weapons, and intelligence, and it was only a matter of time until Qaddafi and his power base would be eliminated. Qaddafi was captured and 89
killed execution-style in October 2011, eight months after the uprising began.

Syria's Minority Dynasty versus the People of Syria

As in Bahrain and Libya, Syria's leaders opted for violent repression against demonstrators inspired by the Arab Spring; at least 9,000 and probably many more died. The Syrian uprising began in January 2011 after another self-immolation.

In Syria, a minority Shiite group (making up 7 percent of population) called the **Alawites** and led by president Bashir al-Assad, son of the late president Hafez al-Assad, ruled the majority (74 percent) Sunni. Resentment of Alawite rule was nothing new in Syria. Four decades of the repressive al-Assad regime rankled the Sunni majority, the Christian minority, and others. In 1982, a brutal army assault against a Sunni Muslim Brotherhood uprising in the city of Hama resulted in tens of thousands of deaths. The 2011/12 revolt involved much broader participation in more cities (•**Figure 6.38**). Arab Spring demonstrations in most of the Arab countries were loudest in the capital cities, but Syria was an exception. The fighting came to Damascus rather late.

The challenge of taking down the Alawite regime in Syria was great. The Alawites controlled the key unit of the Syrian military, so it would be difficult to secure a broad military uprising against the regime. NATO and the United States would be far less inclined to become involved than they did in Libya; the military challenge would be much greater. Western countries stepped up their sanctions against Syria. The United Nations tried to pass resolutions against Syria, but Russia and China vetoed these. The Arab League moved to oust Syria from the organization and to impose its own sanctions. Turkey was more involved, providing shelter to forces of the Free Syrian Army, made up mainly of Sunni defectors

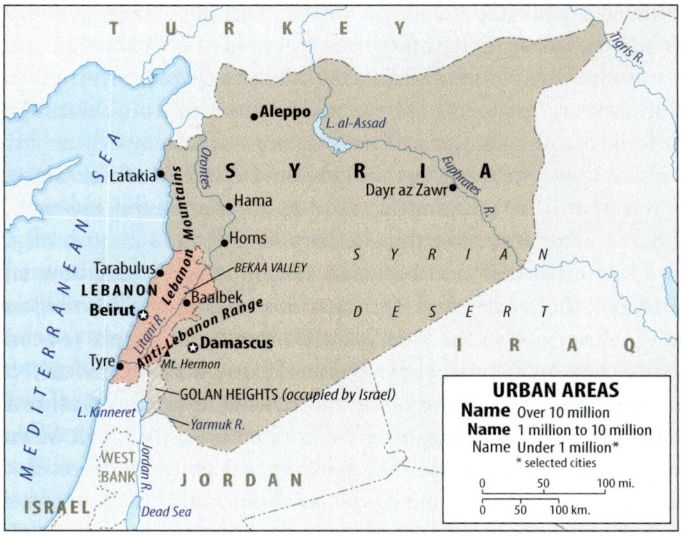

© Cengage Learning 2013

• **Figure 6.38** Principal features of Syria and Lebanon.

from the Syrian army. Israel did not become involved. Iran had much at stake: it would lose its only Arab ally should the Syrian regime fall.

Syria tried to keep foreign reporters out of the country, but a few sneaked in and were able to verify the authenticity of the video footage posted daily on YouTube. The savagery with which the Syrian army took on armed and unarmed civilians was extraordinary. Soldiers seemed to have no compassion even for children and women. It was unrestrained carnage, and there appeared to be no way to bring it to a stop.

Bahrain: A Pearl Is Crushed

In Bahrain, a small but oil-rich Gulf island shaikhdom (Bedouin monarchy) linked to eastern Saudi Arabia by a causeway, there is a Shiite majority of 70 percent that has long been ruled
174 by a Sunni monarchy (•**Figure 6.39**). Like Egyptians, Bahrain's Shiites were inspired by the Jasmine Revolution to challenge King Hamad bin Isa al-Khalifa.

Bahrain was the Arab Spring's best example of how repression can backfire by fueling resentment. The mainly Shiite protestors began by simply expressing wishes for democracy, public participation, and justice. They asked that King Khalifa take his place in a new constitutional monarchy. The monarch could have responded as a benevolent despot, yielding some ground to his subjects.

Instead, he ordered government forces to crush the rebellion with deadly force. As the casualties mounted, demonstrators' cries changed from reasonable petitions to "Death for Khalifa!" Worried that Bahrain's Shiites might overthrow the monarchy and put Bahrain into the embrace of his adversary Shiite Iran, the king called on Saudi Arabian forces to drive across the causeway to Bahrain and helped put down the uprising. 203
Government forces destroyed the demonstrations' epicenter, Pearl Square, with the country's iconic pearl sculpture in the capital city of Manama, and turned it into a traffic intersection to ensure that crowds would not gather again. King Khalifa's opponents retreated, their hopes unrequited. Below we will consider Bahrain's future, especially in its geopolitical context.

Revolt in Yemen's Mountainous Redoubt

Although located on the Arabian Peninsula (**see Figure 6.39**), Yemen has only a small bounty of oil reserves. A stunningly beautiful country graced by mountains and deserts, and a distinctive urban architecture looking like gingerbread houses,

• **Figure 6.39** Principal features of the Arabian Peninsula.

Danita Delimont/Getty Images

• **Figure 6.40** The palace of the Imam in Wadi Dhahr, Yemen.

Yemen is cursed by poverty (•**Figure 6.40**). Loyalties are to clan and tribe, and Yemen has no natural sense of cohesion as a nation state. The country was previously divided into North and South Yemen. Al-Qa'ida has a strong foothold here; Yemen is in fact the ancestral homeland of Osama bin Laden's clan.

With the Arab Spring, a widespread revolt against the longtime ruler Ali Abdullah Saleh erupted. After surviving an assassination attempt, and just before resigning office, President Saleh offered concessions to Yemeni protestors. The country's established opposition political parties pleaded with the young people to stop their protests. Yemen's *shabaab* ignored that advice. They are a force to be reckoned with; the demographic profile of Yemen's *shabaab* is very large. With a population of 23.8 million, Yemen has one of the world's highest population growth rates, a staggering 3.1 percent (with a doubling time of 22 years). Poverty, unemployment, and repression of dissent made Yemen ripe for revolt. Its geopolitical context is also discussed here.

Hallmarks of the Revolution

In the Arab Spring's aftermath, Middle East observer Fareed Zakaria recommended a book on his Sunday morning "GPS" show on CNN. It was the United Nations' *Arab Development Report 2002*. Why recommend a decade-old book? To make the point that little had changed in those years. The chapter titles that so well summarized the region's conditions in 2002 did the same a decade later: "Development not engendered is endangered ... Bridled minds, shackled potential ... As development management of stumbles, economies falter ... The curse of poverty: denying choices and opportunities, degrading lives."[8]

The Arab Spring rumbled through Tunisia, Egypt, Libya, Bahrain, Syria, and Yemen—described in some detail here—and in Sudan, Algeria, Mauritania, Morocco, Jordan, the Palestinian Territories, Lebanon, and Oman. Many of Egypt's problems that we have considered—population growth, unemployment and underemployment, lack of political representation, and oppression by authorities—were shared by these countries. Here are some of the other notable characteristics of the uprising and its aftermath in the countries of the Arab Spring:

- The revolutions were much facilitated by Facebook, Twitter, 24/7 satellite television news (notably from Qatar's *al-Jazeera* network), and other social media. Governments tried to crack down, bringing down the Internet and suspending cell phone service, but to no avail. By the time the authorities woke up, detailed instructions on how, where, and when to gather and protest in public spaces had been widely posted and tweeted.
- Although new media made their debut as tools of revolution in the Arab Spring, the more traditional geography of revolt in public spaces played a prominent role. In the traditional Middle Eastern *medina*, with the exception of the mosque there was no large public space in which large numbers of people could gather. However, spacious traffic hubs that doubled as public squares were established in the post-colonial modern Middle Eastern cities. In the Arab Spring, Cairo's Tahrir Square, Tripoli's Green Square in Libya (renamed Martyrs' Square in recognition of the lives lost there in protest), and Bahrain's Pearl Square, served as central places where social media called upon demonstrators to gather. Many of the key events leading to success (and in Bahrain's case, the failure) of the revolutions took place in these locations.
- Women had an unprecedented strong role in the Arab Spring. In the conservative Muslim societies of the Middle East and North Africa, there is strict segregation of men and women (in some countries more so than in others) in public space. But in Tahrir Square and elsewhere, that convention was thrown aside as women and men stood shoulder to shoulder in protest. In a second round of revolt against Egyptian rule—this one against the Egyptian army late in 2011—the army struck back against women protestors with particular savagery.
- The traditional Islamic classification of sacred time played an important role in the Arab Spring. As often in Muslim history, Friday noon prayers and sermons incorporated politics with prayer. Prayer time gave way to protest time, often on such a huge scale that the authorities dreaded Fridays. This traditional day of prayer was often tweeted, Facebooked, and otherwise announced as the "Day of Rage" for which demonstrators should prepare themselves.
- Remarkably, religion and militant Islamism did not otherwise feature prominently in the Arab Spring. Mainstream and radical Islamists alike were conspicuous by their absence. In Egypt, the Muslim Brotherhood, a well-organized political group that could have promoted itself and its agenda, kept a low profile. Throughout the region,

no demonstrators evoked Osama bin Laden or al-Qa'ida. In fact, the Arab Spring succeeded nonviolently in doing what al-Qa'ida aimed to accomplish through terrorism: to overthrow the autocratic, Western-oriented regimes of the Middle East and North Africa. Bin Laden must have been dismayed by what he saw in the last months of his life.

What Now? The Arab Fall

The Arab Spring transitioned into an autumn, a period of danger and uncertainty that is likely to persist for years. What will take the place of the relative stability that the region's repressive, autocratic regimes guaranteed? What will happen to the dominoes that have yet to fall?

Amidst the hope and such remarkable scenes as Tunisians turning out in October 2011 for their first-ever democratic elections, there were worrisome trends and incidents. Divisiveness and polarization based on major faith, minority sect, and tribal affiliation were of special concern.

Euphoria gave way to hard realities in Egypt. Instability reigned while the army held power in anticipation of democratic elections. Violent sectarian clashes broke out between Muslims and Coptic Christians, leaving scores of people dead. The economy tanked. At least 10 percent of Egypt's economy relies on tourism, and for awhile almost no one wanted to visit Egypt. Foreign investment dried up. The terrible economic conditions that brought people out on the street to begin with only became worse. People began talking about going back to the streets for another revolution, and by late November they were again in Tahrir Square in great numbers, this time challenging the army. Egyptians began voting in truly democratic parliamentary elections that brought members of the Muslim Brotherhood and other "modern Islamists" to power. These peaceful, democratic elections were a first for Egypt and might be an inspiration for many in the Arab World. Egypt's critical peace treaty with Israel was tested on several fronts: Egyptian rioters sacked the Israeli embassy in Cairo, skirmishes broke out between Egyptian and Israeli forces on the Sinai frontier, and Libyan weapons began making their way overland through Egypt into the Palestinian Gaza Strip.

What about Uncle Sam?

The problems between Egypt and Israel raise an intriguing geopolitical question about the Arab Spring's aftermath: How much instability in this troubled, oil-rich region will the United States tolerate without intervention? With economic aid, the United States had effectively "bought" the peace between Egypt and Israel (not counting Iraq for military expenditures, these two countries consistently rank as the top two countries in the value of American foreign aid). Throughout the region, the United States made "deals with the devil," helping to prop up authoritarian regimes like Egypt's that would maintain strategic and political interests in Middle Eastern oil, the region's chokepoints, and Israel's security, to name a few. Another interest was to prevent the ascendance of militant Islamists. Many of these one-party states appealed to the United States for aid, saying that the alternative to their rule was that al-Qa'ida or the like would find a footing. That was an effective ploy in Yemen, Egypt, and elsewhere. Certainly these are valid concerns:

- The well-organized Muslim Brotherhood will grow much stronger in post-Mubarak Egypt, but the Brotherhood does not advocate violence.
- The United States would be loath to "lose" Bahrain to the rule of the majority Shiites, who could help Iran gain traction just miles from Saudi Arabia. Bahrain is also home to the U.S. Fifth Fleet, whose ships patrol the Gulf to ensure safe passage of ships carrying oil through the Strait of Hormuz and to confront Iran if necessary.
- An anti-American Islamist regime, or Islamist insurgents, could find home in post-Qaddafi Libya, or more likely and ominously in Yemen, on Saudi Arabia's doorstep. It is difficult to envision anything other than at least an ongoing low-level war waged by the United States against al-Qa'ida affiliates in Yemen. Those predator drones will be busy.
- With its oil wealth, Saudi Arabia represents the ultimate red line in the sand. Would the United States "allow" the Saudi Arabian monarchy to fall in a popular uprising? That scenario of revolution in Saudi Arabia, perhaps beginning with an uprising by the Shiites in the Eastern Province (**•Figure 6.41**) is the ultimate nightmare for U.S. foreign policy decision makers.

You have witnessed the Arab Spring, one of the region's most momentous events in centuries. These issues will be prominent in world affairs for a long time to come.

Joe Hobbs

• Figure 6.41 Saudi Arabia's strict interpretation of Sunni Islam (Wahhabism) casts the Shiite minority of the country's Eastern Province as heretics. These Shiites have long been the country's poorest and most marginalized group. Already perceived as potentially destabilizing sympathizers with Iran, these Shiites could in fact be among the first to challenge the monarchy. These Shiites are at a market in the Eastern Province town of Qatif.

The Gulf Oil Region

GIS Helps Turn an Arabian Mirage into Reality

A futuristic city is taking shape on the desert sands of Abu Dhabi, one of the United Arab Emirates (location in Figure 6.39). Called **Masdar**, meaning "the Source" in Arabic, this city has the extraordinary goal of being "carbon-neutral." That means that unlike any other settlement in the world, Masdar will minimize the carbon dioxide it produces and neutralize or offset the small emissions it does make (•Figure 6.42). Automobiles will be banned, and the city's 45,000–50,000 residents will be required to use low-emissions public transport. Residents will work in nearby Abu Dhabi city or in the clean technology companies based in Masdar itself. Abu Dhabi's government is bearing the $20 billion price tag to build the city, which is supposed to become a magnet for domestic and foreign investment in high-tech research and development firms. Masdar's first residents are students and faculty of the Masdar Institute, an educational center staffed by professors from the Massachusetts Institute of Technology (MIT). Its graduates are meant to help meet the world's growing demand for scientists and visionaries with knowledge of green technology, green cities, "smart" cities, and renewable energy.

Energy for Masdar City is to be provided by renewable resources, especially photovoltaic energy, a solar tower, and other techniques of solar power that, as you might imagine, has abundant potential there in Arabia. Wind farms, geothermal energy, and the world's largest hydrogen power plant will contribute energy as well. Masdar will be zero-waste, with its waste products reused, recycled, or turned into yet more energy. Water will be provided by desalinization of the very salty Gulf waters, an energy-intensive process that will also use solar power. Eighty percent of the water will be recycled, and greywater—water that has been used but does not contain human waste—will be used for irrigation.

• **Figure 6.42** Artist's rendition of Masdar, the carbon-neutral city under construction in Abu Dhabi, UAE.

In 2007, I was employed by a foundation in the UAE to come up with the first draft of an environmental master plan for the country. I visited each of the seven emirates and spoke with government officials, environmental nongovernment organizations, CEOs, foreign workers, citizens, and others to identify the country's environmental challenges and then proposed specific methods for dealing with them. One of most memorable visits was with Masdar's Chief Executive Officer, Dr. Sultan al-Jaber. While we sipped tea, Dr. al-Jaber spoke at length about his vision of the futuristic city. As I prepared to leave, he said, "Come with me—I have something to show you." He led me to a conference room and opened the door. Inside, seated around a long table, were about 20 persons: engineers, urban planners, GIS technicians, and professors from a variety of fields at the Massachusetts Institute of Technology (MIT). They were on a site visit to see whether they would commit to a five-year leave from MIT, relocating to Abu Dhabi with their families to help build Masdar. You know the expression "the smartest guy in the room." This was a room full of the smartest guys in the room, and Dr. al-Jaber needed them to turn his zero-carbon vision into reality.

As I learned, GIS has an important role in Masdar's urban planning and management. Using ESRI's ArcGIS software (see page 15), Masdar's GIS crew was able to carefully consider the city site's geography of sun angles, wind patterns, street widths, building density, and building heights. "Building a city like this has never been done before," GIS consultant Shannon McElvaney said, "and GIS is proving to be an absolutely critical tool."[9] ESRI's Derek Gliddon wrote:

> Data layers contained in the geodatabase include information such as transportation, vegetation, drainage, structures, boundaries, elevation, biodiversity, buildings, and utilities, as well as terrain elevation, bathymetric data, and remotely sensed imagery. A sophisticated Web browser–based virtual city visualization and navigation tool is used to visualize the construction of the city over time. Construction managers can navigate anywhere in the city; "play" the project timeline; and identify spatiotemporal clashes, accessibility problems, and other logistical issues. On a fast-paced, high-density development, these issues are very important. ArcGIS introduced the spatial analysis and modeling necessary for the most efficient placement of facilities at the city. Water and sewage treatment plants, recycling centers, a solar farm, geothermal wells, and plantations of various tree species were placed using traditional planning principles modeled with ArcGIS. Questions—Is there enough physical space available? How much are the buildings shading each other? How much space is needed between a facility and the residents?—are modeled, and the best answer chosen through GIS.[10]

The global financial crisis has derailed the city's ambitious construction schedule. When Phase I is completed—tentatively in 2015—Masdar will open for settlement. It won't be cheap to live there. Some critics have decried Masdar as a luxury development and the ultimate gated community. Whatever you call it, there is nothing quite like Masdar.

Iraq and the United States: A Deadly Dance

Iraq (population 32.7 million) occupies a broad, irrigated plain drained by the Tigris and Euphrates rivers, together with fringing highlands in the north and deserts in the west (•**Figure 6.43**). Known since ancient times as Mesopotamia, meaning "the land between the rivers," this landscape of today's Iraq has a complex geopolitical history. It has been a cradle of civilization, a seat of empires, a target of conquerors, and in the twentieth and early twenty-first centuries, a focus of oil development and political and military contention.

A critical feature of modern Iraq's geography is its ethnic and religious composition, with three major groups present: the Shiite Arabs, mainly in the south, who make up about 60 percent of the country's population; the Sunni Arabs, about 35 percent of the total and living mainly in the center (especially in the so-called Sunni Triangle); and the Kurds, mainly in the north, most of whom are Sunni Muslims and who represent 15 to 20 percent of the total (•**Figure 6.44**; **see also the discussion of the Kurds on page 203**). There are numerous smaller minorities, including Yazidi and Sunni Turkoman.

On the map of Iraq in **Figure 6.43**, you can see a fundamental geographic disadvantage of this country. It has a tiny, 35-mile (55 km) window on the Gulf. Exports of Iraqi oil by oil tanker are hampered by poor port facilities; this narrow access to the sea does not include natural deepwater ports that would facilitate exports. One of the main reasons Iraq attached Kuwait in 1990 and Iran in 1980 was to widen Iraq's access to the Gulf.

Gulf War I Iraq and the United States became entangled in a deadly dance in 1990, when Iraqi forces invaded Kuwait—a New Jersey-sized country that has the world's sixth largest oil reserves. Claiming that it was historically part of Iraq and calling it Iraq's "19th province," Saddam Hussein's regime attempted to annex Kuwait. Saudi Arabia and other Gulf oil states seemed in danger of invasion. Worldwide opposition to Iraq's

• **Figure 6.43** Principal features of Iraq.

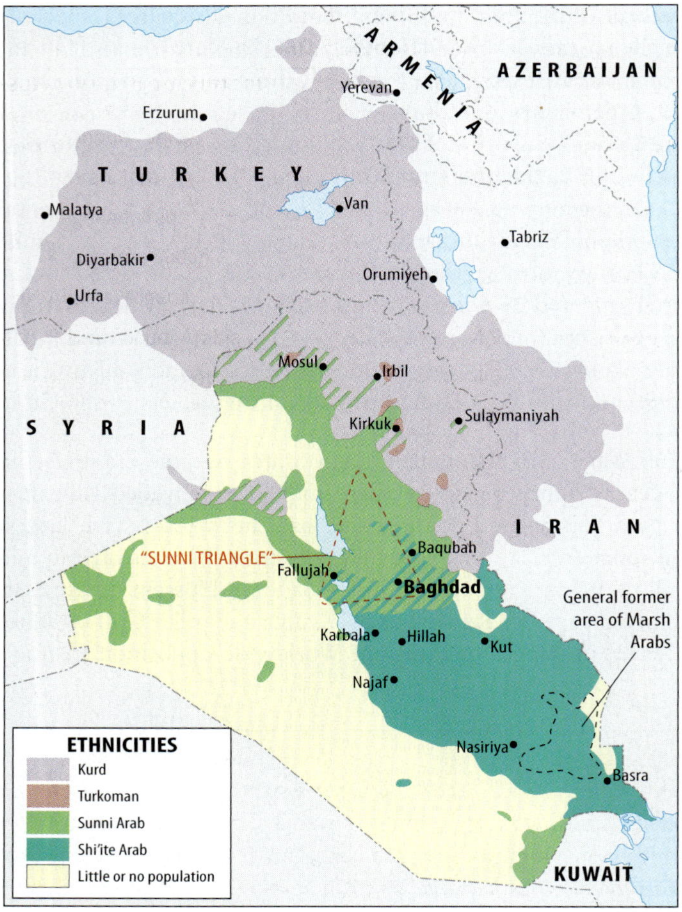

Hassan Partow, UNEP/EROS/NASA

• **Figure 6.44** Ethnicity in Iraq and neighboring regions.

aggression resulted in a sweeping embargo on Iraqi foreign trade imposed by the Security Council of the United Nations. Led by the United States, a coalition of 28 UN members (including several Arab states) staged a rapid buildup of armed forces in the Gulf region. The stage was set for what would come to be known as "the Gulf War," and later as "Gulf War I."

Beginning in January 1991, the coalition's crushing air war drove the Iraqi air force from the skies, severely damaged Iraq's infrastructure, and hammered the Iraqi forces in Kuwait and adjacent southern Iraq with intensive bombardment. In the ensuing ground war, the Iraqis were driven from Kuwait within 100 hours. Overall, an estimated 110,000 Iraqi soldiers and tens of thousands of Iraqi civilians were killed, while the technologically superior coalition forces suffered 340 combat deaths. As they withdrew from Kuwait, the Iraqis also made war on the environment, setting fire to hundreds of oil wells in Kuwait and allowing damaged wells to discharge huge amounts of crude oil into the Gulf.

A formal cease-fire to the Gulf War came in 1991, with harsh conditions imposed on Iraq by the UN Security Council. UN-sponsored economic sanctions created shortages of food and medical supplies, bringing great hardship to the country's population. Iraq agreed to pay reparations for the damage its forces had caused, and it renounced its claims to Kuwait. Saddam Hussein's surviving forces then mercilessly put down rebellions by Shiite Arabs in southern Iraq and by Kurds in the north. To deter Iraq's military and provide a safe haven for the Kurds in the north and the Shiites in the south, the United Nations established "no-fly zones," where Iraqi aircraft were prohibited from operating. But Iraqi forces had no difficulties in mounting ground campaigns against the Shiites and Kurds. Coalition forces did not intervene on behalf of the Shiites and Kurds, who felt betrayed: the coalition had encouraged them to rise up against Saddam Hussein but then abandoned them and left them at the mercy of Saddam's vengeful forces. Iraq's Shiites, both in their urban strongholds, including Basra, and in the countryside, fared especially badly.

One of history's great environmental and cultural tragedies was Iraq's systematic destruction after 1991 of the marshes adjoining the southern floodplains of the Tigris and Euphrates rivers (**•Figure 6.45; see also the ethnicity map in Figure 6.44**). This

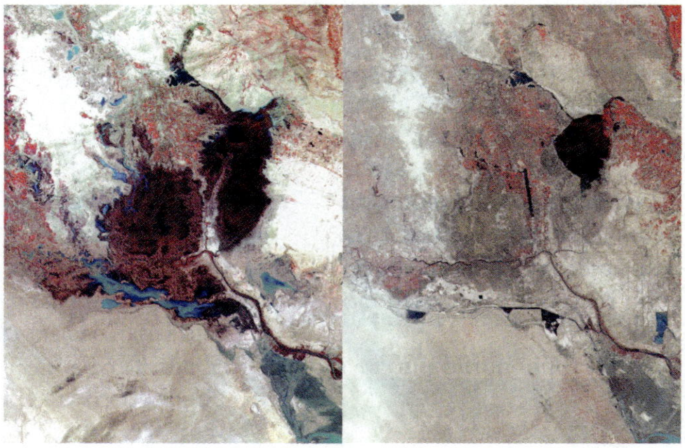

Chris Hondros/Getty Images

(a)

AP Photo/Hameed Rasheed

(b)

• **Figure 6.45** **(a)** Remote sensing of Iraq's southern marshes in 1971 and 2000. This is a "false color" image; the red is the real-life green color of marsh vegetation. Iraqi engineers built embankments to cut off the flow of Tigris and Euphrates waters into these wetlands, laying waste to a unique ecosystem and the way of life of the "Marsh Arabs." **(b)** Marsh Arab life in the marshes in 1974, before their destruction. Note the unique architecture of dwellings and community buildings, all made from marsh reeds. Some Biblical scholars have proclaimed that this marsh was the Garden of Eden, where Adam and Eve begat humankind.

area is the homeland of the ancient culture of the **Ma'adan**, or **Marsh Arabs**. These people established a way of life seen nowhere else in the Middle East, using the resources of the marshland to sustain almost all their needs. To quell any opposition these Shiite Muslims might pose, the Iraqi regime destroyed their marshy habitat and forced some 300,000 Ma'adan to flee. Engineers built massive embankments and canals to drain the marshes, and later set fire to the dry vegetation and to Ma'adan settlements.

After the Iraqi regime was taken down in 2003, the embankments were demolished, and restoration of the marshlands began. More than 50 percent of the marshes have come back to life, albeit with less biodiversity than before. The fires changed the soil chemistry so that many of the soils are now claylike and unsuited for the vegetation that grew there previously. Less than one-third of Ma'adan who fled have returned to restore their traditional way of life. As extraordinary as their adaptation to this environment was, their life in the marshes was one of poverty. The task of rebuilding communities in the marshes is an ongoing social challenge. Much damage was done to a culture whose social fabric and economy was woven into the reeds and waters of the marsh.

Throughout the 1990s, the regime of Saddam Hussein played a cat-and-mouse game with UN inspectors whose duty was to eliminate Iraq's potential to produce weapons of mass destruction. Iraq refused to allow full access by the UN inspectors. United States and allied forces responded with air strikes and more stringent economic sanctions. International support for the sanctions against Iraq weakened greatly, especially in the Arab World. President Saddam Hussein endured, much to the chagrin of his foes, and might have been able to hang on to power indefinitely if not for 9/11.

Gulf War II Immediately after the Gulf War and during the Clinton administration of 1993–2001, the reasoning in Washington was that a unified Iraq under Saddam Hussein was better than the alternative: its possible fragmentation into the three entities of a Shiite south, a Sunni center, and a Kurdish north, some or all of which might be hostile to U.S. interests in the region. Although unsavory, Saddam Hussein could be contained. Under the new Bush administration, and especially after 9/11, the downfall of Saddam Hussein became an urgent priority, ranking alongside the War on Terrorism. The Bush administration asserted that Saddam Hussein had linkages with al-Qa'ida. There was no basis for this association, however, as al-Qa'ida had an unwavering contempt for Saddam.

Publicly, Washington built the case that Iraq was harboring and continuing to develop weapons of mass destruction that might be used against the United States and its allies. The United States was able to prompt further United Nations pressure on Iraq, which finally allowed UN weapons inspectors to resume their work. When these inspections failed to turn up weapons, the United States insisted that Iraq was hiding them and that only the use of force could remove this threat. It was necessary to effect "regime change" in Baghdad. To take on Saddam Hussein, President Bush assembled what he called a "coalition of the willing," consisting mainly of Britain, Australia, Denmark, and Poland. It fell far short of the coalition assembled during the Gulf War. France, Germany, and other traditional allies of the United States opposed the impending war against Iraq, and even within the United States there was considerable dissent against what was perceived as a war of choice rather than one of necessity.

The U.S.-led ground and air assault on Iraq (dubbed **Operation Iraqi Freedom**) that began in March 2003 led quickly and with relatively few casualties to the downfall of the Iraqi regime. With about 130,000 troops, concentrated mainly in the Sunni-dominated center of the country around Baghdad, the United States settled in for a long and troubled occupation of the country. The United States handpicked what it called the Governing Council of Iraqis, broadly representative of the country's ethnic composition, and ran many of Iraq's security and other affairs with its own Coalition Provisional Authority (CPA) until June 2004, when it turned over sovereignty to an interim Iraqi government.

The United Nations, with strong U.S. support, organized a January 2005 election for a 275-member transitional national assembly. In this, Iraq's first-ever free election, Iraq's largest ethnic population unsurprisingly won the lion's share of the seats: the Shiites, who turned out in huge numbers to vote, won 48 percent for their main political party. Anticipating the Shiite success and condemning the election as an illegitimate U.S. action, most Sunnis boycotted the vote. That made it possible for the main political party of the Kurds, a smaller ethnic group, to capture another large bloc of the assembly votes—26 percent. Iraq's Sunnis ended up with just 2 percent of the assembly seats. Having been the dominant force in Iraq for decades under Saddam Hussein (who was convicted of war crimes by an Iraqi court and hanged in 2006), the Sunnis now found themselves deprived of power and increasingly angry about the American occupation. The pro-U.S. government meanwhile faced a perplexing array of challenges in cementing a constitution for the fragile country: how much autonomy the Kurds and Shiites might receive within the country, how oil revenues would be split among the geographic and ethnic regions, and what the role of Islam should be in national affairs, for example.

Sunni resistance against the United States grew. Using roadside bombs, car bombs, small arms, and rocket-propelled grenades, Sunni militants were able to inflict American casualties on an almost daily basis. Predictably, al-Qa'ida used the U.S. occupation of Iraq as a rallying cry for intensified resistance
against the United States. A well-organized insurgency offi- 184
cially linked with al-Qa'ida started up, under the leadership of Jordanian-born Abu Mus'ab al-Zarqawi. His al-Qa'ida in Iraq organization kept up a steady stream of bloody attacks on U.S. targets even after al-Zarqawi was killed in 2006. Also in the crosshairs were the Iraqis perceived to be collaborating with the Americans, particularly those who joined the new Iraqi army. Sunni insurgents relentlessly targeted Shiites in

(Left): Chris Hondros/Getty Images
(Right): AP Photo/Hameed Rasheed

• Figure 6.46 In February and June of 2006, Sunni militants—possibly belonging to al-Qa'ida in Iraq—bombed the holy site of the Imam al-Askari shrine in Samarra, north of Baghdad. Shiite Islam's holiest sites outside Saudi Arabia are in Iraq. Sunni insurgents used the tactic of bombing these sites to sow sectarian strife in Iraq.

particular, even twice bombing one of their holiest sites, the golden-domed Askariya Mosque in Samarra (**•Figure 6.46**).

These Sunni assaults on Shiite sacred space helped push Iraq into civil war. Increasingly, this internecine violence took the form of ethnic cleansing, in which one group systematically murdered members of the other. The more militant anti-Sunni and anti-U.S. Shiite factions, rallying chiefly behind a cleric named Moqtada al-Sadr and his Mahdi Army (Mahdi Militia), ascended. Moderate Shiite leaders such as Ayatollah Ali al-Sistani were no longer able to restrain a Shiite backlash. As Shiites retaliated against Sunnis, a ceaseless cycle of revenge killings ensued.

At home, the Bush administration found itself in an increasingly difficult position. Costly in lives and dollars—with thousands of soldiers and supporting civilians killed and a price tag of over $100 billion per year—the war grew more unpopular among Americans, and the president's approval rating slid. Many opposition politicians, academics, and others spoke increasingly of Iraq as "Vietnam," referring to a previous war
258 that proved untenable and unwinnable for the United States. The Bush administration faced a dilemma: "stay the course," as President Bush insisted, with about 140,000 U.S. soldiers still on the ground in Iraq, or pull out (as the United States did in Vietnam) and risk Iraq's plunge into further chaos. Fearing anarchy, the Bush administration in 2007 opted for a "surge" in U.S. troops to maintain order, even as its main ally, Britain, drew down its military presence (primarily in Shiite southern Iraq). The surge, which appears to have been an effective tactic in assisting Iraqi troops in securing the volatile Anbar province, took the number of U.S. troops in Iraq up to nearly 170,000. By this point in the war, al-Qa'ida in Iraq had become deeply unpopular among Iraqis, including the Sunnis who were intended to become al-Qa'ida's main supporters. "Moderate" Sunnis and U.S. forces joined up to push the militants out of Anbar Province.

As a candidate for the presidential office, Bush's successor Barack Obama promised a speedy withdrawal of U.S. troops from Iraq, effectively handing the country's problems back to the Iraqis. His Iraq policy helped Obama clinch the election in 2008. The troop drawdown began in 2009. In August 2010, President Obama announced that U.S. combat operations in Iraq had ceased. He declared that Operation Iraqi Freedom was over, and that Operation New Dawn was beginning. All but some American military "advisors" were withdrawn from Iraq in 2011. To keep a foothold in the unstable Gulf region, the U.S. military beefed up its forces in Kuwait. It would be especially vigilant toward Iran, which would surely seek to fill the vacuum left with the withdrawal of U.S. forces.

The Iraq War is now being viewed through historical lenses. The conflict was seen in a new light after the controversial release in 2010 of 400,000 Wikileaks documents about the war. These documents reinforced and fleshed out the major findings of the official U.S. government investigation into the 9/11 attacks, establishing the following narrative. The Iraq war was a miscalculation by the United States. The claim that there were weapons of mass destruction in Saddam Hussein's Iraq was a costly intelligence failure. The allegation of Saddam Hussein's connection with al-Qa'ida was false. In the wake of 9/11, the Gulf War squandered American lives and military resources in Iraq, while the United States lost its focus on Osama bin Laden and al-Qa'ida in Afghanistan and neighboring Pakistan. In invading Iraq, the United States lost most of the international sympathy and goodwill that grew out of 9/11. The United States inadvertently fulfilled many of Iran's ambitions in Iraq. And finally, the United States created enemies in Islamic countries who perceived that the United States was at war with Muslims. The United States inadvertently played into the hands of the conspiracies theories created by Osama bin Laden and other terrorists.

About 4,400 U.S. soldiers and perhaps 150,000 Iraqi soldiers died during the war. About 5 million Iraqis became refugees, with about half fleeing abroad (mainly to countries in the Arab World) and half displaced internally.

The Kurds

A mostly Sunni Muslim people of Indo-European origin, the Kurds are the largest of several non-Arab minorities in Iraq and the largest minority group in Turkey. They have long revolted against Iraqi and Turkish authorities. Kurds today remain the world's largest ethnic group without a country. European powers promised an independent state of Kurdistan at the end of World War I, but the governments concerned never took steps to create it. The Kurdish Question—How are the political wishes of these people to be accommodated?—has international implications because the area occupied by an estimated 30 million Kurds extends from Turkey and Iraq into Iran, Syria, and the southern Caucasus region (see Figure 6.44); an estimated 2 million more Kurds live outside the Middle East.

A book about the Kurds is appropriately titled *No Friends but the Mountains.*[11] Turkish officials have long downplayed the identity of Kurds (who make up about 20 percent of Turkey's population), often referring to them as "Mountain Turks," and even banned or restricted Kurdish-language media in Turkey until 2002. A Kurdish rebellion against Turkish rule, rising in part from the great poverty of the Kurds relative to the rest of Turkey's population, has smoldered and sometimes flamed since 1984. The main Kurdish resistance to Turkish rule comes from the **Kurdistan Workers Party**, known by its Kurdish acronym **PKK**.

Under Saddam Hussein, Iraq tried to "Arabize" its Kurds by forcing them to renounce their ethnicity and sign forms saying they were Arabs. However, after a no-fly zone and "safe haven" established after Gulf War I put them beyond Saddam Hussein's reach, Iraq's Kurds came to enjoy relative autonomy and prosperity. Even in the tumultuous years after the U.S.-led invasion in 2003, the Kurdish region of the north escaped most of the bloodshed and economic chaos characteristic of the south.

Hotspot Iran

Iran, the easternmost country of the Middle East, was known formerly as Persia (•**Figure 6.47**). From the city of Persepolis in southern Persia, Persian kings ruled a great empire in the fifth century B.C.E. Alexander the Great destroyed Persepolis, but Persia's grandeur would return. Some of the world's great architectural wonders were commissioned by Shah Abbas the Great in sixteenth-century Esfahan. Persia became a backwater in subsequent centuries and was a poor and undeveloped country a century ago. In the 1920s, a military officer of peasant origin, Reza Khan, seized control of the government and began a program to modernize Iran and free it from foreign domination. He had himself crowned as Reza Shah Pahlavi, the founder of the new Pahlavi dynasty. Influenced by the modernizing efforts of Mustafa Kemal Atatürk in neighboring Turkey, the new shah (king) introduced social and economic reforms. In 1941, his young son took the throne as Moham- 204
med Reza Shah Pahlavi. A decade later, intervention by the U.S. Central Intelligence Agency (CIA) helped to solidify the shah's absolute rule of Iran. The shah then set out to expand the program of modernization and Westernization begun by his father. In the 1970s, Iran played a leading role in raising world oil prices. Mounting oil revenues underwrote explosive industrial, urban, and social development.

• **Figure 6.47** Principal features of Iran.

In the countryside, the shah's government tried to upgrade agricultural productivity and rural standards of living through land reform and better technologies. Despite the fact that Iran was a cradle of plant and animal domestication, dry and rugged habitats make agriculture difficult over large parts of the country. Agricultural reforms generally failed, and poverty- 40
stricken families from the countryside poured into Tehran and other cities. From this devoutly Muslim group of new urbanites came much of the support for the revolution that ousted the shah in 1978–1979. Dissent also grew because the regime spent vast sums on its military, mainly for U.S.-made weapons, thus undercutting benefits to the people.

The revolution that overthrew the shah took Iran abruptly out of the American orbit. The central figure among the revolutionaries was an **ayatollah** ("sign of God") named Ruhollah Khomeini, an elderly critic of the regime whom the shah had forced into exile in 1964. From Paris, Khomeini sent repeated messages to Iran's Shiites that helped spark the revolution. A furious tide of Shiite religious sentiment rose against corruption, police heavy-handedness, modernization, Westernization, Western imperialism, the exploitation of underprivileged people, and the monarchy, which Shiite beliefs regard as illegitimate. 174
Growing numbers of people staged strikes and demonstrations in 1978, forcing the shah to flee the country and abdicate his throne. After hospitalization in the United States and life in exile in Panama and Egypt, the shah died in Cairo in 1981 and is buried there. The Ayatollah Khomeini returned in triumph to Iran in 1979. His followers seized and held 52 U.S. diplomatic personnel as hostages in Tehran for 444 days. U.S. President Jimmy Carter proved unable to resolve this crisis, which contributed to his defeat in his 1980 bid for reelection.

Under Khomeini (who died in 1989), the country became an Islamic republic governed by Shiite clerics, who included a handful of revered and powerful ayatollahs at the head of the religious establishment, and an estimated 180,000 priests called **mullahs**. Khomeini served as the Guardian Theologian, an infallible supreme leader enshrined by the principle of **divine rule by clerics** (*velayet-e-faqih*). These religious leaders continue to supervise all aspects of Iranian life and perform many functions allotted to civil servants in most countries. They base their authority on Shiite interpretations of Islam (about 90 percent of Iran's 78 million people are Shiites).

In contrast to their status during the shah's reign, when they were encouraged to take on larger public roles, women in post-revolutionary Iran have lost many freedoms. Under the new regime, "polluting" influences of Western thought and media were purged. However, recent years have seen a general softening of restrictions on personal freedom of expression, and the Internet and other media have introduced broader worldviews to Iran's youth, a very important cohort, as about 60 percent of the population is under the age of 30. Young people led an unprecedented series of prodemocracy protests in 2002. Five years later, Iranians of different generations took to the streets to protest government hikes on gasoline prices; ironically, this oil-rich country has a low oil-refining capacity and needs to import one-third of its gasoline. Iran imports more gasoline than any country in the world except for the United States. The country is fortunate to have the world's fourth largest reserves of oil, and natural gas reserves that are second only to those of Russia.

Before 2005, there was tug-of-war between the moderate and reform-minded president Muhammad Khatami and the less tolerant supreme religious leader, Ayatollah Ali Khamanei (Khomeini's successor), who has lifetime control over the military and the judicial branch of government. There was speculation that the country's younger and more Western-oriented sectors might win the day and that relations with the wider world—the United States in particular—would improve. But the winner of Iran's 2005 presidential election was Mahmoud Ahmadinejad. This hardliner outspokenly opposed American policies in the region. He goaded Israel by refusing to recognize its existence, insisting that the Holocaust never happened and calling for the country's destruction. He accelerated Iran's
183 nuclear power program. The level of rhetoric and tension between Iran and the United States escalated.

Here are a few things to keep in mind about Iran in its regional context. First, remember that Iranians (Persians) are
170 not Arabs; their ethnolinguistic roots are Indo-European. Iran (and previously Persia) is a historic enemy of most Arabs. Its big Arab neighbors Iraq and Saudi Arabia despise Iran, and vice-versa. Even the Shiite Muslims in Iraq do not naturally ally themselves with Iran; their ethnicity (Arab) seems to outweigh their faith where relations with Iran are concerned. In other contexts, however, Iran's identity as a Shiite state has important effects on its relations with Arabs. Iran claims sovereignty over the Gulf Arab country of Bahrain, where
195 70 percent of the population is Shiite. During the Arab Spring of 2011, while Saudi Arabia came to the aid of the minority Sunni monarchy, Iran appealed to the Shiites to overthrow the regime.

The Sunni/Shiite equation is reversed in Syria, with Sunnis the majority and Shiites the minority. When (Arab) Iraq attacked (Persian) Iran in 1980, the Arab nation of Syria allied itself with Iran. Why? In part because, as you have seen, Syria's Sunni Muslim majority is ruled by a family practicing an offshoot of the Shiite faith known as Alawites. Iran is strongly allied with the militant Shiite Arabs of Hizbullah, in Lebanon, and is able to support Hizbullah logistically because of strong ties with Lebanon's neighbor Syria. 183
Iran also supports Hamas in the West Bank, not because its members are Shiites (they are Palestinian Sunni Muslims), but because Hamas is set on Israel's destruction, which Iran would like to see. 183

Turkey: Where East Meets West

Like Iran, Turkey is a demographic heavyweight (population 74 million people) and has long had an enormously important presence in the region (•**Figure 6.48**). The Turks who organized the Ottoman Empire beginning in the fourteenth century originated as pastoral nomads from central Asia, where Turkic cultures still prevail. From the sixteenth 142
century to the nineteenth, their empire, based in what is now Turkey, was an important power. Modern Turkey was created from the wreckage of the old empire after World War I. Its founder, Mustafa Kemal Atatürk, was determined to Westernize the country, raise its standard of living, and make it a strong and respected national state. He inaugurated social and political reforms designed to break the hold of traditional Islam and open the way for modernization and Turkish nationalism.

Islam had been the state religion under the Ottoman Empire, but Atatürk divorced church from state, and to this day Turkey is the only Muslim Middle Eastern country to officially separate them. The wearing of the red cap called the *fez*, an important symbol of piety under the Ottoman caliphs, was prohibited, and state-supported secular schools replaced the all-pervasive religious schools. To facilitate public education and remove further traces of Muslim culture, Latin characters

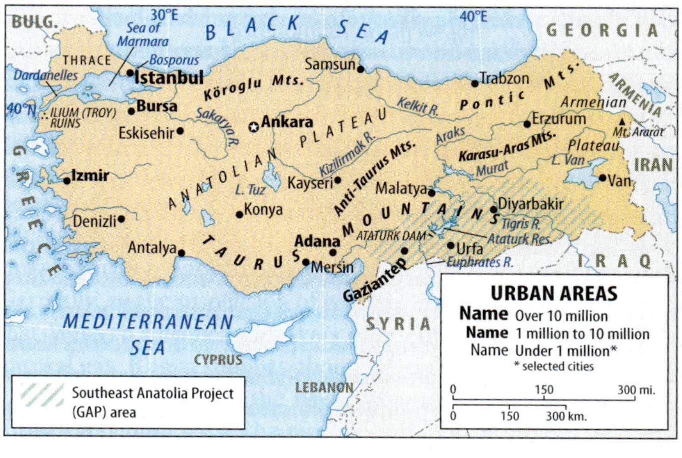

• **Figure 6.48** Principal features of Turkey.

replaced the Arabic script. Slavery and polygamy were outlawed, and women were given full citizenship. Legal codes based on those of Western nations replaced Islamic law, and forms of democratic representative government were instituted, although Atatürk ruled in a dictatorial fashion.

Turkey's economic standing is near the top of the LDCs, close to MDC status. Not long ago, poverty and prominently rural characteristics marked Turkey. But the country embarked on a course of change that is modernizing agriculture, expanding industry, and raising standards of living. Its economic growth has been impressive. The value of output from manufacturing (mainly textiles, agricultural processing, cement manufacturing, metal industries, and automobiles) is greater than that from agriculture, which is considerable in this fertile, well-watered country. Some of Europe's most recognizable brand names, like Bosch, Renault, and Fiat, have factories in Turkey. Turkey does produce some oil.

Early in the 1990s, Turkey began an impressive \$32 billion agricultural effort called the **Southeast Anatolia Project** (known by its Turkish acronym **GAP**). Its aim is to double the country's irrigable farmland, converting Turkey's semiarid southeastern into the "breadbasket of the Middle East," especially with surplus production of cotton, wheat, and vegetables for export. The centerpiece of this massive irrigation project is the Atatürk Dam on the Euphrates River. About 20 other dams on the Turkish Euphrates and Tigris are also part of the ongoing project to provide hydroelectricity as well as water (•**Figures 6.48** and **6.49**).

Like most great dam projects, including the Aswan Dam described earlier, The GAP is controversial. It is inundating historic settlements and priceless archaeological treasures and is compelling people to move. It is being developed in the part of the country where a poor and restive Kurdish population
202 is seeking development, human rights, and autonomy. Many Kurds do not have confidence in government promises that the project will boost their prospects. Tens of thousands of Kurds are being relocated from areas inundated by reservoirs, and there are fears that local people will succumb to waterborne diseases. In addition, Turkey's Tigris and Euphrates waters flow downstream into neighboring states, raising serious questions about downstream water allocation and quality. 182

Joe Hobbs

• **Figure 6.49** Hasankeyf, on the Tigris River. Important archaeological sites are to be inundated by a reservoir that will form behind a dam that is part of the Southeast Anatolia Project.

Is Turkey European?

Turkey is an "in-between" country. Economically, it is near the line between MDCs and LDCs. Culturally, it is between traditional Islamic and secular European ways of living. Part of Turkey—the section called Thrace—is actually in Europe, and Turkey as a whole aspires to become more European. It has been a member of the North Atlantic Treaty Organization (NATO) since 1952. It is an associate member of the European Union and is a candidate for full membership. The European Union, however, insists that it would first like to see Turkey become more European politically, for example, by diminishing the army's influence in government and by improving its human rights record. Turkey insists it has made huge strides 87
toward satisfying the minimal democratic and human rights norms for EU member countries. It has abolished the death penalty and eased restrictions it imposed on the cultural expressions of its minority Kurdish population.

At expansion talks, EU leaders nevertheless have repeatedly rejected Turkey's bid and would not promise a date for membership negotiations. They argued that Turkey has still not gone far enough with democratic and human rights reforms. The Europeans probably fear the costs of absorbing Turkey's mainly low-income population into the union—a fear that is justified, considering the EU's struggles with its own lower income countries like Greece, which have threatened the very existence of the EU. 88

Many Turks feel that their membership in the European Union was denied because they are Muslims. It is indeed possible that behind the scenes, EU decision makers rejected EU membership for Turkey because they feared that Islamist terrorists would easily make their way from an EU Turkey into the heart of continental Europe. The European Union's rejection of Turkey contributes to the sentiment in the Islamic world that the West is waging a systematic campaign against Muslims. Turkish public opinion has become decidedly more anti-European, even as the prospect of joining the troubled European Union, with its sovereign debt crises and other problems, looks increasingly unattractive. A Turkey shunned by Europe may seek alliances eastward in Russia, India, and China.

Or Is Turkey a Great Middle Eastern Power?

Turkey's standing in Middle Eastern affairs has changed dramatically in recent years. Not long ago, diplomatically and otherwise, the Arabs reviled Turkey, just as they had for centuries (remember that Turks are not Arabs and that The Ottoman Empire that subjugated much of the Middle East from the sixteenth century until World War I was Turkish). Turkey's closest ally in the Middle East was Israel, which is also reviled in much of the Arab World. That association made Turkey deeply unpopular among Arabs.

A single incident in 2010 dramatically rearranged Turkey's regional priorities and its standing in the Arab Middle East. Israel has a blockade that prohibits any unauthorized access to the Palestinian-controlled Gaza Strip. On May 10, 2010, a so-called "Gaza Freedom Flotilla" of ships carrying humanitarian aid and construction materials to the Gaza Strip was confronted at sea by Israeli commandoes. The Israeli forces boarded several of these ships belonging to the Free Gaza Movement and the Turkish Foundation for Human Rights and Freedoms and Humanitarian Relief. An altercation ensued, and a number of Turkish civilians as well as Israeli commandoes were killed. Turkey condemned Israel and demanded that Israeli apologize, pay compensation, and prosecute those responsible for the Turkish fatalities. Israel refused.

The incident accelerated an ongoing reshuffling of the Middle East's political landscape. Turkey's Prime Minister, Recep Tayyip Erdoğan, denounced the raid as Israeli "state terrorism." Turkey recalled its ambassador to Israel, suspended military trade with Israel, and otherwise indicated that Israel was no longer an ally. Turkey had long promoted a national policy of "zero problems with the neighbors," but Prime Minister Erdoğan now portrayed Israel as a regional problem and offered himself and Turkey as champions of Arab causes. Amidst the tumult of the Arab Spring in 2011, Turkey appealed to the United States to recognize it, rather than Israel, as the best ally of the United States in the Middle East. Prime Minister Erdoğan visited the countries most transformed by the Arab Spring—Egypt, Tunisia, and Libya—and vowed to assist in their reformations. He spoke of a new axis of power, based on Turkey's alliance with Egypt, the Arab World's strongest country. He indicated that Turkey would use military force if needed to advance the interests of its Arab allies and of itself—for example, in preventing the Republic of Cyprus from working with Israel to tap oil and gas reserves in the Mediterranean Sea. 102

Domestically as well, Prime Minister Erdoğan shook up Turkey. After coming to power in 2003, he began restraining the powers of the military. Turkey's constitution invests the military with the responsibility of protecting the country's secular government, and the army has used its powers to overthrow four elected governments since 1960. Erdoğan's growing power and popularity blunted any prospect that he would be ousted by the military. Turkey's economy improved markedly under his rule—most notably, it became a powerhouse of automobile manufacturing—and trade, transport, and communication with the Arab countries accelerated. As the leader of secular Turkey, Prime Minister Erdoğan has had to restrain his personal convictions. Prior to his election, he had described himself as a political Islamist and an advocate of Islamic law or *shari'a*. Some analysts worry that he will violate Turkey's rigid separation of church and state, and advocate Islamist domestic and foreign policies.

This concludes our journey through the Middle East and North Africa. We now turn eastward into the realm of South and East Asia.

summary

- Misleading stereotypes about the environment and people of the Middle East and North Africa are common, as people outside the region often associate the region solely with military conflict and terrorism.
- The region has bestowed on humanity a rich legacy of ancient civilizations and the three great monotheistic faiths of Judaism, Christianity, and Islam.
- Middle Easterners include Jews, Arabs, Turks, Persians, Berbers, people of sub-Saharan African origin, and other ethnic groups who practice a wide variety of ancient and modern livelihoods.
- Arabs are the largest ethnic group in the Middle East and North Africa, and there are also large populations of ethnic Turks, Persians (Iranians), and Kurds. Islam is by far the largest religion. Jews live almost exclusively in Israel, and there are minority Christian populations in several countries.
- Population growth rates in the region are moderate to high. Oil wealth is concentrated in a handful of countries, and as a whole, this is a developing region.
- The Middle East has served as a pivotal global crossroads, linking Asia, Europe, Africa, and the Mediterranean Sea with the Indian Ocean. These countries have historically been unwilling hosts to occupiers and empires originating far beyond their borders.
- The margins of this region are occupied by oceans, high mountains, and deserts. The land is composed mainly of arid and semiarid plains and plateaus, together with considerable areas of rugged mountains and isolated "seas" of sand.
- Aridity dominates the environment, with at least three-fourths of the region receiving less than 10 inches (25 cm) of yearly precipitation. Great river systems and freshwater aquifers have sustained large human populations.
- Many of the plants and animals on which the world's agriculture depends were first domesticated in the Middle East.
- The Middle Eastern "ecological trilogy" consists of peasant villagers, pastoral nomads, and city dwellers. The relationships among them have been mainly symbiotic and peaceful, but city dwellers have often dominated the relationship, and both pastoral nomads and urbanites have sometimes preyed on the villagers, who are the trilogy's cornerstone.
- About two-thirds of the world's oil is here, making this one of the world's most vital economic and strategic regions.
- Since World War II, several international crises and wars have been precipitated by events in the Middle East. Strong outside powers depend heavily on this region for their current and future industrial needs. Unimpeded access to Persian-Arabian Gulf oil is one of the pillars of U.S. foreign policy.

- The region of the Middle East and North Africa is characterized by a high number of chokepoints, strategic waterways that may be shut off by force, triggering conflict and economic disruption.
- Oil pipelines in the Middle East are routed both to shorten sea tanker voyages and to reduce the threat to sea tanker traffic through chokepoints, but are themselves vulnerable to disruption.
- Access to freshwater is a major problem in relations between Turkey and its downstream neighbors, Egypt and its upstream neighbors, and Israel and its Palestinian, Jordanian, and Syrian neighbors.
- Al-Qa'ida and affiliated Islamist terrorist groups aim to drive the United States and its allied governments from the region and to replace them with an Islamic empire. Al-Qa'ida is an apocalyptic group that seeks to inflict mass casualties on its enemies, particularly on Americans in their home country.
- The United Nations Partition Plan of 1947 attempted a two-state solution to the dilemma Britain had created by promising land to both Arabs and Jews in Palestine. It envisioned geographically fragmented states, making each side feel vulnerable to the other. War prevented the plan's implementation. The Arab-Israeli conflict has continued from that time, with Israel gaining more territory and the indigenous Palestinians failing to acquire a country of their own. Principal obstacles to peace between Israelis and Palestinians are the status of Jerusalem, the potential return of Palestinian refugees, the future of Jewish settlements in the Occupied Territories, and Israeli construction of a security barrier that penetrates into portions of the Palestinian West Bank.
- Israel's eastern neighbor, Jordan, has a majority Palestinian population.
- The Arab Spring of 2011 was a wave of revolutions against autocratic regimes across the region, notably in Tunisia, Egypt, Libya, Syria, Bahrain, and Yemen.
- Masdar is a "carbon-neutral" city under construction in Abu Dhabi, one of the United Arab Emirates.
- Two-thirds of the world's proven petroleum reserves are concentrated in a few countries that ring the Persian-Arabian Gulf. Saudi Arabia controls more than one-fifth of the world's oil.
- Under Saddam Hussein, Iraq squandered its oil wealth on military misadventures, including invasions of Iran and Kuwait. The Kuwait invasion led to an enormous U.S.-led counterattack that devastated the country's infrastructure and subjected it to years of economic sanctions. The U.S.-led invasion of Iraq in 2003 resulted in Saddam Hussein's downfall and death, and an uncertain future for this ethnically diverse country (with large Shiite Arab, Sunni Arab, and Kurdish populations) engaged in civil war.
- Iran's oil revenues were used to modernize and Westernize the nation during the Pahlavi dynasty. Rapid social change and uneven economic benefits precipitated a revolutionary Islamic movement, which forced the shah to abdicate and flee in 1979. This theocratic state has had difficult relations with the United States, most recently over its alleged nuclear weapons program. Iran's young population is agitating for change, but the ruling clerics are resisting.
- Turkey was formerly the seat of power for the Islamic Ottoman Empire. It is a secular nation with membership in NATO and possible future membership in the European Union. It is blessed with freshwater resources that it has mustered with the hope of creating the "breadbasket of the Middle East."

Key Terms + Concepts

Afro-Asiatic language family (p. 169)
- Berber subfamily (p. 170)
- Semitic subfamily (p. 169)

al-Aqsa *Intifada* (p. 191)
al-Aqsa Mosque (p. 174)
al-Qa'ida (p. 184)
Alawite (p. 194)
Allah (p. 174)
anti-Semitism (p. 172)
Arab (p. 169)
Arab-Israeli War of 1956 (p. 177)
Ashkenazi Jews (p. 188)
ayatollah (p. 203)
bazaar (p. 167)
Camp David Accords (p. 187)
Carter Doctrine (p. 178)
cataract (p. 181)
chokepoint (p. 177)
Church of the Holy Sepulcher (p. 172)
Copts (p. 173)
Crusades (p. 172)
Diaspora (p. 172)
divine rule by clerics (p. 204)
Dome of the Rock (p. 172)
downstream countries (p. 180)
drought avoidance (p. 160)
drought endurance (p. 160)
Druze (p. 188)
energy crisis (p. 176)
environmental perception (p. 164)
Exodus (p. 170)
"facts on the ground" (p. 189)
Falashas (p. 188)
Fertile Crescent (p. 169)
final status issues (p. 190)
First Temple (p. 170)
Foundation Stone (p. 170)
Gaza-Jericho Accord (p. 190)
ghurba (p. 188)
Green Line (p. 191)
Gulf War (p. 178)
Hamas (p. 183)
Hebrews (p. 170)
Hizbullah (p. 183)
hydropolitics (p. 180)
International Atomic Energy Agency (IAEA) (p. 183)
Intifada (p. 189)
Islamic law (*sharia*) (p. 175)
Islamists (p. 183)
Israelites (p. 170)
Jasmine Revolution (p. 192)
Jewish settlements (p. 189)
Jews (p. 169)
jihad (p. 184)
Judaea and Samaria (p. 189)
Ka'aba (p. 173)
Kurdistan Workers Party (PKK) (p. 203)
Kurds (p. 170)
Labor Party (p. 189)
Land for Peace (p. 189)
Law of Return (p. 188)
Likud Party (p. 189)
Ma'adan (Marsh Arabs) (p. 201)
Maronites (p. 173)
Masdar (p. 198)
medina (p. 167)
megacity (p. 168)
Middle Eastern ecological trilogy (p. 164)
- pastoral nomads (p. 165)
- urbanites (p. 165)
- villagers (p. 164)

mullah (p. 204)
"New Rome" (p. 172)
Night Journey (p. 174)
Nile Water Agreement (p. 181)
Noble Sanctuary (*al-Haraam ash-Shariif*) (p. 172)
Occupied Territories (p. 187)
oil embargo (p. 178)
Operation Iraqi Freedom (p. 201)
Organization of Petroleum Exporting Countries (OPEC) (p. 176)
Oslo I Accord (p. 190)
Oslo II Accord (p. 190)
Palestine Liberation Organization (PLO) (p. 189)
Palestinian Authority (PA) (p. 190)
Palestinians (p. 186)
People of the Book (p. 173)
Persian (p. 170)
Pillars of Islam (p. 174)
  almsgiving (p. 174)
  fasting (p. 174)
  pilgrimage to Mecca (hajj) (p. 175)
  prayer (p. 174)
  profession of faith (p. 174)
pre-1967 borders (p. 186)
Promised Land (p. 170)
Red–Dead Peace Conduit (p. 180)
remittances (p. 177)
right of return (p. 186)
risk minimization (p. 165)
"road map for peace" (p. 191)
salinization (p. 181)
sand sea (p. 160)
Second Temple (p. 170)
security fence (p. 191)
sedentarization (p. 165)
Sephardic Jews (p. 188)
Shia, or Shiite (p. 174)
Six-Day War (p. 186)
Southeast Anatolia Project (GAP) (p. 182)
Suez Crisis (p. 177)
Sunni (p. 174)
suq (p. 167)
Temple Mount (p. 172)
terrorism (p. 183)
Turks (p. 169)
two-state solution (p. 186)
underemployment (p. 193)
United Nations Resolutions 242 and 338 (p. 189)
upstream countries (p. 180)
usufruct (p. 167)
War on Terror (p. 183)
weapons of mass destruction (WMD)(p. 178)
Western Wall (p. 170)
youth bulge (p. 158)
zero-sum game (p. 180)
Zion (p. 172)
Zionist movement (p. 172)
1948–1949 Arab-Israeli war (p. 186)
1973 Arab-Israeli war (p. 187)

Review Questions

1. What countries constitute the Middle East and North Africa? Which are the three most populous? Which encourage and discourage population growth?
2. What are the major climatic patterns of the Middle East? Where are the principal mountains, deserts, rivers, and areas of high rainfall?
3. Why can this region be described as a culture hearth? What major ideas, commodities, and cultures originated there?
4. What are the major ethnic groups, and in which countries are they found? What is an Arab? A Jew? A Turk? A Kurd? A Persian? A Muslim?
5. What are the principal beliefs and historical geographic milestones of Jews, Christians, and Muslims?
6. What is the difference between Shiite and Sunni Islam?
7. Where is oil concentrated in this region?
8. What is the Carter Doctrine?
9. What options does the United States weigh in dealing with Iran's suspected nuclear weapons development program?
10. What are upstream and downstream countries, and which are usually the more powerful? What are the exceptions to this rule?
11. What are Hizbullah, Hamas, and al-Qa'ida?
12. What precipitated the various conflicts between Israel and its Arab neighbors? How did each of these conflicts rearrange the political map?
13. What lands did the peace process of the Oslo Accords yield to Palestinian control? What is the current situation in those areas? Who controls the Gaza Strip and the West Bank?
14. According to Lester Brown, what three steps can be taken to avert conflict over Nile waters?
15. What are the pros and cons of Egypt's Aswan High Dam?
16. What were essential elements of the Arab Spring, especially in Egypt, and why is the United States wary about the "Arab Fall?"
17. What are the objectives of the MASDAR project?
18. What happened when Saddam Hussein's troops invaded Kuwait? What happened when they invaded Iran? What critical decisions must the United States make in Iraq?
19. What is the ethnic composition of Iraq? How did that consideration cause the United States to avoid invading the country in 1991?
20. Who are the Kurds, and where do they live?
21. What were some of the grievances against the Pahlavi dynasty that led to revolution in Iran? What changes has the Islamic regime brought to the country?
22. What three options has the United States weighed in response to Iran's nuclear weapons program?
23. In what ways is Turkey an "in-between" country, and how is Turkey's role in the MENA changing?

Notes

1. Ulrich Deil, Heike Culmsee and Mohamed Berriane. 1980. "Sacred Groves in Morocco: A Society's Conservation of Nature for Spiritual Reasons." *Silva Carelica 49*, 185–201.
2. Lester Brown, "When the Nile Runs Dry." *New York Times*, June 2, 2011, p. A23.
3. Fouad Ibrahim and Barbara Ibrahim, *Egypt: An Economic Geography*. (London: I.B. Tauris, 1983).
4. From the U.S. State Department website at http://www.state.gov/s/ct/rls/crt/2000/2419.htm.
5. Quoted in Daniel Dombey, James Blitz and Najmeh Bozorgmehr, "Atomic Agitation." *Financial Times*, January 8 2010, p. 7.
6. Rohan Gunaratna, *Inside al Qaeda: Global Network of Terror* (New York: Cambridge University Press, 2002), p. 188.
7. Suleiman Abu Ghaith, "Al-Qa'ida Statement (October 10, 2001)." In *Anti-American Terrorism and the Middle East,* Barry Rubin and Judith Colp Rubin, eds. (Oxford University Press, 2002), p. 251.
8. United Nations Development Programme Report 2002. http://www.arab-hdr.org/publications/other/ahdr/ahdr2002e.pdf
9. Derek Gliddon, "Building an Oasis in the Desert." *Arc News, 31*(3) Fall 2009, 26–27.
10. Ibid.
11. John Bullock, *No Friends but the Mountains* (London: Oxford University Press, 1993),

Online Resources

CourseMate: Make the most of your study time by accessing everything you need to succeed in one place. Read your textbook, take notes, review flashcards, watch videos, complete activities, take practice quizzes, and more—online with CourseMate. Log in at **www.cengagebrain.com**.

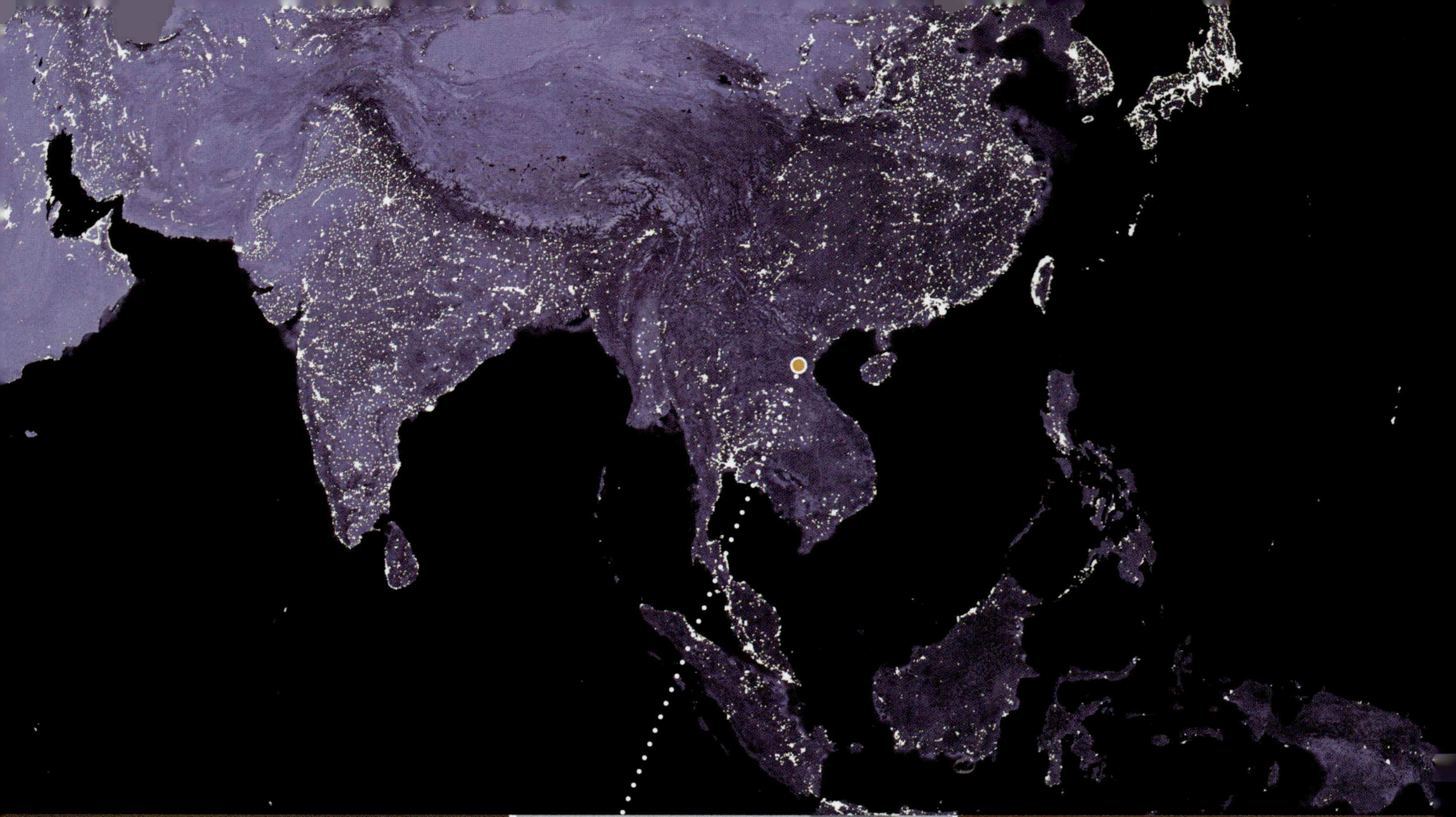

"A book holds a house of gold."

—CHINESE PROVERB

Southern Asia at night. Note especially how the Himalaya separate the densely-populated Indian subcontinent from the Tibetan wilderness; the belts of urban-industrial activity in southeast China and in Japan; and the contrast between the two Koreas. Left: boats on the New Year pilgrimage to the Perfume Pagoda, Vietnam.

C. Mayhew & R. Simmon (NASA/GSFC), NOAA/ NGDC, DMSP Digital Archive

Joe Hobbs

7 South and East Asia

South and East Asia make up a great triangle that runs from Afghanistan in the west to Japan in the northeast and the island of New Guinea in the southeast (•Figure 7.1). In this land area of about 8 million sq mi (21 million sq km—a little more than twice the size of the United States [•Figure 7.2]), or less than one-quarter of the earth's landmass—lives more than half of the world's population (•Figure 7.3). The cultural makeup of these 3.8 billion people is as diverse and expressive as the physical environments they inhabit. The world's highest mountain peaks, some of the longest rivers, and the earth's most highly transformed and densely settled river plains are the settings for some of the oldest civilizations and the most modern economies. Just as this region has played a major role in human development over thousands of years, it is a major force in world affairs in the twenty-first century. Some experts say this will be known as the **Asian Century** for the ways in which the region will reshape the globe.

chapter outline

chapter objectives

This chapter should enable you to

- Recognize the role of seasonal monsoon winds and rains in people's livelihoods and perceptions.
- Appreciate China and India as the demographic giants and surging economies of early-twenty-first-century Asia.
- Learn how unique spatial and spiritual considerations have influenced the traditional layout of Asian settlements.
- Know the basic beliefs of Hinduism, Buddhism, Confucianism, and Daoism.
- Become familiar with the pros and cons of the Green Revolution.
- Understand the geopolitical dimensions of the tensions between India and Pakistan, North Korea and the West, and Islamists and governments in Pakistan and Indonesia.
- Consider how Afghanistan has become known as the "Graveyard of Empires."
- Recognize the tensions between Muslims and Hindus in South Asia that often give rise to violent conflict.
- Appreciate how the partition of India created lasting problems in political relations, resource use and allocation, and industrial development.
- See that food production has kept pace with enormous population growth in India.
- Recognize the serious consequences of political insurgencies in Kashmir, India, Pakistan, and Sri Lanka.
- Appreciate the catastrophic reach of tsunamis and some of the region's other natural hazards.
- Understand some of the major political obstacles and environmental consequences associated with large dams and other water management projects.
- Recognize China as a land empire in which a single ethnic group added vast peripheral areas to its eastern core.
- Appreciate the constraints imposed by China's Communist system on personal, political, and economic freedoms, and the recent substantial easing of those restrictions.
- Balance the pros and cons of China's hugely ambitious plans to transform its natural environment through dams, canals, forest restoration, and other projects.
- Consider the economic discrepancies between China's rural and urban populations and the forces behind large-scale rural-to-urban migration.
- Understand the regional and geopolitical risks of Taiwan's aspirations for, and identity independent of, China.
- Appreciate how countries burdened by a lack of natural resources can become prosperous by marshaling their human resources.
- Understand the unique challenges associated with a postindustrial society that has a prosperous and aging population.
- See how a country's geographic location can make it a target of division and conquest.
- Appreciate how different political and economic systems can produce dramatically different results for almost identical peoples.

The countries of eastern Eurasia were at one time referred to as the Orient, from a Latin word meaning "facing the east" or "facing the sunrise." (The West was then known as the Occident, derived from a word meaning "facing the sunset.") But this terminology reflected the view from Europe and became associated with Western stereotypes and misconceptions about Asia. These terms have fallen out of favor, but in older literature you are sure to read about the Orient.

7.1 Area and Population

South and East Asia encompass the following subregions, which in some geography books are classified as separate regions:

- East Asia, which includes Japan, North and South Korea, China, Mongolia, Taiwan, and many near-shore islands
- South Asia, which includes Afghanistan, Pakistan, India, Sri Lanka, Bangladesh, the mountain nations of Bhutan and Nepal, and the island country of the Maldives
- Southeast Asia, which includes both the peninsula jutting out from the southeast corner of the Asian continent, on which are located Myanmar (Burma), Thailand, Laos, Cambodia, Vietnam, Malaysia, and Singapore, and the island world that rings this peninsula, which includes the countries of Indonesia, the Philippines, Brunei, and Timor-Leste (•**Table 7.1**; see also Figure 7.1)

Demographic Heavyweights of South and East Asia

Perhaps the most critical feature in the geography of South and East Asia is that this region is home to about 55 percent of the world's population (Figure 7.3; see also Figures 3.13 and 3.14). Two countries alone, India and China, together have 2.6 billion people, or 37 percent of the world total. Their extraordinary demographic weight reflects thousands of years of human occupation of productive agricultural landscapes, combined with twentieth-century advances in health technology that reduced death rates. The highest population densities are found in the areas of most abundant precipitation or surface water and with the most productive soils; rivers and coastal plains are especially densely settled. Rugged mountains; almost waterless deserts; and until recently, tropical rain forest vegetation have limited populations elsewhere.

Some of the world's highest urban densities are found in the cities of Hong Kong, Macao, and Singapore, but in Mongolia and Nepal population densities are extremely low. Singapore is one of the few nations to claim a 100 percent urban

• **Figure 7.1** South and East Asia.

population; in contrast, only 17 percent of Nepal's and 15 percent of Sri Lanka's population are urban. A mere 1 percent of Mongolia is arable, whereas farmers in India and Bangladesh cultivate more than 50 percent of their lands.

Population Growth Patterns

It is not possible to generalize about population growth in South and East Asia. Growth rates range from about zero percent growth (ZPG) in Japan to a very high 3.1 percent per year in Timor-Leste. The region reflects the general worldwide trends of poorer countries having higher growth rates and wealthier countries having lower growth rates. The region is demographically and otherwise recognizable as one of mainly less developed countries (LDCs), with some outstanding exceptions. Japan is a classic postindustrial country and, like Italy, must worry about a declining, aging population in which young people will have to shoulder an increasing economic
burden to support their elders. As discussed later in this chap- 69
ter, there are other strong economies in the region. 233

Migration is discussed in many parts of this chapter, but a brief note on the Philippines has a place here. The country is unique in the world for its strong tradition of sending family members to work abroad. One quarter of the country's workforce works abroad. There is an acronym for them: **OFWs**, for **overseas foreign workers**. They send as much money home as possible, so much that these remittances account for more than 10 percent of the Philippines' gross domestic product. There is a price to pay for the many years that a Filipino mother or father spends working abroad: loneliness; isolation; the pain of being far from one's children; broken families; and ultimately, divorce are common.

One of the big anomalies in Asia's population equation is China. As a developing country with a per capita GNI PPP of just $6,890, China might be expected to have a rather high

Table 7.1 East and South Asia: Basic Data

Political Unit	Area (thousands; sq mi)	Area (thousands; sq km)	Estimated Population (millions)	Estimated Population Density (sq mi)	Estimated Population Density (sq km)	Annual Rate of Natural Increase (%)	Human Development Index	Urban Population (%)	Per Capita GNI PPP ($US)
South Asia	**1,984.4**	**5,139.6**	**1,653.7**	**833**	**322**	**1.6**	**0.535**	**29**	**2,990**
Afghanistan	251.8	652.2	32.4	129	50	2.8	0.398	22	860
Bangladesh	55.6	144.0	150.7	2,710	1,046	1.5	0.500	25	1,550
Bhutan	18.1	46.9	0.7	39	15	1.4	0.522	33	5,290
India	1,269.3	3,287.5	1,241.3	978	378	1.5	0.547	29	3,280
Maldives	0.1	0.3	0.3	3,000	1,158	2.0	0.661	35	5,250
Nepal	56.8	147.1	30.5	537	207	1.9	0.458	17	1,180
Pakistan	307.4	796.2	176.9	575	222	2.1	0.504	35	2,680
Sri Lanka	25.3	65.5	20.9	826	319	1.3	0.691	15	4,720
Southeast Asia	**1,735.3**	**4,494.4**	**602.0**	**347**	**134**	**1.3**	**0.619**	**42**	**4,930**
Brunei	2.2	5.7	0.4	182	70	1.3	0.838	72	N/A
Cambodia	69.9	181.0	14.7	210	81	1.8	0.523	20	1,820
Indonesia	735.4	1,904.7	238.2	324	125	1.3	0.617	43	3,720
Laos	91.4	236.7	6.3	69	27	2.2	0.524	27	2,200
Malaysia	127.3	329.7	28.9	227	88	1.6	0.761	64	13,710
Myanmar	261.2	676.5	54.0	207	80	1.2	0.483	31	N/A
Philippines	115.8	299.9	95.7	826	319	1.9	0.644	63	3,540
Singapore	0.2	0.5	5.2	26,000	10,039	0.5	0.866	100	49,780
Thailand	198.1	513.1	69.5	351	135	0.5	0.682	31	7,640
Timor-Leste	5.7	14.8	1.2	211	81	3.1	0.495	22	4,730
Vietnam	128.1	331.8	87.9	686	265	1.0	0.593	30	2,790
East Asia	**4,545.6**	**11,773.1**	**1,573.5**	**346**	**134**	**0.4**	**0.711**	**55**	**9,760**
China	3,696.1	9,572.9	1,345.9	364	141	0.5	0.687	50	6,890
Japan	145.9	377.9	128.1	878	339	−0.1	0.901	86	33,400
Korea, North	46.5	120.4	24.5	527	203	0.6	N/A	60	N/A
Korea, South	38.3	99.2	49.0	1,279	494	0.4	0.897	82	27,240
Mongolia	604.8	1,566.4	2.8	5	2	1.8	0.653	61	3,330
Taiwan	14.0	36.3	23.2	1,657	640	0.1	N/A	78	N/A
Summary Total	**8,265.3**	**21,407.1**	**3,829.2**	**463**	**179**	**1.0**	**0.621**	**42**	**6,080**

Sources: World Population Data Sheet, Population Reference Bureau, 2011; Human Development Report, United Nations, 2011; World Factbook, CIA, 2011.

population growth rate like India, which has a similar economic ranking. But in the twentieth century, China's population was so alarmingly large and rapidly growing that its Communist leaders decided to restrain its growth drastically. China adopted a coercive "one-child policy" that has brought its growth rate to just 0.5 percent per year, representing a doubling time of 140 years (see Geographic Spotlight, page 216). That is a remarkable turnabout from the situation in 1964, when China had a growth rate of 3.2 percent and a doubling time of 22 years.

To reduce its poverty, India has made great strides in reducing its population growth, but not in China's coercive fashion. India is projected to overtake China as the world's most populous country in 2040, when it will have an estimated 1.5 billion people. A wild card in India's population deck is HIV/AIDS. The number of Indians infected with HIV was estimated at 2.4 million in 2011. The potential for a **pandemic** in the world's second largest country is uncertain, and much will depend on how India's government chooses to fight the disease. About 80 percent of India's HIV infections are spread by heterosexual contact, and public health authorities have so far been shy to mount a strong campaign about sexual ethics and practices. In China, the number of people infected was estimated at 740,000 in 2011.

Some Chinese parents want to produce a son if their first child was a girl, and many girls are given to orphanages (China is the world's leading source country of adopted children, about 90 percent of them girls). Female feticide is increasingly associated

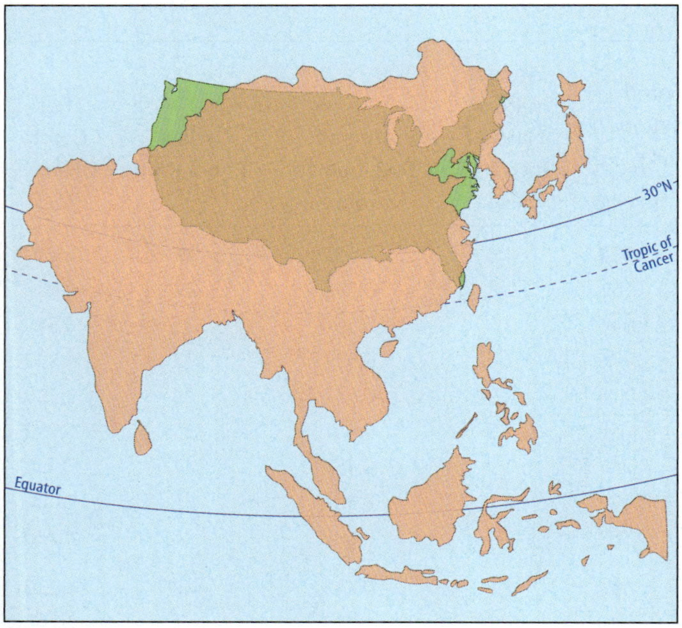

• **Figure 7.2** South and East Asia compared in latitude and area with the conterminous United States.

with the desire to have a son. Couples often insist on an ultrasound test to determine the baby's gender, and about 90 percent of girls detected in this way are aborted. In some provinces where the practice is widespread, three boys are born for every two girls. Countrywide, there are 118 male newborns for every 100 females. As in India, this practice is beginning to have seri-
250 ous repercussions, reflected in a shortage of eligible brides. Rural women have been abducted and trafficked as virtual slaves to
145 other parts of the country, where they become brides. On the positive side, the relative shortage of women in China has helped enhance their status, particularly in urban areas, where they are also becoming more influential in the workplace.

7.2 Physical Geography and Human Adaptations

The physical geography of South and East Asia may be described broadly as consisting of three concentric arcs, or crescents, of land: an inner western arc of high mountains, plateaus, and basins; a middle arc of lower mountains, hill lands, river plains, and basins; and an outer eastern arc of islands and seas (see Figure 7.1).

The inner arc includes the world's highest mountain ranges, interspersed with plateaus and basins. In the south, the great wall of the Himalaya, Karakoram, and Hindu Kush Mountains overlooks the north of the Indian subcontinent. In the north, the Altai, Tien Shan, Pamir, and other towering ranges separate this region of Asia from the countries of central Asia. Between these mountain walls lie the sparsely inhabited Tibetan Plateau, at over 15,000 feet (c. 4,500 m) in average elevation, and the dry, thinly populated basins and plateaus of Xinjiang (Sinkiang) and Mongolia.

The middle arc, between the western inner highland and the sea, is occupied by river floodplains and deltas bordered and separated by hills and low mountains. The major features of this area are the immense alluvial plain of northern India, built up through ages of meandering and deposition by the Indus, Ganges, and Brahmaputra Rivers; the hilly uplands of peninsular India, geologically an ancient plateau; the plains of the Irrawaddy, Chao Praya (Menam), Mekong, and Red Rivers in the Indochinese Peninsula of Southeast Asia, together with bordering hills and mountains; the uplands and densely settled small alluvial plains of southern China; the broad alluvial plains along the middle and lower Chang Jiang (Yangtze River) in central China and the mountain-girded Red Basin on the upper Chang Jiang; the large delta plain of the Huang He (Yellow River) and its tributaries in northern China, backed by loess-covered hilly uplands; and the broad central plain of northeastern China (Manchuria), almost enclosed by mountains.

The outer arc is an offshore fringe of thousands of islands, mostly grouped in great **archipelagoes** (clusters of islands) bordering the mainland. On these islands, high interior mountains (including volcanoes) are flanked by coastal plains, where most of the people live. Most of the islands are in three major archipelagoes: the East Indies, the Philippines, and Japan. Sri Lanka, Taiwan, and Hainan are large, densely populated islands outside these archipelagoes. Between the archipelagoes and the mainland lie the East and South China seas and, to the north, the Sea of Japan (known in Korea as the East Sea). At the southwest, the Indian peninsula projects southward between two immense arms of the Indian Ocean: the Bay of Bengal and the Arabian Sea.

Climate and Vegetation

South and East Asia are characterized generally by a warm, well-watered climate, but there are nine distinct types of climate in the region: tropical rain forest, tropical savanna, humid subtropical, warm humid continental, cold humid continental, desert, steppe, subarctic, and undifferentiated highland (•Figure 7.4, left; see also Figure 2.5). These are generally associated with predictable patterns of vegetation, which have been heavily modified by millennia of human use (Figure 7.4, right; see also Figure 3.4). Remember that discussion of biomes and vegetation considers what ought to be growing in the absence of people. From this baseline, we will look at what people have done to the living landscape (see Perspectives from the Field, page 218).

The tropical rain forest climate zone and vegetation types are typically found within 5° to 10° of the equator and are characteristic of most parts of the East Indies, the Philippines, and the Malay Peninsula. Precipitation is spread throughout the year so that each month has considerable rain. The amount of precipitation is at least 30 to 40 inches (c. 75 to 100 cm) and often 100 inches (c. 250 cm). Average temperatures vary only slightly from month to month; Singapore, for example, has a difference of only 3°F between the warmest and coolest months. High temperatures prevail year-round, although some relief is afforded by a drop of 10°F to 25°F (6°C to 14°C) at

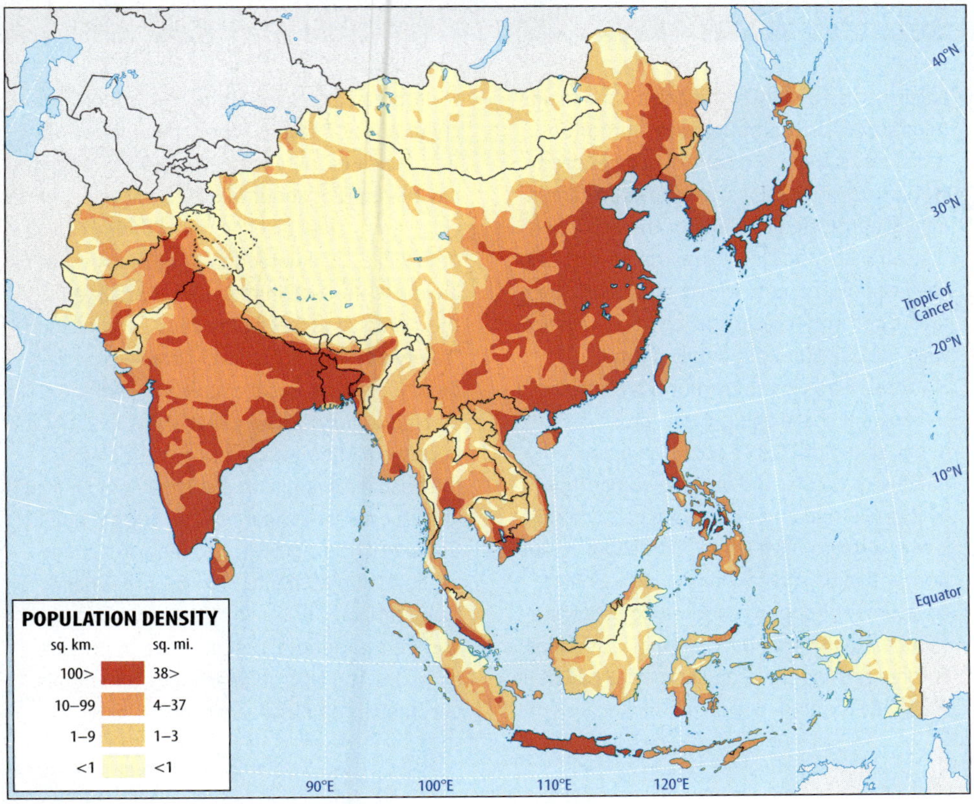

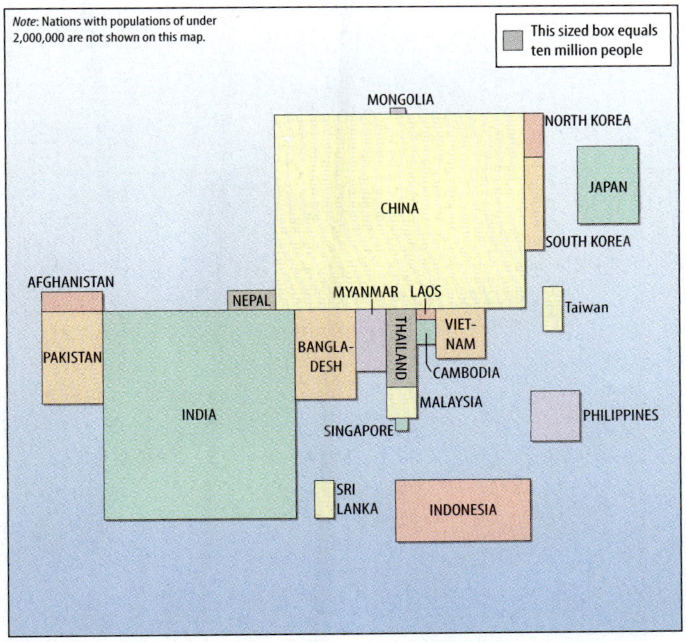

• **Figure 7.3** Population distribution (top) and population cartogram (bottom) of South and East Asia.

night, and sea breezes refresh coastal areas. Winter in climates like this, it can be said, "comes only at night."

The tropical savanna climate, like the tropical rain forest climate, is characterized by high temperatures year-round but is generally found in areas farther from the equator, and the average temperatures vary more from month to month. The savanna has a better-defined dry season, lasting in some areas as long as six or eight months each year, creating problems for agriculture. Tropical savanna climates occur in southern and central India, in most of the Indochinese Peninsula, and on eastern Java and the smaller islands to the east.

In Asia, the characteristic natural vegetation associated with this climate is a deciduous forest of trees smaller than those of the tropical rain forest. Tall, coarse grasses such as those found on African and Latin American savannas grow only in limited areas.

The humid subtropical climate zone prevails in southern China, the southern half of Japan, much of northern India, and a number of other countries. It is characterized by warm to hot summers, mild or cool winters with some frost, and a frost-free season lasting 200 days or longer. The annual precipitation of 30 to 50 inches (c. 75 to 125 cm) or more is well distributed throughout the year, although monsoonal trends produce a dry season in some areas. The natural vegetation is a mixture of evergreen hardwoods, deciduous hardwoods, and conifers.

The humid continental climates characterize the northern part of eastern China, most of Korea, and northern Japan. They are marked by warm to hot summers, cold winters with snow, a frost-free period of 100 to 200 days, and less precipitation than the humid subtropical climate. Most areas experience a dry season in winter. The main natural vegetation is a mixture of broadleaf deciduous trees and conifers, although prairie grasses probably formed the original cover in parts of northern China. Extreme northeastern China and northern Mongolia have a subarctic climate associated with coniferous trees.

Steppe and desert climates are found in Xinjiang and Mongolia and in parts of western India, Pakistan, and Afghanistan. Due to limitations on agriculture, along with very high mountains, they are the region's least inhabited climate zones. The undifferentiated

GEOGRAPHIC SPOTLIGHT

China's One-Child Policy

China's population of 1.3 billion represents nearly one-fifth of the world's people and is increasing by 6.7 million each year. Population is a very serious matter for a less developed country whose area of about 3.7 million square miles (9.4 million sq km) is only slightly smaller than the United States but whose inhabitants outnumber the U.S. population by about five to one. But a glimpse at China's age structure diagram, with its wide middle ages but tapering youth, reveals that China is one of the world's great success stories in terms of population management (•**Figure 7.A**).

The drive for smaller families was motivated by the fact that China's population was surging while per capita food output was increasing at only a very modest rate. In 1976, China's Communist regime instituted one of the most stringent and most controversial programs of birth control ever attempted. It culminated in the **one-child campaign,** which aims to limit the number of children per married couple to one (there are some exceptions, for example, for ethnic minorities) (•**Figure 7.B**). The government tracks individual family birthing patterns—especially in the cities—and maintains surveillance through local authorities. It dispenses free birth control devices, free sterilization operations, free hospital care in delivery, free medical care for the child, free education for the child, an extra month's salary each year for parents who comply, and other favors and preferences to induce compliance with the one-child-per-family norm.

Those who violate the norm are subjected to constant social and political pressure, denial of the privileges accorded one-child families, pay cuts, and fines. A woman who becomes pregnant after having had one child is pressured to have a free state-supplied abortion, with a paid vacation provided. China's government hopes to continue to decrease childbearing so that population will stabilize before reaching the projection of 1.5 billion by 2025. The number may be higher because of widespread disregard of the "one couple, one child" policy. Reasons for noncompliance vary. In some cases, parents feel wealthy enough to pay the fines of about $6,200 or more per extra child. Many rural dwellers think their extra children will escape detection by authorities, who are most effective in the cities. Anticipating their years as elders, some couples want the "safety net" of additional children—especially boys. If a young urban couple has just one child and it is a girl, it is likely that she will marry and become, by Chinese tradition, such a part of the husband's family that she may not be available to care for her parents as they get older. China has traditionally used the son's family as the "old folks' home." On the other hand, rising property prices in cities may affect this preference for boys: parents have to provide a married son with an apartment, and that is becoming prohibitively expensive in cities.

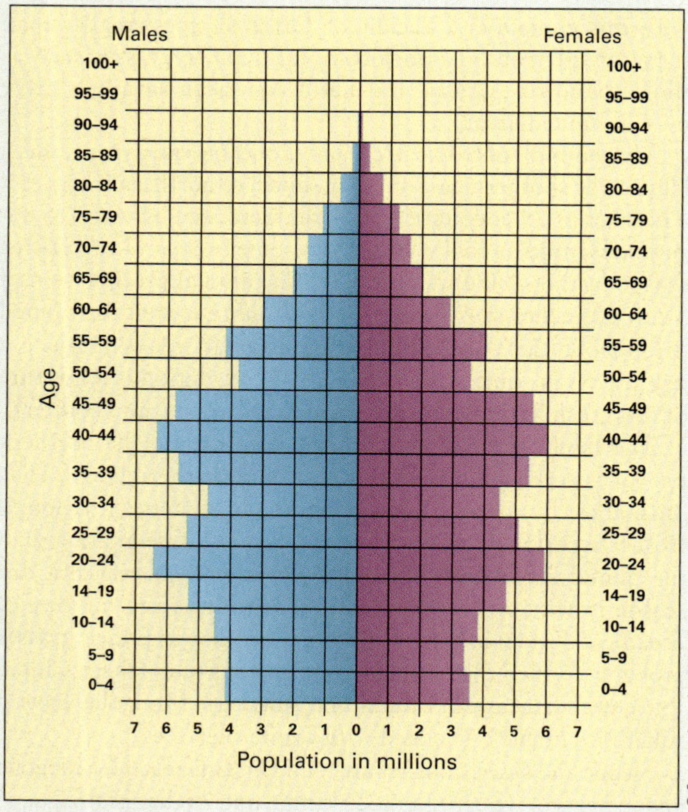

• **Figure 7.A** The age structure diagram of China (2011) reveals the country's success in slowing population growth.

Joe Hobbs

• **Figure 7.B** Strong disincentives mean one child is the norm for Chinese couples.

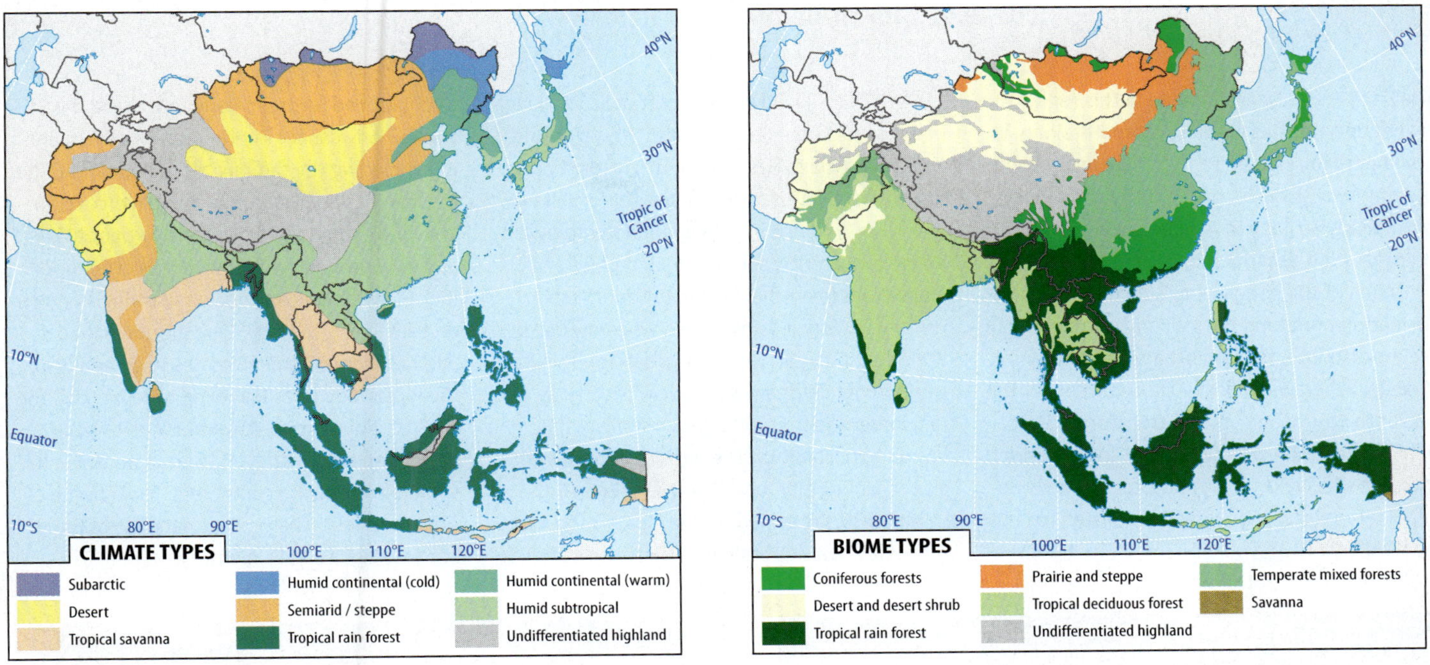

• **Figure 7.4** Climates (left) and biomes (right) of South and East Asia.
Sources: (left) Based on Rand McNally's Classroom Atlas, 2003. (right) Based on World Wildlife Fund ecoregions data, 1999.

highland climate is most extensive in the Tibetan Highlands and adjoining mountain areas. It is made up of a broad range of montane microclimates that limit plant growth and vary from place to place according to elevation, solar exposure, and latitude.

The Monsoons

The **monsoons** are the prevailing sea-to-land and land-to-sea winds that are the dominant climatic concern for people living in this world region (•**Figure 7.5**). They play a significant role in both wet and dry environments and are especially influential in the coastal plains and lowlands of South Asia, the peninsula and islands of Southeast Asia, and the eastern third of China. The ways in which the sea and land absorb **insolation** (heat from the sun) are different, causing the instability in air masses that creates monsoonal winds. If coastal waters and the adjacent coastline receive about equal amounts of warmth from the sun, the land becomes warm or hot much more quickly than the seawater. As a result, air over the land begins to become unstable and to rise. This ascending air creates a low-pressure attraction for the air masses over the water, and marine winds begin to blow toward the land. In Asia, this is the **summer monsoon**, characterized by high humidity, moist air, and generally predictable rains. Because of the moisture carried by such air masses, these wind shifts are the sources of major rainfall in the late spring, summer, and early fall seasons. Agriculture and patterns of human activity are keyed to these incoming rains.

If there is even a little bit of elevation on the landscape—hills or mountain flanks—the moist air is driven higher, where it cools and releases even more rain in the pattern known as *orographic precipitation*. If there is a lot of elevation—as there is with the giant mountain walls of the Himalaya and Karakoram—the orographic effect of wringing out the moisture-laden atmosphere is impressive, with tremendous rainfall and snowfall events.

In the **winter monsoon**, the land loses its relative warmth while the sea and coastal waters stay warm longer. As a result, the wind shifts and air masses begin to flow from the inland areas of Asia toward the sea. However, because there is hardly any moisture in the source areas over land for the winter

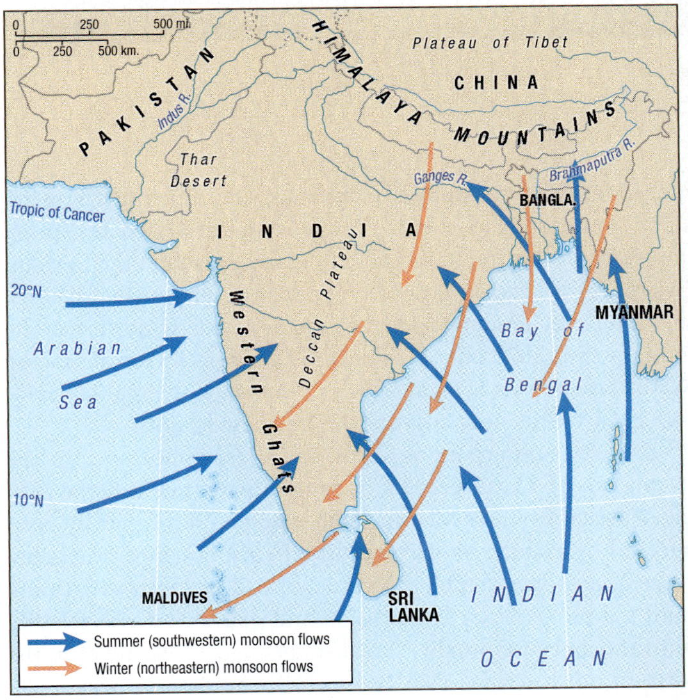

• **Figure 7.5** This map illustrates how the monsoons work, with prevailing winds blowing from the sea (wet summer monsoons) and toward the sea (dry winter monsoons).

PERSPECTIVES FROM THE FIELD The Lost City of Bangkok

The pace of landscape change in this region in recent decades has been truly breathtaking. We could perhaps say the same is true in all world regions except Antarctica, but the change is particularly striking in Asia because of the deforestation described on page 251.

Much of the geographer's focus is on transformation of natural or rural environments, but we are also interested in urban environmental change. When we look at Asian cities, we often see both nature and culture transformed. Nowhere else in the world is nature so prominent in urban life. Next time you watch a television reporter in Karachi or New Delhi, listen for the hooded crows.

As a child, I lived in the desert of Saudi Arabia. In 1968, my parents took my brother, sister, and me across Asia to Bangkok. In this verdant paradise of a city, it was love at first sight for this animal-loving youngster. Behind our hotel sprawled a vast lawn literally seething with reptiles and amphibians (many flew home to Saudi Arabia with me; others did not make it when my dad saw the sides of my carry-on bag pulsating). This was a green, wet city: canals ran every which way, and rain forest and fruit trees grew wherever they had an opportunity. It seemed like a forest with an enchanting city tucked into it (•Figure 7.C).

In 2006, I returned to Bangkok for the first time. How was it possible this was the same city? I looked for trees—any trees. And the canals—any canals. There were some remaining, including the famous "floating market," but most were gone, cut down and paved over in Thailand's rapid modernization, urbanization, and economic growth. Bangkok is often gridlocked, with some of the worse traffic anywhere on Earth (•Figure 7.D). One morning I had a 9:00 A.M. appointment across town and took the locals' advice to leave the hotel at 6:00 for the 30-minute drive. Any later and it would have been a two- or even three-hour drive. In some of those long commutes, I had plenty of time to reflect: The verdant paradise of Bangkok was just a child's memory.

© Dean Conger/CORBIS

• Figure 7.C Bangkok, 1968

Joe Hobbs

• Figure 7.D Bangkok, today

monsoon, little moisture is picked up, and much less rain results. A monsoon climate, overall, is characterized by spring and summer precipitation and a long dry season in the low-sun (winter) cycle. Japan, however, experiences an unusual pattern of winter precipitation related to the monsoon wind shifts. The winter winds blow out of China and East Asia, sweep across the Sea of Japan (East Sea), and pick up moisture, dropping heavy, wet snow on the west coast of the Japanese islands.

The wet summer monsoon is of critical importance to the livelihoods of a large portion of humanity, particularly the people of India. By early May in India, the land has typically been parched by months of drought, and the rising temperatures become almost insufferable. No thought can be given to farming until the rains come. Anticipation and discomfort rise steadily until the monsoon finally "breaks," bringing great relief in the form of much needed rain and lower temperatures (Figure 7.6). Sometimes the onset of the rain comes with great ferocity. India and Bangladesh have experienced particularly heavy early monsoonal rains for the past three decades, with significant loss of

Joe Hobbs

• Figure 7.6 Torrential monsoon rains are regular and usually welcome features of land and life in much of the region. This is a flooded road in south central Sri Lanka.

INSIGHTS | Shifting Cultivation

Many Asian farmers, particularly in tropical rain forest regions, practice **shifting cultivation** (also known as **swidden cultivation** or **slash-and-burn cultivation**), or land rotation between crop and fallow years (•**Figure 7.E**). The procedure begins with people clearing forest cover and then burning the dried vegetation during the dry season. Ash from the burn provides short-lived fertility to the soil in which farmers plant their subsistence crops. Most tropical soils lose their natural fertility quickly when they are cropped, and after two or three harvests must be rested for several years (often 10 to 15 years) before they will again produce a crop. During this **fallow** period, the land reverts to wild vegetation.

The system of shifting agriculture is widespread in the tropics. Although it is not very productive, it does provide a bare living for people too poor to afford fertilizer or farm machinery. Traditionally, this system has minimized soil erosion because most of the land is not in cultivation at any given time. Unfortunately, there has been a recent widespread trend for farmers to reduce the length of the fallow period in the cycle of shifting cultivation or to abandon it altogether. This is a direct result of a surging growth in population: with more mouths to feed, in many places there is simply not enough land to provide the "luxury" of letting it lie fallow.

The tragic consequence is that short-term overuse of the land is eroding the soil. On the **lateritic soils** characteristic of many tropical regions, excessive exposure of the land to the sun's ultraviolet radiation and to repeated wetting and drying turns the ground to a bricklike, iron-rich substance called **laterite** or **plinthite**, which is essentially impossible to cultivate. In fact, this transformed soil is so tough that people used it in the construction of some of Asia's greatest architecture, including the temples of Angkor Wat in Cambodia (•**Figure 7.F**). When the best available lands are exhausted, farmers are often marginalized to places unsuitable for farming, such as semiarid lands or slopes that are too steep to till without causing disastrous
erosion. Such dilemmas of land use are typical 47
of the world's tropics from Indonesia to West Africa and Amazonia.

345, 384

Joe Hobbs

• **Figure 7.E** Shifting cultivation (swidden or slash-and-burn cultivation) is widespread in the world's tropical regions and can be sustained as long as the land is given sufficient time to recover. Part of the cycle is harvesting crop residue (here, cornstalks) to use as fuel or fodder. This Hmong woman's backbreaking work is part of the greater fuelwood crisis scenario described on page 48.

Joe Hobbs

• **Figure 7.F** Laterite or plinthite, created by the transformation of tropical soils, is hard enough to provide foundations for buildings. This is one of the outstanding examples: the twelfth-century temple complex in the Cambodia site of Angkor Wat.

life. Pakistan has had fewer events of this kind, but unusually heavy, sustained monsoon rains in 2010 caused catastrophic flooding all along the Indus floodplain. Inevitably, some scien-
32 tific fingers point to global climate change as the culprit.

Agricultural Adaptations

The wet and warm climates associated with the monsoons might be expected to offer excellent opportunities for agriculture. However, the high temperatures and heavy rains promote rapid leaching of mineral nutrients and decomposition of organic matter, so many soils are infertile. The lush vegetation cover in these low-latitude areas is deceiving because most of the system's nutrient matter is in the vegetation itself rather than in the
underlying soil. When Asian farmers clear the tree cover, they 384
often find that the local soils will not support more than one or two poor harvests (see Insights, this page). Agricultural and other land uses in South and East Asia are depicted in •**Figure 7.7**.

The Importance of Rice

Agriculture in South and East Asia—with the principal exception of Japan—is characterized by the steady input of arduous manual labor. Known as **intensive subsistence agriculture**, it is built around the growing of cereals, especially rice, which is the premier staple of South and East Asia and the crop of choice in areas with adequate rainfall or where irrigation waters are available (see the areas of rice subsistence in Figure 7.7).

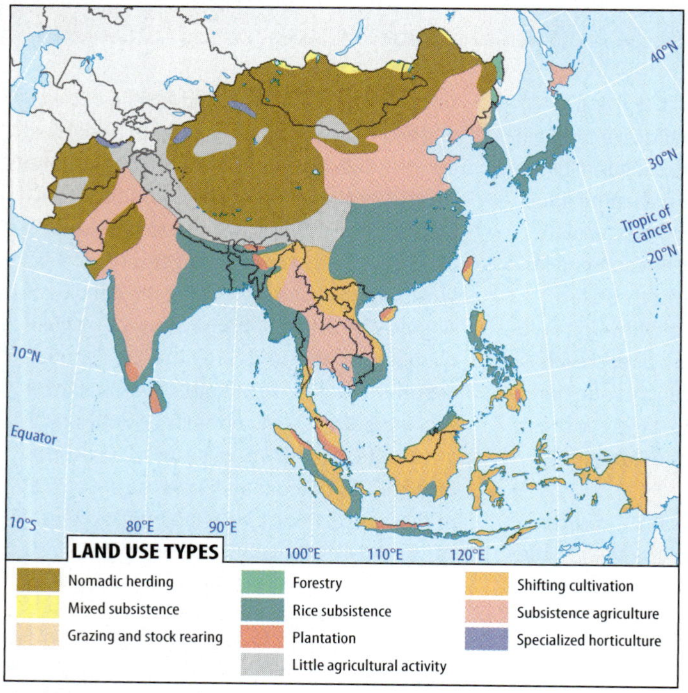

• **Figure 7.7** Land use in Asia.

Source: The Oxford School Atlas edited by Patrick Wiegand (OUP, 1997), copyright © Oxford University Press. Used by permission of Oxford University Press.

Joe Hobbs

• **Figure 7.8** Harvesting wet rice from an irrigated terrace in Ha Giang, Vietnam

Farmers use organic fertilizers from both their animals and their own latrines and irrigate to the extent that they can or need to. Where natural conditions are not suitable for irrigated rice, grains such as wheat, barley, soybeans, millet, sorghum, or corn (maize) are raised. Where possible, farmers grow cash crops for local markets. Large-scale, commercial plantation agriculture is common in Southeast Asia, the subregion of South and East Asia in which cash cropping has had the strongest role.

Shifting cultivation represents a precarious adaptation to local soil and climate conditions, capable of sustaining only small populations for brief periods in particular locales. At the other end of the spectrum of human adaptation to Asian lands is **wet rice cultivation**, capable of producing two or three crops per year and sustaining large populations over long periods of time. Wet rice agriculture takes place both in lowland floodplains and on carefully constructed upland **terraces** requiring enormous inputs of human labor (the term **paddy**, also spelled **padi**, may refer to the rice itself or the field in which it is grown). Terraces have been built almost everywhere in Asia where there is a combination of sloping land, available irrigation water, and productive soil (•Figure 7.8). The dramatic stair-step terraces of the island of Luzon in the Philippines are the most famous, but dramatic cultural landscapes of this type can be found all across the region.

Terraces have the important function of preventing or dramatically slowing soil erosion, as the terrace walls allow heavy rainfall to run off without carrying topsoil away. Doing what the Chinese call "teaching water," people use canals and pipes to direct irrigation water from a source to the terrace, which acts like a kind of dam to hold the water in place. Wet rice is actually immersed in standing water for extended periods, creating an environment unfavorable to many crop pests. The productivity of wet rice agriculture is sometimes exceptionally high, supporting as many as 500 people per square mile (200/sq km) on the volcanic soils of Java and Bali in Indonesia. This extraordinary output is a contributing factor to the high densities and overall populations of some of the Asian countries and regions, including Java. One anthropologist has observed that most of Asia's rice is eaten within walking distance of where it is grown.[1]

Are There Correlations between Agriculture and Culture?

Anthropologists have observed some unique cultural adaptations to opportunities for terraced wet rice. Group welfare takes precedence over individual interests, for example. Individual farmers cannot cultivate at will but must work with the community to ensure that crops are staggered over time. This allows water to be shared fairly and efficiently, and gives some fields the chance to dry out while others are irrigated. This agricultural system also prevents the use of tractors and other mechanical technology. Over much of Asia today, just as in ancient times, wet rice cultivation is the product of intensive work by many human hands, mostly women's. They spend much of their lives in very difficult labors, bent over to plant and harvest in challenging heat and humidity.

Geographers and other social scientists have also observed a remarkable set of cultural adaptations to harsh environments on high elevation steep slopes in the region, especially in Southeast Asia; see the Regional Landscapes feature "Zomia."

REGIONAL LANDSCAPES | Zomia

Stretching from Vietnam in the east to India in the west, at elevations over 1,000 feet (300 m) is a region known as **Zomia** (•**Figure 7.G**). Evocative of imaginary Asian realms like Shangri-La, the name *Zomia* comes from a word meaning "highlander" in Burma, Bangladesh, and India. Although first used in the scientific literature by a Dutch scholar, Zomia was popularized by Yale University professor James C. Scott in a groundbreaking book entitled *The Art of* Not *Being Governed.* Scott's thesis is that in China, Vietnam, Cambodia, Laos, Thailand, and Burma, upland peoples including the Hmong, Akha, Karen, Lahu, Mien, and Wa have found something perhaps all of us yearn for: an escape from tax collection and other unpleasantries of government rule. They have done so by choosing to live in the highest, most difficult, and least appealing lands of Southeast Asia—cold, steep, infertile places where no one else wants to live. They wish to be left alone, ungoverned, and undisturbed.

Scott's thesis is controversial, but after reading his book it is impossible to visit these places, as I did on a journey along the China-Vietnam frontier, without thinking, "I am in Zomia!" In the foothills below Zomia, the ethnic Tay people enjoy a prosperous livelihood growing wet rice (they are depicted in Figure 7.8). Higher in the mountains, ethnic Yao and Nung are clearly poorer than the Tay, growing dry rice and maize (corn) on slopes that they must maintain carefully with terraces to ward off erosion. Finally, on the steepest and most forbidding slopes near the mountaintops are the ethnic Hmong. The Hmong are the poorest of these ethnic groups, and it is easy to see why: it is all but impossible to productively farm these lands. The Hmong must resort to slash-and-burn farming, cutting the forest and planting maize without the benefit of terracing, which is impossible because the slopes are too steep. Without terraces, soils wash away and leave a desert-like environment in their wake. The Hmong move on and clear more forest. The result is breathtaking: an impoverished, lunar landscape where these people scour the hillsides for the last patches of forest to clear and plant (•**Figure 7.H**).

At first glance, it appears that the Hmong are the ultimate illustration of marginalization, where more powerful groups push weaker ones to the least hospitable lands. If Scott is 47
correct, however, the Hmong are marginalized by choice: they prefer to live in Zomia because no one else—including governments and other powerful entities—would want to live in and control such daunting places.

Many Hmong have resettled from Southeast Asia to the United States, especially in rural California and urban Minnesota and Wisconsin. Here too the Hmong have largely succeeded in being left alone, at least culturally. Their American neighbors are often baffled why Hmong have not assimilated as other immigrants have. There is an exceptional account of the culture clash between the traditional Hmong and modern America—especially in the realm of medicine. (The book's focus is on a Hmong girl afflicted with epilepsy, which the Hmong attribute to the work of evil spirits. They struggle to keep her out of the care of doctors with their powerful anti-seizure medications.) This is Anne Fadiman's *The Spirit Catches You and You Fall Down.*

• **Figure 7.G** Map of Zomia.

Source: Based on James C. Scott, 2009, The Art of Not Being Governed: An Anarchist History of Upland Southeast Asia. New Haven: Yale University Press, p. 17.

Joe Hobbs

• **Figure 7.H** Landscapes of the Hmong up to an elevation of about 6600 feet (2000 m) in the Ha Giang province, northern Vietnam near the Chinese border.

In some cases, population growth has outpaced the ability of even the most productive wet rice farming to feed the people, creating the classic scenario of "people overpopulation." In Nepal, for example, high rates of population growth on limited areas of moderately sloped, terraceable land have been a
56 push factor behind migration. After the 1950s, when malaria began to be eradicated in the lowland Terai region, millions of Nepalis migrated there, cleared the forests, and planted crops. For decades, the Terai served as a virtual "safety valve" for overpopulation in Nepal. Now that the Terai itself is mostly settled, there is nowhere for landless Nepalis to go but up. They clear and cultivate increasingly steep lands on which it is impossible to build terraces. Without terraces in place to hold the topsoil, heavy monsoon rains cause relentless erosion of their farmlands. Locally, the consequences are that the eroded plots cannot be cultivated again, and landslides occur downslope, often causing loss of life.

Much farther away, in the lowlands of India and Bangladesh, the increased sediment load contributed by erosion in Nepal causes rivers to swell out of their banks and bring extensive flooding—again, with losses of crops and lives. This sequence of events, beginning with overpopulation in Nepal and ending with flooding in Bangladesh, has been described as the **theory of Himalayan environmental degradation.**[2]

Where Asians Live

Many of the world's largest cities are in South and East Asia, including the greatest of all, Tokyo, with a metropolitan population of 38 million. About 1.6 billion people, or 42 percent of the region's total, live in urban settlements (•Table 7.2).

Table 7.2 South and East Asia: Metropolitan Populations Data

1	Tokyo, Japan	37.7
2	Seoul, South Korea	22.7
3	Mumbai, India	21.9
4	Manila, Philippines	20.6
5	Shanghai, China	19.9
6	Jakarta, Indonesia	19.2
7	Delhi, India	18.9
8	Osaka, Japan	17.4
9	Kolkata, India	15.6
10	Dhaka, Bangladesh	14.3
11	Karachi, Pakistan	13.2
12	Beijing, China	13.2
13	Bangkok, Thailand	10.1
14	Hong Kong, China	9.5
15	Nagoya, Japan	8.9
16	Taipei, Taiwan	8.5
17	Kuala Lumpur, Malaysia	8.1
18	Chongqing, China	7.8
19	Chennai, India	7.4
20	Lahore, Pakistan	7.1
21	Bandung, Indonesia	6.7
22	Shenyang, China	6.6
23	Bangalore, India	6.5
24	Tianjin, China	6.4
25	Hyderabad, India	6.4
26	Singapore	6
27	Gunagzhou, China	5.7
28	Pune, India	5.5
29	Ahmedabad, India	5.4
30	Xi'an, China	5.3
31	Ho Chi Minh City, Vietnam	5.3
32	Hangzhou, China	5.1
33	Yangon, Myanmar	5

Population in millions.

In Villages

The main unit of Asian settlement is the village. Most of Asia's roughly 2 million villages are essentially groups of farmers' homes, although some villages are home to other occupational groups, such as miners or fisherfolk. Clusters of houses bunched tightly together are typical, and inexpensive and simple structures—often made of local clays and other building materials—are characteristic. Indoor plumbing continues to be the exception in village homes, but electricity lines have been extending steadily throughout Asia. So have cell phones and satellite dishes; it is not unusual today to see what appear to be very "traditional" and even quite poor Asians enjoying some of the latest innovations in global communications (•Figure 7.9). In South and East Asia, as elsewhere in the developing world, many people are making the jump directly from nineteenth- to twenty-first-century technology. Wireless technology is especially appropriate where landlines and other infrastructure are difficult and expensive to install and where bureaucracies often slow down applications for access to utilities.

The original site selection for Asian villages was usually closely related to natural conditions and to perceived spiritual circumstances as well (see Geography of Sacred Space, page 224.). With consideration always given to the possibility of flooding monsoon rains, lowland villages tend to be situated on natural **levees** (raised riverbanks built up by deposition of sediments during floods), dikes, or raised mounds. Early villages in Indonesia were often built in defensible mountain sites. European colonists sometimes rejected traditional settlement patterns to serve their own interests. Dutch colonial administrators in Indonesia, for example, required that villages be built along main roads and trails in the lowlands, making it easier to exercise

Joe Hobbs

• **Figure 7.9** Across South and East Asia, satellite dishes sprout incongruously from traditional dwellings.

control, collect taxes, and draft soldiers or laborers for road-
45 work and other projects.

The most infamous example of foreign intervention in traditional rural settlements came during the Vietnam War in the
258 1960s. The American military wanted to make it more difficult for insurgent Viet Cong forces to infiltrate South Vietnamese villages, where they might indoctrinate and blend in with civilians in their fight against the Americans. The military therefore forced millions of farming families to relocate into fortified communities known as **strategic hamlets**. The project was an utter failure. The insurgents managed to infiltrate these settlements. More important, the Americans violated one of the strongest precepts of Vietnamese culture: to live close to the graves of
228 their ancestors. The U.S. military lost the "hearts and minds" of Vietnamese civilians in the strategic hamlets program. Today, U.S. intelligence agencies are still trying to learn from the lessons of Vietnam by trying to understand the culture of the enemy (for example, the Pashtun of Afghanistan's Taliban) (see A Closer Look, page 244.). In a controversial program known previously as "human terrain" and now remarkably as "human geography", social scientists with expertise in local cultures are imbedded with troops. Early on, some of these scholars met gruesome ends when the enemy captured them and discovered their roles. But now the program is more effective and less risky.

In Cities

The majority of people in South and East Asia are rural, but the region's future, in terms of demographic weight, economic power, and cultural change, is focused on the cities (see the feature on megacities on page 226, and Table 7.2). Some of the countries are already highly urbanized, notably Japan (86%), South Korea (82%), and Taiwan (78%) and entities like Singapore and Brunei that are effectively city-states. The layouts and architectures of the cities span a continuum from the ancient, walled fortress city to the modern Western-style metropolis, and many, like Beijing in China and Lahore in Pakistan, accommodate both. Unfortunately, in the push for urban modernization, many traditional urban structures and communities uniquely adapted to local environments have been sacrificed. Urban conservations are struggling to protect some of the remaining old buildings including those of Dhaka in Bangladesh and of Shanghai in China.

A striking pattern across the region is the steady stream of rural-to-urban migration, with people motivated by both push and pull factors to leave the relative poverty of the countryside in
search of employment or even prosperity in the city. Many Asian 56
cities are hard-pressed to provide accommodation and services to their swelling populations, and in authoritarian countries (notably China), restrictions on internal migrations are imposed.

7.3 Cultural and Historical Geographies

South and East Asia have been home to some of the most important cultural developments of humankind in landscape transformation, settlement patterns, religion, art, and political systems. From Korea (not Gutenberg's Germany, as is widely thought) came the first movable printing type. From China came gunpowder, paper, silk, and china (porcelain). From India came the great faiths of Hinduism and Buddhism. Handmade textiles, artware, bone and leather and paper products, and precious metals and gems have flowed from Asia into global trade for millennia. Spices, foodstuffs, and exotic plants and products from the East Indies have been major commodities in Asian-European trade and commerce for more than five centuries. Some of the world's most important domesticated plants and animals originated in this region, including rice, cabbage, chickens, water buffalo, zebu cattle, and pigs.

These countries, and their cultures, histories and geographies, are a source of endless fascination. It is for example difficult to many people, the present author included, to dispute Ramachandra Guha's claim that "India is the most interesting country in the world."[3]

Ethnic and Linguistic Patterns

As would be expected of such a vast region, in many places fragmented by islands and mountain barriers, the ethnic and linguistic composition is rich and complex (•Figure 7.10). Central and South Asia gave birth to the Indo-European language family, which contains tongues as diverse as English and Hindi. *Hindi* and *Urdu*, which are major languages in India and Pakistan, respectively, are in the **Indo-Iranian subfamily** that includes such other major South Asia languages as *Pashto* (closely related to Persian, and spoken in western Pakistan and Afghanistan; Afghanistan's official language is *Dari*, or Afghan Persian). Indo-Iranian tongues also include *Sanskrit* (which is extinct as a vernacular language but is still used in Hindu ritual), *Sinhalese* (the majority language of Sri Lanka), *Punjabi* (spoken on both sides of the northern frontier between Pakistan and India), Nepal's national language of *Nepali, Bengali* (the language of Bangladesh), and
Romany (the tongue of the Roma, or Gypsies, of Europe). 79

GEOGRAPHY OF SACRED SPACE

The Korean Village

In North America, people are accustomed to communities laid out on a perfect grid of streets intersecting at right angles, with their homes and businesses symmetrically fronting those streets. But in large parts of Asia, settlement and residence considerations are far more varied, especially because they are tied to spirituality, aesthetics, and topography. Villages, and the homes in particular, tend to be arranged for maximum harmony with the natural and spiritual environments—a practice known as **geomancy**—and constitute a special realm that may be considered sacred space. The traditional Korean village illustrates these unique Asian qualities.[4]

From ancient times, Koreans developed ways of determining the most auspicious village sites. The principles of *feng shui* and of Confucianism played strong roles in how they adapted settlements and homes to geographic conditions. **Feng shui** (pronounced "fung-*shway*") is a traditional theory for selecting favorable sites for buildings, homes, and cities and for guiding many other urban and architectural planning decisions.[†] It is essentially an ordering device for environmental planning and for daily living. *Feng shui* (the words mean "wind" and "water") attempts to balance **yin** and **yang** to achieve harmony. *Yin* (conceived of as negative, dark, and feminine) and *yang* (representing the positive, bright, and masculine) stand for all the dualistic, opposing forces that may be reconciled harmoniously, including female and male, night and day, moon and sun, cold and hot, and right and left. With cultural diffusion and growing interest in Asian lifestyles, *feng shui* principles are spreading in American suburban home architecture, as in houses built to conceal overhead beams and sharp protrusions (which are disharmonious) and provided with windows opening to the east (a harmonious situation).

The theory of *feng shui* and the Confucian view of nature have some common characteristics that affect site, settlement, and residence designs. Confucianism reflects the sociopolitical philosophy of the Chinese philosopher known in the West as Confucius (see page 231). One of the main Confucian principles is that architecture and the lived environment should be in harmony with nature. This can be seen in the Confucian institutes for higher learning, called *seowon,* which in Korea peaked in the Joseon Dynasty, around 1400 C.E. Each *seowon* had a lecture hall where Confucian scholars gathered to learn philosophy. The buildings were laid out in such a way that from the inside of the complex, the scholars could see distant landscapes. The scholars' minds and spirits were enhanced as they gazed at the mountains and contemplated nature's grandeur. They believed that the earth's power was an invisible force that would help them become men of dignity and eminence.

The principles of *feng shui* and Confucianism are also apparent in the design of Korean villages. Each village is laid out to achieve harmony with the mountains—an important consideration in a land that is 70 percent mountainous. Rising behind many villages is a mountain, called *jinsan,* which is a spiritual focus for the villagers. *Jinsan* serves as the village symbol and guardian. Typically, a small stream flows in front of the village, and beyond it, there is a flat expanse of farmland. Beyond those fields is another mountain, called *ansan,* facing the village (•**Figure 7.1**).

These geographic conditions are economically advantageous to the community, providing it with productive fields and a good supply of water. But just as important, this situation is psychologically, spiritually, and symbolically beneficial to the villagers. The natural features are meant to embrace the village, as Koreans say a mother would hold a child. Korean historical records describe the traditional village as a place of comfort and serenity that safeguards against negative natural forces.

Symbolic guardians are associated with the topography. Ideally, there is the Green Dragon to the east, the White Tiger to the west, the Black Tortoise to the south, and the Red Bird to the north. These guardians, named for figures in Chinese astronomy, could have varied manifestations. The Green Dragon, for example, could be a hill or a river to the east of the village, while the White Tiger could be a mountain or a road on the west side. One enters the village through a symbolic portal. For many villages, a pair of "spirit posts" stands at the foot of a nearby mountain. The posts mark the preliminary boundary to the outside, and they ward off evil spirits.

Not only is the village site selection process different from that in the West, but the settlement pattern is uniquely Asian too. Unlike in Western settlements, no more than two houses should be built side by side, they should not stand parallel to each other, and their gates should never face each other. The side street should curve naturally between the home lots. Overall, then, Koreans choose to emphasize the surrounding topography rather than geometric order in the construction of their communities and homes.

The traditional Korean home is adapted to the natural environment, embracing its surroundings while offering protection, comfort, and order. As in the *seowon,* the layout brings outdoor scenery into the house itself. Homes are also adapted to the environment in their construction materials. A thatched roof and earthen walls effectively block heat in the summer and insulate in winter, a perfect combination for Korea's sweltering summers and raw winters. The mulberry paper covering lattice windows diffuses sunlight, providing a pleasant natural light while absorbing sound and allowing some ventilation. The home's eaves are built so that sunlight penetrates deep into the house during the cold weather but never during summer—a rather recent innovation in the West, where it is known as "passive solar" design.

Joe Hobbs

• **Figure 7.1** The Korean village of Yangdong is nestled between mountains that are separated by a river and lush rice fields.

• **Figure 7.10** Languages of South and East Asia.
Source: Based on Bernard Comrie, et al., The Atlas of Languages, Revised Edition (New York: Facts on File, 2003).

When the first of the Indo-European languages was carried from central Asia into India—apparently by the ethnic group known as the **Aryans**, between 2000 and 1000 B.C.E., languages in the ancient and indigenous **Dravidian language family** were being spoken there by peoples who had much darker skin than the newcomers. The Dravidian languages, today found almost exclusively in southern India and northern Sri Lanka (with the exception of a small pocket in Pakistan), include *Tamil, Malayalam*, and *Telegu*. Along the eastern side of the Himalayan mountain wall separating South Asia from China, people speak

A closer LOOK

Asia, Motherland of Megacities

A **megacity** is a city with more than 10 million inhabitants. We are accustomed to living in a world of such gigantic cities. Yet as recently as 1975, there were only three of them: New York City, Mexico City, and Tokyo. Today there are 23, including 6 in the Americas, 3 in the Middle East, 2 in Europe, and 1 in Africa. The rest are in South and East Asia. India has 3, and China and Japan have 2, including the world's greatest city, Tokyo, with its 38 million souls. South Korea, the Philippines, Indonesia, Pakistan, Bangladesh, and Thailand each have 1, and each of
95 these is that country's primate city.

South and East Asia have a lot of urban growth ahead of them. On a worldwide scale, urban dwellers surpassed rural dwellers for the first time in history in 2008. Only about 40 percent of Asians are urbanites, so just to catch up with the global trend Asians will be moving into cities. Three-quarters of the world's people will live in cities by 2050—again, with the greatest numbers in Asia.

It is impossible to make any generalizations about "the Asian megacity." They range from prosperous Tokyo to Bangladesh's poor Dhaka, and from planned, vertical, and tidy Shanghai to unplanned, sprawling, and squalid Mumbai. In site and situation too, they span the continuum from the auspicious, chosen layout of the original Forbidden City in Beijing to the unfortunate but necessary placement of Tokyo atop a network of active tectonic faults.

Having described the Asian village as a sacred nurturing cradle, it is tempting to contrast it with secular fleshpots of the modern Asian city. But such a depiction of the Asian megacity would be wrong. As Edward Glaeser writes of the Asian city in his book *Triumph of the City,*

> Cities are at the apex of human endeavour. High-density cities are creative, thrilling and less environmentally destructive than sprawling car-based suburbs typical of America. Cities are passports from poverty. They attract poor people, rather than creating them. They are where humans are at their most artistically and technologically creative.[5]

As for the sacred, one need not wander far from any spot in the Asian city to find people praying, whether in the neighborhood mosque or below the towering *goporums* of a South Indian Hindu temple complex. At the larger pilgrimage sites, many of the devotees are villagers who have travelled far to petition for better harvests or harmonious marriages of daughters and sons. Many of those daughters and sons will eventually move into the city.

---

languages in the **Tibeto-Burman subfamily** of the **Sino-Tibetan language family**. *Tibetan,* spoken farther north in Tibet, and *Burmese,* spoken to the east in Myanmar (Burma), are also in the Tibeto-Burman subfamily. So is *Koro*, a language spoken by only 800 people in India's northeasternmost state. Koro was unknown to the outside world until 2008, when it was discovered by a team of researchers working for National Geographic's Enduring Voices Project.

With so few speakers, Koro is an endangered language, and like so many other languages, seems doomed for what linguists call **language death**, which occurs when a tongue's last speaker dies. The last fluent speaker of a language dies, on average, every two weeks. Of the world's 6,909 known languages, about half are expected to die in this century. The main reason is that young people are drawn by globalization and other forces to learn more dominant languages. Koro-speaking youth, for example, are going to boarding schools where they study Hindi or English.

In other cases, there are stringent efforts to protect languages. Indian languages, for example, have served as the basis for the drawing of many of the country's state boundaries. India has 14 official languages, and the country's political map is still in the process of being redrawn along ethno-linguistic lines. There are at least five ongoing efforts to create new states, with the main justification being that these boundaries will protect linguistic and cultural identities. These languages are in many cases mutually unintelligible; neither speaker understands the language spoken in an adjacent state. One economic benefit of British colonialism was that it bestowed English as a *lingua franca* on India. This has given India a tremendous advantage 78
in the marketplace of outsourced call centers. 235

The greatest numbers of people in China speak languages in the **Sinitic subfamily** of the Sino-Tibetan languages, overwhelmingly *Chinese*, and its variants, including *Mandarin, Xiang, Min,* and *Yue*. The Chinese government has adopted Mandarin (Putonghua) as the country's official dialect—making it the national tongue of one-fifth of the world's population—but only about half the population can speak it. Almost all of these more than 1 billion speakers of about 1,500 Chinese dialects belong to the **Han Chinese** ethnic group, which is culturally, politically, and economically the dominant ethnic group of China. Northern and northwestern China, along with Mongolia, are home to speakers of languages in the Altaic language family including *Uighur, Kazakh, Turkic, Mongolian,* and *Tungus*. Cut off from their Altaic-speaking kin are the *Korean* speakers of the Korean Peninsula and the *Japanese* speakers of Japan. Japan's indigenous **Ainu** language has no known linguistic relatives and is close to language death.

The people of Taiwan include Chinese-speaking migrants from China and indigenous speakers of an unrelated language. Most linguists assign this tongue to the **Austronesian subfamily** of the more widespread **Austric language family**. Extreme southern China and adjacent portions of Vietnam and Laos are linguistically complex, reflecting the presence of many minority ethnic groups. Several are kin to the peoples of peninsular Southeast Asia, including the *Hmong* of southern China, Laos, and Vietnam, who speak a language in the **Hmong-Mien subfamily** of the Austric language family. Most people in Laos and neighboring 221
Thailand speak *Lao* and *Thai* in the **Tai-Kadai subfamily** of the Austric family. The majority languages of Cambodia and Vietnam, respectively, are *Khmer* and *Vietnamese,* in the **Mon-Khmer** branch of the **Austro-Asiatic subfamily** of the Austric languages.

Southward and eastward into peninsular Malaysia, Indonesia, and the Philippines, the linguistic map is exceptionally rich and fragmented. The eight languages of the Philippines belong to the Austronesian group. Also in the Austronesian branch is a multitude of languages spoken in Indonesia, including *Malay, Dayak, Kayan-Kenyah, Barito, Javanese, Bugis, Tomini,* and *Timorese*. In many cases, a single language is spoken almost exclusively by an ethnic group inhabiting a single island. In other cases, as will

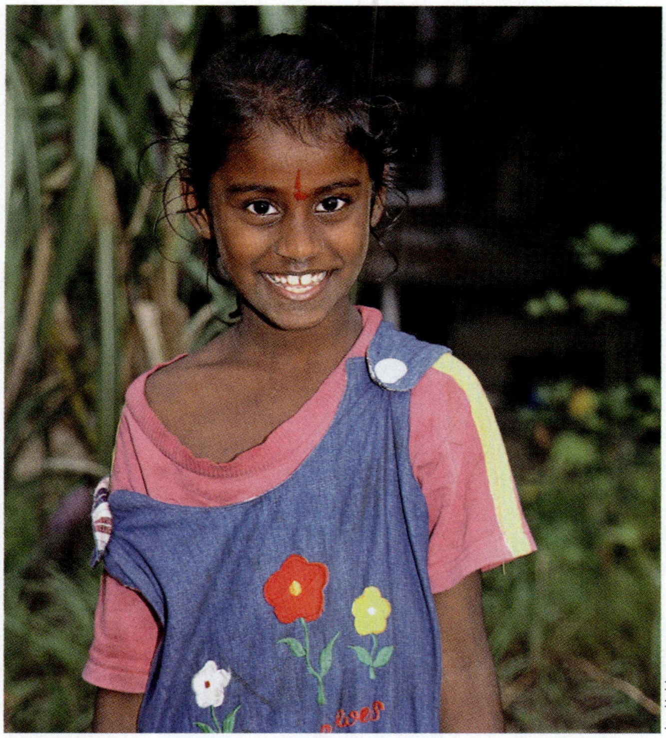

Joe Hobbs

• **Figure 7.11** Tamil youth near Ipoh in West Malaysia. Making up about 10 percent of Malaysia's population, the Tamils are descendants of laborers brought from India to (what was then called) Malaya by the colonial British.

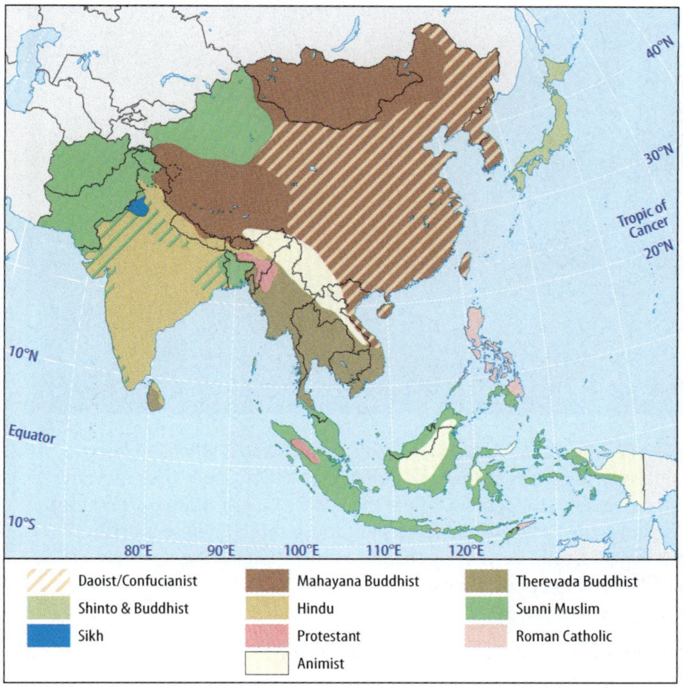

• **Figure 7.12** Religions of Asia.

be seen later, different ethnolinguistic groups sometimes live in conflicted proximity to one another on a single island. The large island of New Guinea and some of its smaller island neighbors are home to the 750 different **Papuan languages**, which are not
291 related to Austronesian tongues.

Colonialism and regional economic enterprises led to the infusion of distant tongues into many parts of East and South Asia. As well as bringing English to India, British colonists introduced Tamil-speaking Dravidians from south India into what is now Malaysia, where they worked the rubber plantations and tin mines and where their descendants remain as an important mi-
295 nority today (•Figure 7.11). Another small but economically prosperous minority in several countries (and a majority in Singapore) is ethnic Han Chinese. Some people in Indonesia speak Dutch, reflecting that country's former colonial status, and the Philippines' former status as a U.S. commonwealth is reflected in the fact that English remains one of that country's two official languages.

Religions and Belief Systems

South and East Asia are, along with the Middle East, one of the world's two great hearths of religion. Hinduism, Buddhism, and related traditions of Confucianism and Daoism—collectively practiced or observed by an enormous 2 billion people (or more than a quarter of the world's population) originated in this region. The basics of these belief systems are introduced here. The precepts of Islam and Christianity, also practiced in the region, are described in Chapter 6. The faiths with small numbers of adherents, like Zoroastrianism and Sikhism, are described later in the contexts of particular countries. Religion is sometimes, unfortunately, at the root of domestic or international conflict, and these cases are also discussed in the appropriate places.

The geographic distribution of the region's faiths is depicted in •Figure 7.12. As this map reveals, Hinduism is dominant in India, although many other religions are practiced there as well. Islam prevails in Afghanistan and Pakistan, Bangladesh, most parts of the East Indies, parts of the southern Philippines, parts of western China, and among the Malays of western Malaysia. Indonesia, in the East Indies, has the world's largest number of Muslims; 90 percent of the 238 million Indonesian people are Muslim. However, Hinduism and Buddhism thrived in Indonesia until Islam arrived in the fifteenth century, and many aspects of these religions permeated Indonesia's unique Islamic culture. Various forms of Buddhism are dominant in Myanmar (Burma), Thailand, Cambodia, Laos, Tibet, and Mongolia. Most people in Sri Lanka and Nepal are Buddhists or Hindus. The religious patterns of the Chinese, Vietnamese, and Koreans are more difficult to describe. Among them, Buddhism, Confucianism, and Daoism have been influential, often in the same household.

Growing prosperity among the Chinese has been accompanied by a renewed interest in spiritual matters. China's government officially recognizes and even encourages five faiths (Buddhism, Islam, Daoism, Catholicism, and Protestantism) but discourages or forbids some sects and belief systems, including Tibetan Buddhism; the quasi-Buddhist movement known as **Falun Gong** (or Falun Dafa, a practice of calisthenics and meditation, that the government calls an "evil cult") and "house church" Protestantism (•Figure 7.13). In Japan, religious affiliations often overlap, with an estimated 84 percent of the Japanese population sharing Buddhism with the strongly nationalistic religion of **Shintoism**, itself a blend of nature reverence, Japanese folklore, and Chinese ritual. The Philippines and Timor-Leste, with large Roman Catholic majorities, are

Joe Hobbs

• Figure 7.13 Falun Dafa (Falun Gong) is outlawed in China. In many countries you will see followers sharing information about their movement with passersby. This posting is near the Sydney Opera House. After I took this photo, the Falun Dafa followers eagerly greeted Chinese tourists stepping off a tour, bus, telling them about the human rights abuses against their compatriots back home in China.

the only predominantly Christian nations in Asia, although Christian groups are found in all other countries in this region. About one-third of South Koreans are Christian, and Korean missionaries work in many countries of the world.

Other customs include **ancestor veneration**, which is especially prominent among Chinese, Japanese, Vietnamese, and Koreans. Most homes in these countries contain small shrines with photographs and other memorabilia of deceased ancestors at which family members place fresh fruits and other offerings on a daily basis. It is crucial in these cultures that the ancestors never be forgotten (**•Figure 7.14**). Many members of hill tribes in Southeast Asia believe that natural processes and objects possess souls, a doctrine
322 known as **animism**. Such indigenous animistic beliefs have been incorporated into a number of mainstream religions, creating vibrant and highly varied belief systems by Western standards. This

Joe Hobbs

• Figure 7.14 Ancestor veneration is not a religion per se but an important spiritual tradition that requires the living to remember and pay tribute to their forebears. This is a shrine to the deceased parents of an elderly woman in a southern Thai village.

Joe Hobbs

• Figure 7.15 Worship at the Main Temple of the Holy See of the Cao Dai religion in Tay Ninh, southern Vietnam. Note the "All-Seeing Left Eye of God" at the upper right. Tay Ninh is like the Vatican City of the Cao Dai faith.

blending, known as **syncretism**, has parallels elsewhere in the
world, for example, among the Maya of Central America. 365

The ultimate syncretic faith is the **Cao Dai** religion, which originated in Vietnam during the French colonial era and persists today despite constraints imposed by the communist Vietnamese government (**•Figure 7.15**). It recognizes and blends the precepts and iconography of Buddhism, Daoism, Confucianism, Judaism, Christianity, and Islam. Prophets and saints of the Cao Dai faith include Moses, Louis Pasteur, Victor Hugo, and Albert Einstein.

This kind of layering and mixing in the spiritual realm is typical of most belief systems in South and East Asia. Most of the faithful put stock in what many Westerners might dismiss as "superstitions." Numerology, shamanism, amulets, horoscopes, and fortune telling have an important place in peoples' lives. Marriage, travel, site selection for homes and businesses, and countless other aspects of life are frequently influenced by auspicious and inauspicious omens. In Thailand, Laos, Cambodia, and Myanmar, the most visible sign of beliefs in a multifaceted spiritual realm is the "spirit house" (**•Figure 7.16**). The spirit house is placed in an auspicious spot of a home of business, where it serves as an abode for spirits that people petition for special favors or appease to avoid bad fortune.

Hinduism is a regionally varied faith and a belief system so complex that it is extremely difficult to summarize. However, it

Joe Hobbs

• **Figure 7.16** Spirit House in Laos.

has the following characteristic elements. It lacks a definite creed or theology. It is very absorptive, encompassing an unlimited pantheon of deities (all of which, most Hindus say, are simply different aspects of the one God, Brahman) and an infinite range of types of permissible worship (•**Figure 7.17**). Most Hindus recognize the social hierarchy of the caste system, deferring authority to the highest (Brahmin) caste. They practice rituals to honor one or another of the principal deities (Brahman the creator, Vishnu the preserver, and Shiva the destroyer) and thousands of their manifestations (*avatars*) and lesser deities, attending temples to worship them in the form of sanctified icons in which the gods' presence resides. They believe in reincarnation and the transmigration of souls and revere many living things (**see A Closer Look**).

Joe Hobbs

• **Figure 7.17** Hinduism has a complex and very colorful pantheon. This is Vishnu, preserver god, and his consort, Lakshmi, goddess of wealth and prosperity.

A closer LOOK

The Sacred Cow

Many attitudes, beliefs, and practices associated with cattle make up the world-famous but often poorly understood "sacred cow" concept of India. There are nearly 200 million cattle in India, representing about 15 percent of the world total and the largest concentration of domesticated animals anywhere on earth. India's dominant religion of Hinduism forbids the slaughter of cows but allows male cattle and both male and female water buffalo to be killed. Reverence for the cow is well founded. Indians favor cow's milk, ghee (clarified butter), and yogurt over dairy products from water buffalo. They value cows as producers of male offspring, which serve as India's principal draft animals. Both cows and bullocks provide dung, an almost universal fuel and fertilizer in rural India.

Reverence for the cow in particular and cattle in general pervades Hindu religion and mythology. The bull Nandi is associated with the Hindu god Shiva, the cow is linked to the goddess Lakshmi, and Krishna (an incarnation of the god Vishnu) was a cowherd. People allow cattle to roam the streets of Indian cities, and simply accommodate them in vehicle and pedestrian traffic (•**Figure 7.J**). As a symbol of fertility, cows are associated with several deities. The mother of all cows, Surabhi, was one of the treasures churned from the cosmic ocean. Hindus honor cows at several special festivals. They use cow's milk in temple rituals. They believe that the "five products of the cow"—milk, curds, ghee, urine, and dung—have unique magical and medicinal properties, particularly when combined. In India, there are about 4,000 *gaushalas,* or "old folks' homes," for aged and infirm cattle.

Remarkably, however, the subcontinent countries of Pakistan, Bangladesh, and India are major exporters of leather and leather goods. In India, the leather industry is mainly in the hands of Muslims, who do not share Hindus' restrictions on killing or eating cows. Muslims are forbidden to eat pork; therefore, pigs, a major food resource in many developing countries, are of little importance here. They are eaten mainly by Christians, very low-caste Hindus (some of whom are pig breeders), and tribal peoples.

Joe Hobbs

• **Figure 7.J** People and cows share street space throughout India.

They are supposed to be tolerant of other religions and ideas. They participate in folk festivals to commemorate legendary heroes and gods. To earn religious merit and to struggle toward liberation from the bondage of repeated death and rebirth, they make pilgrimages to sacred mountains and rivers.

The Ganges is a particularly sacred river to Hindus, who believe it springs from the matted hair of the god Shiva, and who call it "Mother" and "Goddess." The pilgrimage to celebrate the festival of Ardh Kumbh Mela at Allahabad, at the confluence of the Ganges and Yamuna rivers, drew 75 million pilgrims in January 2007. It was the largest pilgrimage, and probably the largest gathering of any kind, in human history. The most enduring Hindu pilgrimage destination is the Ganges city of Varanasi (Benares) in the state of Uttar Pradesh. Many elderly people go to die in this city and be cremated where their ashes may be strewn in the holy waters. Indian authorities have passed laws, so far ineffective, to clean up the Ganges, polluted in part by incompletely cremated corpses; many of the faithful poor cannot afford to buy the fuel needed for thorough immolation. Scavenging water turtles released into the river at Varanasi help dispose of the cadavers.

Buddhism is based on the life and teachings of Siddhartha Gautama, known as the Buddha or the Enlightened One (•**Figure 7.18**). The Buddha was born a prince in about 563 B.C.E. in northern India.[6] Although presumably born a Hindu, he came to reject most of the major precepts of Hinduism, including caste restrictions. When he was about 29 years old, he renounced his earthly possessions and became an ascetic in search of peace and enlightenment. The Buddha eventually settled for a "middle path" between self-denial and indulgence, and he meditated through a series of higher states of consciousness until he attained enlightenment (*bodhi*). From that point, he traveled, preached, and organized his disciples into monastic communities. In his sermons, he described the "Four Noble Truths" revealed in his enlightenment: life is suffering; all suffering is caused by ignorance of the nature of reality; suffering can be ended by overcoming ignorance and attachment; and the path to the suppression of suffering is the Noble Eightfold Path, made up of right views, right intention, right speech, right action, right livelihood, right effort, right mindedness, and right contemplation.

An important Buddhist concept is the doctrine of **karma**, a person's acts and their consequences. A person's actions lead to rebirth, in which good deeds of the previous life are rewarded (for example, by which one could be reborn a human) and bad deeds punished (for example, by being reborn as a resident of hell). The goal of the practicing Buddhist is to attain **nirvana**, a transcendent state in which one is able to escape the cycle of birth and rebirth and all the suffering it brings.

Buddhism evolved into two separate branches: **Theravada** ("Way of the Elders") and **Mahayana** ("Great Vehicle"). (Mahayanists sometimes apply the term *Hinayana*, "Lesser Vehicle," in a derogatory fashion to the Theravada.) Theravada Buddhism is strongest in Sri Lanka, Myanmar, Laos, Cambodia, and Thailand. Theravada Buddhists claim to follow the true teachings and practices of the Buddha. Mahayana Buddhism originated in India and then diffused along the Silk Road to

Joe Hobbs

•**Figure 7.18** The Buddha has many manifestations. This temple of the Theravada faithful is in Chiang Mai, Thailand.

central Asia, Tibet, and China and eventually into Vietnam, Korea, Japan, and Taiwan. Mahayana Buddhists accept a wider variety of practices than those espoused by the Theravada, have a more mythological view of the Buddha, and are interested in broader philosophical issues.

Confucianism is not as much a belief system as it is a sociopolitical philosophy that serves on its own or is blended with other religions, particularly among ethnic Han Chinese. It is based on the writings of Confucius (Kung Fu-tzu, or "Master Kung," who lived from 551 to 479 B.C.E.), collected primarily in his *Analects*. Mencius (c. 371–288 B.C.E.) later became a major force in the widespread diffusion of this belief system throughout China. Following a period in which Communist authorities disavowed them, Confucian ideals are making a major comeback in China today.

The major tenets of Confucianism are embodied in a system of ethical precepts for the proper management of society, emphasizing honor of elders and other authorities, hierarchy, and education. Confucius never proclaimed his beliefs to constitute a religion; he was simply attempting to create a social contract between different classes central to Chinese government and society. He viewed his philosophy as secular, and its diffusion to Korea, Japan, and Vietnam has been part of the cultural baggage taken abroad by a steady stream of Chinese emigrants to those places.

Another important Chinese school of thought, second only to Confucianism and also widely blended with Buddhism among ethnic Han Chinese, is **Daoism** (sometimes spelled **Taoism**). Its philosophy comes from a body of work known as the *Daodejing (Tao-te Ching)*, or in English, *Classic of the Way and Its Power*, ascribed to Chinese Lao-tzu (Laozi), who lived in the sixth century B.C.E. Daoism encourages the individual to reject Confucian-style social conformity and seek to conform only to the underlying pattern of the universe, the "Way" (Dao, or Tao). To follow that way, the individual should "do nothing," meaning nothing unnatural or artificial. One can become rid of all doctrines and knowledge to achieve unity with the Dao and thereby gain a mystical power. With that power, the individual can transcend everything ordinary, even life and death. Daoists hold the simple earthly life of the farmer in high esteem.

There are about 25 million Christians in India, most of whom live in the south of the peninsula; Buddhists, Jains, Parsis, and members of a variety of tribal religions make up the numerous smaller remaining religious minorities. Most of the estimated 7 million Buddhists are recent converts from among India's lowest castes. Numbering perhaps 75,000 and concentrated mainly in Mumbai, the famously entrepreneurial **Parsis** have attained wealth and economic power far out of proportion to their number. Their religion is the pre-Islamic Persian faith of **Zoroastrianism**, a monotheistic faith known mainly for its reverence of fire. Zoroastrians follow the teachings of the prophet Zarathustra, who was born around 600 B.C.E. in Persia (modern Iran). He preached that people should have good thoughts, speak good words, do good deeds, and believe in a single god, Ahura Mazda ("Wise Lord"). Persia's Zarathustras, as the Zoroastrians are also known, fled Persia in two waves, first during the conquest by Alexander the Great around 300 B.C.E. and then after Islam swept into Persia in the seventh century. They settled in China, Russia, and Europe, where they became assimilated with other cultures. In India, they maintained a separate existence until quite recently; now about one in three takes a spouse of another religion and abandons Zoroastrianism. Many scholars consider theirs an endangered religion, with only 124,000 to 190,000 followers worldwide.

India's 6 million **Jains** also have influence beyond what their numbers suggest, as they control a significant share of India's business. They live mainly in the western state of Gujarat. **Jainism**, founded on the teachings of Vardhamana Mahavira (a contemporary of the Buddha, c. 540–468 B.C.E.), is renowned for its respect for geographic features and animal life. Jains believe that souls are in people, plants, animals, and nonliving natural entities such as rocks and rivers. Jainism has taken the principle of nonviolence (*ahimsa*, also present in Hinduism) to mean they should not even harm microbes. Therefore, Jain worshipers often wear masks to prevent inhalation of microscopic organisms, and Jains cannot be farmers because they would have to destroy plant life and living organisms in the soil.

Effects of European Colonization

As in much of the developing world, Western colonialism reshaped many traditional geographic patterns in South and East Asia (•Figure 7.19). By the end of the fifteenth century, Portugal and Spain had begun to extend economic and political control over some islands and mainland coastal areas of South and Southeast Asia. In the eighteenth and nineteenth centuries, the pace of colonization and economic control quickened, and large areas came under European domination. By the end of the nineteenth century, Great Britain was supreme in India, Burma (modern

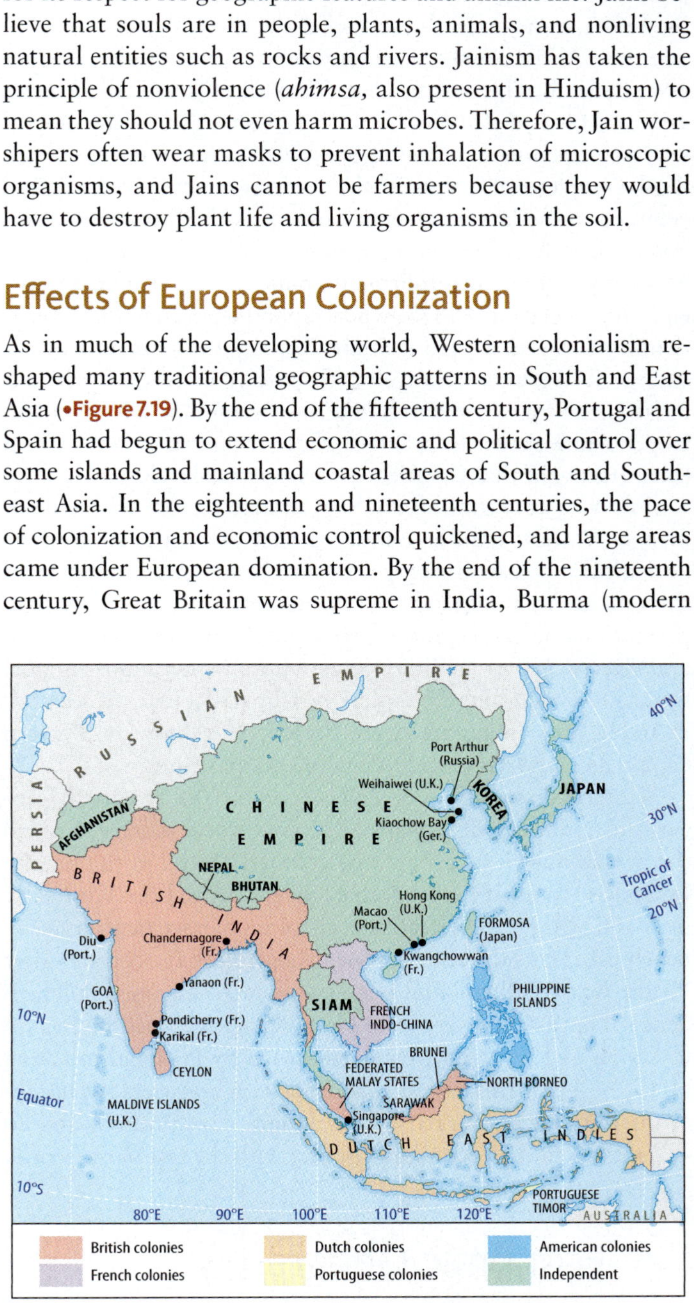

• **Figure 7.19** Colonial realms and independent countries in South and East Asia at the beginning of the twentieth century.

Source: Based on 'The World in 1900,' in R. R. Palmer, ed., Atlas of World History (Chicago: Rand McNally, 1957), pp. 169–171.

Myanmar), Ceylon (today's Sri Lanka), and Malaya and northern Borneo (now parts of Malaysia); the Netherlands possessed most of the East Indies (Indonesia); France had acquired Indochina (the region of modern Vietnam, Laos, and Cambodia) and small holdings on the coasts of India; and Portugal had Goa and Diu in India, Macau in China, and Timor (today's Timor-Leste) in the East Indies. Although retaining a semblance of territorial integrity, China was forced to yield possession of strategic Hong Kong to Britain in the 1842 Treaty of Nanking and to grant special trading concessions and extraterritorial rights to various European nations and the United States through the latter half of the nineteenth century. In 1898, the Philippines, held by Spain since the 1500s, came under the control of the United States.

A few Asian countries escaped domination by the Western powers during the colonial age. Thailand (historically known as Siam) formed a buffer between British and French colonial spheres in peninsular Southeast Asia. Japan withdrew into almost complete seclusion in the mid-seventeenth century but emerged in the late nineteenth century as the first modern, industrialized Asian nation and soon acquired a colonial empire of its own. Korea also followed a policy of isolation from foreign influences until 1876, when Japan began to colonize it. It is noteworthy that Japan, South Korea, and Thailand would emerge as very or relatively prosperous countries, whereas most of the former colonies lagged in economic development. Dependency theorists point to these discrepancies to affirm
44 their argument that colonization hampered development.

Asia was an extraordinarily profitable region for the colonizers. Western nations extracted vast quantities of tropical agricultural commodities and in turn found large markets for their manufactured goods. Westerners also invested heavily in plantations, factories, mines, transportation, communication, and electric power facilities. Some of the region's most important cities, including Shanghai in China, Kolkata (Calcutta) in India, and Singapore, were developed mainly by Western capital as seaports serving Western colonial enterprises.

Western domination of Asia ended in the twentieth century for a variety of reasons. The two world wars weakened the West's ability to conduct its colonial affairs in the region; Japan rose to great power status and challenged the West early in World War II; and effective anti-colonial movements arose in nearly all areas subject to European control. In the decades after World War II, all colonial possessions in Asia gained independence. The last to revert were Hong Kong (returned by Britain to China in 1997) and Macau (returned by Portugal to China in 1999). Until this era of independence, however, twentieth-century Asia was marked by revolution, war, turmoil, and an inability to gain an economic footing with the rest of the world.

7.4 Economic Geography

With some exceptions, this is a dynamic region with some of the world's strongest and fastest-growing economies. Their economic performance is often described as miraculous, given the region's tumultuous history. At the end of World War II, South and East Asia were poorer than Africa and Latin America. "The depleted states of East Asia, including Korea, Taiwan and Singapore, with few sources of money and practically no existing industry, seemed especially hopeless," writes journalist Michael Schuman in his book *The Miracle*. "Yet it was in these countries, supposedly the bottom of the international economic barrel, that the Miracle was born."[7]

If not literally a product of divine intervention, what explains Asia's extraordinary economic success? Many of the answers—government policies that nurtured the growth of industries, and government investments in industrial growth rather than in public amenities and services, for example—are explained in the following pages, in studies of the individual countries, including Japan and South Korea. It is also useful to look at the level of the individual and family to appreciate Asia's economic geography and economic success. To generalize, Asians are savers (whereas Americans are spenders!). These individual traits arguably played a significant role in the recent "Great Recession:" Americans went deeply into debt to finance consumption, while Asians stepped up their savings and exported more goods to Western consumers.[8] Savings are also important because most of the Asian countries provide few "safety nets" such as retirement pensions and health care. And most important of all, savings help to take care of the family, which is virtually sacred in many Asian cultures. It is always impressive to see how young people working even for low wages manage to save a significant share of those earnings to send home as remittances to Mom and Dad and Grandma. Needless to say, one of the great hopes in the West is that Asians will cut back on savings, loosening purse strings especially to buy Western exports—thus pulling Western economies and (the West argues) the world economic system out of the abyss.

The world's second and third largest economies are in East and South Asia: China and Japan, respectively. The recent global economic downturn slowed them down somewhat, but the economies of China and India have still been surging, with both goods and services inundating the global economic system. Their clout, their accomplishments, and their ambitions are enormous. Some economists and equities marketers have taken to conflating the two countries and calling them "Chindia," a double-barreled engine of economic growth.

It must be noted, however, that this growth is occurring against the backdrop of extraordinary poverty, and many hundreds of millions of people remain desperately poor. Although the region as a whole has taken great strides at reducing poverty rates, the gap between rich and poor is growing in many countries. This is particularly true in India, where the overall economy has grown about 8 percent annually since 2000; yet in that same period, there has been almost no reduction in childhood malnutrition, which afflicts almost half of India's young. Three-quarters of India's people live on less than $2 per day, which is the World Bank's benchmark for poverty. About 70 percent of Indians live in rural areas, but agriculture produces only about 20 percent of India's growth domestic product; no wonder then that rural people are poor, as they are a majority who live off a small portion piece of the economic pie. Poverty

tends to be deepest in the rural areas, fueling the classic push
56 factor of rural-to-urban migration. China overtook Japan to become the world's second largest economy in 2010 (and is projected to be the largest by 2020), yet on a per capita basis the country remains poor, with over a third of the people living on less than $2 per day. "For the first time, you have this odd combination," an economist in Hong Kong wrote as China reached this milestone. "One of the world's largest economies is also one of the world's poorest economies."[9]

The milestone of China becoming the world's second largest economy is one of the many that is described as a "China moment." With its galloping growth, China reaches one of these milestones after another. The world awaits the China moment of the country becoming the world's largest economy. There was a China moment recently when the country consumed so much coal that for the first time it began to import coal. There was the recent China moment of China becoming the world's second largest oil importer, after the United States. And so on.

Despite being generally a region of less developed countries, South and East Asia are home to the strong, industrialized, export-oriented economies of South Korea, Taiwan, Hong Kong, and Singapore, which late in the twentieth century came to be known as the **Asian Tigers** (or "Little Dragons"). Some economists consider the rapidly industrializing Southeast Asian countries of Thailand, Malaysia, Indonesia, the Philippines, and Vietnam the **New Asian Tigers**, or **Tiger Cubs**, following the path of rapid economic growth blazed by their neighbors. Indonesia's economy has been doing so well that some economists would like to include it as a second "I" in the exclusive "BRIC" club, which would then be Brazil,
133 Russia, India, Indonesia, and China, or "BRIIC."

Japan was the first Asian country to develop modern cities and modern types of manufacturing on a large scale. It has been a major industrial power for more than a century. China and India are much better supplied with mineral resources than is Japan. With the world's largest populations, they also have cheaper labor. But Japan's labor force is, at least for the time being, generally more skilled, and it leads Asia in high **value-added manufacturing**—the process of refining and fabricating more valuable goods from raw or semi-processed materials. Now, however, there are signs that Japan's preeminence in innovation is waning as China's intellectual resources grow. The biggest economic story in East and South Asia is the rise of China. Some highlights of that story are told elsewhere in Section 7.6.

China's Surging Economy

The statistics portray China as a steamroller. Among the China moments, China became the world's largest exporter, overtaking Germany in 2010. China overtook the United States as the world's largest energy consumer in 2010. China overtook Japan to become the second largest economy that same year.

China's economic growth soared at an average annual rate of 10 percent between 1990 and 2011. China's exports (along with imports) increased about 1,000 percent during those

Joe Hobbs

• **Figure 7.20** Mechanized silk production in China. Note the white cocoons in the vats below the worker's hands; the silk is spun from these.

years, growing an average of 50 percent a year. China's share of world exports of goods grew 8 percent between 1981 and 2010 (compared with a decline of 3.5 percent in the United States during the same period).

Just a few years ago, China was notable for making low-value-added products—cheap toys, for example—that the richer countries really had no interest in making. There were shirts and shoes too, most of them manufactured in more than 20,000 factories in the Special Economic Zone (SEZ) of Shen-
zhen, adjoining Hong Kong. Now China is making at least a 267
little bit of almost everything, becoming the "workshop of the world" (•**Figure 7.20**). The country is making especially rapid progress in manufactures and exports of information technology (IT) hardware. Already China is the world's third largest manufacturer of personal computers, sold under such brand names as Dell and IBM. The roots of China's boom in IT may be traced to the dot-com bubble that burst in the United States in 2000. Many of the talented Chinese who worked in California's Silicon Valley at that time went back to China,
taking their skills with them in a kind of "reverse brain drain." 56

Advances in China's global status and in government economic policy have fueled the boom. China joined the World Trade Organization in 2001, prompting a surge in foreign investment. China sweetens the deal for investors with incentives like subsidized loans, tax exemptions, and 50 percent discounts on land prices. The addition of Chinese brawn and brains creates a nearly perfect investment climate. Chinese labor is much less expensive than is typical of most more developed countries, and China has a rapidly growing pool of university-trained engineers and other innovators.

China's Economic Impact

China's economic boom is having enormous consequences for other countries, particularly its Asian neighbors. Japanese electronics companies are cutting costs by moving increasing amounts of production to China. Surging investment in China is linked to disinvestment elsewhere, especially in Southeast Asia. That region's Tigers and Tiger Cubs—notably Malaysia, Singapore, and Thailand—dominated the 1990s with manufactures and exports of medium- and high-tech industrial products. With its growing consumer appetite, China was buying such goods from these countries. But recently the manufacturing shifted to China from those countries. South Korea's Samsung made China the main base for production of its flat screens and computers, for example. Japan's Toshiba makes televisions only in China. Dell recently moved some of its computer-making facilities from Malaysia to China.

With China's increasing gravity, Southeast Asian economies may at least temporarily be relegated to supplying low-end products, especially food and raw materials, to China. In turn, they may end up buying cheap Chinese manufactured goods, harming their own fledgling industries. They may try to counter this trend by developing higher-value niche products to meet demands in China. Thailand, for example, wants to boost manufactures of health care products, while Singapore focuses on biomedical products and financial services. Poorer Indonesia will probably be able to count on helping satisfy China's growing appetite for natural gas to fuel its southern industrial cities. Meanwhile, the large economies of Japan, Taiwan, and South Korea have lost ground to China in the market of foreign investment, as investors have been lured away from those countries by China's lower labor costs and its potential 1.3 billion consumers.

Cheap labor is unlikely to make China a permanent magnet for foreign investment, however. As affluence rises there, so will wages, and China's less prosperous neighbors will again attract investment. These figures are telling: in 2011, the average factory worker in Vietnam made $136 per month, in Indonesia $129 per month, and in China $413 per month. China is no longer cheap. China's one-child policy also means it will have fewer industrial workers in the future. Mindful of this demographic problem, farsighted multinational companies are boosting manufacturing investment in India, where the pool of young workers is likely to remain large and growing for decades to come.

In the long-dominant United States, there are rising fears of China's economic ascendance. Many Americans feel their country is becoming subservient to China because Chinese investments are funding U.S. federal deficits and because the United States is losing ground to China's industrial might. China's rise has been steadily chipping away at U.S. economic dominance in the region. In one country after another (including the most critical, Japan and South Korea), China has eclipsed the United States as Asia's most essential trading partner. The U.S.-China relationship is growing in China's favor. Of China's exports, about 18 percent go to the United States (its largest single trading partner), but its imports from the United States lag far behind. This U.S. trade deficit with China threatens to become a sore point in relations between the countries. Textile and other industries in the United States are pressuring the U.S. government to pass punitive tariffs on Chinese imports, a move that could precipitate a
costly trade war. Problems with tainted food and toy imports 420
have already led to bans of some Chinese products in the United States.

The United States is also furious at China's reluctance to stop the counterfeiting or "piracy" of American products, particularly computer software and DVDs of Hollywood films. China is the epicenter of prolific Asian manufacture and trade in pirated products. In some countries like Malaysia, authorities try to crack down on sales of such goods on street corners, but the vendors are usually prepared to disappear at a moment's notice with their goods in tow, only to reappear moments later when police leave. In other countries like Vietnam, no one needs to run: there are countless permanent shops selling the latest counterfeited music, movies, watches, handbags, and more (**•Figure 7.21**).

Another great concern in the United States is how many jobs and profits are lost to outsourcing, especially to India (**see Insights, next page**).

Joe Hobbs

• Figure 7.21 Counterfeited goods made in China and sold throughout Asia bear the brand names of U.S. and other manufacturers but are sold to happy consumers at a fraction of the cost of the genuine product. This $4 digital audio player claims to be an Apple iPod, but you can be sure it is not.

INSIGHTS | Outsourcing

Outsourcing, also known as **offshore outsourcing** or **offshoring**, is the flight of technology and other jobs from countries with high manufacturing and service costs to countries with low manufacturing and service costs. India is the world's largest recipient of these jobs, and the United States is the leading outsourcer. India's outsourcing industry was a natural development after 1990, when the country abandoned its Soviet-style, centrally planned economic model of self-sufficiency and opened its doors to world markets and economic liberalization.

In the United States, there is growing worry and even alarm over the trend of outsourcing jobs to India. Many of these jobs are in computer programming and a wide variety of technical support, from computer problem phone calls to X-ray diagnosis. Many are mundane, like bill processing and order taking. Increasingly, however, high-paying white-collar employment is being outsourced to India, in such fields as investment banking, aircraft engineering, and pharmaceuticals research. Top Western corporations like the United States' Cisco Systems, the world's leading maker of communications equipment, are beginning to relocate substantial numbers of their senior executives to India, where much of the firm's work is actually done. The top destination is the south-central city of Bengaluru (Bangalore), which has come to be known as "India's Silicon Valley."

India, which has earned a reputation of being the West's "back office," is an especially attractive venue for information technology outsourcing for several reasons. It has a large population of well-educated people, most of whom speak excellent English. Those who work the telephone call centers typically receive training in vernacular American English and strive to perfect their American accents while shedding their Indian and British pronunciations. The bottom line is another major advantage: Indian skilled labor is cheap. The average computer programmer in the United States earns $75,000 per year, while his or her counterpart in India earns $9,000 per year.

Finally, the Internet and superb telephone communications, ironically combined with India's physical distance from the United States, provide some remarkable opportunities. While people in the United States sleep, the day shift is on in India. An American doctor can X-ray a patient in the afternoon and by early the next morning have the analysis—performed overnight in India. On the downside, when an American consumer places a technical service call for a computer problem, he or she may be speaking with a very tired Indian employee working the graveyard shift. There are many accounts of fatigue and burnout among Indian call center employees.

What about the loss of American jobs due to outsourcing? There is an expression that one's job may be "Bangalored." Some projections are alarming. A former official of the U.S. Federal Reserve described outsourcing as a "Third Industrial Revolution," predicting it would take jobs from as many as 42 million American workers in the coming years.[10]
Outsourcing companies generally argue that 420
reducing the costs of their services reduces inflationary costs for their consumers and improves efficiency and productivity. This rationale is, of course, cold comfort to the American worker whose job has moved offshore. Another trend that Americans may fear is that with its well-educated work force (including many graduates of U.S. universities), India is moving further up the value chain. There is new investment in research and development so that the country's information technology workforce will be innovating, and not just producing goods that Western thinkers invented.

There are surprising new twists in the economy of outsourcing. One is "reverse outsourcing," in which Indian and other companies hire American workers because their qualifications cannot be matched in India and other countries. Secondly, the pride in "made in America" products has grown great enough, and the costs of intellectual capital, wages, raw materials, real estate, and energy in India and China have risen high enough, that many American firms have begun to "onshore" work they had previously offshored. This countertrend, known as **insourcing**, onshoring, inshoring, or homeshoring, is especially popular in the United States at a time when jobs are scarce.

The Green Revolution

Most Asian countries do not want to improve their economies only by enhancing their outputs of services and manufactured goods. They also want to boost agricultural self-sufficiency and crop exports. Revolutionary changes in crops are at the heart of efforts to reshape Asia's agricultural economies. One of the most significant agricultural innovations in Asia during the past century is the development of new seed types and innovations in planting, cultivating, harvesting, and marketing of crops, efforts known collectively as the **Green Revolution**. These developments are not exclusive to South and East Asia, but much of the growth and investment have taken place here.

Ever since 1962, when the International Rice Research Institute (IRRI) was founded in the Philippines, there have been efforts to use science to increase food yields (particularly of rice) to stave off hunger and generate export income. Notable success has been achieved in breeding the new high-yield varieties of seed stock, and there has been a large upsurge of production in certain areas where the new strains have been widely introduced. Many of the new varieties are
bioengineered or genetically modified (GM) crops. This means 86
that precise manipulation of the genetic components of a crop can not only produce a higher yield but also a crop more resistant to drought, flood, or pests and which generates a higher amount of a desired nutritional component such as vitamin A. China, second only to the United States in the global biotechnology industry, has produced genetically engineered rice, corn, cotton, tobacco, sweet peppers, petunias, poplar trees, and more.

Because the underlying premise of the Green Revolution is economic—more crops will be produced at lower cost for higher revenue—most countries in the region are racing to

increase their biotechnological output, especially in an effort to catch up with China and ensure they do not lose positions in the global marketplace. Malaysia hopes to build Biovalley, a biotechnology research park near Kuala Lumpur, and Indonesia is planning a similar Bioisland. One of Biovalley's goals is to create palm oil trees genetically modified to produce the raw
253 material for specialized plastics used in medical devices. South Korea, Japan, and India have large and well-funded biotechnology research programs. Japan has so far resisted production of GM food crops out of fears of possible health hazards, a concern that most Europeans and their governments share.

For Asian farmers to capitalize fully on the Green Revolution, they must overcome many obstacles. Success requires levels of capital that are often beyond the means of peasant farmers, landlords, and governments. Such expenditures are needed for water supply facilities (for example, the tubewells that now number hundreds of thousands in the Indo-Gangetic Plain of the northern Indian subcontinent and that are severely depleting the groundwater), chemical fertilizers, and chemicals to control weeds, pests, and diseases. And as agriculture becomes more mechanized, considerable increases in the costs of machinery, fertilizers, and fuel have to be borne by farmers who have, in many cases, had little experience with the cash economy. Governments must improve transportation so that the large quantities of fertilizer required by the new seed varieties can be delivered in a timely fashion. Not only must there be these associated infrastructural changes to support this "revolution," but also the crop calendar of the farmer becomes much less forgiving because many of the newest seed grains demand more precise water, fertilizer, and cultivation requirements than traditional grains. Grain storage facilities, now subject to plundering by rats, must be improved.

There are other problems associated with the Green Revolution. Economic dislocations result when rice-importing countries become more self-sufficient, causing hardships for rice exporters. Large infusions of the agricultural chemicals associated with high-yield varieties have negative repercussions on natural ecosystems. India's government subsidizes urea fertilizer, and farmers are naturally tempted to overuse this cheap product. Unfortunately, the overuse of urea degrades the soil, and fertilized soil demands much more water. There is also concern that development of a limited number of high-yield crop varieties grown in vast monocultures will dramatically reduce the genetic variability of crops, in essence interfering
29 with nature's ability to adjust to environmental changes.

7.5 Geopolitical Issues

In the Middle East and North Africa, and in Russia and the Near Abroad, principal geopolitical concerns focus on the production and distribution of fossil fuel energy resources. In South and East Asia, by contrast, some of the most serious geopolitical issues are prospects for what may be done with weapons created from a particular energy source: nuclear energy. There are also concerns about seafloor resources (see Geography of Energy, next page), Islamist terrorism, and the security of shipping lanes. Of these, seabed resources may prove to be the most critical. There are several discussions in this chapter of seemingly insignificant small islands that are extremely critical because of the fossil fuels and other resources that may lie in nearby waters. 23

The earlier discussion of the mushrooming economic clout 27
of China and India raises a big picture geopolitical reality: Asia is emerging as a center of gravity that will seriously challenge the century-long primacy of the United States in world affairs. Japan long stood alone as a pillar of strength in Asia, but with China and even India surging ahead, regional might has been reshuffled. Eyes in the region and around the world will be focused especially on China, which some analysts fear may use its fast-increasing military power to enhance its fast-increasing economic clout.

Nationalism and Nuclear Weapons

As we will see in Section 7.6, after its independence from Britain in 1947, India emerged as an avowedly secular democratic state. But after the 1998 victory of the Hindu nationalist Bharatiya Janata Party (BJP), hopes for new privileges emerged among India's Hindu majority. Among BJP supporters, there was hope for a "Hindu bomb," a counterweight to a long-feared "Islamic bomb" in neighboring Pakistan.

On May 11, 1998, much to the surprise of U.S. and other Western intelligence agencies, India conducted three underground nuclear tests in the Thar Desert. With the blasts, India's government seemed to be staking India's claim as a great world power, exhibit its military muscle to Pakistan and China, and increasing political support for the BJP. Initial reactions among India's vast populace were highly favorable; 91 percent of residents polled within three days of the event supported the tests. Outside India, there was alarm. India had defied an informal worldwide moratorium on nuclear testing that went into effect in 1996, when 149 nations (not including India and Pakistan) signed the **Comprehensive Test Ban Treaty** (also known as the **Nuclear Nonproliferation Treaty**, or **NPT**), which prohibits all nuclear tests.

The world's eyes turned quickly to India's neighbor to the west. Governments pleaded with Pakistan to refrain from answering India with nuclear tests of its own, arguing that Pakistan would have a public relations triumph it if exercised restraint: it would appear to be a mature and responsible power, whereas India would seem to be a dangerous rogue state. But Pakistan's leaders felt obliged by their population's demands for a tit-for-tat response to India's blasts, and soon followed with six nuclear tests. India and Pakistan thus joined just five other nations—the United States, Russia, China, the United Kingdom, and France—in acknowledging that they possess nuclear weapons. Israel neither confirms nor denies that it has nuclear weapons, but is thought to have about 250.

There is disagreement about what this recent regional and global shift in the balance of power means. Some analysts fear an escalating nuclear arms race that, perhaps ignited by a border skirmish in Kashmir or by a terrorist attack like the

GEOGRAPHY OF ENERGY | The Spratly Islands, Geopolitical Hot Spot in the South China Sea

About 60 islands make up the Spratly Island chain, which lies in the South China Sea between Vietnam and the Philippines (for its location, see Figure 7.37). They are an idyllic tourist destination where divers can hire luxury boats to explore the coral reefs and palm-lined beaches of remote atolls. However, the islands are much more significant for their strategic location between the Pacific and Indian Oceans. During World War II, Japan used the islands as a base for attacking the Philippines and Southeast Asia. Still more significant, as much as $1 trillion in oil and gas may lie beneath the seabed around the Spratlys.

Not surprisingly, many nations covet control of the Spratlys. Six countries claim some or all of the islands, and international law is not equipped to sort out these conflicting claims (see the UN Convention on the Law of the Sea, page 272). China, Vietnam, and Taiwan claim sovereignty over all of them, while Malaysia, Brunei, and the Philippines claim some of them. During the Cold War, the competing nations felt it was too hazardous to push their claims on the islands. As the Cold War drew to a close, however, the situation became more volatile. All the contenders except Brunei placed soldiers, airstrips, and ships on the islands. In 1988, the Chinese navy invaded seven of the islands occupied by Vietnam, killing about 70 Vietnamese soldiers. In 1992, China again landed troops in the islands and began exploring for oil in a section of the seabed claimed by Vietnam. In 1995, China moved to expand its territorial designs on islands already claimed by the Philippines and has since built what it calls "shelters," which look like fortifications to the Filipinos. Late in the 1990s, Malaysia occupied two disputed Spratly Islands and began building on one of them. In 2004, Vietnam began renovations of an airport on one of the islands, saying it was necessary to boost tourism there. China condemned the construction as a violation of its territorial sovereignty, an accusation it repeated as Vietnam negotiated natural gas and oil development with Indian and other foreign companies in the Spratlys. The situation moved to the brink of all-out conflict in 2011, when Chinese ships sabotaged Vietnamese oil exploration vessels operating within what Vietnam claimed was its own economic zone in the South China Sea (which Vietnam calls the East Sea—the same name the Koreans call the Sea of Japan).

As China's need for energy grows, so does its assertiveness over the Spratlys, the nearby Paracel Islands, and the Senkaku Islands, among others (see •**Figure 7.K**).

"China has indisputable sovereignty over the South China Sea islands and adjacent waters," a Chinese Foreign Ministry spokesman asserted in 2011.[11] Such a bold claim makes smaller powers in the region fear that China could turn small islands into "permanent aircraft carriers" that China could use to dominate them militarily. China even used a tiny submarine to plant a Chinese flag on the bed of the South China Sea. Not surprisingly, the United States took a position in this geopolitical chess match: in 2010, Washington declared that the South China Sea was of strategic importance to the United States. The United 248
States is solidifying its position to offset China: it has an alliance with the Philippines, and has drawn closer militarily to Vietnam. Clearly, an incident in these remote islands could trigger a 276
much wider and more serious conflict in Asia.

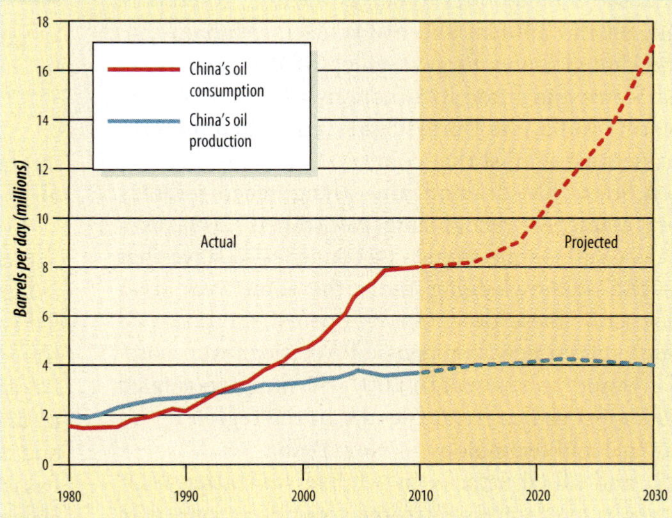

• **Figure 7.K** With China's oil appetite expected to grow enormously, every potential source of oil—like the Spratly Islands'—takes on unprecedented importance.

Source: Based on 'The World in 1900,' in R. R. Palmer, ed., Atlas of World History (Chicago: Rand McNally, 1957), pp. 169–171.

one in Mumbai in December 2008, could lead to the **mutually assured destruction (MAD)** of Pakistan and India. Others, particularly in South Asia, argue that the weapons represent the best deterrent against such a MAD conflict, as they did for decades between the countries of NATO and the Warsaw Pact. However, few people anywhere argue against the view that the nuclear rivalry between India and Pakistan has taken an enormous economic toll in two nations that need to wage war on poverty.

To the United States, both India and Pakistan are **pivotal countries**, defined by several influential historians as countries whose collapse would cause international migration, war, pollution, disease epidemics, or other international security problems.[12] The disposition of their nuclear arsenals is one of the main reasons for their being considered pivotal. If the government of Pakistan were to fall, Pakistan's nuclear weapons could come into the possession of a rogue element or a new government hostile to the West. One of those influential historians, Francis Fukuyama, declared that Pakistan is "the most dangerous place in the world."[13] India has much to fear from Pakistani scenarios. Either a right-wing or an Islamist government in Pakistan would be far more likely than the current regime to take India on in the disputed Kashmir region. Such a confrontation could set the stage for a nuclear exchange between Pakistan and India, not only decimating those countries but also sending economic shock waves around the world. India's recently surging economy has been a major force in India's recent diplomatic overtures to Pakistan; nothing discourages investment like war or the fear of it.

U.S.-Pakistan Relations since 9/11

Pakistan became even more pivotal with the events of 9/11. Up to that point, the United States had slapped tough economic sanctions against Pakistan (and less severe ones against India) for the nuclear weapons tests. The United States pursued a diplomatic courtship with India, both as a counterweight against China (India's longtime foe in the region) and as a means of expressing displeasure with Pakistan (China's ally) for helping Afghanistan's Taliban harbor Osama bin Laden. But the United States and Pakistan did an abrupt diplomatic about-face and embraced one another after 9/11. Pakistani President Pervez Musharraf instantly dropped support for the Taliban and quietly allowed the United States to use its territory to prepare for the assault on the Taliban and al-Qa'ida in Afghanistan. In return, the United States forgave much of Pakistan's debt to the United States and lifted its sanctions against Pakistan. To avoid isolating India, the United States also lifted its post-nuclear test sanctions against that country.

The United States has dramatically strengthened military ties with both countries, recognizing Pakistan in 2004 as a major non-NATO ally—one of just 14 countries to have that designation—and acknowledging India the same year as a "strategic partner." These designations opened the doors of both countries to major new shipments of American war material. But India would be favored: in 2007, Washington singled out India as its most critical strategic ally in the region by offering nuclear fuel and technology to New Delhi.

Under U.S. and allied attacks in Afghanistan after 9/11, Taliban and al-Qa'ida fighters retreated to and regrouped in western Pakistan, particularly in the semiautonomous federally administered tribal areas (FATA; see Figure 7.29). Here the populace is mainly Pashtun and is sympathetic to the causes of their Taliban ethnic kin and their al-Qa'ida spiritual kin (•Figure 7.22). This is an area where Pakistani government authority has long been kept at bay and where it was considered extremely difficult to insert Pakistani forces on a sustained basis. Bowing to American pressure, Pakistan's government periodically undertakes military operations in the FATA but considers a permanent or ongoing presence there as potentially detrimental to its survival.

The United States is deeply unpopular in Pakistan, so the sustained insertion of U.S. troops in Pakistan (in search of al-Qa'ida, for example) would be very destabilizing, perhaps even result in rebellion or coup against Pakistan's government. This could lead to the rise of an anti-American administration in Pakistan and raise questions about the disposition of Pakistan's nuclear weapons. To avoid that scenario, the U.S. military has relied heavily on predator drones to stalk and kill al-Qa'ida and Taliban suspects in Pakistan.

Joe Hobbs

• Figure 7.22 The Pashtun of western Pakistan and Afghanistan are a tight-knit ethnic group with a long history of defiance against foreign rule. Most are strongly opposed to American interests in the region.

Even those unmanned missions are politically risky for the United States, so imagine the risk involved in sending U.S. troops on the mission that killed Osama bin Laden, deep in Pakistani territory. The fact that bin Laden had lived not far from Pakistan's capital, and just around the corner from a Pakistani military academy, raised many questions on the quality and the loyalty of Pakistani intelligence forces. Many analysts and politicians in the United States asserted that elements within Pakistan's intelligence service, the ISI (Inter-Service Intelligence agency), had helped protect bin Laden. There are also suspicions that Pakistani intelligence and armed forces also continue to support the Taliban both in Afghanistan and Pakistan. The American military and intelligence communities often perceive Afghanistan and Pakistan as a single "theater" of operations, and had a neologism for it until the Pakistanis objected: *AfPak*.

What Does North Korea Want?

There are also major concerns about nuclear weapons in Northeast Asia. Ever since suffering Hiroshima and Nagasaki, Japan has had an official policy never to develop or use atomic weapons. But Japan worries about the potential nuclear threat from three adversaries, all of which possess nuclear weapons: Russia, China, and North Korea. The Japanese feel that the West dismissed potential threats from Russia too readily when the Soviet Union dissolved. Japan still has territorial disputes with Russia, particularly involving the four Kuril Islands of Kunashiri, Etorofu, Shikotan, and Habomai (lying just off the

northeastern coast of Hokkaido; see Figure 7.56), which the government of Josef Stalin seized at the end of World War II. To show that Russia was intent on maintaining its control, Russian President Medvedev visited one of the islands in 2010 and talked
147 about raising living standards there. The Japanese were furious.

Japan and China also have a territorial dispute concerning the East China Sea, and if, as expected, major oil reserves are discovered there, relations between the two countries could deteriorate. China continued to test nuclear weapons through 1996, adding to tensions in Japan. Finally, Japan fears reunification in Korea, which, as a former Japanese colony from 1910 to 1945, has a historical dislike of Japan. There is speculation that Japan may do the unthinkable and the officially disavowed—develop nuclear weapons—to counter the perceived threat of North Korea's nuclear weapons program.

The world looks nervously at the troubled relations between North and South Korea and between North Korea and the West. A crisis flared in 1994 when North Korea refused to permit full inspection of its nuclear facilities by the International Atomic Energy Agency (IAEA). The country was suspected of separating plutonium that could be used in making nuclear bombs. Over time, the United States, South Korea, and Japan worked out an agreement in which North Korea would agree to freeze development of nuclear weapons in exchange for the others providing fuel oil and assistance in building nuclear power plants. Those nuclear reactors would be of the light water variety, much less likely than North Korea's plutonium-based reactors to be "dual use" for both military and civilian purposes. The Clinton administration trumpeted the agreement as a success, but it had several flaws. One was that North Korea did not have to dispose of its existing nuclear fuel; it simply had to stow it safely away—meaning that it could quickly be reactivated.

President George W. Bush in 2002 proclaimed that North Korea was one of three countries making up an "axis of evil," along with Iraq and Iran. Late that year, North Korea dropped a virtual bombshell on the United States by admitting—after being confronted with evidence collected by U.S. intelligence agencies—that it did indeed have an active nuclear weapons program (developed, as it turns out, with Pakistani technical assistance). This admission followed earlier strident denials of such a program by North Korean officials.

North Korea's admission could have brought an end to the economic assistance in the form of food and fuel that the United States, South Korea, and Japan had been providing for a decade to the beleaguered country. What, then, was North Korea hoping to accomplish by revealing its nuclear weapons program? There are several possibilities. The least likely is that North Korea was instigating a military confrontation with the United States and its allies South Korea and Japan. In any war scenario, those allies would obliterate North Korea. The United States is obliged to defend South Korea in the event of a war with the North and would effectively wield its huge military advantage (•Figure 7.23). However, that victory would come at an enormous price for South Korea and possibly Japan. North Korea has a standing army of nearly 1 million soldiers, an impressive military arsenal including short- and medium-range missiles, a stockpile of chemical and biological weapons that it might not be shy to use, and presumably at least a few nuclear weapons that it could deploy. The casualties in South Korea would be enormous, whether from a conventional or an unconventional assault from the North.

Joe Hobbs

• **Figure 7.23** A little humor eases the tension for U.S. troops serving along what may be the world's tensest border, the "no-man's-land" separating North and South Korea. Even a minor incident here could touch off a conflagration that might take a huge toll in human lives.

It is much more likely that by disclosing its nuclear weapons program, North Korea sought guarantees that it could avoid war and also gain even more assistance from the West. For years, North Korea had only to mention the prospect of reactivating its nuclear weapons program to get more concessions from the United States and its allies—especially food aid during a succession of droughts and floods in the 1990s. In effect, North Korea was extorting money from the United States and its allies, which have been happy to ante up as a means of containing the rogue nation.

A series of on-again, off-again negotiations known as the Six Party Talks (held between North and South Korea, the United States, China, Japan, and Russia) culminated in 2007 with an agreement: The United States would help unfreeze North Korean funds in a Macau bank, remove North Korea from its list of countries supporting terrorism, lift trade sanctions in place since the Korean War ended in the 1950s, and along with its negotiating partners restart the flow of fuel oil to North Korea; in return, North Korea would close its nu-
274 clear plants and allow them to be inspected by international monitors. In the deal, North Korea did not promise to dismantle its nuclear facilities at Yongbyon, but only "abandon" or "disable" them, which it did do by the end of that year. North Korea is not required to destroy any existing stocks of nuclear weapons, and much uncertainty remains. The United States apparently hopes that the involvement of China and the other countries will make North Korea less likely to restart its nuclear program this time. North Korea watchers did not know how the policies of the country's new leader, Kim Jong-Un, might differ from those of his deceased father Kim Jong-Il.

Islands, Sea Lanes, and Islamists

As will be seen in Section 7.6, Indonesia—another pivotal country from the U.S. perspective—is an ethnically complex nation whose integrity has at times been threatened from a variety of secessionist movements. These disturbances range all the way from Aceh (pronounced "a-chay") in the country's far west to Papua (formerly Irian Jaya) in the east (see Figure 7.37). A possible **domino effect** scenario for Indonesia concerns foreign powers: If one province falls or proclaims its independence, many others will, and the largest country in the region will break apart. The United States and other countries fear what would happen to vital international shipping lanes should several new and perhaps militant states emerge from Indonesia's fragmentation. Washington is concerned that many states, some of them anti-Western, could emerge in Indonesia's place and threaten American interests in the country—notably its oil, natural gas, and copper resources and its location astride vital shipping lanes.

There have been scores of pirate attacks on vessels plying the strategic Strait of Malacca (see Figure 7.37), a critical chokepoint through which a quarter of the world's trade passes, including two-thirds of the world's shipment of liquefied natural gas and half of all sea shipments of oil (it ranks second only to the Strait of Hormuz as an oil shipping lane). Security in 177
the Strait of Malacca is a source of great concern to Japan, which imports 80 percent of its oil on ships that pass through the strait. Aceh's vast natural gas deposits and Papua's copper and gold are also seen as critical in the global economy, so like Indonesia, many world powers are anxious to see that they stay in the "right" hands. The United States gives generous amounts of foreign aid to Indonesia, but some Indonesians resent the United States, perceiving it as anti-Muslim in the wake of 9/11.

U.S. intelligence agencies have kept a watchful eye on Southeast Asia, and Indonesia in particular, fearful that it might emerge as an Islamist terrorist hearth. The country with the largest number of Muslims, Indonesia is a secular rather than a religious state, and a moderate form of Islam prevails in politics and in everyday life. What concerns Western intelligence authorities and Indonesia's moderate leadership are the more militant Islamist organizations that gathered strength after 9/11. One of these is **Laskar Jihad**, created in 2000 to mobilize Indonesian Muslims to fight Christians in Indonesia's Sulawesi and Muluku islands and Americans throughout the country. Another is **Jemaah Islamiah (JI)**, led by Abu Bakar Bashir, whose nominal role was preacher in an Islamic school. Western authorities are much concerned about such religious schools, known as *pesantrens* in Indonesia. Like the *madrasas* of Pakistan and the Arab heartland of the Middle East, these are venues for both religious and secular learning, and some of the schools' curricula are imbued with virulent anti-Western content. Baasyir admitted to having tutored 13 of the men arrested in connection with a plot to blow up Western targets in Singapore.

This concludes an introduction that has set the stage for further exploration of land and life in the subregions and countries of South and East Asia. The journey continues in the region's far west.

7.6 Regional Issues and Landscapes

South Asia

South Asia has an enormous range of ethnic groups, social hierarchies, languages, and religions. This is the most culturally complex area of its size on earth. Its cultures have roots that are deep in antiquity. Its modern engagements with the world are very important in the realms of security, economics, and population, among others.

Afghanistan: Graveyard of Empires

Like Sudan and Turkey, Afghanistan (•Figure 7.24) is a country that is transitional between regions; some geographers place it in Central Asia or the Middle East. It is a land of limited resources, poor internal transportation, and little foreign trade. Afghanistan is one of only five landlocked nations in South and East Asia. As we have seen, most landlocked countries are disadvantaged by their location and are poor. Afghanistan is 45

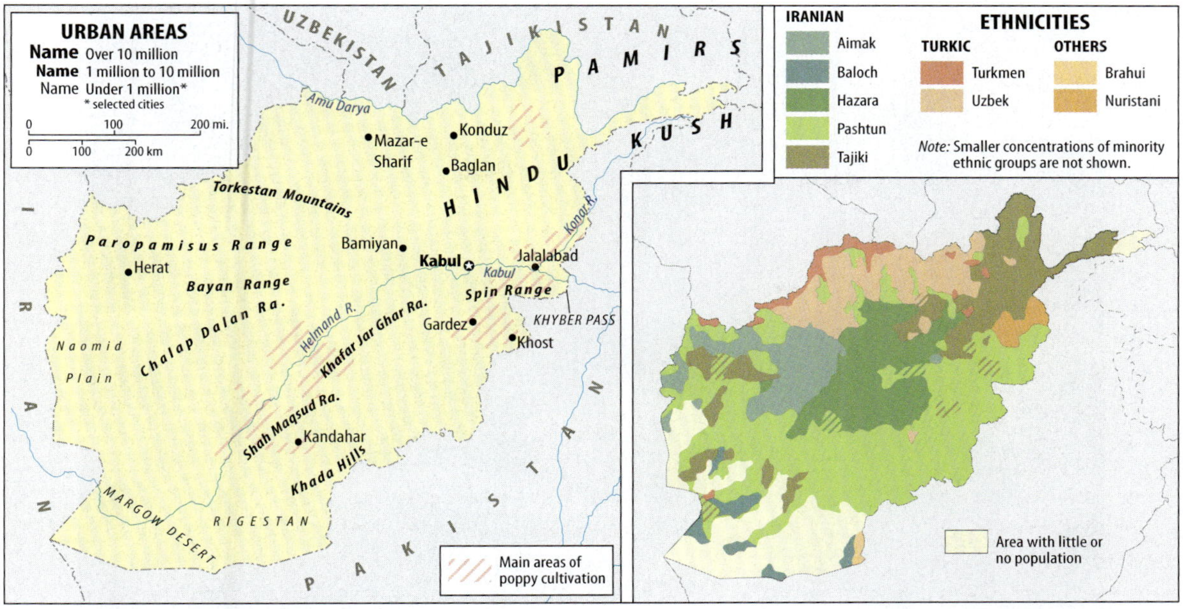

• **Figure 7.24** Principal features and ethnic groups of Afghanistan.

indeed one of the world's poorest countries and also ranks high as a failed state (seventh on that list).

Afghanistan is also a land of much conflict. It has repeatedly drawn the world's most powerful countries into conflict. Why? In large part because of its geographical situation: historically, it has occupied an important strategic location between India and the Middle East, and between India and Central Asia. Major caravan routes crossed Afghanistan, and a string of empire builders sought control of its passes. And although poor in natural resources, Afghanistan is neighbor to oil-rich Iran. This fact was critical to the Carter Doctrine discussed in Chap-
178 ter 6 and below.

High and rugged mountains dominate Afghanistan. Most of the country's 32 million people live in irrigated valleys around the fringes of the mountains. Northern Afghanistan borders three of the five central Asian countries, and millions of people on the Afghan side are related to peoples of those countries. The most populous area is the northeast, particularly the fertile valley of the Kabul River, where the capital and largest city, Kabul (population 2.3 million; elevation 6,200 ft [1,890 m]), is located. Most of the inhabitants of the southeast are Pashtuns (also known as Pushtuns or Pathans). Their language, Pashto, is
170 related to Persian. The Pashtuns are the largest and most influential of the numerous ethnic groups that make up the Afghan state; the next largest are the Tajik and Hazara, respectively.

The independent-minded tribal Pashtun people have always been loath to recognize the authority of central governments. In 1893, during the days of Britain's Indian Empire, British authorities and the Afghan King drew the "Durand Line" to separate their respective areas of influence. The line sliced right through traditional Pashtun lands, and Pashtun peoples have never accepted it. This border area saw warfare among Pashtun tribes, tribal raids on British-controlled areas, and British punitive expeditions against the tribes. In those days, the city of Peshawar—on the Indian (now Pakistani) side of the Khyber Pass into Afghanistan—became the hotbed for British and Russian agents playing what came to be known as the **Great Game** of vying (and spying) for strategic influence in Central Asia. The British were especially fearful that the Russians might use Afghanistan as a staging area to invade India, the "jewel in the crown" of Britain's colonies. To this day, Peshawar feels like a wild, lawless frontier city. It plays an important role in regional trade of weapons and drugs, especially heroin. For generations, the people of one of its outlying communities, Daraa, have specialized in replicating whatever foreign weapons a buyer desires, from old British muskets to modern anti-aircraft guns (•**Figure 7.25**).

Joe Hobbs

• **Figure 7.25** Pashtun *mujahadiin* in an Afghan refugee camp outside Peshawar, Pakistan. When I took this photo, these men were taking a break from fighting Soviet troops occupying Afghanistan. They and their sons could well have become Taliban fighting American troops in Afghanistan. Their bought their weapons in Daraa, the nearby factory and market for arms.

THOMAS J. ABERCROMBIE/National Geographic Stock

• **Figure 7.26** The Bamiyan Valley and the Koh I Baba Range of north-central Afghanistan. This photo conveys some of the country's rugged beauty and pastoral character.

Afghanistan is overwhelmingly agricultural and pastoral. Its agriculture relies on traditional methods and produces low yields. The country is so mountainous and arid that only about 12 percent is cultivated (•**Figure 7.26**). The most successful crop has been the opium poppy, which routinely accounts for between one-third and two-thirds of the country's gross domestic product, and more than 90 percent of the world's opium (!). In the context of the global drugs trade, portions of Afghanistan, Pakistan, and Iran have come to be known as the **Golden**
257 **Crescent**. A succession of rebel armies and then Afghanistan's government under the Taliban used revenues from opium and heroin to obtain their arms. Most of the production takes place in the southern province of Helmand, where the Taliban have retrenched themselves most successfully. Principally to reduce funding to the Taliban, and also to crack down on illicit drugs, Britain and the United States have stepped up efforts to eradicate opium. A policy split has developed: the NATO mission in Afghanistan worries that in the absence of crop alternatives, elimination of the poppy crop could turn destitute farmers into vengeful fighters. One option rarely called for but perhaps most appropriate in Afghanistan's future would be to allow the country to produce opium legally for pharmaceutical needs, as a handful of countries, including Turkey and India, are allowed to do.

Through most of the twentieth century, this highland country was remote from the main currents of world affairs. After the Islamic revolution in Iran in 1979, however, Afghanistan's location next to that oil-rich country made it once again the
178 target of foreign interests. The Soviet military intervention of 1979 and the ensuing devastation catapulted the country into world prominence. The USSR's motives for its invasion of Afghanistan may have included a desire to prevent by force the spread of Iranian-style Islamic fundamentalism into Afghanistan (which is 80 percent Sunni and 19 percent Shiite) or into the central Asian Soviet republics deemed vital to the superpower's security. For its part, with the Carter Doctrine the United States warned the Soviet Union that it would not tolerate further Soviet expansionism.

The Soviet war in Afghanistan brought widespread killing and maiming of civilians, destruction of villages, burning of crops, killing of livestock, sabotage of irrigation systems, and sowing of land mines over vast areas. Soviet ground and air forces caused several million Afghan refugees to flee into neighboring Pakistan and Iran. Arms from various foreign sources filtered into the hands of the *mujahidiin*, the anti-Soviet rebel bands that kept up resistance in the face of heavy odds. The United States was one of the powers supporting the rebels and in that sense waged a proxy war against the Soviet Union in Afghanistan. In its waning years, the USSR recognized that it could not win its "Vietnam War," and its troops withdrew
from Afghanistan by 1989. It is estimated that of the 15.5 mil- 258
lion people who lived in Afghanistan when it was invaded in 1979, at least 1 million died, 2 million were displaced from their homes to other places within the country, and 6 million fled as refugees into Pakistan and Iran. About two-thirds of these refugees have returned to Afghanistan.

Support that the United States and moderate Arab states such as Egypt and Saudi Arabia lent to the *mujahidiin* and sympathetic Arab fighters soon came back to haunt those countries in a phenomenon that came to be known as **blowback**.

Emboldened by their victory against the Soviets, the most militant Islamists among the fighters turned their attention to the United States and its Middle Eastern allies. Chief among them was Osama bin Laden, who developed "the base" (*al-Qa'ida* in Arabic) of thousands of Afghans, Arabs, and other anti-Soviet war veterans he could call on to wage a wider *jihad*. Bin Laden's organization trained an estimated 10,000 fighters in al-Qa'ida camps in Afghanistan and from this unlikely, remote setting devised spectacular acts of terrorism: an assassination attempt on Egyptian President Hosni Mubarak, the bombing of U.S. military barracks in Saudi Arabia, the bombing of U.S. embassies in Kenya and Tanzania, the attack on the *U.S.S.*
184 *Cole,* and the 9/11 attacks.

The *mujahidiin* succeeded in overthrowing the Communist government of Afghanistan in 1992, but after that, rival factions among the formerly united rebels engaged in civil warfare. During this period, most other countries utterly neglected Afghanistan, inadvertently promoting the growth of militant movements (which tend to thrive in remote, underdeveloped, and ungoverned regions). By 1996, one of the rebel factions, the **Taliban** (backed by Saudi Arabia and Pakistan), gained control of most of the country, including the capital of Kabul. Proclaiming itself the sole legitimate government of Afghanistan, the Taliban imposed a strict code of Islamic law in the regions under its control and gained international notoriety for its austere administration. The Taliban removed almost all women from the country's workforce, forbade public education of girls, and outlawed "un-Islamic" practices such as dancing, flying kites, watching television, keeping birds, and trimming beards.

The Taliban continued to make advances against its opponents inside Afghanistan (particularly the **Northern Alliance**, whose leader, the Ahmed Shah Massoud—to his people a hero known as "the Lion of Panjshir"—was assassinated just days before and in apparent preparation for the 9/11 attacks) and, by 2001, controlled 95 percent of the country's territory. The neighboring central Asian countries, Russia, and even Iran grew increasingly fearful of the spread of the Taliban's extreme interpretation of Islam into their nations. Russia ended up supporting some of the rebels it had fought because they were now fighting the Taliban. Russia's once unlikely ally in supporting those rebels was the United States, which successfully lobbied the United Nations to levy economic sanctions against Afghanistan in 1999. The United States hoped that economic losses would pressure the Taliban into turning over bin Laden for prosecution, and it also announced a $5 million reward for information that would lead to bin Laden's capture or death.

The bin Laden bounty rose into the tens of millions of dollars after September 11, 2001. Named as the mastermind of the attacks against New York and Washington, bin Laden became the "world's most wanted" as U.S. President George W. Bush evoked a Wild West vow to have him captured "dead or alive." In crafting its pronounced war against terrorism (which it promised to carry anywhere in the world necessary), the U.S. administration identified the Taliban as al-Qa'ida's mentor and
184 targeted both organizations for elimination.

Within a month of the attacks on the United States, American warplanes and special forces struck in **Operation Enduring Freedom** against Taliban and al-Qa'ida facilities and personnel around Afghanistan. The British military assisted in the air campaign, and most of the ground fighting was carried on by the Northern Alliance, the Taliban's rival and nemesis. The United States successfully persuaded Pakistan to drop its support of the Taliban and join the effort, and much of the military campaign was based on U.S.-Pakistani cooperation. One after another of the Taliban's urban strongholds fell, including Kabul and the Taliban spiritual capital, Kandahar. Withering assaults from U.S. warplanes crushed the tunnels and other hideouts used by al-Qa'ida guerrillas, and many al-Qa'ida prisoners were taken to the U.S. naval base in Guantánamo
Bay, Cuba. The elusive bin Laden, however, along with the Tal- 375
iban leader Mullah Omar, slipped away, presumably seeking safe harbor in Pakistan.

The reconstruction of Afghanistan began almost as soon as the Taliban were driven from power. Billions of dollars in aid went into the building of highways and other infrastructure, hospitals, and schools. The numbers of Afghans enrolled in primary schools rose from 25 percent in 1999 to over 90 percent, and girls and women are again being educated. Progress is slow because the country is so devastated, the security situation so tenuous, and the Afghan government so corrupt. Of all those billions of dollars in aid, some unknown amount has been siphoned off to line the pockets of those responsible for distributing it. U.S. Secretary of State Hillary Clinton said it was "heartbreaking" to see that billions of dollars of U.S. aid to the country had been wasted.

American forces and the Afghan government headed by Hamid Karzai effectively took control of only a small area around Kabul, and NATO troops were stationed in the troubled south of the country. Much of the country remains in the hands of warlords whose allegiance is up for grabs. Al-Qa'ida and Taliban fighters operate in the rugged countryside, sometimes carrying fighting into Kabul and other cities.

The mission of this war—the longest in U.S. history—changed substantially over the years. Initially, its purpose was clear: rid Afghanistan of al-Qa'ida and its Taliban supporters. With al-Qa'ida effectively displaced into Pakistan and elsewhere, the focus fell on the Taliban. American and NATO ground troops fought seemingly endless numbers of Taliban fighters, who fought with classic insurgent hit, run, and hide tactics. Western forces became engaged in a conflict that paralleled the long war of Soviets against the *mujahidiin* in Afghanistan. Although the Taliban and other Afghans, including civilians, bore heavy losses, the conflict also cost more than 1,800 American lives and over $400 billion (through 2011). Nearing the end of his first term in office, U.S. President Barack Obama steadily drew down the number of American forces in Afghanistan and announced that in 2014 the mission there would change from one of combat to one of support. U.S. military analysts came to the conclusion that the fight against the Taliban was unwinnable. For the losses of life and retreats suffered in earlier conflicts by the British and Russians

A closer LOOK

American Strategic Interests in Your Education

Arguably one of the most important answers to the question *Why has it been difficult for the United States to be involved in Afghanistan?* is something that cannot be seen in an atlas: the fact that the United States has been short on knowledge about the cultures, tribal allegiances, and other social complexities of Afghanistan, and in people fluent in some of the peoples' languages. The U.S. government is playing catch-up, in part by funding undergraduate and graduate-level college students to study the languages and become immersed for long periods in countries deemed critical to national security. See for example the Boren Scholarships, which "provide American undergraduate students with the resources and encouragement they need to acquire skills and experiences in areas of the world critical to the future security of our nation" (http://www.borenawards.org/). Boren Scholarships include language study of four of the five languages of Afghanistan depicted in Figure 7.10: Pashtu, Persian (Dari), Uzbek, and Turkmen. Scholarships like this can open many doors for you. They are also available for graduate study.

In the educational context, national security goes far beyond concerns with military thinking. Consider this definition of national security on the Boren Scholarship program overview: "It draws on a broad definition of national security, recognizing that the scope of national security has expanded to include not only the traditional concerns of protecting and promoting American well-being, but also the challenges of global society, including: sustainable development, environmental degradation, global disease and hunger, population growth and migration, and economic competitiveness." These concerns are virtually a definition of the field of geography. In a sense, your study of geography in this book is contributing to national security.

in Afghanistan, the country had come to be known as the "Graveyard of Empires." Historians began to add America to the list of empires bloodied by Afghanistan.

We can now check our answers to the geographic question that opened the book: *Suggest two or three reasons why it has been difficult for the United States to be involved in Afghanistan.* Even if you knew very little or nothing about Afghanistan, you might have looked at it in this book or in an atlas and proposed that it is mountainous, making supply logistics difficult, and that it is landlocked, also complicating those logistics. Now that you have read this section, you can supply many more details about its geopolitical and other contexts (see A Closer Look, this page).

Faith, Sectarianism, and Strife

The Indian subcontinent is one of the world's culture hearths, rich in a historical legacy and modern diversity of ethnicities, social practices, and faiths. Ironically, this cultural wealth often threatens the well-being and security of the subcontinent's peoples.

There are major social and political divisions within the subcontinent that relate to religion. Religious differences between the two largest religious groups—Hindus (whose religion is described on pages 228–230) and Muslims (whose religion is described in Chapter 6)—are often troublesome. Hinduism was the dominant religion at the time Islam made its appearance in this region. Muslims are significant minorities in the region's non-Muslim countries, including 13 percent of the population of India and 4 percent of Nepal. Muslims make up a seemingly small percentage of India's population, but they number over 160 million people, making India home to the world's third largest Muslim population, behind Indonesia and Pakistan. Muslims are majorities in Pakistan (97 percent) and Bangladesh (90 percent).

Seldom in history have two large groups with such differing beliefs lived in such close association with each other as Hindus and Muslims in the subcontinent. Islam holds to an uncompromising monotheism and prescribes uniformity in religious beliefs and practices. Hinduism is monotheistic for some believers but polytheistic for others, and asserts that a variety of religious observances is consistent with the differing natures and social roles of humans. The exuberant and loud celebrations of the Hindu faith are a striking contrast to the austere ceremonies of Islam. Islam has a mission to convert others to the true religion, whereas most Hindus regard proselytizing as essentially useless and wrong; one is born or reborn as a Hindu, so conversion is not an issue. Islam's concept of the essential equality of all believers is in total contrast to the inequalities of the caste system endorsed in Hinduism (see page 248). Islam's use of the cow for food is an anathema to Hinduism. And then there are religion-based politics, which often turn differences into discord and conflict.

During the colonial era, the colonizing power frequently favored one ethnic or socioeconomic group over another,
sowing the seeds for eventual discord. This was the case in 339
British India, where the formerly subordinate Hindus came to dominate the civil service and most businesses. Many Muslims feared the results of being incorporated into a state with a Hindu majority, and their demands for political separation led in 1947 to the creation of two independent countries: the secular but predominantly Hindu nation of India (•Figure 7.27) and the Muslim nation of Pakistan (Figure 7.29). The respective secular and religious identities of both countries remain critical issues today. Pakistan had two parts, West Pakistan and East Pakistan, separated by 1,000 miles (c. 1,600 km) of Indian territory. In 1971, an Indian-supported revolt in East Pakistan led to the birth there of the new independent country of Bangladesh.

Immediately preceding and following partition in 1947, violence broke out between Hindus and Muslims, and hundreds of thousands of lives were lost in wholesale massacres. More than 15 million people migrated between the two countries. Pakistan today has a large population known as **Mohajirs**, or "migrants," the Muslim immigrants and their descendants who poured into the two Pakistans from India when partition occurred. Particularly in Karachi and elsewhere in the southern province of Sind, violent conflict frequently occurs between rival factions among the Mohajirs and between the Mohajirs and other ethnic groups, including Pashtuns (Pathans) and

• **Figure 7.27** Principal features of India.

Biharis. Large-scale migrations of ethnic groups like these are part of humanity's troubled experience.

The partition of colonial India precipitated all kinds of problems for the new countries. Some of them are still prominent today. The largest problem is the disputed Kashmir region. Since 1947, India and Pakistan have been in conflict over the status of Kashmir (called Jammu and Kashmir in India), a disputed province straddling their northern border (•**Figure 7.28**). Before independence, Kashmir was a princely state administered by a Hindu maharaja. About three-fourths of Kashmir's estimated population of 15 million is Muslim, which is the basis of Pakistan's claim to the territory. But under the partition arrangements, the ruler of each princely state was to have the right to join either India or Pakistan, as he preferred. Kashmir's Hindu ruler chose India, which is the legal basis of India's claim.

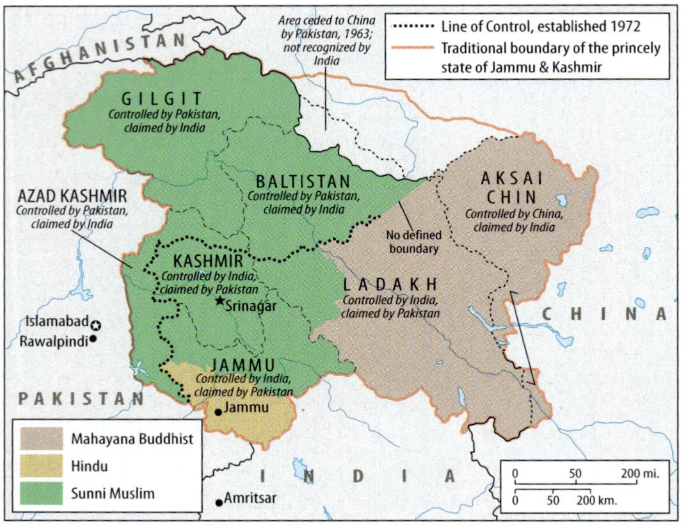

• **Figure 7.28** Political boundaries of Kashmir.

After partition, fighting between India and Pakistan led to the demarcation of a cease-fire line—the "line of control" that still separates the forces—leaving eastern Kashmir, with most of the state's population, in India and the more rugged western Kashmir in Pakistan (see **•Figure 7.29**). Beginning in the mid-1950s, China pressed its own claims to remote northern mountain areas of Kashmir and occupied some of the border territories. Pakistan ceded part of the occupied territory to China and established friendly relations with its giant northern neighbor, but India rejected China's claims (relations between these countries remain chilly). India's subsequent small-scale military actions failed to dislodge Chinese forces. India now holds about 55 percent of the old state of Kashmir; Pakistan, 30 percent; and China, 15 percent. India has another dispute with China over China's control of 34,750 square miles (90,000 sq km) in India's northeastern state of Arunachal Pradesh.

Three wars between India and Pakistan have effected little change in Kashmir. In 1965, conflict began in Kashmir, spread to the Punjab, and escalated to a brief but indecisive full-scale war involving tanks, airborne forces, and widespread air raids. Renewed hostilities in Kashmir in 1971, as part of the war in which India supported the revolt of Bangladesh against Pakistan, again did not alter the political landscape of Kashmir. During that conflict, the Siachen Glacier in the Karakoram Mountains, at about 20,000 feet (2,800 m), earned the title of "world's highest battlefield." As recently as 2011, negotiations to demilitarize the glacier have failed. Without this move, it is proving difficult for the two countries to normalize relations.

Kashmir's Muslim majority escalated the campaign for secession from India in 1989, and the strife has continued ever since. India and Pakistan came perilously close to the brink of war over Kashmir in 2002. Late in 2001, armed insurgents attacked India's parliament in New Delhi in a failed attempt to assassinate Indian politicians. India's government identified the assailants as members of two Muslim rebel movements (Lashkar-e-Taiba and Jaish-e-Muhammad) based in Kashmir. A far more spectacular attack—also attributed to these two groups, and with possible links to al-Qa'ida—took place in December 2008. In India's commercial capital of Mumbai, scores of civilians were killed and injured in multiple terrorist strikes against a Jewish center, transport stations, and hotels and restaurants frequented by Westerners.

The upper portion of the Indus River and many of its tributaries are in Kashmir (**•Figure 7.30**; see location in Figure 7.29), and both India and Pakistan seek as much control as possible over

• **Figure 7.29** Principal features of Pakistan.

Joe Hobbs

• **Figure 7.30** Western Kashmir, now in the hands of Pakistan, is a land of massive mountains and grinding glaciers. The world's second highest mountain, K2, is here, as is the "world's highest battlefield" (see page 246). People manage to eke out a livelihood almost all the way to timberline at 13,000 feet. This is the mighty Indus River, not far from K2 and from its source.

these waters. The geopolitical contest for this water threatens to become very dangerous. The militant Lashkar-e-Taiba has threatened to blow up India's dams in its portion of Kashmir. The **Indus Waters Treaty** of 1960 prescribed exactly how much water each side could use (80% to Pakistan, 20% to India), and India's dams have so far not violated that treaty. Pakistan fears, however, that India could use its dams as a weapon to deprive downstream Pakistan of water. Pakistan is sending chilling warnings to India, for example, in a newspaper editorial that read, "Pakistan should convey to India that war is possible on the issue of water and this time war will be a nuclear one."[14]

In the 2002 and 2008 incidents, Indian citizens and many government authorities laid blame on Pakistan's government. Foreign diplomats exerted tremendous efforts to avert war between the South Asian giants. Pakistan banned the militant groups after the 2002 assault (causing many Pakistanis to blame their government for "selling out" to the West following the 9/11 attacks), but the Mumbai attacks revealed how incapable Pakistan might be of preventing dangerous provocations against India. There are also questions about how much control Pakistani leaders have over their own intelligence service, the ISI, which may contain rogue elements intent on continued
238 support of the militants fighting India.

The Kashmir conflict has been very costly to both sides, with more than 65,000 people killed since 1965, about half of those since 1989. India's international reputation suffered because of the country's unyielding position on the region. Pakistan has steadily spent about a third of its budget on defense, mostly focused on a potential engagement with India over Kashmir. There would be economic benefits for both India and Pakistan if lasting peace were achieved. "Kashmir is a Paradise on Earth," wrote Samsar Kour.[15] The stunning snow-crowned peaks and flower-laden valleys of this western Himalayan region (more habitable than the Pakistani-held portion of Kashmir) once lured millions of Indian tourists and ranks of Western trekkers each year, and the tourism development potential is enormous.

Sectarian violence (known in the region as **communal violence**) between Hindus and Muslims is a frequent threat to India's social and political fabric. Contention over places sacred to both faiths sometimes ignites widespread violence. The sixteenth-century Babri Mosque in the city of Ayodhya in Uttar Pradesh (location in Figure 7.27) was a place of prayer for Muslims on a spot revered by the Hindus as the birthplace of the god-king Ram, an incarnation of the god Vishnu. Backed by Hindu nationalists in the provincial government, a mob of about 250,000 Hindu fundamentalists demolished the mosque in December 1992 with the intention of building a Hindu temple on the site. The ensuing communal violence between Hindus and Muslims throughout India, even among the cosmopolitan and usually tolerant urbanites of Mumbai, left thousands dead.

Ever since this incident, as a means of garnering votes from Hindu Indians, Hindu politicians have made veiled promises to proceed with construction of the temple. In 1998, the BJP platform included a pledge to build a Hindu temple atop the ruins of the mosque at Ayodhya. His apparent support for the temple helped sweep the BJP leader, Atal Vajpayee, into power as prime minister that year. A 2010 high court ruling divided the site, with two-thirds going to Hindus and one-third to Muslims. The conflict over this multireligious sacred space in India has parallels with the Temple Mount dispute between Israel and the Palestinians. Politicians in India today are increasingly aware of the threat 171
that Hindu-Muslim discord poses to the country's development.

In India's Punjab, where Sikhs make up 60 percent of the population and Hindus, 36 percent, the 1980s and early 1990s were violent years during which about 20,000 people died in armed clashes. Sikh factions were intent on establishing their own homeland of **Khalistan**, prompted to do so in part by New Delhi's plans to divert water from the Punjab. They challenged Indian authority and turned Amritsar's Golden Temple, the holiest site in the Sikh faith, into their military stronghold. In a controversial move to quell the revolt, Prime Minister Indira Gandhi ordered troops to storm the Golden Temple in June 1984. The resulting deaths and desecration led directly to Gandhi's assassination by her own Sikh bodyguards later that year. Indian authorities accused Pakistan of arming the Sikh militants.

Other trouble spots for India are the northeastern states of Assam, Manipur, Nagaland, and Arunachal Pradesh, an extraordinarily diverse region with more than 200 ethnic groups and followers of many different religions. Several rebel groups have been active in this region for years, agitating for autonomy and independence from India. The government has begun investing more resources in the area, especially infrastructure to boost trade ties with adjacent Myanmar. There are also troubles in half of India's states from Maoist insurgents, known as **Naxalites**, who claim to champion the causes of the rural poor and promise to carry violent rebellion into India's cities and ultimately expand their "compact revolutionary zone" to include the entire country. Some Maoists have sought support among India's roughly 700 tribal groups, whose population is about 85 million.

A conflict over land use has arisen in the eastern state of Orissa, where ancient tribal traditions are in a deadlock with modern industry. Steel companies want to tap the state's deposits of iron ore and coal and develop mills near them. But tribal people living in the forests atop those mineral deposits insist that their identity is tied to the land, forest, and water. If they were displaced by mining, they say, they would lose their cultural and social identities. 383, 404

Sri Lanka has two major ethnic groups, distinguished from each other by religion and language: the predominantly Buddhist **Sinhalese**, making up about 75 percent of the population, and the Hindu **Tamils**, about 10 percent (•Figure 7.31). The light-skinned Sinhalese are an Indo-Aryan people who settled in Sri Lanka about 2,000 years ago. The dark-skinned Tamils, whose main area of settlement is the Jaffna Peninsula and adjoining areas in the north, are descendants of early invaders and more recent imported tea plantation workers from southern India. The migrations of Tamil tea workers from India to Sri Lanka, and their subsequent travails there, are part of the legacy of British colonial rule. Later in this chapter, we will see how the British also relo- 295
cated Tamils to work the rubber plantations of Malaya. Ethnic discord in modern Malaysia is a legacy of British colonialism.

Sri Lanka's civil conflict reflected antagonisms between ethnic groups and discontent with economic and political conditions,

© Cengage Learning 2013

• **Figure 7.31** Principal features of Sri Lanka.

especially among the minority Tamils. Between 1983 and 2009, more than 70,000 people were killed and many more made homeless as the Tamils of the country's north have fought for autonomy or independence from Sri Lanka's majority Sinhalese government. Tamils complained that the Sinhalese treated them as second-class citizens and deprived them of many basic rights. Tamil fighters called the **Tamil Tigers**, officially the Liberation Tigers of Tamil Eelam (LTTE), wanted to establish their own homeland, Tamil Eelam, along the east, north, and west of the country. The rebels succeeded in holding wide swaths of northern Sri Lanka (see Figure 7.31). Unrest in rebel-held areas made it particularly difficult to rebuild northern sections of the area hit hard by the December
254 2004 tsunami. Government troops gradually pushed the Tamil forces back, and declared victory over them in 2009 (•**Figure 7.32**).

Joe Hobbs

• **Figure 7.32** The war between Tamils and Sinhalese has ended in Sri Lanka, but Tamil protests against Sinhalese rule continue where they can be practiced. This is a demonstration in Sydney, Australia. The depiction if of the Tamil Tiger founder and leader Velupillai Prabhakaran, who was killed just weeks before this demonstration. The red on the map on the right depicts parts of Sri Lanka held by the Tigers at their maximum advance. The Tamil flag, not fully in view, depicts a tiger jumping through a halo and crossed bayonets.

Arguably, one of the victors in the Sri Lankan civil war was China, which sided with the government and financed the construction of a port in the country's south. Some analysts argue that China will use this port to project its power into the Indian Ocean, where much of the world's oil and other valuable cargo is in transit. They see Sri Lanka as one pearl in China's **"String of Pearls"** strategy to gain critical allies in a line across the region. The largest "pearls" in this part of the string are China's allies Pakistan and Myanmar. India, which wants to see itself projecting power across these sea-lanes, views this as encirclement by China. Keep in mind that China and India are enemies and have been ever since they fought a high-altitude border war in 1962 (which India lost).

Known to early Arab seafarers as Seredib, the "island of Serendipity," Sri Lanka is a breathtakingly beautiful country. The return of peace to Sri Lanka should reopen the doors to international tourism.

The Caste System

One ancient Hindu belief is that every individual is born into a particular **caste**, a social subgroup that determines the individual's rank and role in society. Castes form a hierarchy, with the Brahmin caste at the top (comprising about 5 percent of all Hindus) and three others (Kshatriya, Vaisya, and Sudra) below. Caste membership is inherited and cannot be changed. Particular castes are associated with certain religious obligations, and their members are expected to follow traditional caste occupations. Marriage outside the caste is generally forbidden, and meals are usually taken only with fellow caste members.

At the bottom of the social ladder are the **Dalits**, or *scheduled castes,* once known as **untouchables**, accounting for about 20 percent of all Hindus. They are not part of the caste system but are "outcastes," and according to Hindu belief, they are not reincarnated. The untouchables were so called because they traditionally performed the worst jobs, such as the handling of corpses and garbage; therefore, their touch would defile caste Hindus. I will never forget the horrified look on my taxi driver's face when I shook hands with a man whose job was to cremate bodies on the "ghats" of Mumbai.

This ancient, rigid social system is changing. India's constitution outlawed the caste system in 1950, but a single decree could not undo thousands of years of tradition. Brahmin privileges are now being increasingly challenged and restricted by more laws. Indian law explicitly forbids recognition of Dalits as a separate social group. Confronting the fact that upper castes account for less than one-fifth of India's population but command more than half of the best government jobs, the Indian government has instituted an affirmative action program to allocate more jobs to members of lower castes. Indian leaders of high caste have championed the Dalits' cause for both moral and practical reasons. Politicians use caste divisions to their advantage to bring out voters, promising if elected to act in the interests of their respective large social blocs. In 1997, a Dalit was elected president of India for the first time in the country's history. Since then, Dalits have won more regional and local offices.

The Dalits are voting in greater numbers than ever before, with the conviction they can use India's democratic system to overcome ancient discrimination against them. They are making progress. Disintegration of the caste system has been especially rapid in the cities, where people of different castes must mingle in factories, in public eating places, and on public transportation. In India's vast rural areas, the caste system is more entrenched. There is also a north–south difference: lower caste southern Indians began agitating against the caste system early in the twentieth century and kept up their struggle. In northern India, by contrast, there was less effort to tackle the caste system. Caste is so important in northern politics that it is said that voters don't cast their vote; they vote their caste. Remarkably, partly because of the loosening of caste restraints, economic growth is much stronger in southern than in northern India.

Keeping Malthus at Bay

Poverty and human health are serious problems in South Asia, and the prospect of continued high population growth raises the question of whether growth in food supplies can avert the proverbial Malthusian crisis (see Chapter 3).

India's population has surged since independence, from 352 million in 1947 to 1.2 billion in 2011. In recent decades, the overall trend has been toward lower population growth rates. Fifty years ago, the average Indian woman had six children; now she has three. That rate is still well above the natural replacement rate, however, that would keep the population steady. The population base is already so vast, and the base of its age structure diagram so wide (about 50 percent of the population is younger than 25), that even modest growth will
53 add huge numbers. India is predicted to overtake China as the world's most populous country by 2040. There is uncertainty about how high India's population will grow. Estimates range from 1.5 to 1.9 billion. Most Indian decision makers think that even the low end of that range is too high. Unlike China, India is a democracy and cannot simply declare how many children a couple can have. Indian states rather than the central government establish laws and policies, and there is a wide range of them in the realm of population. Some states offer financial incentives, such as a $100 bonus if couples delay their first child's birth.

For many people, the name *India* invokes an image of grinding poverty, and with good reason: it is estimated that fully 40 percent of the population is "abjectly poor," defined as living on less than $1.25 per day, and almost half of India's children are malnourished. But perhaps in defiance of Malthus, an increased population has so far succeeded in producing more economic resources overall. A few people in India are very wealthy, and there is an emerging middle class of as many as 300 million people—the world's largest middle class (**•Figure 7.33**). India's economy has grown impressively in recent years. An apparent problem is that the benefits of the growing economy are unevenly distributed. Most of the economic growth benefits those who are already better off, widening the gap between rich and poor.

The issue of population growth is also critical for Pakistan, which has never mounted an effective family planning program, mainly because of opposition from Muslim religious authorities, who regard birth control as an intervention against God's will

Joe Hobbs

• Figure 7.33 This Indian family visiting the Red Fort in Agra is among the growing ranks of the subcontinent's middle class. Note the Taj Mahal gleaming in the distance.

(Pakistan's birth rate in 2011 was 25 per 1,000, compared with 21 per 1,000 in India). Muslim Bangladesh, along with Nepal and Bhutan, also have high birth rates, but they are much lower in Bangladesh than they once were. In the 1970s, the average woman had seven children, whereas now she has two, and contraception use nationwide has increased from 4 percent then to more than 50 percent now. South Asia's success story in bringing population growth under control is Sri Lanka, with a birth rate of just 17 per 1,000 and an annual population growth of 1.3 percent.

Agricultural output in South Asia has increased since independence. Most notably, despite its huge and growing population, India had achieved self-sufficiency in grain production, with a fourfold increase in output since 1950. Almost half a million "ration shops" sell subsidized food staples to the country's poorest people. The successes of agriculture in the subcontinent have been due mainly to the increased use of artificial fertilizers, the introduction of new high-yield varieties of wheat and rice associated with the Green Revolution, more labor provided by the growing rural population, the spread of education, the development of government extension institutions to aid farmers, and better irrigation. "Mother Nature," however, has the last word. Sixty percent of India's farmland is rain-fed rather than irrigated, and when summer monsoons fail to deliver rainfall, as they did in 2009, the yields of valuable rice, sugarcane, and other crops plummet. This causes immediate difficulties for people (in the form of higher food prices) and the nation (in terms of trade and economic growth).

Many experts in agriculture and economics are calling for a "Second Green Revolution" in India to address agriculture in the context of climate change. The focus would be more nuanced than the Green Revolution of the 1960s. Now there is a need for nationwide cooperation to improve irrigation, capture rainwater and conserve groundwater while boosting crop yields. There is much at stake, not just food security but human well-being and equity. During droughts like the one in 2009, debt-ridden farmers resort to the dire options of selling their
wives or daughters (for $83 to $248) or committing suicide. 145

As the sale of wives and daughters suggests, one of the challenges confronting both rural and urban India is to improve the

status of women. Despite a 1961 ban on dowries (money and gifts given by a bride's parents to the groom), the practice is continuing. So is the rate of killing brides who do not provide enough dowry. The burden placed on the bride's family is one reason that many prospective parents choose to abort female fetuses, which
213 can be detected with ultrasound technology. India banned this use of ultrasound in 1996, but the practice continues. Female feticide—particularly prevalent among India's more educated and affluent, who can afford the ultrasound screening—and infanticide have taken tens of millions of young and unborn girls. In some of India's states, especially Haryana, Punjab, Delhi, and Gujarat, one result is a serious shortage of local brides; for every 1,000 men, there are 900 women. The gap is being filled by marriages
276 arranged with women from distant Indian states and Bangladesh.

Low-Lying Bangladesh and Maldives: Canaries in the Climate Change Coal Mine?

Formerly East Pakistan, Bangladesh is about the size of the state of Iowa or the nation of Greece—not an important fact until one considers that it is home to 150 million people (•**Figure 7.34**). With its limited resources, people-overpopulated Bangladesh is South Asia's third-poorest country on a per capita GNI PPP basis, after mountainous Bhutan and Nepal.

The Bangladesh portion of the Ganges-Brahmaputra Delta ranks, along with the Indonesian island of Java, as one of the two most crowded agricultural areas on earth. This area is subject to catastrophic flooding. Massive tropical cyclones (hurricanes) ravage Bangladesh regularly in September and October. Deforestation upstream on the steep slopes of the Himalayas has increased water runoff and sediment load in the Ganges, also contributing to Bangladesh's flood problems (the region's principal rivers may be seen in Figure 7.34).

Ironically, although there are concerns about climate change–related flooding in Bangladesh, there are even more worries about climate change–related water shortages and droughts. Climate change models point toward accelerating depletion of the

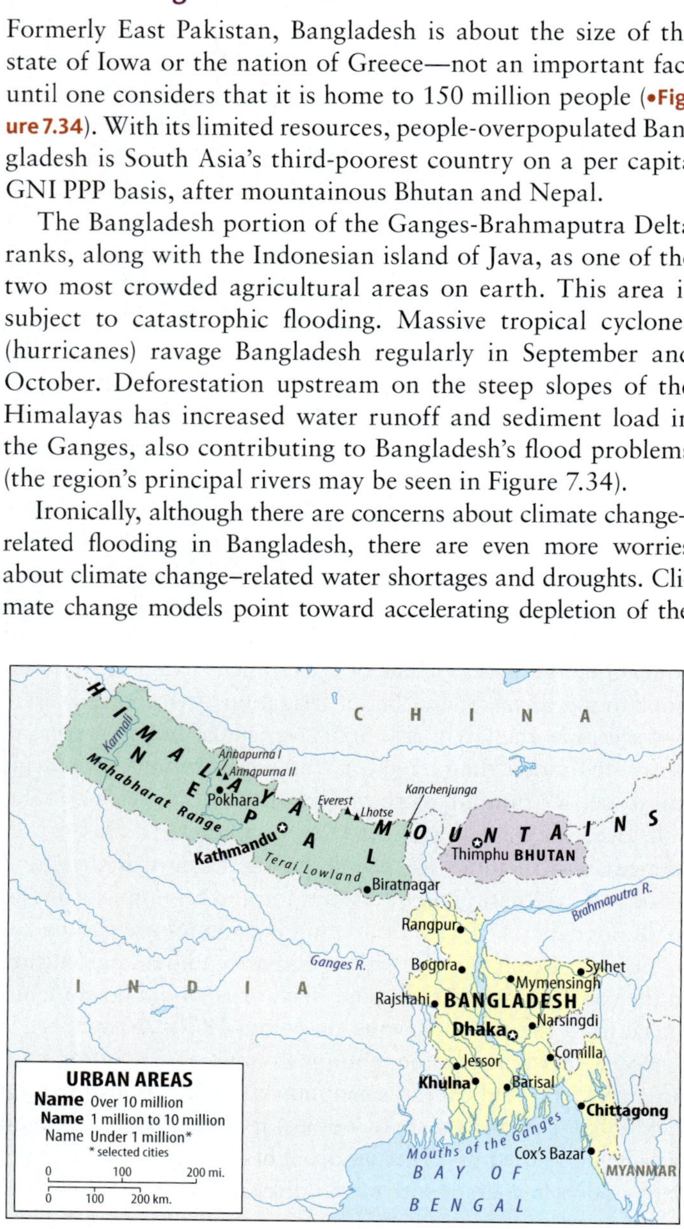

• **Figure 7.34** Principal features of Bangladesh, Nepal, and Bhutan.

• **Figure 7.35** Melting of glacial ice in the Himalaya and other great mountain ranges of South Asia threatens future water supplies for people downstream. This is Gokyo Glacier, close to Nepal's Mt. Everest.

so-called **"Water Towers of Asia"**—the glaciers and snow cover of the Himalaya, Karakoram, and other great mountain ranges that feed the region's 10 great rivers and scores of smaller ones. The perennial ice mass here is exceeded in volume only by the ice of the Arctic and Antarctic, earning the Tibetan Plateau and environs the name "The Third Pole." The region's roughly 18,000 glaciers, like most others in the world, are melting at a steady, relentless pace (•**Figure 7.35**). With temperatures on the Tibetan plateau rising much faster than the world average, a respected Chinese glaciologist predicts that two-thirds of them will be gone by 2050. 76,
These glaciers supply water to about 40 percent of the world's 358,
population! The prospect of their melting away is so alarming that 428
longstanding enemies India and China are putting away their differences and collaborating in scientific research on climate change.

Bangladesh and adjacent portions of India must also keep a wary eye on the potential for the sea level to rise if the world's temperatures increase, as predicted by many climate change models. The country's environment minister has forecast that 20 per- 33
cent of Bangladesh will be underwater by 2020 if nothing is done to control global warming. The Intergovernmental Panel on Climate Change recognizes this as one the world's most endangered areas, predicting a rise in sea level there of 23 inches (c. 58 cm) by 2100. On the Indian side of the delta, scientists are documenting the shrinkage of the river delta islands known as the Sundarbans. Renowned as habitat for the Bengal tiger, the Sundarbans are succumbing to a combination of sea level rise and deforestation that makes them very prone to flooding and erosion.

The roughly 1,100 islands that make up the Maldives (population 300,000; capital, Male, population 90,000), an independent country since 1965, appear to be the very essence of tropical paradise (•**Figure 7.36**). So prized are its palm-blessed and coral-fringed beaches that more than half a million tourists, mainly Europeans, visit the country each year. More than 60 percent of the country's foreign-currency earnings come from tourism, with fishing and clothing providing most of the rest. Some sources list the Maldives as South Asia's wealthiest country on a per capita basis. The greatest wealth is in the

Peter Essick/Getty Images

• **Figure 7.36** This aerial view of Male (pronounced *mah*-lay), the Maldives' capital, reveals how vulnerable the site is to storms and sea level rise. This is the world's most densely populated city.

tourist sector, where it is concentrated in so few hands that a potentially dangerous rich–poor divide has developed.

There may be other troubles in paradise. More than 80 percent of this country's very limited land area consists of limestone atolls less than 3 feet (90 cm) above the sea. If sea levels were to rise, as predicted by many common climate change scenarios, the entire country could rapidly be submerged. This prospect led to the Maldives' president's famous cry at a climate change conference: "We are an endangered nation!"

Southeast Asia

Deforestation of Southeast Asia

Southeast Asia today is composed politically of 11 countries: Myanmar (formerly Burma), Thailand, Laos, Cambodia, Vietnam, Malaysia, Singapore, Indonesia, Timor-Leste (formerly East Timor), Brunei, and the Philippines (•**Figures 7.37 and 7.38**). With the exception of Thailand (which was never colonized) and

• **Figure 7.37** Principal features of Myanmar, Thailand, Cambodia, Laos, Vietnam, Malaysia, Brunei, Singapore, and the Philippines.

• **Figure 7.38** Principal features of Indonesia and Timor-Leste.

Timor-Leste (which gained independence first from Portugal and then from Indonesia), all of these states became independent from colonial powers after 1946. These 11 nations are fragmented both politically and topographically. Many of the countries are also culturally fragmented and have experienced strife among different ethnic groups. Outside intervention has sometimes complicated and worsened local discord, producing enormous suffering in the region and lasting trauma among many of the foreign soldiers who fought the determined inhabitants of Southeast Asia. The region also must contend with environmental problems related both to its economic progress and its lingering poverty.

Near areas of dense settlement in Southeast Asia, there are still some large areas covered in primary forest, but they are shrinking as population expands and development progresses. Many scientists view the destruction of the region's tropical rain forest as an international environmental problem, especially because of its contributions to climate change (•**Figure 7.39**). In 2012, only 35 percent of Thailand's original forest cover remained. Poorer neighbor Myanmar had about a 45 percent forest cover. Vietnam lost more than one-third of its forest between 1985 and 2012, by which time only about one-third of the original forest cover remained. Deforestation is advancing most rapidly in Malaysia. Officially, 63 percent of the land area is still forested. However, official statistics include palm oil plantations, so these figures should be evaluated carefully. Only 11 percent of Malaysia's remaining forests are considered "pristine". Deforestation is also progressing quickly in Indonesia, where about half the land is forested. Malaysia is now the world's largest exporter of tropical hardwoods. Virtually all of the primary forests in Peninsular Malaysia have already been cleared, and it is projected that commercial logging will decimate those remaining outside protected areas in Malaysian Borneo quite soon.

Although Southeast Asia's tropical forests are smaller in total area than those of central Africa and the Amazon Basin, they are being destroyed at a much faster rate. The main forces of deforestation are commercial logging for Japanese and Chinese markets, and clearing of forest to plant oil palm trees, from which palm oil is rendered (•**Figure 7.40**). The demand for palm oil is soaring for use in biofuels and in household

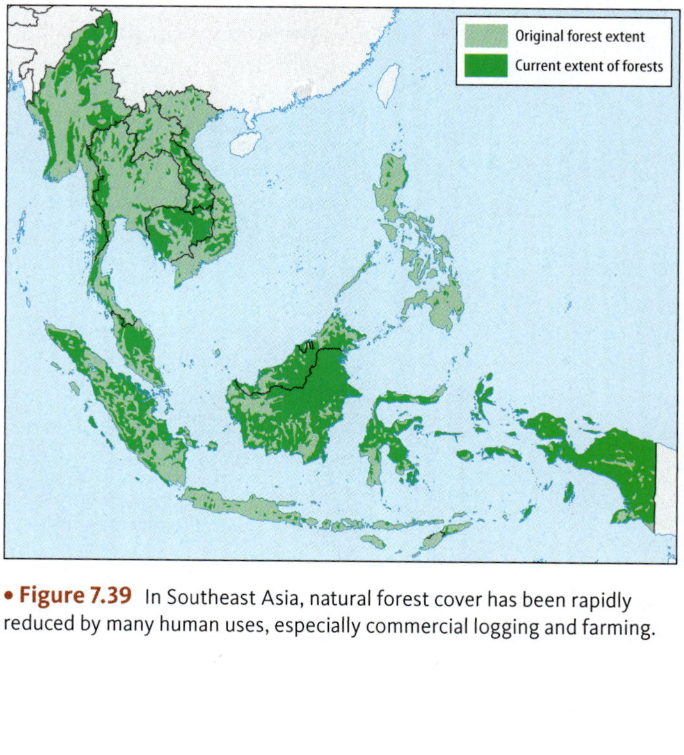

• **Figure 7.39** In Southeast Asia, natural forest cover has been rapidly reduced by many human uses, especially commercial logging and farming.

• **Figure 7.40** Over much of the island of Borneo, tropical rain forest has been replaced by palm oil plantations.

products including Vaseline and Dove soap, and food including Kit Kat candy bars. Environmentalists fear the irretrievable loss of plant and animal species and the potential contribution to global warming that deforestation in this region may cause.

Remarkably, the destruction and burning of forests on the islands of Borneo and Sumatra have made Indonesia the world's third largest emitter of carbon dioxide, after China and the United States. There is a new scientific consensus that a fifth of the world's greenhouse gas emissions is due to land use change, with Indonesia and Brazil the most responsible. Indonesia accounts for 40 percent of that; thus the cutting and burning of rain forest mainly to produce palm oil there is responsible for 8 percent of global greenhouse gas emissions. The conversion of peat bogs—some inhabited by Sumatran tigers and bears—to tree plantations for paper and pulp also releases carbon dioxide. As deep as 50 feet, and made up of decomposed trees and plants, the peat stores billions of tons of carbon dioxide.

Due to its large contributions to greenhouse gases through forest conversion, Indonesia is negotiating many projects for the Reducing Emissions from Deforestation and Forest Degradation (REDD) program described in Chapter 2 (see page 35). One major paper and pulp company working in Indonesia is negotiating a REDD deal in which by protecting forests it will receive carbon credits that it can sell on the global carbon exchange for large sums of money. 34

Environmentalists have put pressure on large companies like Cargill, Unilever, and Nestlé to stop buying palm oil produced unsustainably. Nestlé has responded favorably, promising a "zero deforestation" policy. Malaysia and Indonesia have passed laws to slow the destruction, with Indonesia taking the aggressive step of banning the use of fire to clear forests. However, enforcing such legislation has so far proved impossible. Hundreds of Indonesian and Malaysian companies, most of them large agricultural concerns (for example, palm oil plantations and pulp and paper companies) with close ties to the government, continue
to use fire as a cheap and illegal method of clearing forests. The 382
World Wildlife Fund estimates that 70 percent of Indonesia's exported trees are illegal, many of them removed by "timber barons" in the country's national parks while bribed park and police officials look the other way. The urgent need for construction materials after the devastating tsunami of 2004 (see p. 254) further increased the pressure on the country's forests.

There is a growing movement in the consuming countries to boycott the tainted timber by selling only certified "good wood" and only palm oil that has been produced from sustainably harvested oil palms. The U.S. company Home Depot, for example, sells only imported woods bearing the Forest Stewardship Council (FSC) logo. Even the source countries are coming on board the certification bandwagon. Malaysia began work on an agreement with the European Union in 2011 to become the first country to sign a voluntary agreement guaranteeing that all tim-
ber exports are harvested legally. Malaysia would promise the 330
European Union that its trees would not be cut from protected forests and that endangered tree species would not be harvested.

Deforestation in Southeast Asia has many actual and potential transboundary consequences. In 1997, a widespread

drought attributed to the warming of Pacific Ocean waters by the phenomenon known as El Niño turned the annual July–
358 October burning season into a conflagration of human origin. Deliberately set fires raced out of control over large areas of Sumatra and Borneo, resulting in a choking haze that shut down airports, closed schools, deterred tourists, and caused respiratory distress to millions of people in Indonesia, Malaysia, Singapore, the Philippines, Brunei, and Thailand. Less widespread but still severe smog crises struck Malaysia and Indonesia, particularly the island of Sumatra, in subsequent years, and Indonesian authorities have fought back by revoking the licenses of plantation companies.

Many of the plants and animals in these forests are **endemic**
344 **species** (found nowhere else on earth). Indonesia, which contains 10 percent of the world's tropical rain forests, is known, after Brazil, as the world's second most important **megadiversity country**, with about 11 percent of all the world's plant species, 12 percent of all mammal species, and 17 percent of
377 all bird species within its borders. The Southeast Asian region is also particularly significant in biogeographic terms because of the so-called **Wallace's Line**—named for its discoverer, the English naturalist Alfred Russel Wallace—that divides it. Nowhere else on earth is there such a striking local change in the composition of plant and animal species in such a small area on either side of this divide. East of the line (separating the Indonesian islands of Bali and Lombok, for example), marsupials are the predominant mammals, and placental mammals prevail west of the line. Similarly, bird populations are remarkably different on either side of the line.

The Great Tsunami of 2004

On December 26, 2004, the most cataclysmic natural event of modern times afflicted a vast portion of the earth. Beneath the Indian Ocean, just off the northwestern coast of the Indonesian island of Sumatra, a gargantuan "megathrust" earthquake measuring magnitude 9.3—with the force of more than 26 megatons of TNT, or 1500 times the power of the nuclear weapon that devastated Hiroshima—struck at 6:58 A.M. local time. In the Java (Sunda) Trench, as much as 750 miles (1,200 km) of fault line slipped 60 feet (over 18 m) along the subduction zone where the Indian Plate is sliding beneath the Burma Plate (part of the greater Eurasian Plate; see Figure 2.1).

As the seabed of the Burma Plate instantly rose several meters vertically above the Indian Plate, the massive displacement of seawater created a series of tsunamis (*tsunami* is Japanese for "harbor wave"; the phenomenon is sometimes popularly but incorrectly referred to as a "tidal wave") that pulsed across the Andaman Sea and the Indian Ocean at up to 500 miles (800 km) per hour. On the open ocean, such waves are barely perceptible, but they break on coastlines with the force of giant storm surges. The tsunami roared ashore, at heights up to 45 feet (14 m), in 14 countries: Indonesia, Malaysia, Thailand, Myanmar, Bangladesh, India, Sri Lanka, the Maldives, Seychelles, Somalia, Kenya, Tanzania, Yemen, and South Africa (•**Figure 7.41**).

These were the deadliest tsunamis in recorded history. The total number of dead exceeded 200,000. By far the greatest number of deaths (over 130,000) was in Indonesia, where Aceh's principal city of Banda Aceh was obliterated (•**Figure 7.42**). Sri Lanka lost 35,000; India, 12,000; and Thailand, over 5,000. Almost all the deaths occurred within a mile of the shoreline and were caused by blunt force from debris and by drowning. An estimated one-third of the fatalities were children. There were reports of children running out to investigate fish stranded by retreating seas—a characteristic precursor to a tsunami—only to be struck seconds later by the wall of water.

As many as 2 million people were made homeless by the disaster. An unprecedented international relief effort, designed to feed and house these refugees and to contain the spread of epidemic diseases, began within hours. In many cases, it was clear that recovery efforts were profoundly successful, sometimes in unexpected ways. In India's Tamil Nadu State, for

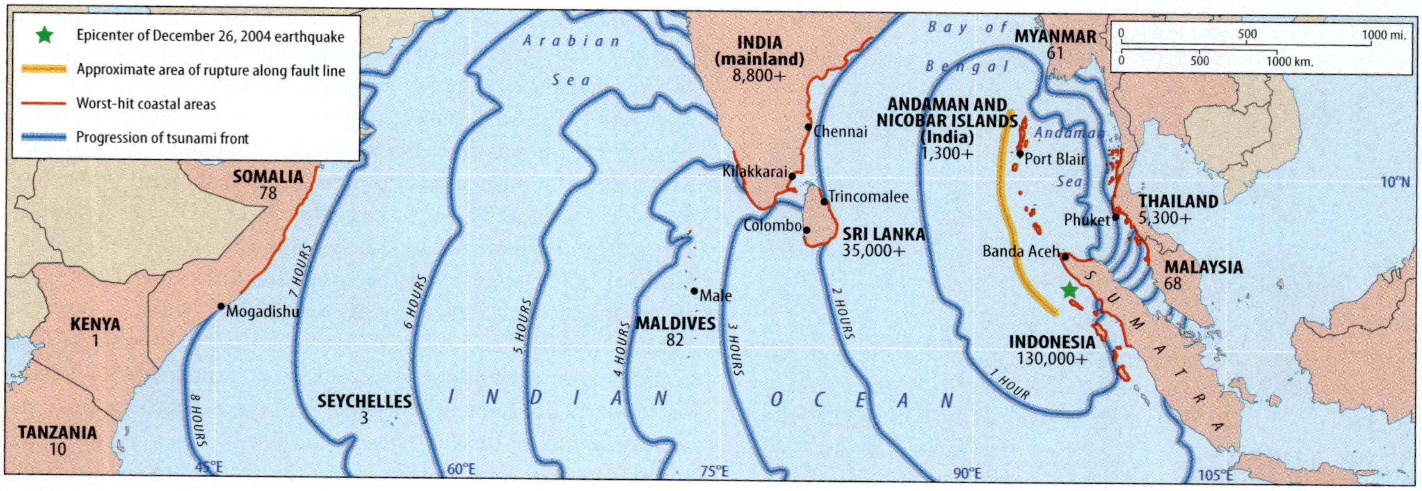

• **Figure 7.41** The epicenter of the 2004 earthquake, which registered magnitude 9.3, was located off the coast of Sumatra, Indonesia. It created a tsunami of tremendous force and scope.

• **Figure 7.42** Banda Aceh before and after the tsunami.

example, there were new roads, housing and utilities, and jobs training. New homes were titled in women's names (because women were deemed more trustworthy than men), giving women an unprecedented rise in status as concerns respect. In the tsunami's wake, the social fabric was transformed.

Tsunami experts concur that countless thousands of lives would have been saved had the affected Indian Ocean region been equipped with the same kind of tsunami early warning system now in place in the Pacific Ocean.[16] That investment had not been made because historically about 90 percent of the world's tsunamis have occurred in the Pacific and because many of the nations afflicted lacked the financial resources for the wave sensors and other infrastructure required by the system. Installing an early warning system in the Indian Ocean basin became an urgent priority, and a system was up and running by 2006. Had this system been operational in 2004, Indonesia would have had just a few minutes' warning, but Thailand would have had two hours, and Somalia, seven hours to prepare for the disaster.

Japan's tragic tsunami in 2011 is discussed on page 16. Its mechanism was similar: an upthrust earthquake along a section of underwater plates. Its magnitude was slightly less (9.0 versus 9.3), and the loss of life in wealthy, tsunami-prepared Japan far less (15,000 versus 200,000). It was such a massive tsunami, however, that Japan's engineering works were simply overwhelmed.

Misrule in Myanmar

Myanmar (population 54 million, Figure 7.37) is centered in the basin of the Irrawaddy River and includes surrounding uplands and mountains.

Britain conquered what was then called Burma in three wars between 1824 and 1885. It became an independent republic outside the British Commonwealth in 1948. Civil war has been almost constant since independence in a country where ethnic minorities make up about 40 percent of the population. Communism, targeted persecution, and ethnic separatism have motivated rebellions by ethnic Karens and Shans, a number of less numerous hill peoples, and Muslims of the Rohingya ethnic group who fear the establishment of a pure Buddhist state. Beginning around 1999, however, Myanmar's ruling military government reached cease-fire agreements with most of these ethnic groups (except the Karens) and gave these groups what it called **contingent sovereignty**, offering them more civil rights and economic opportunities. The government also encouraged profits made by some of these groups in the drug business to be "invested" in national development.

The economy has been badly damaged by fighting and for years was mismanaged under a form of rigid state control promoted as the **"Burmese Way to Socialism."** Myanmar became the poorest non-Communist country in Southeast Asia, a sad contrast with its reputation in early independence years for having the best health care, highest literacy rate, and most efficient civil service in Southeast Asia. It used to be the world's largest rice exporter; it now exports less than one percent of the world total. Myanmar's government spends comparatively less on health care for its citizens than all but three countries in the world.

In the 1990s, Myanmar began to abandon socialism in favor of a free-market economy and enjoyed considerable economic growth, with China its main trading partner and military ally. Most of the fruits of this economic growth, however, lined the pockets of the ruling military. The **State Law and Order Restoration Council (SLORC)**, the military government that seized power in Burma in 1988 (and changed the country's name to Myanmar), has yielded little to popular pressure for democratization. Myanmar remains one of the world's most repressive places to live. Access to the Internet was prohibited until 1999 and is still strictly regulated. Foreign journalists are banned, and citizens may not allow foreigners into their homes. It is illegal to gather outside in groups of more than five. Giant green billboards across the country read: "Crush all internal and external destructive elements as the common enemy."

In an apparent move to gain strategic depth from its own population and from any outside threats, the government in 2005 established a new capital at Naypyidaw, in the country's remote central region. The move did nothing to quell popular discontent with the regime. Widespread anti-government protests, with broad participation of Buddhist monks, broke out all over the country late in 2007. Authorities responded with deadly force and the insurgency faded—at least temporarily.

In the wake of this political turmoil, Myanmar's isolated leaders turned a natural disaster into a humanitarian catastrophe. On May 2, 2008, a category 4 strength typhoon (hurricane) named Nargis slammed ashore across the Irrawaddy Delta. In this most agriculturally productive and densely populated region of Myanmar, more than 135,000 people lost their lives, and an estimated two and a half million lost their homes and livelihoods. The country's junta largely banned foreign aid agencies from providing medical, food, and other relief supplies. Foreign journalists were also barred entry, so the real impacts of the tragedy may never be known.

The Lady of Burma

A much publicized contest, followed around the world, has been raging between the government and one of its citizens, the Nobel Peace Prize laureate Aung San Suu Kyi, who had been placed under house arrest after leading her antimilitary political party to an overwhelming victory in the 1990 elections (which the military government nullified). In 1995, the government released "The Lady," as Suu Kyi is popularly known. She then called for dialogue with the military junta and appealed for the parliament that was elected in 1990 to be convened. The government responded by again placing her alternatively under house arrest and in prison, triggering strong economic sanctions by the United States.

Aung San Suu Kyi has called for would-be tourists to avoid Myanmar, which has abundant sites of interest but few visitors (•**Figure 7.43**). She argues that tourism revenue will help strengthen the repressive government. She has also been successful in gaining support abroad for boycotts of U.S. and other Western companies doing business with Myanmar. American corporations including PepsiCo and ConocoPhillips have withdrawn from Myanmar, and the U.S. government has banned new investments. The United States regards the country's government as illegitimate and therefore rejects the change of the country's name to Myanmar.

Gail Ludwig

• **Figure 7.43** The huge golden Shwedagon Pagoda is the holiest Buddhist shrine in Yangon, Myanmar, attracting multitudes of the faithful and the few tourists visiting the country. It is said that the relics or bones of three earlier buddhas and the eight hairs of Gautama Buddha are located within this spectacular golden architectural masterpiece.

The U.S. government was displeased when the American firm Unocal built a pipeline from the offshore Yadana natural gas field to a port west of Bangkok, Thailand, where the gas is used to generate electricity. Human rights groups complained that slave labor was used to build the pipeline and that profits from gas sales help prop up Myanmar's repressive government. The pipeline has to be well defended in view of the government's many domestic opponents who might try to destroy it as a means of weakening the regime. Other pipelines will serve China's energy appetite. One will carry Myanmar's offshore natural gas across the country to southern China. A parallel pipeline will carry Middle Eastern oil offloaded on the coast of Burma and piped to southern China. This will increase China's energy security, in part by avoiding the congested chokepoint
of the Strait of Malacca. India is vying with China for access to 240
Myanmar's natural gas, copper, and other natural resources. 251

The United States has been reevaluating its policies toward Myanmar, especially for an important geographical reason: its geographic *situation* is extremely important, as the country sits right between the two great powers of China and India. By withdrawing from Myanmar and imposing sanctions against it, the United States has been missing out on important commercial and strategic engagements with the country, pushing Myanmar into the arms of its rival China and enemy North Korea, and with sanctions effectively punishing the people rather than the government.

The long isolation of Myanmar may be ending, perhaps because Nargis forced the country to open its doors a little bit to allow food aid. Many changes came in 2011, when the 10 countries of the regional organization known as the **Association of Southeast Asian Nations** (ASEAN) agreed to allow its member state Myanmar to chair the bloc in 2014. In 2011, Myanmar made changes to its economy, released hundreds of political prisoners, eased some restrictions on the Internet, media, and political parties, stopped a dam project that was backed by China, and released Aung San Suu Kyi from house arrest again. In 2012, the ruling party allowed Aung San Suu Kyi to run in national elections. She won a seat in Parliament. In the same year the U.S. and Myanmar restored diplomatic relations; they had broken in 1990.

Sex, Drugs, and Health in Southeast Asia

Southeast Asia has long been one of the world's main source areas for opium (sometimes eclipsing Afghanistan) and its
highly addictive semisynthetic derivative, heroin. For decades, 242
global concern focused on combating the drug trade originating in this region and on dealing with the problem of heroin addiction in the consuming nations. More recently, owing to the emergence of HIV/AIDS, the Southeast Asian industries of prostitution and tourism have woven drugs into a complex problem of global concern.

The region's drug production is centered in the so-called **Golden Triangle**, a vernacular region comprised of the

• **Figure 7.44** Raw opium oozes from slices in a poppy pod. Heroin, which is derived from opium, has a hearth in Southeast Asia's Golden Triangle region.

borderlands where Laos, Thailand, and Myanmar historically exercised little control over their territories (see Figure 7.37). As in Afghanistan, for a long time the absence of strong government presence together with ideal growing conditions facilitated ex-
242 plosive growth in drug production (•**Figure 7.44**). But as Thailand and Laos exercised increasing authority in the area, drug production plummeted. Today, Myanmar is the only one of the three countries with a strong opium poppy economy, but even there, production has declined. The main reason is that neighboring China, with an eye to safe, conventional trade with Myanmar, is pressuring authorities there to crack down near their common border. As poppy cultivation slumped in the Golden Triangle, it spiked in the Golden Crescent of Afghanistan. This phenomenon, in which cracking down on a problem in one place causes it to pop up in another, is called the balloon effect and applies to
374 terrorism as well as to drugs.

In Asia, the U.S. Drug Enforcement Administration and other international antidrug authorities focus their concerns on Myanmar, where refining of heroin from opium is done in remote laboratories. In northeastern Myanmar, armed ethnic minority groups have produced both heroin and a potent methamphetamine known as *ya baa*. They smuggle raw opium into both China and India, and amphetamines into Thailand, providing investment capital to contribute to Myanmar's "development." Laundered drug money has paid for the country's recent construction boom, and the government encourages such investment from the drug sector to make up for shortfalls in legitimate export revenues.

Not all of the associated problems are exported. A flood of cheap and potent heroin hit the market in Myanmar after the current regime came to power in 1988. Heroin addiction has soared ever since. Primary delivery is by needles shared in tea stalls. The number of users continues to grow; perhaps 1 percent of the adult population is addicted. Accompanying the heroin use is an AIDS epidemic, with about 1 percent of the adult population infected. The best available figures indicate that the HIV infection rate among heroin users in Myanmar is around 37 percent.

The epidemic that began among heroin users spread quickly to Myanmar's sex industry. Recent estimates put the HIV infection rate among prostitutes at 26 percent, three times the rate in neighboring Thailand. Many young men with AIDS are shunned by their families and go to monasteries to die. Tragically, their virus has spread among the populations of Buddhist monks (there are now 400,000 in Myanmar), who share razors to shave their heads. AIDS prevention is virtually unheard of. The government had outlawed condoms until 1993. They are now legal, and over 30 million have been distributed by the government since 2001, but this is still short of what is needed. Practically no anti-HIV drugs find their way to Myanmar.

Myanmar's prosperous, cosmopolitan neighbor Thailand also has the scourge, with a 1.3 percent HIV infection rate among adults in 2011. This is a substantial reduction from the peak infection rate the country recorded in 1994. Thailand's liberal social climate has given rise to the unique phenomenon of sex tourism, with package tours catering to an international clientele (•**Figure 7.45**). Thailand's overall annual income from prostitution is estimated to be up to $4 billion, with prostitutes sending large sums as remittances each year to relatives in the villages from which they migrated. However, the spread of HIV infection among intravenous drug users and among prostitutes in the red-light districts of Bangkok and the popular beach resorts of Pattaya and Phuket threatens both the country's public health and its lucrative tourist trade. The Thai government mounted an all-out war on the drug trade in 2002 that resulted in the deaths of more than 2,000 people, most of them suspected drug dealers.

There are striking differences in how this region's countries are dealing with HIV/AIDS. Thailand, regarded globally as

• **Figure 7.45** Bangkok's red-light district was a breeding ground for HIV/AIDS until the government and nongovernmental agencies aggressively promoted a safe-sex campaign.

a leader in efforts to combat the problem, has run an aggressive and increasingly successful anti-AIDS public awareness campaign, resulting in a dramatic decrease in the number of new HIV infections. Thailand distributes antiretroviral drugs to all AIDS patients at extremely low cost. Other Southeast Asian countries lag far behind. The government of the Philippines has been unsuccessful in pursuing an anti-AIDS campaign because of opposition from the country's powerful and anti-contraception Catholic clergy. Church officials have denounced the government's anti-AIDS program as "intrinsically evil" and have set boxes of condoms on fire at anti-government demonstrations. Similarly, in the conservative and largely Islamic nations of Indonesia and Malaysia, Muslim clerics denounce their governments' anti-AIDS campaigns as efforts to encourage promiscuity.

Vietnam Then and Now

The Vietnam War, which is such an important chapter in the American experience, had roots that preceded U.S. interests in the region. This conflict profoundly affected Southeast Asia and has a lingering legacy there today.

France conquered Indochina between 1858 and 1907. The French first extinguished the Vietnamese empire, which covered approximately the territory of today's Vietnam, and defeated the Chinese, whom the Vietnamese called on for help. Later, the French took Laos (in 1893) and western Cambodia (in 1907) from Thailand. France's administration of Indochina, centered in Hanoi and Saigon, left a big cultural imprint. Its major economic success lay in opening the lower Mekong River area to commercial rice production, converting both Vietnam and Cambodia into important exporters of rice.

Japanese forces overran French Indochina in 1941, initiating five decades of warfare in the area. During World War II, a Communist resistance movement led by a Vietnamese nationalist named Ho Chi Minh carried on guerrilla warfare against the Japanese. When French forces attempted to reoccupy the area after the end of war, these forces fought to expel them. In 1954, Ho's forces destroyed an entrenched French garrison at Dien Bien Phu, in the northwestern mountains. With this defeat, France withdrew from Indochina, ending a century of colonial occupation.

Four countries came into existence with France's departure (**see Figure 7.37**). Laos and Cambodia became independent non-Communist states. North Vietnam, where resistance to France had centered, emerged as an independent Communist state. South Vietnam gained independence as a non-Communist state that received many anti-Communist refugees from the north. Vietnamese Catholics were prominent among those who fled southward.

The partition of Vietnam into two countries was largely the work of U.S. power and diplomacy. South Vietnam was from the outset a client state of the United States, whereas North Vietnam became allied first with Communist China and then with the Soviet Union. Warfare in Vietnam gained momentum. Between 1954 and 1965, the **Viet Cong**, an insurrectionist Communist force supported by North Vietnam, achieved increasing successes in the south in its bid to reunify the country. Increasing intervention by the United States in South Vietnam escalated in 1965 and eventually involved the commitment of half a million U.S. military personnel. Why did the United States get involved on such a massive scale in a small country half a world away? Remember that those were Cold War days, and fighting communism was a foreign policy priority for the United States.

North Vietnam responded to the U.S. military escalation by fielding regular army forces against American and South Vietnamese troops. The United States never invaded North Vietnam, although American air strikes there were devastating, with more ordnance dropped on Vietnam than on all of Europe during all of World War II. American bombs, napalm (an incendiary explosive), and defoliants such as Agent Orange (dioxin) caused enormous damage to the natural and agricultural systems of Vietnam, destroying over 8,400 square miles (nearly 22,000 sq km) of forest and farmland in an effort described in military parlance as **"denying the countryside to the enemy."** Agent Orange also left a legacy of birth deformities among Vietnam's postwar generation.

The United States avoided invading North Vietnam partly because of the risks posed by Chinese and Soviet support of the North. In limiting the theater of ground warfare, however, the United States found itself unable to expel from the South both the Viet Cong and the North Vietnamese army. Those combined forces were determined, skillfully commanded, increasingly better equipped by its allies, accomplished in guerrilla tactics, and willing to bear heavy losses. More than 3 million Vietnamese soldiers and civilians died in the conflict, and about 58,000 U.S. soldiers and support staff perished. In 1973, almost all American forces were withdrawn from the costly war, which was extremely divisive at home as growing ranks of Americans questioned its rationale and protested its conduct. There are still 1,948 Americans listed as missing in action in Indochina.

North Vietnam completed its conquest of South Vietnam in 1975. Saigon was renamed Ho Chi Minh City (although most of its people still call it Saigon), after the country's leader (who died in 1969). Vietnam entered a period of relative internal peace, but there were ongoing conflicts, including Vietnam's 1979 conquest of Cambodia to oust the brutal Khmer Rouge regime, continued warfare against Cambodian and Laotian resistance groups, and a brief border war with China. Within its borders, the Communist government of the reunited Vietnam unleashed a repression against those who had assisted the Americans, including the rebellious hill tribes known collectively as Montagnards and members of the Cao Dai religion. This backlash caused a mas- 228
sive outpouring of refugees, including the "boat people" who risked much to flee. Most of the ethnic Vietnamese living in the United States today are those refugees and their descendants.

Vietnam (population 88 million), with its capital at Hanoi (**•Figure 7.46**), has restored parts of its war-torn landscape through large-scale reforestation, agricultural reclamation, and nature conservation programs, aided by international organizations such as the World Wildlife Fund. Animal species new to science have been discovered in the rugged Annamite Cordillera (Truong Son Mountains). Such natural treasures are attracting international ecotourists who provide much-needed revenue, and tourism in general is booming.

GEOGRAPHY OF THE WORLD'S GREAT RIVERS | The Mekong

The Mekong River, known as "Water of Stone" in Tibet, "Great Water" in Cambodia, "Nine Dragons" in Vietnam, and "Mother of Waters" (Mae Nam Khong, contracted to Mekong) in Laos and Thailand, is Southeast Asia's great river (its course may be seen in **Figure 7.37**). From its source on the Tibetan Plateau in the Chinese province of Qinghai, the Mekong snakes along a 2,600-mile (4,200 km) path through China's Sichuan and Yunnan provinces, forms the boundary between Laos and its neighbors Myanmar and Thailand, winds across Cambodia, and fans out across Vietnam's densely populated Mekong Delta region, before empting into the South China Sea.

Unlike the Chao Praya in Thailand, the Red River in Vietnam, and the Irrawaddy in Myanmar, the Mekong is not a great magnet for cities and civilizations. The biggest city on its banks is Cambodia's Phnom Penh, with just over a million people. The main factor limiting more settlement along the Mekong has been the difficulty in navigating the river above Phnom Penh. Historically, very little trade has been carried along much of its length; the oceans and overland routes have provided superior opportunities.

The Mekong is not economically unimportant, however. It is one of the world's most prolific fisheries, yielding about 2 million tons of fish each year, or twice the annual North Sea catch. There are more species of fish in the river (about 1,200) than in any other rivers but the Amazon and the Congo. Unique creatures inhabit its waters, including about 70 Irrawaddy dolphins and the giant catfish that attain 10 feet (c. 3 m) in length and 660 pounds (300 kg) in weight. Overfishing and the construction of dams have all but eliminated this superlative catfish throughout the Mekong Basin.

A big push is being made to capitalize on the Mekong's potential to produce hydroelectric power, with plans to build about 100 dams on the river and its tributaries. The first dam on the river itself (the Man Wan in China) was completed in 1993. China then went to work on five more. Vietnam is building five dams on the Mekong tributary called the Dak Hoyt, which forms part of its border with Cambodia. Thailand has dammed its main tributaries to the Mekong. Laos plans to build the most dams of all—20—and thereby become mainland Southeast Asia's biggest supplier of electricity. Laotian officials boast that their country will become "the battery of Asia." With funding from the World Bank—despite the criticism this organization has received for supporting large dams—Laos built the giant Nam Theun 2 dam. The World Bank insists it went to extraordinary lengths to minimize environmental impacts from this project.

These dams will nevertheless have enormous impacts, some of them unforeseeable. The fish catch will definitely fall dramatically 181, 265
because the main reason for the Mekong's productive fishery is its regular flooding, which will be dramatically reduced. Fishermen are already complaining of smaller catches since dam building began. There are concerns that dams will ruin the ecosystem of Cambodia's Tonle Sap Lake, which is drained by a river that actually changes direction when downstream Mekong floodwaters inundate its basin. More than a million people make their living by fishing this lake. Inevitably, the downstream countries suffer many ill effects of dams. In this case, for example, Vietnam's Mekong Delta will be deprived of the fertile silt that helps make it an agricultural "breadbasket."

As always with shared river systems, there are serious questions about how the Mekong waters should be divided. Vietnam, Cambodia, 181, 247, 438
Laos, and Thailand have joined to form the **Mekong River Commission**, whose biggest responsibilities are to fix the minimum amount of water each country must discharge downstream on the Mekong and its tributaries, and to agree on rules to ensure water quality. Both quality and quantity of water are threatened by upstream Myanmar and China, which have refused to join the Mekong River Commission. The downstream countries complain that China does not seem to care about what happens to them as a result of China's dam-building and other uses of the Mekong.

Joe Hobbs

• **Figure 7.46** Typical street scene in the Old Quarter of Hanoi, Vietnam's capital. Urban traffic in Vietnam is a cacophony of motorbikes, bicycles, cars, and pedestrian traffic.

Since its embrace in 1986 of free-market reforms in the policy known as *doi moi,* Vietnam's economy has improved markedly. The poverty rate fell by half in the 1990s, and in recent years Vietnam has had Asia's second-fastest-growing economy, after China's. The Communist government halted collectivized farming in 1988 and turned control of land over to small farmers under 20-year lease agreements. The results have been impressive; Vietnam since 1988 has become a major exporter of rice, second in the world after Thailand. This is a vulnerable commodity, however, especially because production is concentrated along the Mekong River, which is prone to flooding in the May–October rainy season, and to the specter of rising sea levels due to climate change (see following discussions of the World's Great Rivers and GIS). Vietnam is also pinning hopes on exports of its proven reserves of 2.5 billion barrels of oil, especially to energy-hungry China.

Vietnam in 1995 restored full diplomatic relations with the United States. Recently, the United States and Vietnam have been boosting military ties, as they share a joint interest in containing China. U.S. businesses have been scrambling for consumers in this large, promising market of mainly young consumers (about 45% of the Vietnamese are under age 25). After the two countries signed a trade agreement in 2000, Vietnam began exporting shoes, finished clothing, and toys to the United States. Both its imports and its exports boomed when the country joined the World Trade Organization (WTO) in 2007. The dual personalities of North and South are being perpetuated in the new economic climate, with most investment and infrastructural improvement focused in the South. The North is a more austere economic landscape, and northerners dominate the civil service. Along with a growing gap between the rich and the poor, the differences between North and South could slow the country's overall development.

Visualizing Climate Change with GIS

In 2010, I was invited by the government of Vietnam to advise the country's decision-makers on how best to deal with the anticipated impacts of climate change. My focus had to be on adaptation rather than mitigation (see page 34); Vietnam produces only a tiny fraction of the world's greenhouse gases, but has one of the world's longest and most densely populated coastlines. It is commonly ranked as one of the top 10 countries most threatened by climate change, particularly in the forms of rising sea levels and stronger typhoons (hurricanes; **•Figure 7.47**). How could Vietnam adapt to these threats?

Joe Hobbs

• Figure 7.47 In recent years, Vietnam has experienced an unprecedented number of "super-typhoons," equating to the category 4 and 5 hurricanes of the western hemisphere. I arrived in the coastal city of Hoi An hours before super-typhoon Xangsane slammed ashore on October 1, 2006, and took this photo 3 days later as the city's residents struggled to recover from extensive flooding and other damage. Many climate change scenarios anticipate more frequent and severe typhoons in the country's future. Just as in Louisiana and in Bangladesh, higher sea levels would magnify the impacts of these storms. Engineering adaptation to climate change in the Mekong Delta must therefore consider sea level rise plus storm surge.

The best way to approach this question is by the basic geographical question: *Where* will the impacts of climate change take place? Using GIS, Vietnam's Ministry of Natural Resources and Environment (MONRE) has developed maps showing exactly where rising sea levels will inundate Ho Chi Minh City and the Mekong Delta (Vietnam's most densely-populated region, and one of the world's "breadbaskets," particularly for rice production). MONRE's Director, Dr. Tran 259
Thuc, told me that IPCC forecasts for sea level rise were far too low, because they did not take into account the melting of the world's great ice sheets in Antarctica and Greenland. Dr. Tran therefore commissioned maps depicting the impacts of 65-, 75-, and 100-centimeter (26-, 30-, and 39-inch) sea level rise by 2100—and told me he "fully expects" Vietnam to experience at least the 100 centimeter (one meter) rise.

MONRE used a GIS method known as "statistical downscaling" to derive large-scale maps of the Mekong Delta and Ho Chi Minh City from small scale global sea level models produced by the IPCC. The maps showing the one meter rise are breathtaking (the Mekong Delta map is shown in **•Figure 7.48**). Almost 40 percent of the Mekong Delta, and 25 percent of Ho Chi Minh City, would be underwater. About nine million people—10 percent of Vietnam's population—would have to be relocated. These GIS visualizations are enormously useful, especially in helping Vietnamese leaders appeal for outside foreign aid to help build sea walls, dikes, and other infrastructure to adapt to climate change, and in illustrating the potential impact of climate change on global agriculture and urban infrastructure.

Much more sophisticated GIS work lies ahead. In order for sea walls to be built in just the right places and according to the correct specifications, very detailed large-scale modeling and mapping must be done. Vietnam can look to the Netherlands,

• Figure 7.48 The impacts of a 1-meter (39-inch) sea level rise on the Mekong Delta, Vietnam. About 40 percent of this extremely productive farmland, one of the world's breadbaskets, would be inundated by the sea by the year 2100.

where Dutch geographers are using a GIS-based process of "spatial planning" to map out exactly where engineering works must be built to stave off the rising waters of the North Sea.

Indonesia: One Country, One People, One Language, 300 Ethnic Groups

The national credo of Indonesia (see map, Figure 7.38), seen on banners and placards throughout the country, is "One country. One people. One language." The government's constitution, in an effort to promote national unity, officially recognizes four faiths: Islam, Christianity, Hinduism, and Buddhism. But the presence of some 300 different ethnic groups, combined with religious frictions, physical fragmentation, and economic problems, has made it difficult to attain peace, order, and unity in Indonesia. There is an official national language (Malay, known locally as Bahasa Indonesian), but more than 200 languages and dialects are in use. The largest ethnic group is the Javanese, who make up 41 percent of Indonesia's population. Eighty-six percent of Indonesians are Muslim, 9 percent are Christian, and 2 percent are Hindu, living mainly in Bali.

Various groups in the outer islands have resented the dominance of the Javanese, who have traditionally asserted their power in colonialist fashion over the larger, more resource-wealthy outer islands. Such animosities have escalated at times to armed insurrections. The Suharto regime (1968–1998) countered these militant expressions with a state ideology called ***Pancasila***. Aimed mainly at suppressing militant Muslim aspirations, *Pancasila* was a pan-Indonesian nationalist ideology designed to neutralize all ethnic identities.

For a time, Indonesia's government vowed to break with the repressive ways of Suharto and appeased would-be separatists with promises of **liberal autonomy**, meaning that the provinces would have greater control over local administration and take more profits from the sales of local natural resources. With growing economic problems, pressure from nationalists, and accusations of corruption leveled against Suharto, however, the government shifted to using an iron fist against every Indonesian province that might aspire to follow the path the former enclave of East Timor took.

Now the independent country of Timor-Leste, East Timor was a Portuguese possession that Indonesia occupied after the collapse of the Portuguese colonial empire in the 1970s. The government carried on a long and bitter struggle against an independence movement led by East Timor's Catholics. In 1998, Indonesia and Portugal reached an autonomy agreement that would give the Timorese the right to local self-government.

Indonesia's incoming Wahid administration was bolder, however, and in 1999 allowed the people of East Timor to vote on whether they wanted outright independence. The result was overwhelmingly in favor of independence.

Since its birth, this tiny young country has had many woes: a breakdown of law and order, interethnic violence, and growing unemployment. The economic future of this poor nation could be brightened by resolution of an international legal wrangle over resources. Between Timor-Leste and Australia, in the portion of the Timor Sea known as the Timor Gap, lies a huge field of petroleum and natural gas. Because the distance between Timor-Leste and Australia is less than 400 miles (640 km)—making it impossible to apply the international standard of a 200-mile (320 km) offshore territorial limit—the two countries must come to
an agreement on how to divide these undersea riches. Any deal is 272
likely to yield significant wealth to Timor-Leste.

From the West ... Of outstanding concern for Indonesia's unity is the province of Aceh (population 4.5 million), at the northernmost tip of Sumatra. In 1976, the inhabitants of Aceh, who are a predominantly Muslim people of Malayan ethnicity, began seeking independence from Indonesia. The central government had made them a promise of autonomy with Indonesia's 1949 independence from the Netherlands, but failed to keep it. The Acehnese insisted that the Jakarta government was too secular for their Muslim tastes and that too little of the revenue from sales of Aceh's abundant natural gas reserves remained in the province (which also has a wealth of oil, gold, rubber, and timber).

The secessionists' main voice was the **Free Aceh Movement** (known by the acronym **GAM**), which wanted to install a member of its own indigenous royal family as president of a free Aceh. Seeing Aceh as vital to the nation's unity and economic viability and fearful that Aceh might inspire separatism elsewhere in the country, the government refused to discuss independence for the province. Violent clashes persisted, and the death toll grew—to 12,000 between 1976 and 2004. But then an unlikely arbiter appeared. The tsunami of December 2004
concentrated its unspeakable sorrow on Aceh (see Figure 7.42). 255
The international community stepped in with a massive relief effort, and Indonesian authorities listened with unprecedented interest to Acehnese concerns. GAM dropped its demands for independence, and Indonesia agreed to allow GAM to participate in local elections. Nominally secular Indonesia allowed Aceh to become the only one of its 33 provinces to adopt Islamic
sharia law. Redefining itself as the Aceh Transitional Commit- 176
tee, GAM laid down its weapons, and Indonesia scaled back its military presence in Aceh (•Figure 7.49). The government gave

HOTLI SIMANJUNTAK/EPA/Newscom

• **Figure 7.49** Following a victory in 2009 local elections, Aceh Party members who are former rebels of the Free Aceh Movement (GAM) hold a press conference in Banda Aceh.

amnesty to the rebels and, critically, pledged that 70 percent of the revenues from Aceh's natural resources would return to the provincial government.

Through a program called Aceh Green, founded by a former rebel, Aceh may emerge as a shining example of sustainable development. In one of Aceh Green's activities, hundreds of former rebels, who know the countryside very well, are being trained by the nongovernmental organization (NGO) Flora and Fauna International as rangers to look out for wildlife poachers and illegal loggers. If the rangers are successful in protecting the Ulu Masen rain forest, Aceh could net $26
35 million in carbon credits from the REDD funding mechanism.

... to the East Progress in Aceh raised hopes for a similar future for Papua, at the extreme eastern end of the Indonesian archipelago. Known formerly as Irian Jaya and West Irian, Papua is home to 3 million people of 200 different tribes and speaking 100 different languages. Most are of Melanesian origin, and most are at least nominally Christian. Collectively known as **Papuans**, they have little in common with the Javanese Muslims who control them from 2,500 miles (4,000 km) away in Jakarta. Their homeland on the western half of the island of New Guinea is of great importance to the economies of Indonesia and the United States because it contains the world's largest copper and gold mines. The American company Freeport-McMoRan, which provides more taxes to Indonesia's treasury than any other foreign source, operates these mines. There are also large assets of oil, natural gas, and timber.

Indonesia has never recognized the province's 1961 unilateral declaration of independence. It is little wonder that Jakarta refuses to let its poorest but most resource-rich province go; the Dutch, coveting this wealth, originally refused to relinquish the area when it granted independence to Indonesia. Pressure from the United States led to a UN-sponsored plebiscite on independence in 1969. The vote in favor of the region's becoming part of Indonesia was widely regarded as rigged. Recently, Indonesia has promised some autonomy and a greater share of resource revenues to Papua, but Papuans complain that these promises are unfulfilled.

China

Han Colonization of China's "Wild West"

China's growth as a land empire has involved both subjugation of people who are not ethnic Han and the colonization
226 of those ethnic areas by ethnic Han. There are at least 56 non-Han ethnic groups in China, in total about 100 million people living mainly in the west and southwest of the country. As the Soviets did in the Soviet Union, China's Communist government granted at least token recognition of the distinctive identity and rights of five large minorities by creating **au-**
127 **tonomous regions**. These are Guangxi, bordering Vietnam; Nei Mongol (Inner Mongolia) and adjacent, small Ningxia in the north; Xizang (Tibet); and Xinjiang in the far west (see **•Figure 7.50**).

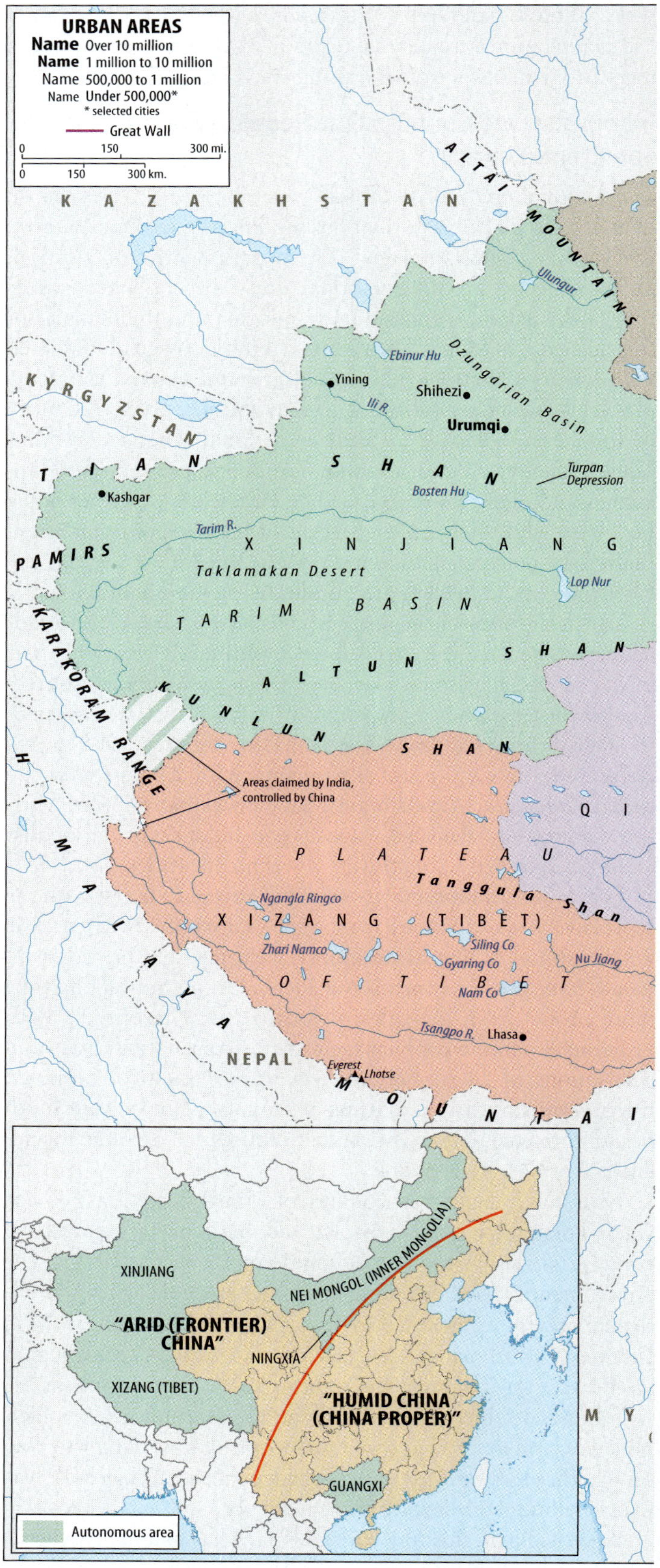

• **Figure 7.50** Principal features of China and Mongolia. On the inset map, the line represents the division between Humid China, or China Proper, to the east and Arid China, or Frontier China, to the west. Humid China has the great majority of the Chinese population and economic activity, and Arid China is home to many minority peoples.

The Hui of Ningxia are Muslims of many different ethnicities who absorbed Chinese language and many Han customs. In contrast, the Uighurs of Xinjiang and the Tibetans have defiantly resisted becoming Chinese—and so Han colonization, or **Hanification**, focuses especially on their territories. The Chinese government aims either to assimilate these minority peoples into a broader, Han-based Chinese culture or to establish a strong enough Han demographic presence among them to dispel any hopes of autonomy or independence.

Chinese authorities have also initiated what they call the **Western Big Development Project** with the stated objective of improving locals' livelihoods enough to diminish their desire for ethnic and political separatism. Focused on the autonomous regions and four other provinces in which there are large minority populations, this project seeks to improve infrastructure like airports and roads and to boost output from factories and farms. The Uighurs, Tibetans, and other minorities generally fear that this "development" is just a cover for demographic, political, and economic domination by the Han.

Buddhism reached Tibet in the middle of the eighth century, and its arrival was followed by decades of conflict between Chinese and Indian Buddhists. The Chinese were defeated in this struggle and expelled from Tibet at the end of the eighth century. China sometimes reexerted loose control over Tibet, but Chinese authority vanished again with the overthrow of the Qing (Manchu) Dynasty in 1911. From 1912 until the Chinese Communists' violent conquest of 1951, Tibet existed as an independent state, with its capital and main religious center at Lhasa.

After the Chinese completed roads to Tibet in late 1954, an increase in restrictive measures by the Communist government contributed to the rise of Tibetan guerrilla warfare. This culminated in a large-scale Tibetan revolt in 1959 and the flight of the **Dalai Lama** (the spiritual and political leader of Lamaism, or Tibetan Buddhism; •**Figure 7.51**) and many other refugees to India. Subsequently, the Chinese drove most of the monks from their monasteries, expropriated the large monastic landholdings, implemented socialist programs, and prohibited organized religion in the rest of China. Large-scale immigration by ethnic Han picked up; about half of Lhasa's population is now Chinese.

HARISH TYAGI/EPA /Landov

• **Figure 7.51** The Dalai Lama is the enduring symbol of the nonviolent Tibetan quest for autonomy from China.

Beijing insists it is helping Tibet with major development investment, including the $4 billion Qinghai-Tibet railway recently constructed between Beijing and Lhasa. But Tibetan resistance to Chinese rule continues. From his place of exile in Dharamsala, India, and on frequent international road trips, the Dalai Lama continues to appeal for Tibetan freedom. He asks China to grant what he calls **genuine autonomy** to Tibet, meaning that Tibetans would have authority over most matters except foreign affairs and defense, which the Chinese government would control. His peaceful cause has widespread support. But some Tibetans want stronger, even violent, action. China's government has so far refused to negotiate with the Dalai Lama. For many years, India had promoted the cause of Tibetan autonomy, but as its trade ties with China have grown more vital, its advocacy for Tibet has quieted.

China's government regards Xinjiang's Muslims, mainly from the 8-million-strong Uighur ethnic group, who speak a Turkic language and are kin to the Kazakhs to the west, as the most problematic minority groups. Many Uighurs would like to pull at least a part of Xinjiang out of China's orbit to create an independent Chinese Muslim state. Two short-lived independent Uighur republics, both known as the Eastern Turkestan Islamic Republic, existed in 1933 and again from 1944 to 1949, and some Uighur separatists are fighting for one again. Beyond execution of those separatists deemed to be terrorists, China's response has been to press ahead with development projects and Han immigration.

At the outset of the Communist takeover of China in 1949, the Han population in Xinjiang amounted to 6 percent of the province's population. Today, it accounts for over 40 percent, and some 250,000 more Han Chinese immigrate to the cities (especially Urumqi and Yining) annually, enticed in part by relatively high salaries and other incentives. Many of these Han migrants are in Xinjiang only a short time; they stay long enough to make some money and then return to eastern China.

Human rights monitors report systematic Han abuses of the Uighurs. Scores of mosques have been torn down; Uighur literature has been burned; Muslim clerics have been forced into "reeducation" training; Islamic schools have been closed; public prayer is now forbidden; head scarves have been banned; and prodemocracy demonstrations are brutally quashed. China's struggle to maintain a grip on its ethnically distinct regions is reminiscent of Soviet efforts in non-Russian areas. 134

The Three Gorges Dam

The Chang Jiang ("Long River"), also known as the Yangtze River, has been central to China's identity and welfare for thousands of years. The river delivers precious water and fertile soils, creating environments that have been intensively settled and farmed, especially for rice and wheat. But this great river has also delivered tragedy in the form of devastating floods. Flooding in August 1998 affected 300 million people along the river and did an estimated $24 billion in damage, for example. Inevitably, questions arose about the advantages to be gained—especially in flood control, drought relief, and hydroelectricity production—if the river were to be dammed. The concept of a giant dam on the river dates to 1919, when it was proposed by Sun Yat-sen. His vision has been realized on a scale far grander than he could have imagined in the massive Three Gorges (Sanzxia) Dam, begun in 1994 and completed in 2009 (•**Figures 7.52 and 7.53**; see also location on Figure 7.50).

As with any large dam project, like the Aswan Dam seen earlier (page 181), the Three Gorges may best be evaluated by considering its pros and cons. On the plus side, this largest dam ever built—1.3 miles (2.1 km) wide and 610 feet (186 m) high—is creating a reservoir 385 miles (620 km) long that will dramatically reduce the threat of downstream flooding. Stored waters will also help alleviate drought. The Three Gorges Dam will also improve navigation and therefore enhance

• **Figure 7.52** Changed landscape of the Three Gorges Dam region; NASA visualization.

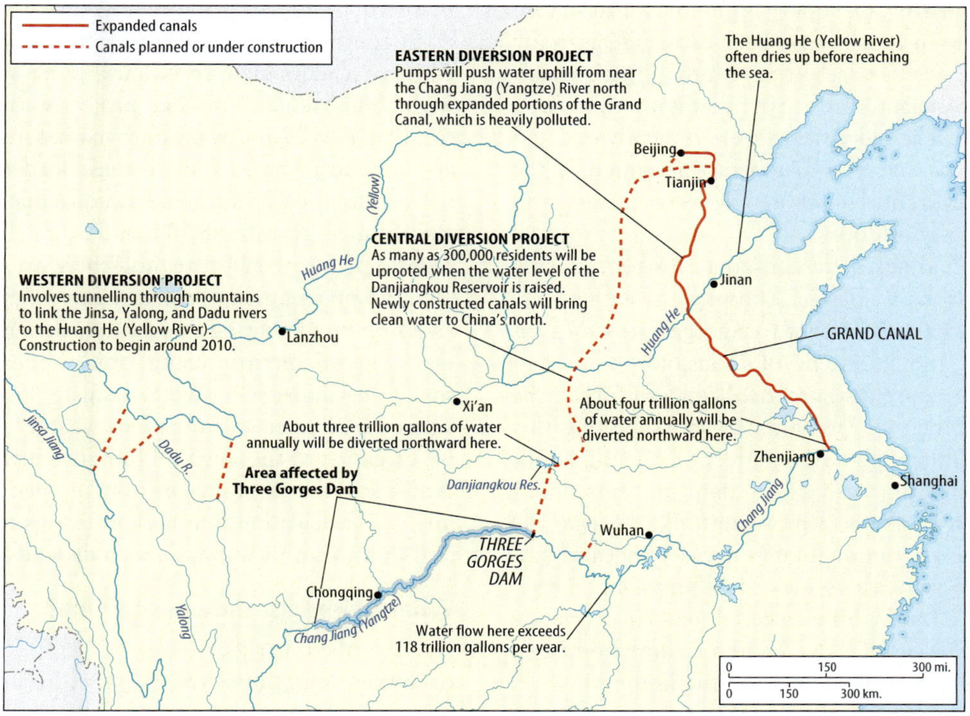

• **Figure 7.53** The Three Gorges Dam and the Chang Jiang Water Transfer Project.

trade, especially by connecting the burgeoning inland city of Chongqing to the world abroad. Historically, the Chang Jiang was able to accommodate freighters of 10-foot (3 m) draft all the way from the East China Sea to Wuhan. Above Wuhan, the shallower river was traditionally plied by smaller freighters to Yichang, situated just below the river's spectacular gorges. Yichang was a classic **break-of-bulk point**: the point at which cargo is unloaded and broken up into smaller units for delivery, thus minimizing transport costs. In this case, freight and passengers were offloaded and put on flat-bottomed junks and sampans drawn upstream through the gorges to Wanxian by human trackers who pulled and rowed the vessels (you can see all these locales in Figure 7.50. With the completion of the Three Gorges Dam project, vessels with much deeper draft are able to sail 1,500 miles (2,400 km) upstream from Shanghai all the way to Chongqing in the Sichuan Basin. Small vessels are able to pass over the dam in a ship elevator, a vast steel box that lifts the ship the height of the dam in 42 minutes. Most ship traffic uses a system of locks with five levels, taking three hours in transit. With these means, shipping to Chongqing has increased fivefold. Some observers say the increased commerce is in the process of transforming Chongqing into a "new Hong Kong." Chinese authorities view Chongqing as a new model for urban growth, as discussed on page 269.

Also on the plus side, the world's largest hydropower plant that is built into the dam provides about 85 billion kilowatt-hours of energy annually, enough to satisfy 15 percent of China's electricity needs. Some of the power is sold to private users, offsetting some of the dam's huge construction costs.

Most of the negatives are related to the vast reservoir forming behind the dam. The new 500 feet (150 m) deep lake has inundated 4,000 villages, 140 towns, 13 cities, numerous archeological sites, and nearly 160 square miles (over 400 sq km) of farmland. Because their homes have been drowned, as many as 1.4 million people have been resettled. Many of the vacated settlements are quite ancient, and their inhabitants had deep cultural roots in them. The old villages were typically situated close to fertile farmlands that were also being inundated. In the new, higher communities into which people were located, soil conditions were generally inferior.

The reservoir has had aesthetic consequences too, erasing the world-class wondrous wild river running through scenic canyons. The natural marvels of the three gorges drew 17 million Chinese tourists and hundreds of thousands of foreign visitors every year. There are also practical concerns about the silt that is building up behind the dam. It is possible that this increased soil load will have the ironic effect of causing flooding upstream from the dam, even as the dam prevents flooding below. There are fears of the almost unthinkable: that the dam could burst, causing unimaginable devastation and suffering downstream. Finally, there may also be seismic consequences of the dam. Central China suffered massive economic and human losses (about 70,000 dead and 5 million homeless) in the magnitude 8.0 Great Sichuan Earthquake of May 12, 2008. There is growing scientific evidence to suggest that this quake was triggered by the weight of a reservoir close to the fault on which the quake occurred. The Chinese government admits that the Three Gorges Dam has created seismic and landslide risks, and that its reservoir has been plagued by algae carpets and waste dumped by communities on its shores.

Although recently admitting some environmental consequences, the government of China pursued the mighty Three Gorges Dam project without remorse and without tolerance for dissent within China. The enormous cost of the project, nearly $30 billion, has been born solely by China. Like mega-dams elsewhere, the Three Gorges Dam is a statement that its builder is a great power to be reckoned with on the world stage. 181

Chang Jiang Water Transfer Project

China has one of the world's lowest per capita water supplies and most uneven distributions of water. More than 40 percent of China's people live in the north, but less than 15 percent of the country's water is there. Chinese engineers see water transfer on a massive scale as the way to redress this imbalance.

One project being considered is the transfer of seawater through a pipeline all the way to the dried up salt lakes and desert basins of Xingiang. The hope is that the water will evaporate to induce rainfall over dry northern and northwestern China.

Another project is underway: as part of the massive **Chang Jiang Water Transfer Project**, the Three Gorges Dam will provide water to arid North China (Figure 7.53). This spectacular engineering project—more ambitious even than the Three Gorges Dam—will move water from south to north through three canals. One of its most difficult challenges is to blast through some of the giant mountains of the Himalayas to divert water from the upper reaches of the Yangtze River to the headwaters of the Yellow River. This feat will be the first time in history that people have merged two river basins.

In contrast with their approach to the Three Gorges Dam, Chinese leaders view the Chang Jiang Water Transfer Project with some trepidation. The project's concept originated with Chairman Mao and was approved before China's present leaders came to power. Many of these leaders feel the project will have unfortunate, unintended consequences, both social and environmental. Like the Three Gorges Dam, this project will require people to relocate, and there are fears of social unrest. And the environmental consequences of changing the course of rivers and reallocating their waters are very difficult to foresee. Dai Qing, an environmental and political activist, had this to say of the Chang Jiang Water Transfer Project: "[It] will be considered a disaster within twenty years. Under Chairman Mao, we often heard the slogan 'Humans must conquer nature,' and that is still the mentality of so many people today—but gradually we are learning that human beings cannot conquer nature and that this philosophy inevitably leads to disaster."[17]

What's Next for Industrial China?

As described earlier (pages 233–234), China's economy has been booming. But as the global economic crisis spread beginning in 2008, it was clear that even this economic dragon had weaknesses.

China's extraordinary recent successes are tempered by several concerns. One is freedom of expression. In 1989, the Chinese government cracked down brutally on prodemocracy demonstrations centered on Tiananmen Square in Beijing. Years later, there continue to be strictures on academic, media, artistic, and other freedoms that some observers believe are conducive to long-term economic well-being. China's government is one of few in the world (including those of Myanmar, Malaysia, Thailand, Singapore, Turkmenistan, Uzbekistan, Saudi Arabia, Syria, the United Arab Emirates, Tunisia, and Iran, Vietnam, Cuba, Sudan, Pakistan, Yemen, and North Korea) to censor the Internet, which is arguably one of the most important tools of economic development today. With what has become known as the "Great Firewall of China," the government requires all computers sold in China to be preinstalled with software blocking political and religious websites and pornography, thus enabling the monitoring of personal communications and the Web sites visited by computer users. The government also monitors mobile phones, the use of which is growing 30 percent annually in China, to ensure that they too do not become instruments of protest (as they did
196 so profoundly in the Arab Spring of 2011). Believe it or not, fearing the potency of a revolutionary symbol, China's government began destroying jasmine plants in the wake of Tunisia's Jasmine Revolution (see page 192).

Another potential problem is that China's leaders have been breathlessly trying to feed what they call the "socialist market economy" with gigantic projects, thereby hoping to create jobs and stimulate economic growth. They reckon that the country needs to register at least 7 percent economic growth annually just to prevent massive unemployment and social unrest. Continuing the infrastructure investment for the 2008 Olympic Games, one of the world's most impressive building booms ever is continuing: new subway systems, railroads, highways, bridges, sewage systems, dams, water diversion projects, a natural gas pipeline, a national electricity project, urban improvements, airports, industrial parks, and more. These efforts cost a lot of money, and the government is drawing down its national treasury to fund them. State banks are pouring billions of dollars into federal projects whose returns are uncertain. (Incidentally, China is the United States' largest foreign lender. Buying up U.S. Treasury Bonds, it helps finance America's national debt.)

A further hazard is that with its galloping growth, China is developing a classic economic bubble that will eventually burst, probably with costly results. Historically, few economies have been able to maintain torrid growth for long, especially when that growth is fueled by speculative cash flows and questionable bank loans, as has been the case recently in China, and as had been witnessed with the 2008 global credit crisis.

There are also environmental costs for such spectacular growth. Due to generally lax environmental quality standards, China's skies and waters are often badly polluted. China has more entries on the World Health Organization's list of most polluted cities than any other nation. China's government acknowledges the problems. The State Environmental Protection Administration (SEPA) reported that one-quarter of the water in China's largest river systems was so polluted that it was dangerous for people and other living things to come into contact with or had "lost the capacity for basic ecological function." Polluted water was killing tens of thousands of people every year and reducing crop yields. SEPA warned that China's approach of "growth through industrialization" was pushing its environment "close to the breaking point."[18]

Having overtaken the United States as the world's largest producer of carbon dioxide, China (which gets 80 percent of its electricity by burning coal) presumably has the largest role in climate change. When the Kyoto Protocol on climate change was negotiated, China was exempted because it was an LDC. Technically this is still true, but China has surprised 35
many by taking some responsibility for climate change. China has also resolved to become the global leader in green technology. China sees this as an opportunity both to confront its own problems with climate change, apparently including both droughts and floods, and to create new jobs. To this end, China has quickly become the world's largest manufacturer of both wind turbines and solar panels, and aims to become the world's biggest builder of nuclear power plants. China has recently passed the United States to become the world's largest market for wind turbines, which are sprouting up on the Gobi Desert and elsewhere.

There is concern that the benefits of China's economic growth have not been distributed evenly throughout the population. Prosperity and consumption have increased enormously, but in a remarkably unequal manner. In 1980, China had one of the world's most even distributions of wealth, but that equation has changed dramatically, with the flamboyant super-rich showing off their wealth to the very poor. These inequalities are comparable to those that existed when the revolution brought Chairman Mao's Communists to power on the promise of redistributing wealth.

Remember that, despite its phenomenal growth, China is still relatively poor (with a per capita GNI PPP of $6,890 (comparable to Turkmenistan) and agricultural (with a 50 percent rural population; see Table 7.1). The agricultural face of China is changing rapidly, though, as the cities boom (see the discussion of megacities on page 226). Most of the economic activity and growth is in the east, and especially in the southeast in the five **Special Economic Zones (SEZs)** designated to attract investment and boost production through tax breaks and other incentives (see •**Figures 7.54 and 7.55**).

One of the results of the clustering of these cities, and the economic power they represented, has been a serious problem of regional imbalance. Most of the country's industrial development and urban growth occurred along its east coast, near the traditional port and river cities that were strongly influenced by Western economic interests from the mid-nineteenth century on. Even the rural areas of eastern China experienced much more growth relative to those in the west, especially because they were well linked to expanding markets and prospering cities. Meanwhile, people in western regions of China, mountain villages all through Humid China, and landscapes

• **Figure 7.54** Major industrial resources, industries, and industrial cities of China.

• **Figure 7.55** The futuristic Pudong section of Shanghai is China's modern face to the world, and is one of the country's SEZs.

lacking in surface transportation in the southwest and parts of upland China were frustrated. They saw relatively little evidence of China's much touted economic prosperity.

Predictably, large numbers of people have migrated from the less prosperous regions of the country to the promising SEZs and other urban centers of the east and south. Millions of migrant workers have left their villages—then populated primarily by children and the elderly—to work in construction or menial labors in the coastal cities, sending much of their meager monthly earnings of $50 to $100 home as remittances. People in the cities often discriminate against these migrants, known derisively as "outsiders," in education, housing, and jobs, treating them like illegal aliens. Many technically are illegal because they fail to sign work contracts that would both register them and require factories to pay fees to local cities.

Chinese cities are characterized by a two-tiered society of legal residents and nonresidents, created by a registration system known as *hukou*. Residency allows people access to better education and health care, and it is financially prohibitive for a nonresident to obtain legal residency. The nonresident migrants provide cheap labor not only to factories but to affluent urbanites who hire them as nannies and servants. Police routinely "shake down" workers without papers, seeking bribes from the frightened migrants. Some of the factories—like the Foxconn complex in Special Economic Zone of Shenzhen known as iPod City because these Apple devices and the iPhone and iPad are assembled here—are world-class, but others are **sweatshops**, where working conditions border on the nightmarish. But even at Foxconn, conditions are so stressful that suicide has become a problem (employees there are now required to sign a contract promising they will not commit suicide!). Management typically suppresses strikes, bans independent trade unions, and fails to enforce minimum-wage laws for workers who toiled through 18-hour workdays and caught brief rest in prisonlike dormitories. It is little wonder that many of the eastern boomtowns see as much as 10 percent annual turnover in the workforce and soaring crime rates.

The Chinese government has a social contract with its people: it pledges that it will create and maintain a "harmonious environment" for them. The basic formula of China's government and economic system is that employment and economic growth needs to be kept high in order to avert social unrest and, in the worst case scenario, revolution. China's leaders watched the Arab Spring with great concern. Although they are often not reported by media, there are growing numbers of labor strikes by workers demanding double-digit salary increases. Recently, the government raised the minimum wage by 20 percent, and employers like Foxconn have granted even higher raises.

Chinese authorities remain convinced that urbanization is the ticket to greater prosperity for China's huge, poor, rural population. But they do not want the magnet cities of the east to absorb all of these people; many would end up unemployed and living in slums. The strategy is therefore that the larger cities of the interior, such as Chongqing, should absorb much of the migrant traffic that has been flowing farther east and south. Already home to 8 million people, Chongqing, for example, is poised to grow much more through its "one-hour economy circle plan." In this plan, by 2017, 4 million rural residents will be moved into newly urbanized urban areas within an hour's drive of the city center. Millions of new jobs are thus being created in the long-neglected urban centers of China's interior. At the same time, wages have been rising sharply all over the country. Once pushed and pulled toward the coast, workers are now finding competitive jobs closer to home. The result is a sudden and surprising shortage of wage laborers in the coastal factories. Once-confident sweatshop managers now worry that the global marketplace will shift jobs to lower-wage countries like nearby Vietnam.

Other trends also point toward the end of the Chinese model. Not only are wages rising, but so too are land prices. A serious "property bubble," partly built up through the same kind of questionable loans that plagued the United States before its credit crisis, is part of the economic bubble building in China. As the population ages, the labor force is shrinking. Cheap capital, cheap labor, cheap energy, and cheap land were at the heart of the Chinese economic miracle, but each of these is becoming pricier. Inflation and unemployment are growing. The danger is, of course, social unrest.

If the bubble does not pop, and if growth continues at about 9–10 percent per year, as many as 70 percent of the Chinese will be middle class or above by 2020. That would represent a huge market of potential consumers. The government wants the Chinese economy to be less reliant on manufacturing and exports, and to become more dependent on domestic consumer spending.

Taiwan and the Two Chinas Problem

Known by Westerners for centuries as Formosa (Portuguese for "beautiful"), Taiwan is separated from South China by the Taiwan, or Formosa, Strait, 100 miles (160 km) wide (see Figure 7.1). The island is nearly 14,000 square miles (c. 36,200 sq km) in area and is home to 23 million people, nearly all of Chinese origin. Driven from the mainland in 1949, the Chinese Nationalist government fled to Taiwan with remnants of its armed forces and many civilian followers. Nearly 2 million Chinese were in this migration. Here, protected by American sea power, the government reestablished itself as the Republic of China with its capital at Taipei. The Nationalists, originally an authoritarian regime but with increasing elements of political democracy, were very successful in fostering industrial development. They united inexpensive Taiwanese labor with foreign capital to build one of Asia's first urban-industrial countries
233 and one of its Asian Tigers. Its major exports are machinery, electronics (notably computers, cell phones, modems, routers, and global positioning systems), metals, textiles, plastics, and chemicals, especially to mainland China and the United States. The population is 78 percent urban; its infant mortality rate is very low; and its birth rate is one of the lowest in Asia (see Table 7.1). The average Taiwanese citizen is four times wealthier than the average mainland Chinese citizen. In almost all respects, Taiwan is clearly recognizable as a more developed country. A major hurdle against even stronger growth has been a lack of energy resources. Some coal and natural gas exist, and some hydropower has been harnessed, but the island depends heavily on imported oil.

Taiwan's Republic of China continues its claim as the legitimate government of China. Meanwhile, the People's Republic of China claims Taiwan as part of its own territory. The United States backed the Nationalist claim for a time, but during the 1970s the United States and the mainland People's Republic developed closer relations. After years of opposition, in 1971 the United States supported the revocation of Taiwan's seat in the United Nations and its replacement by the People's Republic. In 1979, bowing to the wishes of the People's Republic, the United States withdrew its official recognition of Taiwan, instead recognizing China's claim of sovereignty over Taiwan. This **One China Policy** prevents the United States from having formal diplomatic relations with Taiwan. However, the United States opposes the annexation of Taiwan to China by force, vowing to defend Taiwan from attack, just as it opposed Taiwan's call for a Nationalist effort to retake the Chinese mainland during periods of internal weakness there. In the meantime, the United States is a strong backer of Taiwan, supplying it with generous packages of weapons and economic aid.

Taiwan continues to resist mainland China's overtures for reunification, which include promises of broad autonomy, and is moving increasingly away from any type of reunion. That trend reflects Taiwan's ethnic makeup. Less than 15 percent of the island's people descend from the Nationalist refugees, who with their recent ties to the mainland are known as "mainlanders." Native non-Chinese people make up about 2 percent of the population. The vast majority of the island's inhabitants are descendents of Han peoples who emigrated from China as many as four centuries ago, notably the Fujianese, who came from the southeastern coastal province of Fujian, and the Hakka, from Guangdong province. This majority is much less mainland-oriented in political outlook. Despite discrimination by the Nationalists, these "indigenous" Chinese of Taiwan have enjoyed increasing political power.

The two countries must be cautious about their economic and political relations. Taiwan's economy has become very dependent on mainland China, where Taiwanese companies take advantage of low-cost mainland labor (about one-fifth the cost in Taiwan for the manufacture and assembly of products) and where they have a huge market for their products. Any efforts to sever most ties with China and assert Taiwan's independence more forcefully could have large economic consequences and invoke military intervention from Beijing—and an American military response.

Japan and the Koreas

Japan's Postwar and Its Costs

On the Way Up . . . Without colonies or empire, Japan (•**Figure 7.56**) became an economic superpower after World War II. The nation's explosive economic growth after its defeat was one of the most remarkable developments of the late twentieth century and, echoing the broader region's experience, is known widely as the Japanese "miracle." Japan watchers cite 23
different reasons for the country's economic success. Proponents of dependency theory argue that Japan, having never been colonized, escaped many of the debilitating relationships with Western powers that hampered many potentially wealthy countries. Some analysts believe that the country's postwar 44
economic miracle grew from an intense spirit of achievement and enterprise among the Japanese. Notably, many Japanese

• **Figure 7.56** Principal features of Japan.

attribute this industrious spirit to Japan's geography as a resource-poor island nation. To overcome the constraints nature has placed on them, the Japanese people feel they must work harder. Still others attribute Japan's postwar achievements to its association with the United States following World War II. Some of the postwar U.S.-imposed reforms worked well economically, and relations between the two countries since the war have also generally stimulated Japan's industrial productivity.

Economists also point to several unique features of Japanese management and employment to explain the country's meteoric postwar gains. One has been recruitment through an extremely challenging (some say brutal) educational system that emphasizes technical training. Although recent years have seen a move to a more lenient and more comprehensive education, a rigorous and stressful testing system still determines access to higher education. Japanese management strategies emphasize benevolence toward employees, encouragement of employee loyalty, and participation of workers in decision making. About 20 percent of Japanese workers enjoy guarantees of lifetime employment in their firms, and many large Japanese companies help provide housing and recreational facilities to their employees.

One essential factor in Japan's postwar economic growth was a high level of investment in new and efficient industrial plants. Investment capital was made available when government policies cut expenditures on amenities and services such as roads, antipollution measures, parks, housing, and even higher education. Japanese industrial cities grew explosively and became highly polluted areas of dense and inadequate housing, with few public amenities and snarled transportation. The money "saved" by not being spent on amenities and services was made available for investment in industrial growth, but the costs to Japanese society were high.

Some analysts cite elements of Japan's political culture to explain the country's economic successes. One political party, the **Liberal Democratic Party (LDP)**, has with few exceptions been elected to power since Japan regained its sovereignty in 1952. It is a conservative and strongly business-oriented and business-connected organization. The party has promoted Japanese exports with policies to keep the yen (the Japanese unit of currency)—and thus Japanese goods—inexpensive. The party has also cooperated closely with Japanese business in the development of new products and new industries. However, that very coziness between government and business—the "crony capitalism" seen in several other East and Southeast Asian countries—was partly to blame for the economic malaise that spread over Japan in the 1990s, and support for the LDP began to erode.

... and on the Way Down Despite all of these factors in Japan's economic favor, the Japanese miracle did not last. The peak of Japan's postwar success came in the 1980s. A powerful economic boom led to speculative rises in stock prices and land prices. At the end of the 1980s, the total value of all land in Japan was four times greater than that in the United States. The Tokyo Stock Exchange was the world's largest, based on the market value of Japanese shares. Then the **bubble economy** burst, and real estate and stock prices fell by more than 50 percent. Japan had near-zero economic growth through 70,
most of the 1990s and into the early 2000s. Japanese indus- 72, 88,
tries whose growth had seemed unstoppable experienced an 421
increasing loss of market share to U.S. and European producers. Part of the reason the economy could not regain its footing was official Japanese commitment to prop up faltering industries and farms with huge subsidies. Because of the 420
disproportionate political influence of small towns and rural areas, the Japanese tried to modernize remote villages and islands with expensive and underutilized public works projects. Due to such projects, Japan is now the world's most indebted country. Overall, the system succeeds in subsidizing its past at the expense of its future.

Attitudes toward ethnic groups play a significant role in Japan's economic development, or more precisely the lack of development. Japan has one of the world's most homogeneous populations; it is 99.5 percent ethnic Japanese. This lack of racial and ethnic diversity has had mixed results for Japan. Many observers believe that it has helped the country achieve a sense of unity of purpose, allowing the Japanese to persist through periods of adversity, especially the postwar years of reconstruction. However, the Japanese have also earned a reputation for intolerance of ethnic minorities. This attitude may be costly for Japan's future economic development. One way to increase productivity in the face of Japan's declining population would be to import skilled workers from abroad, but most people in the country remain averse to immigration. 70,
Inward-looking tendencies have also left Japan short of people 276, 393
capable of and interested in learning English, a language routinely sought by aspiring businesspeople in nearby South Korea and China.

There is definitely another upside to Japan's homogeneity. Capitalist Japan has had a remarkably egalitarian society, with about 80 percent of the population solidly middle-class. The richest third of the population has a total income just three times as great as that of the poorest third (compared with a fivefold differential in the United States). The Japanese have historically embraced the concept of ***wa***, or harmony, based in part on the principle of economic equality. However, the recessions of the 1990s and early 2000s, followed by an economic rebound until the global financial crisis of 2008, resulted in growing joblessness and homelessness in Japan and a widening gap between haves and have-nots. The country's official unemployment rate stood at 4.7 percent in 2011, compared with 2.1 percent in 1990. There are growing ranks of "permanent temporary workers," mainly young people who live with their parents to keep their expenses low.

The government has balked at many suggestions for economic reform because of the threat they pose to a socially harmonious Japan. There have been loud international appeals for Japan to reform its economic system by allowing inefficient businesses to collapse (rather than prop them up with expensive subsidies) and to resist the temptation to satisfy marginal

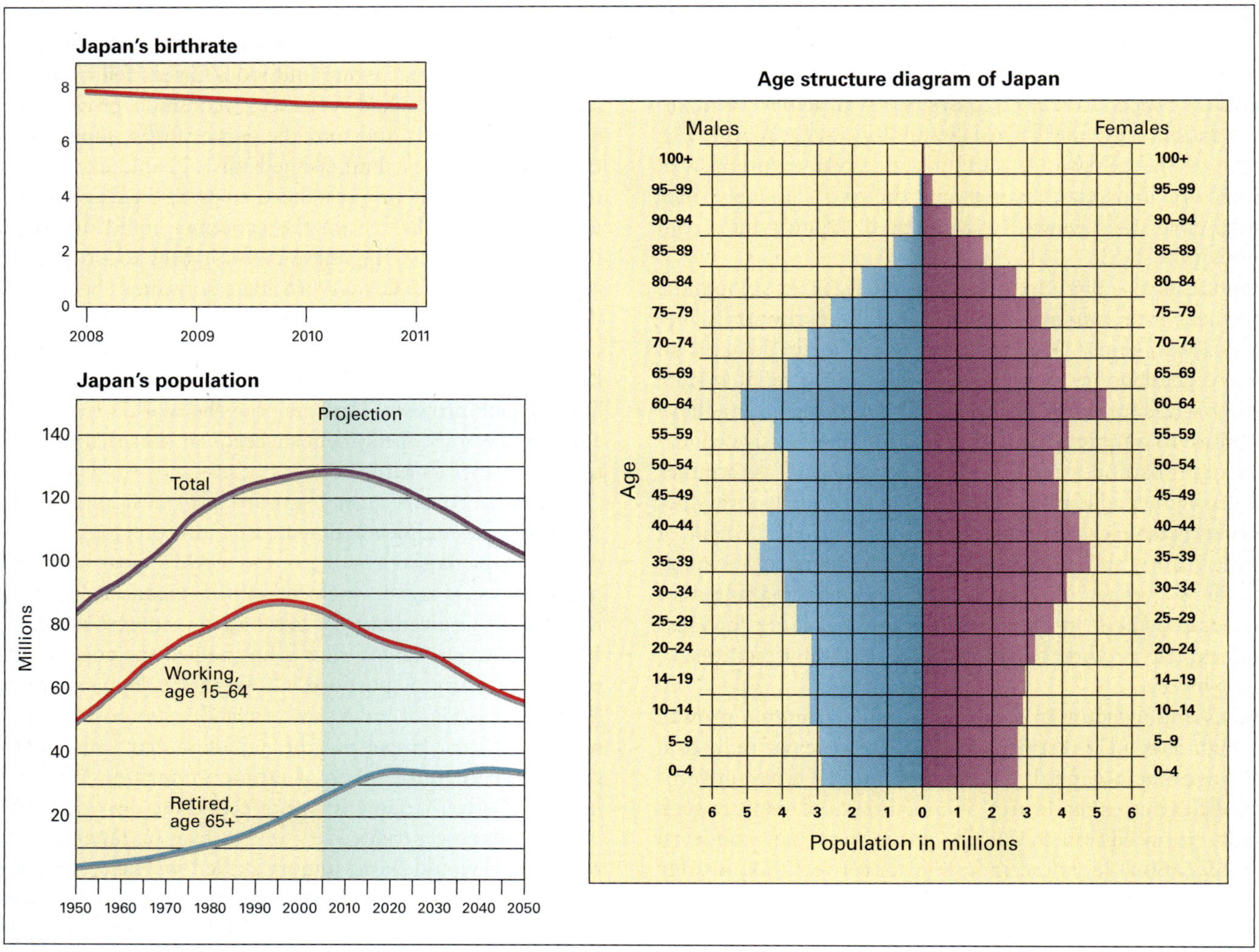

• **Figure 7.57** Many countries face problems of overpopulation, but Japan is concerned about not having enough people to keep its economic engine running in the future.

rural populations with costly education and public works projects.

The legendary Japanese work ethic has had its advantages and drawbacks. It has helped the Japanese create a prosperous country, but it has also created a nation of workaholics beset with the same problems—stress, suicide, depression, and alcoholism—experienced by workers in the world's other MDCs. There is a growing incidence of what Japanese call ***karoshi***, or death by overwork. The educational system encourages children to be highly successful but also to conform; critics say this stifles creativity and innovation. College graduates who are not hired during the annual recruiting season face difficult obstacles to entering the job market. And for those it affects, the practice of lifetime employment makes it difficult to change jobs. The country's welfare system, pensions, and social security are inadequate, compelling workers to work harder for savings. In their long hours at work, men are accustomed to spending little time with their spouses and children.

Women have not achieved parity with men in the workplace, and they complain increasingly of discrimination and sexism. In struggling to increase their footing in Japanese society, growing numbers of Japanese women are both working longer hours and marrying later, contributing to Japan's remarkably low birth rate (7 per 1,000 annually, one of the lowest in the world). This trend and the others make Japan's declining population stand out clearly in **•Figure 7.57**; note especially Japan's chimney-shaped age structure diagram, such a remarkable contrast with the diagrams of LDCs. Ironically, 53
Japan's falling birth rates will only increase the burdens of its workers. The population is both shrinking (it is expected to go from 128 million in 2011 to 100 million in 2050 and 67 million in 2100) and aging (already one in five people were over age 65 in 2011, and 1 million people a year join those ranks). The Japanese of working age will face an increasing load of taxes and family obligations to meet the needs of older citizens.

The Law of the Sea

Some 90 miles (145 km) between the shores of Japan and South Korea in the Sea of Japan (East Sea) lie 34 small, inhospitable islands known to outsiders as the Liancourt Rocks, as Takeshima to the Japanese, and as Dokdo (or Tokdo) to the Koreans (see Figure 7.56). South Korea has controlled the islands since 1956,

but with only a handful of Korean civilians and some Korean coast guard personnel living on the islands; Japan has periodically asserted its right to them. Since the 1990s, these remote rocks have been the focus of a dispute between Japan and Korea, not because of any riches they contain but because of a 1970s United Nations treaty known as the **Convention on the Law of the Sea**. This treaty would permit their sovereign power
37, 27 to have greater access to surrounding marine resources.

The Law of the Sea began as an effort to apportion ocean resources as equitably as possible and to avoid precisely the kind of dispute that developed between Japan and South Korea. The treaty gives a coastal nation mineral rights to its own continental shelf, a territorial water limit of 12 miles (19.3 km) offshore, and the right to establish an **exclusive economic zone (EEZ)** of up to 200 miles (320 km) offshore (in which, for example, only fishing boats of that country may fish). The power that controls offshore islands such as Takeshima and Dokdo can extend its EEZ even further.

By early 1996, some 85 nations had ratified the treaty. South Korea ratified it late in 1995. Japan was preparing to ratify it in 1996, when news reached Tokyo that South Korea had plans to build a wharf on the islands. To avoid provoking Japan, North Korea, and China, South Korea avoided declaring an exclusive economic zone off its waters that would include the islands. Japan ratified the treaty. But fears that South Korea may build facilities and place more people on the islands as a step toward establishing an EEZ, and thereby excluding Japanese fishermen from the area, caused Japan to restate its claim to the islands in 1996. South Korean officials answered with military exercises near the islands, and South Korean civilians staged loud demonstrations outside Japan's embassy in Seoul. South Korea declared the islands a national monument in 2002. Long simmering, the issue boiled up again in 2006 when Japan announced it would send ships to survey the area and South Korea answered that it would form a naval blockade against the Japanese vessels. The problem may have to be resolved in the international legal arena.

Even more important than Japan's competition with South Korea over territorial waters and islands is its contest with China. Both lay claims to the islands in the East China Sea that the Japanese know as the Senkakus and the Chinese as the Diaoyu Islands (see the inset in Figure 7.56). Again, both are interested in staking out an EEZ in which oil reserves might be found. China does not accept the UN definition of the EEZ beginning from a country's coastline; it insists on a wider zone that starts at the continental shelf.

The Senkaku or Diaoyu Islands are uninhabited, but their control is championed by nationalists on both sides and is therefore a potentially explosive issue. In 2004, Chinese activists landed on one of the islands to stake China's claim there, but Japanese authorities promptly arrested them, and Japan took formal possessions of the islands the following year. Since then, there have been several encounters between Japanese and Chinese ships near the islands. The worst was the 2010 incident known as the "Senkaku Shock," which began when a Chinese fishing vessel rammed a Japanese patrol boat near the islands. Japan detained and refused to release the Chinese captain. China responded by halting exports of "rare earth" minerals needed in Japanese high-tech industrial products. By geological accident, and good fortune for China, a very large proportion of the world's supply of 17 rare earth elements are found exclusively or mainly in China. These are indispensable in cell phones, laptops, and many "green" technologies, like wind turbines and hybrid vehicles. They are also important in military technology, and the United States regards access to them as a matter of national security.

Japan's formal claim to the islands dates to 1895, when Japan declared that the islands had previously been *terra nullius*, or no one's property. But China insists that there is Chinese documentation about control of the islands dating to the Ming Dynasty in the fifteenth century. The islands are close to Taiwan, which also lays claim to the islands, so any move by mainland China to secure access to them could also precipitate conflict between the two Chinas.

North and South Korea: Night and Day

Just 110 miles (177 km) from westernmost Japan is the Korean Peninsula, roughly the size of Minnesota or Portugal (•Figure 7.58). The two Koreas have occupied an unfortunate

• **Figure 7.58** Principal features of the Korean Peninsula.

location in historic geopolitical terms. North Korea adjoins China along a frontier that follows the Yalu and Tumen Rivers; it faces Japan across the Korea Strait; and in the extreme northeast, it borders Russia for a short distance. These small countries are thus located near larger and more powerful neighbors—China, Russia, and Japan—that have frequently been at odds with one another and with the Koreans. China and Russia have traditionally feared Japanese rule in Korea because it might serve as a springboard for invasion of their countries. Meanwhile, to Japan, Korea has been seen as a Chinese or Russian dagger aiming at its heart. Japan therefore wanted to occupy or block Korea. For many centuries, the Korean Peninsula has served as a bridge between Japan and the Asian mainland. From an early time, both China and Japan have been interested in controlling this bridge, and Korea was often a subject or vassal state of one or the other.

Despite external ambitions of conquest and division, from the late seventh century to the mid-twentieth century, Korea was a unified state, sometimes invaded and forced to pay tribute but never destroyed as a political entity. The decline of Chinese power in the nineteenth century was accompanied by the rise of modern Japan, whose influence grew in Korea, and from 1905 until 1945, Korea was firmly under Japanese control. In 1910, it was formally annexed to the Japanese empire. A legacy of hostility and occupation continues to cast a shadow over relations between the Koreas and Japan, and disputes periodically emerge between them over issues such as control of islands and fishing rights (see page 272).

Japan lost Korea at the end of World War II, setting the stage for the division between North and South Korea. To understand Korea's split, it is important to consider the Korean War, sometimes in the United States called "America's Forgotten War." The conflict cost the lives of an estimated 2 million North Koreans, perhaps 300,000 South Koreans, and 33,629 Americans.

In the closing days of World War II, the Soviet Union entered the Pacific war as an ally of the United States against Japan. The two sides drew up plans to accept Japan's surrender on the Korean Peninsula, arbitrarily drawing a line at the 38th parallel. The Soviet Union would accept Japan's surrender north of that line and the United States south of that line. This was not meant to be a permanent boundary, but it became one. On either side of the line, the Soviet Union and the United States moved to set up governments that would be friendly to them. By 1948, the Soviet-style, Soviet-backed, Soviet-armed Democratic People's Republic of Korea was established in the north, while in the south, the Western-oriented Republic of South Korea was created under the auspices of the United Nations.

War came soon. On June 25, 1950, North Korean troops invaded the south and quickly took Seoul. In those Cold War days, the United States feared Soviet expansionism into and beyond South Korea. President Truman sent American forces in, and they soon fought under the United Nations flag with support from 16 other countries. Truman selected General Douglas MacArthur to serve as supreme commander of the effort. By September, MacArthur's forces had pushed the North Korean advance all the way back to the Yalu River on the Chinese border. The general proposed using nuclear weapons to press forward from there, but the president refused, and his protests over the issue cost MacArthur his job.

China, seeing enemy forces on its doorstep, reacted by sending huge numbers of forces across the border, driving the allies south across the 38th parallel and taking Seoul on January 4, 1951. Seoul was recaptured on March 15, and the battlefront stabilized generally along the 38th parallel. An armistice was signed at the border site of Panmunjom on July 27, 1953, by the Chinese, the North Koreans, and the United Nations command. It was only a cease-fire, and to this day, there has not been a full treaty or reconciliation between the Koreas.

The border between the two Koreas, often cited as the tensest boundary on earth, still follows this armistice line. The **demilitarized zone (DMZ)** between them, 150 miles (240 km) long and 2.5 miles (4 km) wide, is a virtual no-man's-land of mines, barbed wire, tank traps, and underground tunnels. The world's largest concentration of hostile troops faces off on either side of it. Remarkably, with the near absence of human activity, rare birds like the Manchurian crane and other endangered animal species have taken refuge in this narrow strip. The DMZ thus has the potential to become an international "peace park" like that taking shape along the old Iron Curtain in Europe. 89

The Korean armistice line divides one people, united by their ethnicity and language, into two very different countries (•**Figure 7.59**; also see Table 7.1). South Korea is a republic that has fluctuated between attempts at democracy and a repressive military dictatorship. It has a capitalist economy heavily dependent on relationships with the United States and Japan. North Korea is a rigid and very tightly controlled Communist

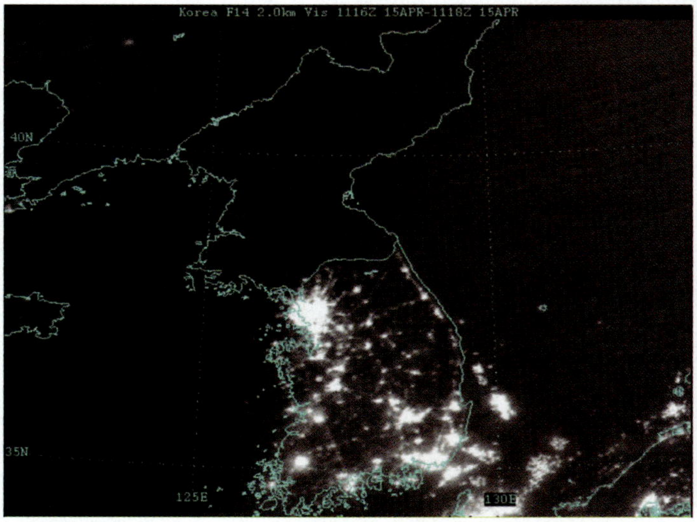

NASA

• **Figure 7.59** The two Koreas are like night and day. A satellite image taken at night shows the profound economic distinctions, with prosperous South Korea awash with urban and industrial lights while North Korea is nearly dark.

state that promotes what it calls ***juche*** (pronounced "*djoo'-*cheh"), or self-reliance. North Korea's leadership has used this political philosophy's three principles of political independence, economic self-sustenance, and national self-defense to limit economic and other exchanges with the outside world. North Korea's principal Cold War ties shifted between China and the Soviet Union. With the collapse of the USSR, China became North Korea's main ally and benefactor, and the Russian Federation is emerging as one of North Korea's few trading partners.

From World War II until his death in 1994, the dictator Kim Il Sung, whom his citizens called "The Great Leader," governed North Korea. His son and successor, Kim Jong Il (dubbed "The Dear Leader"), perpetuated the country's militaristic character. The same is widely expected of Kim Jong-Un, "The Great Successor" and son of Kim Jong Il, who died in 2011. North Korea spends a staggering 23 percent of its gross domestic product on its military, compared with South Korea's expenditure of only 2.7 percent, China's 4.3 percent, and Japan's 0.8 percent. North Korea's lavish expenditures on defense have been one cause of its severe economic decline. North Korea's economy is only 3 percent the size of South Korea's. No data are available for North Korea, but South Korea's GNI PPP is $27,240, similar to that of Israel.

Beginning in 1995, successive waves of flood and drought brought famine to North Korea. The natural calamities only contributed to the already plummeting economic health of the country that came in the wake of the collapse of the Soviet Union, formerly North Korea's main benefactor and trading partner. North Korea's almost impenetrable veil of secrecy made it difficult to calculate the losses, but estimates of the number of people who died in the famine range from 900,000 to 2.4 million, or up to 10 percent of the country's prefamine population. As one crop after another failed, North Korea gradually and reluctantly sought food aid from abroad. It coaxed emergency supplies of wheat from the United States in part by threatening to withdraw from its 1994 agreement with the United States to halt its nuclear weapons program (see the discussion on pages 239–240). Although food aid from the United States, China, and South Korea helped alleviate the immediate crisis, the poor conditions of North Korea's health system, drinking water supplies, and electrical generation have perpetuated malnutrition and disease.

Another physical contrast between the two Koreas is that most of the nonagricultural natural resources are in North Korea. All of the peninsula's major resources—coal, iron ore, some less important metallic ores, hydropower potential, and forests—are more abundant in the North than in the South. North Korea was thus originally the more industrialized state, featuring a typical Communist emphasis on mining and heavy industry, together with hydroelectric production and timber products. Viewed strictly in terms of resource potential for development, North Korea should have become the more prosperous power. Its decision to close its doors to the outside world and insist on strict state control of industry and agriculture provides a fascinating opportunity to appreciate how different political systems produce different economic results.

While the North focused on defense and strict socialism, the South took over industrial leadership in the 1970s and 1980s by developing a dynamic and diversified capitalist industrial economy.[19] South Korea's economy first began to boom between 1961 and 1980 under a state capitalism approach rather than free-market and free-trade capitalism. This was a time during which the government favored and strongly funded a few small companies that it nurtured into giant conglomerates called ***chaebols***. Hyundai is a good example. Originally a small rice milk company, it diversified into trucks and buses. The government believed in its potential and supported Hyundai financially, supervising its growth into a world-class corporation renowned for the manufacture of automobiles and ships. Samsung, now one of the best electronics companies in the world, followed a similar path. Altogether about 15 of these gigantic, interlocking, family-controlled conglomerates came to dominate the economy. Large investments from Japan and the United States, as well as access to markets in those countries, aided in the explosive development of *chaebol* industries. Also important was the availability of inexpensive and increasingly skilled Korean labor. Finally, the factor South Koreans most frequently point to in their success has been an emphasis on education. They sell valued possessions if need be to afford their children's educations at home or abroad, and their culture prizes teachers and professors. South Korea has one of the world's highest numbers of Ph.D.s per capita.

With the world's 13th largest economy, South Korea is now among the top 10 countries worldwide in the production of automobiles, ships, steel, computers, and electronics. The development of its high-tech industries has concerned competitors in Japan and the United States; for example, Samsung has established an increasing market share in an industry dominated by Japanese firms. South Korea is the seventh largest trading partner of the United States, which has long subscribed to a "domino theory" that if South Korea's economy fell, so would Japan's and perhaps even its own. In 2007, the United States and South Korea signed a free-trade pact that will even more strongly enmesh their economies. The United States ratified this agreement in October 2011, and in November Korea did the same.

In sum, South Korea is one of Asia's success stories, an Asian Tiger that made an extraordinary postwar recovery, especially by using brainpower to overcome its resource limitations, and has enjoyed periods of rocketing economic growth. Although South Korea is an MDC by the strict technical GNI PPP definition, it is still something of an "in-between" country, not entirely an MDC or an LDC, not all urban but not all rural either, and both industrial and agricultural. The government still lavishes huge subsidies to protect its rice growers, for example, while also promoting the virtues of free trade in industry.

The transition from LDC to MDC has brought some growing pains, seen especially in the social changes of more women

in the workforce, higher divorce rates, and falling birth rates. Like Japan, South Korea wants to maintain its cohesive society and does not want to invite foreigners to fill its impending
271 shortage of productive young people. The government once encouraged family planning but is now asking its people to have more babies and to stop giving up their own for adoption abroad. Due to a traditional preference for boys, there are fewer women than men in the Korea, and due to young women's migration to the cities for work, there is a pronounced shortage of eligible Korean women in the countryside. The result is a full-blown industry of bride-shopping outside Korea, especially in China, Vietnam, and other parts of Asia.

The bride-seeking Korean farmer typically pays a broker for the full suite of services, from selecting his bride to the wedding ceremony. Prospective brides are lined up for his consideration—in Vietnam's Ho Chi Minh City, for example. He picks one and with no courtship marries her and takes her home to Korea. The Vietnamese and other brides are aliens in the homogeneous Korean society and are commonly ostracized by the grooms' extended families and by society at large. Most husbands and wives do not even speak one another's languages. There are many accounts of the wife's loneliness, depression, physical abuse, and even suicide. Mixed-race babies often suffer domestic abuse and bullying. Happily, there are also reports of successful unions and harmonious families, against the odds.

Despite having some experimental free-trade zones, North Korea is an economically isolated, poor country almost frozen in time, intent on pursuing the command economy model that has failed in every other nation that employed it. Despite its motto of self-reliance, it depends on huge imports of food and energy, especially from Russia. Its principal exports are apparently missiles and other military hardware; heroin, amphetamines, and other synthetic and semisynthetic drugs; and counterfeit money. Its people, whose sentiments and hopes are unknown because they are not allowed contact with the outside world, periodically hover close to famine. Only a select few are allowed access to the Internet, although the infrastructure is there: much of the country is wired with a fiber-optic intranet system. The North Korean leadership has a dilemma: If it gives up on *juche* and opens its doors, the economy might improve, but the people of this country often called the **Hermit Kingdom** will discover the bounties just across the border and might revolt.

Could the Koreans simply reunite as the two Germanys
96 did? The prospect for Korean reunification actually appears to be very remote, even as relations between the two countries improve. North Korea continues to argue that the two countries should unite peacefully and without foreign influence. Outside observers generally believe, however, that North Korea is not really interested in reunification because it would probably doom the country's regime. For its part, South Korea is fearful of the enormous economic cost it would have to pay for absorbing its much poorer neighbor, just as West Germany did in absorbing East Germany. Reunification would probably begin with a huge stream of poor northerners moving south. The nearby great powers are also quietly pleased with Korea's lingering division. The two Koreas provide dual buffer zones between the historical adversaries China and Japan. And if the two Koreas united, finally taking advantage of North Korea's strong natural resource base, China and Japan could suddenly have a rival great power to contend with.

For now, there is little thought of Korean unification. Even with the progress of the Six Party Talks and North Korea's pledge to disable its nuclear weapons facilities (see page 240), there is much more concern with simply avoiding war. Some incidents between the Koreas have been so serious that they threaten to spill over into war. For example, in 2011 North Korean forces sank a South Korean ship claimed to have crossed the Northern Limit Line (drawn hastily by the UN at the end of the Korean War) into North Korean waters. Forty-six South Koreans died. Soon after, North Korean artillery shelled the South Korean island of Yeonpeong, killing four. Because of treaty obligations, incidents in these places or in the DMZ could escalate tension between the United States and North Korea, and even be a tripwire for a great conflict (•**Figure 7.60**). U.S. troops could not avoid being involved, and Chinese troops might be brought in. Some analysts fear that Korea may yet become a nuclear battlefield. Japan worries that it would also be drawn into any conflict between the Koreas. The economic and political stability of these three small countries on the western Pacific Rim is critical to the well-being of the global system.

The book now takes us from that rim out to sea, into the vast Pacific realm.

Joe Hobbs

• **Figure 7.60** Two South Korean soldiers keep a wary eye on their North Korean counterpart, just steps away in the Joint Security Area of the Panmunjom "truce village" in the DMZ. The low concrete slab running left-to-right on the left of the photo marks the border between the two countries. From time to time, a North Korean defector runs across the border, sometimes pursued by North Korean soldiers. Firefights involving both U.S. and South Korean troops against North Korean soldiers have accompanied some of these incidents.

summary

- South and East Asia include the countries of Japan, North and South Korea, China, Taiwan, Pakistan, India, Sri Lanka, Bangladesh, Bhutan, Nepal, Maldives, Myanmar (Burma), Laos, Thailand, Cambodia, Vietnam, Malaysia, Singapore, Indonesia, the Philippines, Brunei, Timor-Leste, and numerous islands scattered along the edges of this major continental bloc.
- This is the most populous world region, with 54 percent of the world's people. It includes the world's most populous countries, China and India. Population growth rates in the region vary widely, from zero growth in Japan to over 3 percent per year in Timor-Leste.
- Three concentric arcs make up the broad physiography of the region: an inner arc of the high mountain ranges, the Himalaya, Karakoram, and Hindu Kush; a middle arc of major floodplains, deltas, and low mountains; and an outer arc of thousands of islands, including the archipelagoes of the East Indies, the Philippines, and Japan.
- Major rivers include the Indus, Ganges, and Brahmaputra in South Asia; the Irrawaddy, Chao Praya (Menam), Mekong, and Red in Southeast Asia; and the Chiang Jiang (Yangtze) and Huang He (Yellow) in East Asia.
- The region's climate types and biomes include tropical rain forest, savanna, humid subtropical, humid continental, steppe, desert, and undifferentiated highland. Shifting cultivation and wet rice cultivation are important forms of agriculture. Wet rice cultivation produces very high yields and is associated with dense human populations.
- Although South and East Asia have some of the world's largest cities, about 60 percent of the region's population is rural. Many uniquely Asian traditions helped shape village settlement planning and home design.
- South and East Asia's ethnic and linguistic compositions are diverse. Major language families are Indo-European, Dravidian, Sino-Tibetan, Altaic, Austric, and Papuan.
- Major religions and sociopolitical philosophies of Asia are Hinduism, Islam, Buddhism, Confucianism, Daoism, and Christianity.
- Great Britain, the Netherlands, France, and Portugal were the most important colonial powers in this region. Most of these domains were relinquished by the middle of the twentieth century, and the later British return of Hong Kong and the Portuguese return of Macao (both to China) closed the colonial period.
- China has the region's strongest economy, and globally is second only to that of the United States. Recent decades have seen strong economic growth among the Tigers and Tiger Cubs of the region, including South Korea, Taiwan, Singapore, Thailand, and Malaysia. The most important emerging economic power is China, whose large and inexpensive labor force has recently been attracting investment away from other parts of the region. India's well-educated and inexpensive labor force is contributing to its strong economic growth. There are fears in the United States about the outsourcing of American jobs to India.
- The Green Revolution is a broad effort to increase agricultural productivity in dominant crops. Biotechnology has produced crops that are more drought- and pest-resistant and capable of creating much higher yields, but genetic engineering of crops has perceived risks. Profits and other benefits from the Green Revolution are not uniformly spread.
- There are several major geopolitical issues in South and East Asia. The traditional enemies Pakistan and India (both pivotal countries from the U.S. point of view) now possess nuclear weapons. There are fears that destabilization in Pakistan might allow the weapons to fall into the wrong hands. Pakistan's support of the United States in its war on terrorism is very risky because a popular backlash could bring down the government. North Korea has nuclear weapons and is apparently trying to use them as bargaining chips to get more food aid and other assistance from the West. Anti-government and anti-Western Islamists are active in Indonesia, another country viewed as pivotal by the United States. Strife and fragmentation in Indonesia could threaten vital oceanic shipping lanes.
- Afghanistan is one of the world's poorest nations. It is landlocked and has limited resources, poor internal transportation, and little foreign trade. Opium is its main export. Its location along the ancient caravan routes and adjacent to large oil reserves has made it the target of stronger powers. Backed by the United States, rebels succeeded in driving out a Soviet occupation force in the 1980s. That costly conflict was followed by a period of civil war during which outside countries neglected Afghanistan. The militant Islamist Taliban came to power and protected the presence and training of al-Qa'ida militants. Osama bin Laden and others planned anti-Western attacks, including 9/11, from Afghanistan. Al-Qa'ida was driven into hiding from there in U.S.-led attacks following 9/11, and the Taliban were deposed from power. Reconstruction of the country is progressing, but a Taliban insurgency continues.
- The subcontinent is home to many different faiths, including Hinduism, Sikhism, Christianity, Islam, Buddhism, Zoroastrianism, and Jainism. Religious, ethnic, and other differences underlie serious and often violent conflicts. The main ones have been Hindus versus Muslims in India, Sikhs versus the government of India, and Tamils versus Sinhalese in Sri Lanka.
- Until 1947, what are now Pakistan, Bangladesh, and India formed the single country of India. For over a century, it was the most important unit in the British colonial empire—the "jewel in the crown."
- With independence, India was partitioned between avowedly Muslim Pakistan (which later divided into two countries, Pakistan and Bangladesh) and secular, primarily Hindu India. The partition had many consequences, including violence between

Hindus and Muslims and their large-scale migrations, disputes over the sharing of Indus River waters, and resources and industries stranded on either side of the new partition lines.

- The most severe product of the partition has been lasting conflict in Kashmir. Several wars have been sparked or fought in the still volatile frontier province, which, despite its Muslim majority, was joined with India in 1947.
- Although it is an overwhelmingly poor and rural country, India has many modern industries that, along with privatization of former state-run firms, have contributed to its substantial rate of economic growth in recent years. These include the software and computer industries of Bengaluru and the film industry of Mumbai. Pakistan and Bangladesh are much less industrialized and rely heavily on textile exports. India and Bangladesh have been successful in lowering birth rates, but population growth is a serious issue for both countries. India has managed to feed its huge population.
- South Asia provides large numbers of workers to other countries, ranging from professional to semiskilled and unskilled. Remittances sent home by these workers are important to the South Asian economies. Many of the professionals do not return, contributing to the region's brain drain.
- Some of India's large minorities have had serious difficulties with the majority Hindu population. Hindu nationalism is a problem in India, fanning the flames of communal violence between Hindus and Muslims. In the 1980s, some of the Punjab's Sikhs agitated for an independent country.
- Pakistan's government faces challenges from its Pashtun population in the country's west and northwest and from strong Islamist sentiment throughout the country.
- Bangladesh and the Maldives are threatened by serious natural hazards from the sea. Strong typhoons (hurricanes) regularly devastate large parts of low-lying Bangladesh, and most of the Maldives would be inundated if sea levels were to rise by 3 feet (90 cm).
- Sri Lanka, the world's largest tea exporter, has a Buddhist Sinhalese majority and a Hindu Tamil minority. For more than 20 years, Tamil factions have led a violent struggle to establish an autonomous homeland on the island's east, north, and west periphery. The violence has been punctuated by peace talks that have not yet arrived at a permanent solution.
- A megathrust earthquake off the northeastern coast of Sumatra caused the great tsunami of 2004, which killed more than 200,000 people in 14 countries. A similar earthquake in 2011 triggered the tsunami that caused massive devastation on the coast of northeast Japan.
- These are more Southeast Asian villagers than urbanites.
- Southeast Asia's tropical forests are being destroyed at a faster rate than others of the world, mostly for commercial logging for Japanese markets. In biodiversity terms, Indonesia is a megadiversity country whose forests are under intense pressure.
- Southeast Asia as a whole is among the world's poorer regions. The poorest countries—Myanmar, Cambodia, Laos, and Vietnam—have suffered from warfare and, in the case of Myanmar, from unwise political management.
- Drug production, drug abuse, and HIV/AIDS are serious problems in Myanmar. Neighboring Thailand has some of these problems but has dealt more effectively with them.
- Prodemocracy efforts in Myanmar have been thwarted by official repression.
- U.S. forces succeeded the French withdrawal from Vietnam in the 1950s. American efforts to win a war against communism in Vietnam failed. Normal relations now exist between the United States and Vietnam, and although Communist, Vietnam encourages private enterprise and has seen recent strong economic growth.
- Numerous dams are being constructed on the Mekong River. These will produce hydroelectricity but will damage the productive fishery of this basin.
- Much unrest in Southeast Asia is related to tensions between ethnic groups. Indonesia is in danger of fragmentation. Catholic Timor-Leste has gained impendence from Indonesia, and the provinces of Aceh and Papua are seeking self-rule or independence. Muslim insurgents in the southern Philippines and Communist rebels throughout the country have stepped up their campaigns against the government.
- China may be physically divided into the arid west and the humid east. An arc drawn from China's border with Myanmar northeast to Harbin is the rough dividing line between the arid west and the humid east. The majority ethnic Han are associated with the more densely populated humid east, and non-Han realms are primarily in the arid west. The Huang He (Yellow River) and the Chang Jiang (Yangtze River) descend from the Tibetan highlands.
- The Three Gorges Dam, under construction on the Chang Jiang, is designed to control floods, provide hydropower, and allow transfer of water to the more arid north. The project has been both celebrated and lamented because of the scale of environmental change and social dislocation associated with it.
- China has modeled much of its most dynamic economic growth after capitalist rather than Communist economic and political models.
- Inexpensive labor helped fuel the booming industries of the coastal cities. There is still poverty in China, with the cities generally wealthier than rural areas. The most prosperous areas are in the southeast, along the east coast (including the Special Administrative Regions of Hong Kong and Macau, and several Special Economic Zones), and in farmlands near large cities, whereas regions in the west, southwest, uplands, and Arid China have had less economic development. Jobs and wages are now growing in more areas, and labor shortages have developed in the coastal boomtowns.
- Taiwan (the Republic of China) continues to claim to represent the true government of China and has a tense relationship with the mainland People's Republic of China. Taiwan is heavily invested in mainland China's development. The United States

does not have diplomatic relations with Taiwan but supports it militarily and economically.

- Japan experienced phenomenal economic growth after its crushing defeat in World War II, but it has recently suffered economic recession.
- Due to its relative poverty in natural resources, Japan must import most of its raw materials, most energy supplies, and a large share of its food.
- The quest for possession of offshore resources near islands is a source of tension between Japan and Korea and between Japan and China.
- The hardworking Japanese are experiencing many symptoms of workaholism, and many complain about their low level of amenities relative to that of other prosperous countries.
- The Korean Peninsula has an unfortunate location in geopolitical terms, sandwiched between the greater powers of Russia, China, and Japan. Korea has had a turbulent recent political history, first as a colony of Japan and then as a land divided between the Communist North and the more dynamic and democratic South.
- There are marked physical and economic contrasts between North and South Korea, with North Korea being much more rugged and having more natural resources for industry. Capitalist South Korea is far more prosperous than Communist North Korea.
- North Korea has suffered famine recently but has been reluctant to open up to trade and assistance because North Koreans might revolt.
- There are prospects for reconciliation and, more remotely, reunification, but also for war between North and South Korea. Reunification is not favored by the South because it would be too costly. China and Japan would apparently be happier to have a weaker divided Korea than a stronger unified Korea.

Key Terms + Concepts

Ainu (p. 226)
ancestor veneration (p. 228)
animism (p. 228)
archipelago (p. 214)
Aryans (p. 225)
Asian Century (p. 210)
Asian Tigers (p. 233)
Association of Southeast Asian Nations (ASEAN) (p. 256)
Austric language family (p. 226)
  Austro-Asiatic subfamily (p. 226)
  Austronesian subfamily (p. 226)
  Hmong-Mien subfamily (p. 226)
  Tai-Kadai subfamily (p. 226)
autonomous regions (p. 262)
blowback (p. 242)
break-of-bulk point (p. 266)
bubble economy (p. 271)
Buddhism (p. 277)
"Burmese Way to Socialism" (p. 255)
Cao Dai (p. 228)
caste (p. 248)
chaebols (p. 275)
Chang Jiang Water Transfer Project (p. 266)
communal violence (p. 247)
Comprehensive Test Ban Treaty (p. 236)
Confucianism (p. 231)
contingent sovereignty (p. 255)
Convention on the Law of the Sea (p. 273)
Dalai Lama (p. 264)
Dalits (p. 248)
Daoism (Taoism) (p. 231)
demilitarized zone (DMZ) (p. 274)
"denying the countryside to the enemy" (p. 258)
domino effect (p. 240)
Dravidian language family (p. 225)
endemic species (p. 254)
exclusive economic zone (EEZ) (p. 273)
fallow (p. 219)
feng shui (p. 224)
Free Aceh Movement (GAM) (p. 261)
genuine autonomy (p. 264)
geomancy (p. 224)
Golden Crescent (p. 242)
Golden Triangle (p. 256)
Great Game (p. 241)
Green Revolution (p. 235)
Han Chinese (p. 226)
Hanification (p. 264)
Hermit Kingdom (p. 276)
Hinduism (p. 228)
Indo-Iranian subfamily (p. 223)
Indus Waters Treaty (p. 247)
insolation (p. 217)
insourcing (p. 235)
intensive subsistence agriculture (p. 219)
Jainism (p. 231)
Jains (p. 231)
Jemaah Islamiah (JI, or the Islamic Group) (p. 240)
juche (p. 275)
karma (p. 230)
karoshi (p. 272)
Khalistan (p. 274)
Laskar Jihad (p. 240)
laterite (p. 219)
lateritic soils (p. 219)
levee (p. 222)
liberal autonomy (p. 261)
Liberal Democratic Party (LDP) (p. 271)
Mahayana Buddhism (p. 230)
megacity (p. 226)
megadiversity country (p. 254)
Mekong River Commission (p. 259)
Mohajirs (p. 244)
Mon-Khmer (p. 226)
monsoon (p. 217)
  summer monsoon (p. 217)
  winter monsoon (p. 217)
mutually assured destruction (MAD) (p. 237)
Naxalites (p. 247)
New Asian Tigers (p. 233)
nirvana (p. 230)
Nuclear Nonproliferation Treaty (NPT) (p. 236)
offshore outsourcing (offshoring) (p. 235)
one-child campaign (p. 216)
One China Policy (p. 269)
Operation Enduring Freedom (p. 243)
overseas foreign workers (OFW) (p. 212)
paddy (padi) (p. 220)
Pancasila (p. 261)
pandemic (p. 213)
Papuan languages (p. 227)
Papuans (p. 262)
Parsis (p. 231)
pivotal countries (p. 237)
plinthite (p. 219)
sectarian violence (p. 247)
shifting cultivation (p. 219)
Shintoism (p. 227)
Sinhalese (p. 247)
Sino-Tibetan language family (p. 226)
  Sinitic subfamily (p. 226)
  Tibeto-Burman subfamily (p. 226)
slash-and-burn cultivation (p. 219)
Special Economic Zones (SEZs) (p. 267)

State Law and Order Restoration Council (SLORC) (p. 255)
strategic hamlet (p. 223)
"String of Pearls" (p. 248)
sweatshops (p. 268)
swidden cultivation (p. 219)
syncretism (p. 228)
Taliban (p. 243)
Tamils (p. 247)
Tamil Tigers (p. 248)
Taoism (p. 231)
terraces (p. 220)
theory of Himalayan environmental degradation (p. 222)
Theravada Buddhism (p. 230)
Tiger Cubs (p. 233)
untouchables (p. 248)
value-added manufacturing (p. 233)
Viet Cong (p. 258)
Wallace's Line (p. 254)
"Water Towers of Asia" (p. 250)
Western Big Development Project (p. 264)
wet rice cultivation (p. 220)
yang (p. 224)
yin (p. 224)
Zomia (p. 221)
Zoroastrianism (p. 231)

Review Questions

1. What countries make up South and East Asia?
2. Use the concept of the three arcs to trace the outlines of South and East Asia's geography. What are the largest plains and river valleys of the region? What are the most important mountain ranges? What are the major island groups?
3. What are the region's most populous countries? Where are the highest and lowest population densities? Which countries have the highest and lowest population growth rates?
4. What are the monsoons? What roles do they play in climates and human activities?
5. What types of environments are shifting cultivation and wet rice cultivation practiced in? What are some of the basic methods and land use considerations of both types of agriculture?
6. What is the geographical concept of Zomia, and what is James Scott's argument about the people who live in Zomia?
7. What roles have *feng shui* and Confucianism played in the site selection and design of homes, schools, and villages in parts of Asia?
8. What are some of the innovations and products that originated in South and East Asia and diffused from there?
9. What are the major linguistic and ethnic groups of South and East Asia? What are the principal religions and sociopolitical philosophies?
10. What were the major colonial powers in South and East Asia? What regions did they colonize?
11. What are the region's most significant economic trends?
12. Why are India, Pakistan, and Indonesia considered pivotal countries for U.S. interests? Has the United States been forced to take sides in the traditional enmity between India and Pakistan?
13. Why are there far fewer women than men in some Indian provinces? What problems are associated with this phenomenon?
14. What is the caste system? How does it affect employment and social relations in India? What evidence of change in this ancient system is there?
15. Where is Ayodhya? What site there is critical in relations between Hindus and Muslims in India?
16. What ethnic and political differences have given rise to strife in Sri Lanka?
17. What particular problems do Bangladesh and the Maldives have in terms of climate and climate change?
18. What caused the great tsunami of 2004, and what were some of its effects?
19. What countries are the largest producers and consumers of Southeast Asian tropical hardwoods?
20. What major changes are coming to the Mekong River basin?
21. What political events hampered development in Vietnam?
22. What are the major political and resource-related issues faced by Indonesia?
23. What factors may threaten the continuation of China's economic boom?
24. How can China's economy be characterized? What is its "average" standard of living based on per capita GNI PPP? How effectively is the wealth distributed? Where is it concentrated geographically, and how is the government hoping to change that pattern?
25. Where are China's Special Economic Zones? What does their distribution suggest about China's economic geography?
26. What are the major political attributes of Taiwan?
27. What is Japan's ethnic makeup? What relations have the Japanese majority had with internal minorities and foreign groups at various times?
28. How has Japan been able to overcome its poverty in natural resource assets by commercial, military, or other means?
29. What are some of the features of Japanese society today, especially those related to the country's economy?
30. What precipitated the division of the Korean Peninsula into North and South Korea?
31. What are the major differences between North and South Korea in natural resources, political systems, and economic development?
32. What are the prospects for reunification of the Koreas? How do some outside powers and the Koreans themselves view reunification?

Notes

1. John Reader, British anthropologist and traveler, cited in Fred Pearce, "Terraces: The Other Wonders of the World," *Eurozine,* March 2001, http://www.eurozine.com/articles/2001-03-01-pearce-en.html. Accessed July 11, 2007.
2. Jack D. Ives and Bruno Messerli, *The Himalayan Dilemma: Reconciling Development and Conservation* (New York: Routledge, 1989).
3. Quoted in Pankaj Mishra, "Caste Adrift." *Financial Times*, January 23, 2011, p. 13.
4. This discussion is based largely on two unpublished manuscripts, "Living Spaces of Korean Architecture" and "In Tune with Nature: Rural Villages and Houses in Korea," by the Korean geographer Sang-Hae Lee, of Sungkyunkwan University, and personal talks with him. I am grateful for his contributions.
5. Quoted in David Pilling, "Megacities." *Financial Times*, November 6, 2011, p. HH-1.
6. Descriptions of Buddhism and Daoism are based on the *Microsoft Encarta Reference Library,* 2004.
7. Michael Schuman. 2009. *The Miracle: The Epic Story of Asia's Quest for Wealth*. New York, NY: HarperBusiness.
8. Patrick Barta, July 27, 2009, "Asian Nations Revisit Safety Net in Effort to Bolster Spending." *The Wall Street Journal*, p. A2.
9. Quoted in Bob Davis, "Massive Population Lifts Nation's Growth." *The Wall Street Journal*, p. A12.
10. Alan S. Blinder, quoted in Anand Giridharadas, "India's Edge Goes beyond Outsourcing." *New York Times,* April 4, 2007, p. C4.
11. Quoted in Edward Wong, "China Navy Reaches Far, Unsettling the Region." *New York Times*, June 15, 2011, p. A11.
12. Robert Chase, Emily Hill, and Paul Kennedy, "Pivotal States and U.S. Strategy." *Foreign Affairs,* January–February 1996, pp. 33–51.
13. Fukuyama, quoted in Martin Wolf, "The End of History Man." *Financial Times*, May 29, 2011, p. 3.
14. James Hookway, "Tensions Flare over Disputed Asian Sea." *The Wall Street Journal*, June 10, 2011.
15. Quoted in "Unquenchable Thirst." *The Economist*, November 19, 2011, p.. 27.
16. Samsar Chand Kour, *Beautiful Valleys of Kashmir and Ladakh* (Mysore, India: Wesley Press, 1942), p. 2.
17. See UNESCO, *Assessment of Capacity Building Requirements for an Effective and Durable Tsunami Warning and Mitigation System in the Indian Ocean: Consolidated Report for the Countries Affected by the 26 December 2004 Tsunami* (Paris: UNESCO, 2005), http://unesdoc.unesco.org/images/0014/001445/144508e.pdf. Accessed July 15, 2007.
18. Quoted in Jamil Anderlini, "A Blast from the Past." *Financial Times*, December 15, 2009, p. 11.
19. Jamil Anderlini and Mure Dickie, "Taking the Waters: China's Gung-Ho Economy Exacts a Heavy Price." *Financial Times,* July 24, 2007, p. 11.
20. I thank the Korea Society for sponsoring my field study of Korea, on which much of this discussion is based. Professor Byong Man Ahn of the Hankuk University of Foreign Studies and Professor Taeho Bark of the Graduate School of International Studies at Seoul National University graciously provided much of the information on Korean economies and politics.

Online Resources

CourseMate: Make the most of your study time by accessing everything you need to succeed in one place. Read your textbook, take notes, review flashcards, watch videos, complete activities, take practice quizzes, and more—online with CourseMate. Log in at **www.cengagebrain.com.**

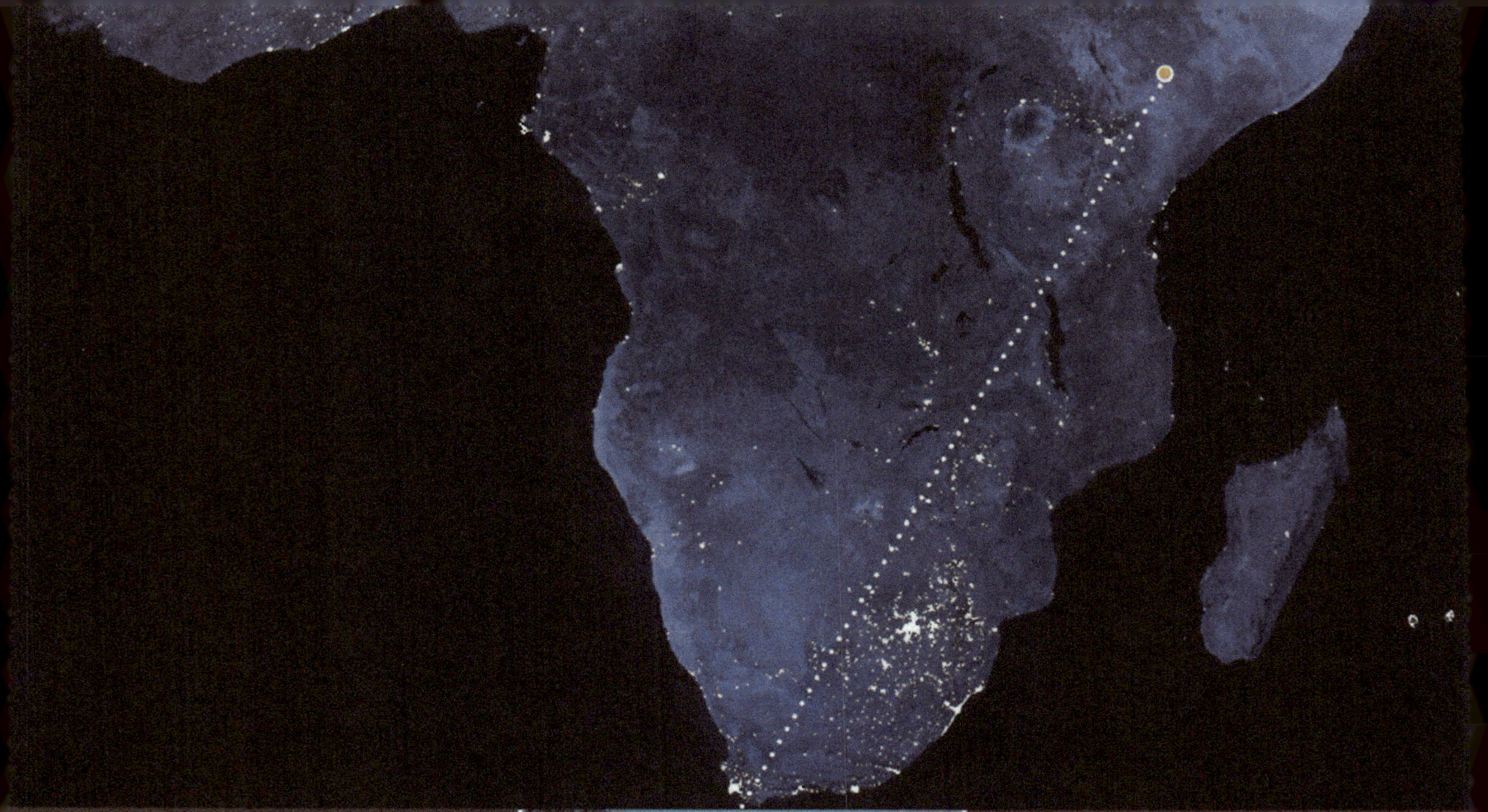

"I dream of the realization of the unity of Africa, whereby its leaders combine in their efforts to solve the problems of this continent. I dream of our vast deserts, of our forests, of all our great wildernesses."

—Nelson Mandela

Africa at night. Note the lights of relatively prosperous South Africa on the world's poorest continent. Left: Doum palms rise from the banks of the Uaso Nyiro River in Kenya's Samburu National Park

C. Mayhew & R. Simmon (NASA/GSFC), NOAA/NGDC, DMSP Digital Archive

Joe Hobbs

Sub-Saharan Africa

9

Geographers typically recognize the continent of Africa as having two major divisions: North Africa (the predominantly Arab and Berber realm of the continent, described in Chapter 6), and ethnically diverse Sub-Saharan Africa (•**Figure 9.1**). This chapter sketches the broad outlines of the geography of Sub-Saharan Africa and examines issues in the major African subregions.

The region of Sub-Saharan Africa is culturally complex, physically beautiful, and problem ridden. This introduction to Africa surveys the region's diverse environments, peoples and ways of living, population distributions, European colonial legacies, and major current issues.

chapter outline

chapter objectives

This chapter should enable you to

- Understand what caused Sub-Saharan Africa to become and remain the world's poorest region.
- Know how the region came to have the highest HIV/AIDS infection rates in the world and how the epidemic could be reversed there.
- Appreciate the pressures on African wildlife and some unique approaches taken to protect the animals.
- Know what is uniquely African about African cultures.
- See how Africans have leap-frogged communications challenges with mobile phones, and use phones to overcome the "digital divide."
- Recognize why, after more than a decade at the sidelines, Sub-Saharan Africa is considered important again in geopolitical affairs.
- Recognize how European colonial favoritism of some ethnic groups over others sowed seeds of modern strife and warfare.
- Appreciate what corrupt leadership has done to impoverish people in countries with enormous oil and other natural resource wealth.
- Understand what causes the "resource curse."
- Understand how drought can trigger a process of environmental degradation.
- Consider the causes of the genocide and "ethnic cleansing" that marred Africa in the 1990s, particularly in East and East-Central Africa.

For many people, Sub-Saharan Africa is the most poorly understood or misunderstood of the world's regions. It seems to have so many countries, so much violence and disease, and so much poverty that it is simply too difficult to think about. Hollywood still perpetuates stereotyped images of an Africa rich in wildlife but with dangerous or backward peoples. This chapter will show that getting to know the real Africa is manageable, rewarding, and interesting. This region's diverse cultures and natural environments are sources of endless wonder. Conditions for many Africans are improving in a number of ways. There are many reasons for hope and much to celebrate in the study of this fascinating region. Just ask the people: a Gallup survey of 50,000 citizens around the world found that Africans are the most optimistic.[2]

9.1 Area and Population

Sub-Saharan Africa (including Madagascar and other nearby Indian Ocean islands, and the newest country of South Sudan) has the second largest land area of all the major world regions described in this book. It covers 8.7 million square miles (22.4 million sq km) and so is more than twice the size of the United States (•**Figure 9.2**).

Sub-Saharan Africa's population was 847 million in 2011. Even with the loss of population due to AIDS, the rate of natural population increase in Sub-Saharan Africa is 2.5 percent per year, or about five times that of the United States. The region's population is expected to double by 2050. As in most LDCs, African parents generally want large families for several reasons: to have extra hands to perform work; to be looked after when they are old or sick; and in the case of girls, to receive the "bride wealth" a groom pays in a marriage settlement. Large families also convey status. However, there are signs of significant change 50
in this pattern of population growth: birth rates have been dropping in every country in this region over the past two decades.

In spite of the large population, much of the region is sparsely populated. The region's average population density is 312
slightly more than that of the United States. Most of the people in this region live in a few small, densely populated areas (•**Figure 9.3**). The main areas are the coastal belt bordering the Gulf of Guinea in West Africa, from the southern part of Africa's most populous country, Nigeria, westward to southern Ghana; the savanna lands of northern Nigeria; the highlands of Ethiopia; the highland region surrounding Lake Victoria in Kenya, Tanzania, Uganda, Rwanda, and Burundi; and the eastern coast and parts of the high interior plateau of South Africa.

Sub-Saharan Africa is the world's most rural region (68%), with rural populations of most countries between 40 and 80 percent. The most rural are two East African nations with very fertile soils: Burundi (89% rural) and Malawi (86%). The most urbanized are Gabon (86%), where people are flocking to partake in an oil boom that is facilitating urbanization, and Djibouti (76%), where there is virtually no arable land and a small number of people are clustered in a port city. The region's major cities and large towns are magnets that attract many poor, rural people. The rate of urban growth is high; in 2011 there were twice as many cities with more than a million residents as there were in 1990 (•**Tables 9.1 and 9.2**). It is projected that 50 percent of Africans will live in cities in 2025. But for now, life in villages is the rule for the majority. A typical rural home is a small hut made of sticks and mud, with a dirt floor, thatched roof, and no electricity or plumbing (•**Figure 9.4**).

Africa has the world's youngest population, with 41 percent of the people under 15 years old (compared with about a third of the populations of Latin America and Asia). The median age is now 20 (compared with 30 in Asia and 40 in Europe). There are different ways of viewing this young, growing population. Some economists see the youthful median age as a "demographic dividend," as large numbers of workers will be able to contribute to the region's economic growth. Others fear that the Malthusian scenario will come about because of the gulf between population growth and agricultural output in these mainly agrarian societies. Since the 1960s, the population of Sub-Saharan Africa has grown at a rate of about 5 percent annually, whereas food production in the region has grown at only about 2 percent annually. This is the only world region where per capita food production declined in that period.

• **Figure 9.1** Principal features of Sub-Saharan Africa.

The wild card in Africa's population deck is the **human immunodeficiency virus (HIV)**. The disease it causes—**AIDS (acquired immunodeficiency syndrome)**—has inevitably been identified by some as the Malthusian "check" to the region's population growth (see Medical Geography, page 316).

9.2 Physical Geography and Human Adaptations

Sub-Saharan Africa is both rich in natural resources and beset with environmental challenges to economic development. It is home to some of the world's greatest concentrations of wildlife and some of the most degraded habitats.

The Landscapes of Africa

Most of Africa is a vast plateau, actually a series of plateaus, with a typical elevation of more than 1,000 feet (c. 300 m). Near the Great Rift Valley in the Horn of Africa and in southern and eastern Africa, the general elevation rises 2,000 to 3,000 feet (c. 600 to 900 m), with many areas at 5,000 feet (1,520 m) and higher (see Regional Perspective, page 317). The highest peaks and largest lakes of the continent are located in this belt. The loftiest summits lie within 250 miles (c. 400 km) of Lake Victoria. They include Kilimanjaro (19,340 ft/5,895 m) and Kirinyaga (Mount Kenya, 17,058 ft/5,200 m; •**Figure 9.5**), which are volcanic cones, and the Ruwenzori range (up to 16,763 ft/5,109 m), a nonvolcanic massif produced by faulting. Lake Victoria, the largest lake in Africa, is surpassed in area

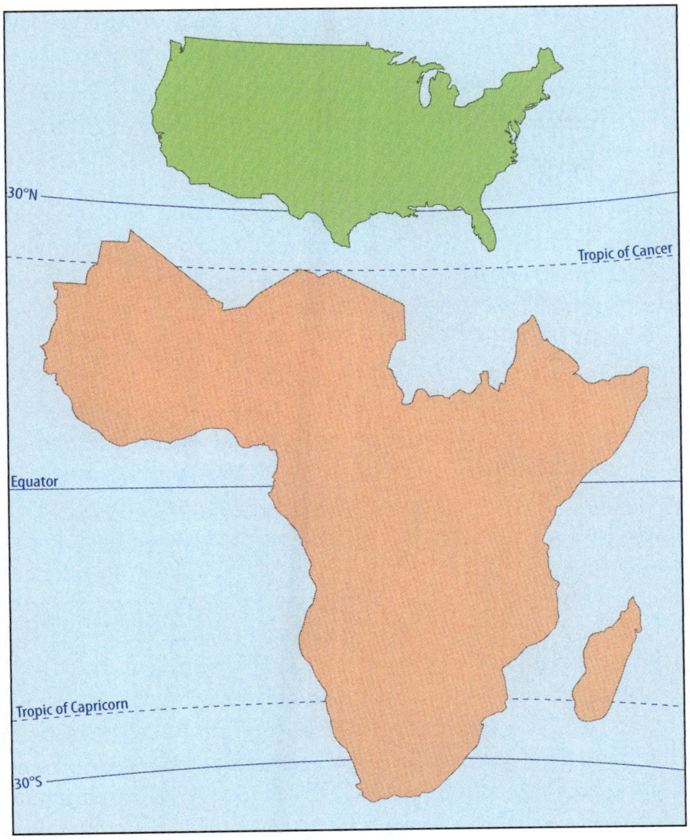

• **Figure 9.2** Sub-Saharan Africa compared in area and latitude with the conterminous United States.

among inland waters of the world only by the Caspian Sea and Lake Superior. It is relatively shallow, however, and the large numbers of people living on its shores are taxing its resources. Other very large lakes in East Africa include Lake Tanganyika and Lake Malawi.

The physical structure of Africa has influenced the character of African rivers. The main rivers, including the Nile, Niger, Congo, Zambezi, and Orange, rise in interior uplands and descend by stages to the sea. At some points, they descend abruptly, particularly at plateau escarpments, with rapids and waterfalls interrupting their courses. These often block navigation a short distance inland. The many waterfalls and rapids do have a positive side: they represent a great potential source of hydroelectric energy. There are major power stations on the Zambezi, Congo, Niger, Volta, and Blue Nile Rivers. But only about 5 percent of Africa's hydropower potential has been realized (compared to about 60% in North America). Many of the best sites are remote from large markets for power. In some cases, geopolitical considerations pose obstacles to dam construction. Downstream Egypt, for example, has expressed concern and even hostile rhetoric about dams and water diversions of the Nile and its tributaries by upstream Sudan, Ethiopia, Uganda, Kenya, and Tanzania (see page 180 in Chapter 6).

Africa's Biomes and Climates

The equator bisects Africa, so about two-thirds of the region lies in the low latitudes, having tropical climates and biomes; Africa is the most tropical of the world's continents (•**Figure 9.6**). Areas of tropical rain forest climate center on the great rain forest of the Congo Basin in central and western Africa. The forest merges gradually into a tropical savanna climate on the north, south, and east. This is the climatic and biotic zone supporting most of the magnificent large mammals of Africa. The savanna areas in turn trend into steppe and desert on the north and southwest. A broad belt of drought-prone tropical steppe and savanna bordering the Sahara on the south is known as the Sahel. There is desert on the coasts of Eritrea, Djibouti, and Somalia in the Horn of Africa. In South Africa and Namibia, a coastal desert, the Namib, borders the Atlantic. The Kalahari Desert, which lies inland from the Namib, is better described as steppe or semidesert than as true desert. Along the northwestern and southwestern fringes of the continent are small but productive areas of Mediterranean climate, while eastern coastal sections and adjoining interior areas of South Africa have a humid subtropical climate. Bordering the subtropical climate region is an area of marine west coast climate.

Total precipitation in the region is high but unevenly distributed; some areas are typically saturated, whereas others are perennially bone dry. Even in many of the rainier parts of the continent, there is a long dry season, and wide fluctuations occur from year to year in the total amount of precipitation. One of the major needs in Africa is better control over water. In the typical village household, women carry water from a stream or lake or a shallow (and often polluted) well. Use of more small dams would help provide water storage throughout the year.

Drought is a persistent problem in most of the countries. Although all droughts create problems, some last for years, with devastating effects in this heavily agricultural region. Droughts have been particularly severe in recent decades in the Sahel and in the Horn of Africa. The number of food emergencies, most of them tied to drought, has tripled since the mid-1980s in Sub-Saharan Africa.

The consensus among scientists is that this region will pay the highest price for global climate change, even though it contributes the fewest greenhouse gases to the atmosphere. United Nations studies predict even more severe droughts and crop losses, along with the destruction of coastal infrastructure that will accompany rising sea levels. Such dangerous environmental changes may be contributing to conflict in the region as well. Pastoralists and farmers already compete for scarce land and water resources in many places—this was one of the problems in Sudan's Darfur conflict, for example—and droughts intensify those struggles.

The patterns of Africa's land use (•**Figure 9.7**) reveal that the most productive lands are on river plains, in volcanic regions (especially the East African and Ethiopian highlands), and in some grassland areas of tropical steppes (notably the High

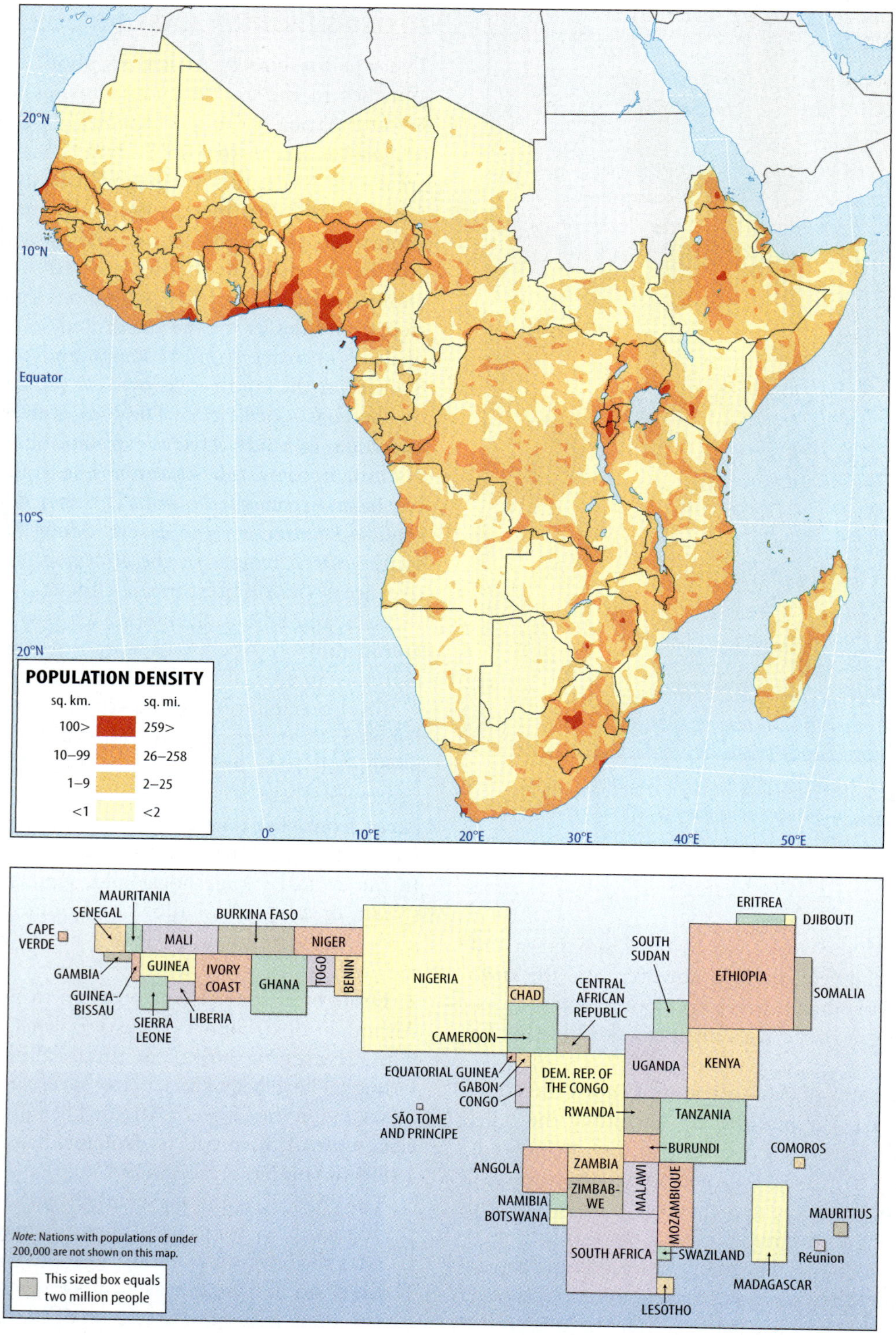

• Figure 9.3 Population distribution (top) and population cartogram (bottom) of Sub-Saharan Africa.
Source: The Oxford School Atlas edited by Patrick Wiegand (OUP, 1997), copyright © Oxford University Press. Used by permission of Oxford University Press.

Veld in South Africa). Soils of the deserts and regions of Mediterranean climate are often poor. In the tropical rain forests and savannas, there are reddish, lateritic tropical soils that are
219, infertile once the natural vegetation is removed and can sup-
345, port only shifting cultivation. Agriculture is discussed in more
384 detail beginning on page 328.

Africa's Wildlife

Although cattle raising is widespread throughout the savannas, cattle are largely excluded from extensive sections both north and south of the equator by the disease called **nagana**, which is carried by the tsetse flies that also transmit sleeping sickness

Table 9.1 Sub-Saharan Africa: Basic Data

Political Unit	Area (thousands; sq mi)	Area (thousands; sq km)	Estimated Population (millions)	Estimated Population Density (sq mi)	Estimated Population Density (sq km)	Annual Rate of Natural Increase (%)	Human Development Index	Urban Population (%)	Per Capita GNI PPP ($US)
The Sahel	**2,296.3**	**5,947.6**	**86.8**	**38**	**15**	**2.8**	**0.324**	**27**	**1,110**
Burkina Faso	105.8	274.0	17	161	62	3.1	0.331	24	1,170
Cape Verde	1.6	4.1	0.5	313	121	1.7	0.568	62	3,530
Chad	495.8	1,284.1	11.5	23	9	2.9	0.328	28	1,160
Gambia	4.4	11.4	1.8	409	158	3.0	0.420	59	1,330
Mali	478.8	1,240.1	15.4	32	12	3.1	0.359	33	1,190
Mauritania	396	1,025.6	3.5	9	3	2.4	0.453	42	1,940
Niger	489.2	1,267.0	16.1	33	13	3.6	0.295	17	680
Senegal	76	196.8	12.8	168	65	2.8	0.459	43	1,810
South Sudan	248.7	644.3	8.2	33	13	N/A	N/A	N/A	N/A
West Africa	**812.4**	**2,104.1**	**246.1**	**255**	**98**	**2.5**	**0.450**	**49**	**1,820**
Benin	43.5	112.7	9.1	209	81	2.9	0.427	43	1,510
Ghana	92.1	238.5	25	271	105	2.3	0.541	52	1,530
Guinea	94.9	245.8	10.2	107	41	2.7	0.344	28	940
Guinea-Bissau	13.9	36.0	1.6	115	44	2.3	0.353	30	1,060
Ivory Coast	124.5	322.5	22.6	182	70	2.3	0.400	51	1,640
Liberia	37.2	96.3	4.1	110	43	3.1	0.329	47	290
Nigeria	356.7	923.9	162.3	455	176	2.5	0.459	51	2,070
Sierra Leone	27.7	71.7	5.4	195	75	2.2	0.336	39	790
Togo	21.9	56.7	5.8	265	102	2.8	0.435	37	850
West Central Africa	**1,576**	**4,081.8**	**99.4**	**53**	**20**	**2.6**	**0.347**	**43**	**1,140**
Cameroon	183.6	475.5	20.1	109	42	2.2	0.482	59	2,190
Central African Republic	240.5	622.9	5	21	8	2.1	0.343	39	750
Congo Republic	132	341.9	4.1	31	12	2.5	0.533	63	3,040
Congo, Democratic Republic of	905.4	2,345.0	67.8	75	29	2.8	0.286	36	300
Equatorial Guinea	10.8	28.0	0.7	65	25	2.3	0.537	40	19,330
Gabon	103.3	267.5	1.5	15	6	1.8	0.674	86	12,450
São Tomé and Principe	0.4	1.0	0.2	500	193	2.7	0.509	63	1,850
East Africa	**703**	**1,820.8**	**143.4**	**155**	**60**	**2.9**	**0.460**	**20**	**1,300**
Burundi	10.7	27.7	10.2	953	368	3.2	0.316	11	390
Kenya	224.1	580.4	41.6	186	72	2.7	0.509	18	1,570
Rwanda	10.2	26.4	10.9	1,069	413	2.1	0.429	19	1,130
Tanzania	364.9	945.1	46.2	127	49	2.9	0.466	27	1,360
Uganda	93.1	241.1	34.5	371	143	3.4	0.446	15	1,190
Horn of Africa	**727**	**1,882.9**	**103.8**	**152**	**59**	**2.7**	**0.363**	**20**	**920**
Djibouti	9	23.3	0.9	100	39	1.9	0.430	76	2,480
Eritrea	45.4	117.6	5.9	130	50	2.6	0.349	22	580
Ethiopia	426.4	1,104.4	87.1	204	79	2.7	0.363	17	930
Somalia	246.2	637.7	9.9	40	16	2.8	N/A	38	N/A

(continued)

Table 9.1 Sub-Saharan Africa: Basic Data (*continued*)

Political Unit	Area (thousands; sq mi)	Area (thousands; sq km)	Estimated Population (millions)	Estimated Population Density (sq mi)	Estimated Population Density (sq km)	Annual Rate of Natural Increase (%)	Human Development Index	Urban Population (%)	Per Capita GNI PPP ($US)
Southern Africa	**2,310.8**	**5,985.0**	**142.4**	**52**	**20**	**1.9**	**0.486**	**45**	**5,440**
Angola	481.4	1,246.8	19.6	41	16	2.8	0.486	59	5,190
Botswana	224.6	581.7	2	9	3	1.2	0.633	59	12,840
Lesotho	11.7	30.3	2.2	188	73	1.2	0.450	23	1,800
Malawi	45.7	118.4	15.9	348	134	2.7	0.400	14	780
Mozambique	309.5	801.6	23.1	75	29	2.8	0.322	31	880
Namibia	318.3	824.4	2.3	7	3	1.7	0.625	39	6,350
South Africa	471.4	1,220.9	50.5	107	41	0.6	0.619	62	10,050
Swaziland	6.7	17.4	1.2	179	69	1.5	0.522	22	4,790
Zambia	290.6	752.7	13.5	46	18	3.1	0.430	36	1,280
Zimbabwe	150.9	390.8	12.1	80	31	1.9	0.376	29	N/A
Indian Ocean Islands	**229.7**	**595**	**24.6**	**107**	**41**	**2.6**	**0.493**	**34**	**1,740**
British Indian Ocean Territory (U.K.)	0.02	0.05	0.003	150	58	N/A	N/A	100	N/A
Comoros	0.9	2.3	0.8	889	343	2.8	0.433	28	1,180
Madagascar	226.7	587.2	21.3	94	36	2.8	0.480	31	990
Mauritius	0.8	2.1	1.3	1,625	627	0.5	0.728	42	13,270
Mayotte (Fr.)	0.1	0.3	0.2	2,000	772	3.0	N/A	50	N/A
Réunion (Fr.)	1	2.6	0.9	900	347	1.1	N/A	94	N/A
Seychelles	0.2	0.5	0.1	500	193	1.0	0.773	56	16,790
Summary Total	**8,655.2**	**22,417.2**	**846.5**	**98**	**38**	**2.5**	**0.423**	**32**	**2,080**

Sources: World Population Data Sheet, Population Reference Bureau, 2011; Human Development Report, United Nations, 2011; World Factbook, CIA, 2011.

to humans (see Insights, page 321). In tropical rain forests, tsetse flies are even more prevalent and few cattle are raised, but goats and poultry are common (as they are in tsetse-frequented savanna areas).

Africa has the planet's most spectacular and numerous populations of large mammals. The tropical grasslands and open forests of Africa are the habitats of large herbivorous animals, including the elephant, buffalo, zebra, giraffe, and many species of antelope, as well as carnivorous and scavenging animals, such as the lion, leopard, and hyena. The tropical rain forests have fewer of these "game" animals (as Africans call them); the most abundant species here are insects, birds, and monkeys, with the hippopotamus, the crocodile, and a great variety of fish in the streams and rivers draining the forests and wetter savannas.

Although documentary films promote a perception outside Africa that the continent is a vast animal Eden, the reality is less positive. Human population growth, urbanization, and agricultural expansion are taking place in Africa, as elsewhere in the world, at the expense of wildlife. Hunting and competition with domesticated livestock also take their toll. Many species are safeguarded by law, especially in the protected areas that make up about 15 percent of the region. Such laws are difficult to enforce, however, and poaching has devastated some species. Still, Africa remains home to some of the world's most extraordinary and successfully managed national parks, including South Africa's Kruger National Park, Tanzania's Ngorongoro Crater National Park, and Kenya's Amboseli National Park (at the foot of Kilimanjaro; see Figure 9.5). International tourism to these parks is a major source of revenue for some countries, particularly South Africa, Tanzania, Kenya, Namibia, and Botswana.

Elephants and rhinoceroses are Africa's largest herbivores and the most vulnerable to global market demand for wildlife products. In 1970, there were about 2.5 million African

Table 9.2 Sub-Saharan Africa: Metropolitan Populations Data

1	Lagos, Nigeria	14.9
2	Kinshasa, D.R.C.	10.9
3	Johannesburg, South Africa	8.7
4	Abidjan, Ivory Coast	6.7
5	Ibadan, Nigeria	5.8
6	Cape Town, South Africa	5.4
7	Luanda, Angola	5.3
8	Kano, Nigeria	4.9
9	Accra, Ghana	4.6
10	Nairobi, Kenya	4.5
11	Durban, South Africa	3.7
12	Harare, Zimbabwe	3.5
13	Addis Ababa, Ethiopia	3.4
14	Dar es Salaam, Tanzania	3.4
15	Lusaka, Zambia	3.3
16	Dakar, Senegal	2.6
17	Pretoria, South Africa	2.6
18	Benin City, Benin	2.5
19	Bamako, Mali	2.5
20	Yaoundé, Cameroon	2.4

Population in millions.

• **Figure 9.4** Homes in African villages are typically made of local materials including thatch, which if replaced regularly do a fine job of keeping rain out. They are usually elevated slightly above to stave off the threat of flooding. Village roads are seldom paved. People live in close proximity with their livestock, in part to protect the animals from predators. This village is in northwestern Madagascar.

elephants living on the continent. Poaching and habitat destruction had reduced their numbers to less than 500,000 in 2012. Poaching for ivory was having a catastrophic impact on African elephants when delegates of the 112 signatory nations of the Convention on International Trade in Endangered Species (CITES, now with 175 members) met in 1989. The organization succeeded in passing a worldwide ban on the ivory trade, and since then the precipitous decline has eased.

• **Figure 9.5** Africa's highest mountains, both volcanic, are Tanzania's Kilimanjaro (left; photographed from Kenya's Amboseli National Park) and Kenya's Kirinyaga (right), also known as Mt. Kenya.

MEDICAL GEOGRAPHY

HIV and AIDS in Africa

AIDS is a global problem. The earth has not seen a **pandemic** (very widespread epidemic) like this since the bubonic plague devastated fourteenth-century Europe and smallpox struck the Aztecs of sixteenth-century Mexico. AIDS had killed over 25 million people worldwide by 2012, and there are almost 3 million new infections around the world each year. Authorities say that worldwide 7,400 people are infected every day. Sub-Saharan Africa is where the virus originated (apparently among chimpanzees; the first human case was reported in the Belgian Congo—today's Democratic Republic of Congo—in 1959) and is today the epicenter of this health crisis. The statistics are startling. In 2012, two-thirds of the world's estimated 33 million people infected with HIV lived in Sub-Saharan Africa.

The epidemic is most severe in southern Africa. In some countries, including Botswana, Lesotho, and Swaziland, as much as one-quarter of the adult population is HIV positive (•Figure 9.A). South Africa has more HIV cases (about 5.7 million) than any other country in the world. This scourge has been causing sharp reductions in life expectancy in Africa and, unless contained, will dramatically alter the region's population and development trends.

The good news is that HIV/AIDS is starting to be contained. The rate of new infections is down about 25 percent in the worst affected countries, mainly in southern Africa. Death rates, which were rising precipitously, are falling. This is brightening the prospects for development in this least developed of the world's regions. The incidence of HIV infection is particularly high among the region's most educated, skilled, and ambitious young urban professionals, including teachers, white-collar workers, and government employees. These are the people, in addition to truckers and merchants, who travel the most, have higher incomes, and are therefore the most likely to have many sexual liaisons.

What can be done to fight the AIDS epidemic? Two countries that have had remarkable success in stemming the AIDS tide provide some answers. Uganda and Senegal have used relatively inexpensive public health tools, especially education, in the fight against the disease. To reduce the huge stigma associated with AIDS and to encourage counseling, testing, and condom use, the countries' political and religious leaders have been outspoken about AIDS. The results have been impressive, with infection rates dropping in Uganda (from 18% to 6%) and staying low in Senegal.

In addition, wealthier countries outside Africa could invest more in combating and preventing the disease. The United Nations in 2001 established the Global Fund to Fight AIDS, Malaria, and Tuberculosis. Bono and other celebrity activists continue to pressure governments to pledge more monies to the Global Fund. But their efforts are flagging due to the global Great Recession and "donor fatigue."

The United States has its own effort to combat AIDS in 12 countries in Africa and 2 (Haiti and Guyana) in the Americas. In a program called **PEPFAR**, the President's Emergency Plan for AIDS Relief, the George W. Bush administration pledged over $45 billion to the effort. The United States gives more money to combat AIDS than all other donors combined. This is an enormous amount of money, and yet it is dwarfed by the scale of the problem: it is enough to treat only about one in four people infected with HIV. The Bush and Obama administrations spent most of those funds to put patients on life-lengthening **antiretroviral (ARV) drugs** and other drugs. About 5 million people worldwide were on these drugs in 2012, and another 10 million needed but could not get them.

The Obama Administration could not afford the escalating costs of treatment, and shifted the focus to prevention and mother/

CITES member states won the ivory ban over the strong objections of several southern African states, led by Zimbabwe, that were experiencing what they saw as an elephant overpopulation problem. At the 1989 CITES meeting, these countries appealed for an exemption from the ivory ban so that they could earn foreign export revenue from a sustainable yield of their elephant populations. The majority of CITES members rejected this appeal, arguing that any loophole in a complete ban would subject elephants everywhere to illegal poaching. Since 1989, CITES has periodically allowed Zimbabwe, Namibia, Botswana, and South Africa one-time sales of ivory under rigorous monitoring as a reward for their positive wildlife policies. All continue to cull (kill) their "excess" elephants, with the meat going to needy villagers and crocodile farms.

If elephant ivory is like gold, rhinoceros horns are like diamonds. Men in the Arabian Peninsula nation of Yemen prize daggers with rhino horn handles (•Figure 9.10). Although Western scientists deny the medicinal efficacy of powdered rhino horn, traditional medicine in East Asia makes wide use of it, including as an aphrodisiac and a cure for cancer. These demands, and the current black-market value of tens of thousands of dollars per horn ($45,000 per pound, more than gold, cocaine, or heroin), have led to a precipitous decline in population of black rhinoceroses in Africa. There were an estimated 65,000 black rhinos in Africa in 1982; in 2012, there were an estimated 3,600. Despite aggressive anti-poaching efforts, the Western Black Rhino became extinct in the wild in 2011, leaving three other subspecies with critically low numbers. Even rhinos mounted on display in European museums have had their horns poached!

9.3 Cultural and Historical Geographies

Many non-Africans are unaware of the achievements and contributions of the cultures of Sub-Saharan Africa. The African continent was the original home of humankind. Recent DNA studies suggest that the first modern people (*Homo sapiens*)

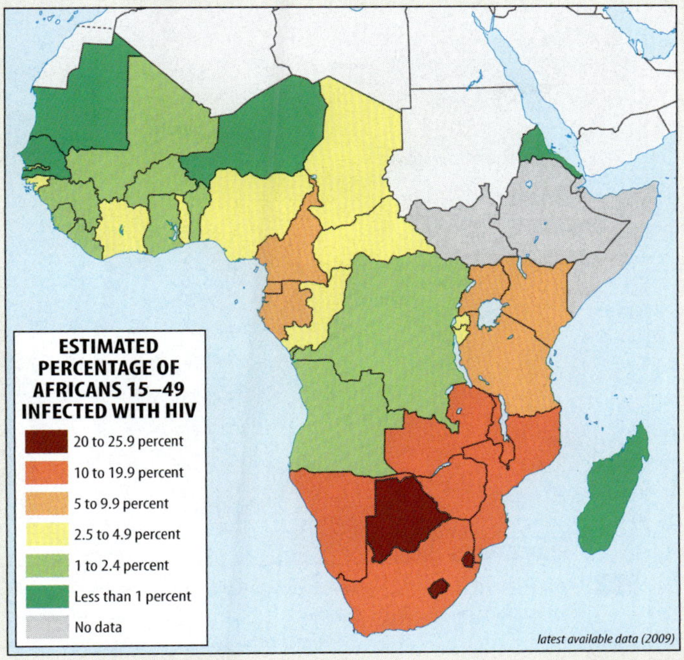

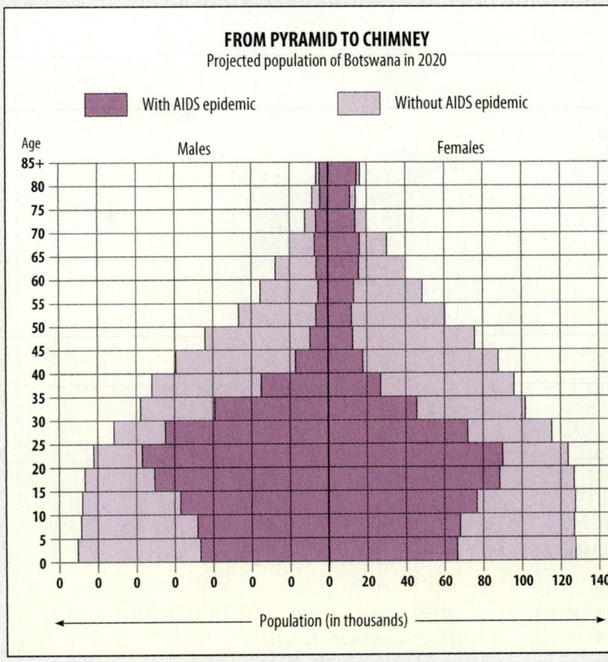

• Figure 9.A HIV/AIDS has had a devastating impact in Sub-Saharan Africa and threatens to redraw the demographic profiles of some countries in unprecedented ways.

child health. For sexual transmissions, the main weapons are male circumcision, a vaginal microbicidal gel, and of course, condoms. Needle-exchange programs are most important in preventing transmission among drug users. So far though, the math is heartbreaking. In the best-case scenario, the Global Fund and PEPFAR jointly will allow 10 million people to be put on treatment by 2014. An estimated thirty-six million will be infected by then. For every 100 people put on treatment, 250 more are newly infected. The overwhelming majority of the newly infected will die untreated. AIDS activists are angry about the spending focus of existing programs, saying that children will be saved now only to die later of AIDS.

to inhabit Asia, Europe, and the Americas were descendants of a small group that left Africa via the Isthmus of Suez about
49 100,000 years ago. After about 5000 B.C.E., indigenous people were responsible for agricultural innovations in four culture hearths: the Ethiopian Plateau, the West African savanna, the West African forest, and the forest-savanna boundary of West Central Africa. Africans in these areas domesticated important crops such as millet, sorghum, yams, cowpeas, okra, watermelons, coffee, and cotton. From Africa, these diffused to populations in other world regions. 82

REGIONAL PERSPECTIVE The Great Rift Valley

One of the most spectacular features of Africa's physical geography is the **Great Rift Valley**, a broad, steep-walled trough extending from the Zambezi Valley (on the border between Zimbabwe and Zambia) northward to the Red Sea and the valley of the Jordan River in southwestern Asia (see Figure 9.1 and also Figure 6.9, page 163 and Figure 2.1, page 22). Its relationship to the tectonic movement of crustal plates is still poorly understood. However, most earth scientists believe it marks the boundary of two crustal plates that are rifting, or tearing apart, causing a central block between two parallel fault lines to be displaced downward, creating a linear valley. This movement will eventually cut much of southern and eastern Africa away from the rest of the continent and allow seawater to fill the valley.

The Great Rift Valley has several branches. Lakes, rivers, seas, and gulfs already occupy much of it. It contains most of the larger lakes of Africa, although Lake Victoria, situated in a depression between two of its principal arms, is an exception. Some, like Lake Tanganyika—at 4,823 feet (1,470 m), the world's second-deepest lake, after Russia's Lake Baikal—are extremely deep. Most have no surface outlet. Volcanic activity associated with the Great Rift Valley has created Kilimanjaro, Kirinyaga, and some of the other great African peaks, along with lava flows, hot springs, and other thermal features. Faulting along the Great Rift Valley in Ethiopia, Kenya, and Tanzania has also exposed remains of the earliest known ancestors of *Homo sapiens*. 39

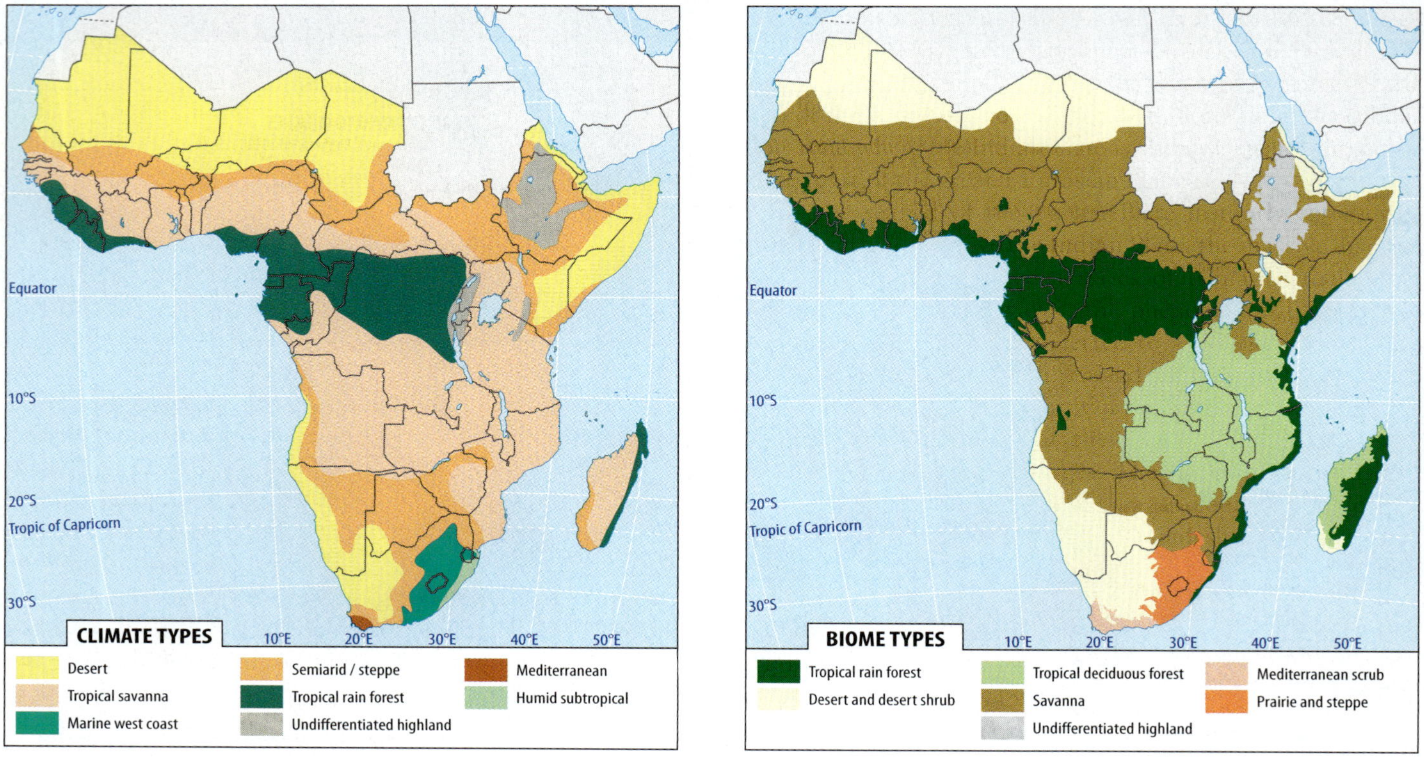

•Figure 9.6 Climates (left) and biomes (right) of Sub-Saharan Africa.
Sources: (left) Based on Rand McNally's Classroom Atlas, 2003. (right) Based on World Wildlife Fund ecoregions data, 1999.

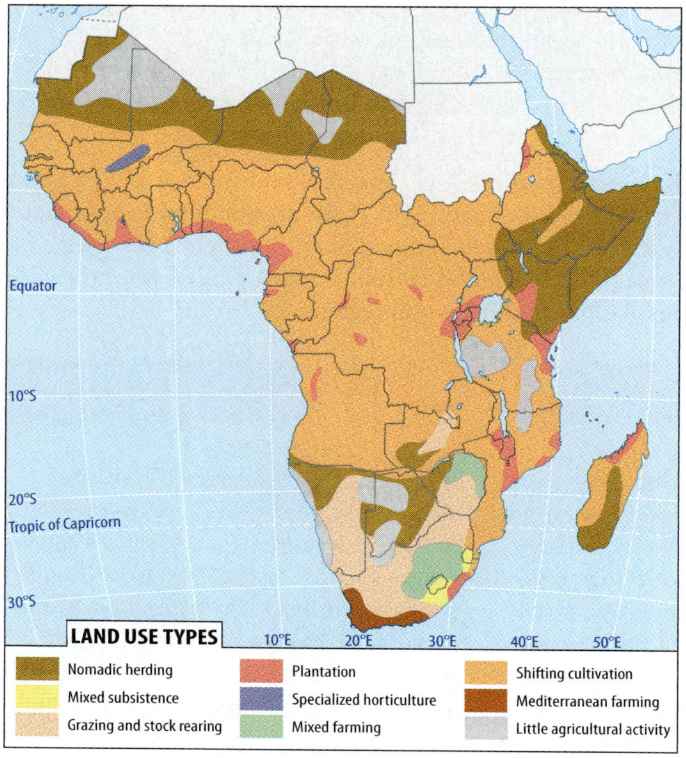

•Figure 9.7 Land use in Sub-Saharan Africa.
Source: The Oxford School Atlas edited by Patrick Wiegand (OUP, 1997), copyright © Oxford University Press. Used by permission of Oxford University Press.

Joe Hobbs

•Figure 9.8 Mother and child in Zimbabwe. Women plant and harvest most of Africa's food and care for most of its children.

Joe Hobbs

• **Figure 9.9** Zebu cattle are extremely important in the cultures and household economies of rural Madagascar. There are about 11 million cattle in this country of 21 million people.

Civilizations and empires emerged in Ethiopia, West Africa, West Central Africa, and South Africa. In the first century C.E., a Christian empire based in the Ethiopian city of Axum controlled the ivory trade from Africa to Arabia. Ethiopian tradition holds that a shrine in Axum still contains the biblical Ark of the Covenant and the tablets of the Ten Commandments, which
170 disappeared from the Temple in Jerusalem in 586 B.C.E. Several Islamic empires, including the Ghana, Mali, and Hausa states, emerged in West Africa between the ninth and nineteenth centuries. All of these agriculturally based civilizations controlled major trade routes across the Sahara. They profited from the exchange of slaves, gold, and ostrich feathers for weapons, coins, and cloth from North Africa. Three kingdoms arose between the fourteenth and eighteenth centuries in what are now the southern Democratic Republic of Congo and northern Angola. These included the Kongo kingdom, which had productive agriculture and was the hub of an interregional trade network for food, metals, and salt. In what is now Zimbabwe, the Karanga kingdom of the thirteenth to fifteenth centuries built its capital city at the site known as Great Zimbabwe. Its skilled metalworkers mined and crafted gold, copper, and iron, and merchants traded these metals with faraway India and China.

Joe Hobbs

• **Figure 9.10** Daggers are a nearly universal dress accessory for men in the Arabian Peninsula nation of Yemen. The most prized dagger handles are of rhino horn, a custom that has had a devastating impact on African rhinos thousands of miles away.

The Languages of Africa

Recent scientific studies have concluded that humans first began using language in southwestern Africa between 50 and 100 thousand years ago. Today, there is great linguistic diversity in Sub-Saharan Africa (•**Figure 9.11**). Africans speak about 2,000 of the world's 6,000 languages, belonging to four broad language groups.

The **Niger-Congo language family** (sometimes considered a subfamily of the Niger-Khordofanian family) is the largest. It includes the many West African languages and the roughly 400 Bantu subfamily languages that fall into seven branches: **Benue-Congo, Kwa, Atlantic, Mandé, Gur, Adamawan,** and **Khordofanian**. Most of these are spoken south of the equator. The *Bantu* language of the Benue-Congo branch is the most widespread.

The Afro-Asiatic language family includes Semitic languages (such as the *Amharic* language of Ethiopia and *Arabic*) and tongues of the **Cushitic** (*Oromo* and *Somali* of the Horn of Africa, for example) and **Chadic** (especially the *Hausa* of northern Nigeria) branches. People living in the area adjoining the Sahara, from West Africa to the Horn of Africa, speak these languages. Even some of the Niger-Congo languages originating south of the Sahara, such as the *Swahili (Kiswahili)* tongue spoken widely in East Africa, have borrowed much from Arabic and other languages with roots elsewhere. The prominence of Arabic words in Swahili reflects a long history of Arab seafaring along the Indian Ocean coast of Africa; in fact, Swahili means "coastal" in Arabic.

The **Nilo-Saharan language family** of the central Sahel region, the northern region of West Central Africa, and parts of East Africa includes about 100 languages in three branches: *Songhai* (a single language, spoken mainly in Mali) and the **Saharan** and **Chari-Nile** language groups.

The **Khoisan languages** of the **San** and related peoples in the western portion of southern Africa are nearly extinct. The

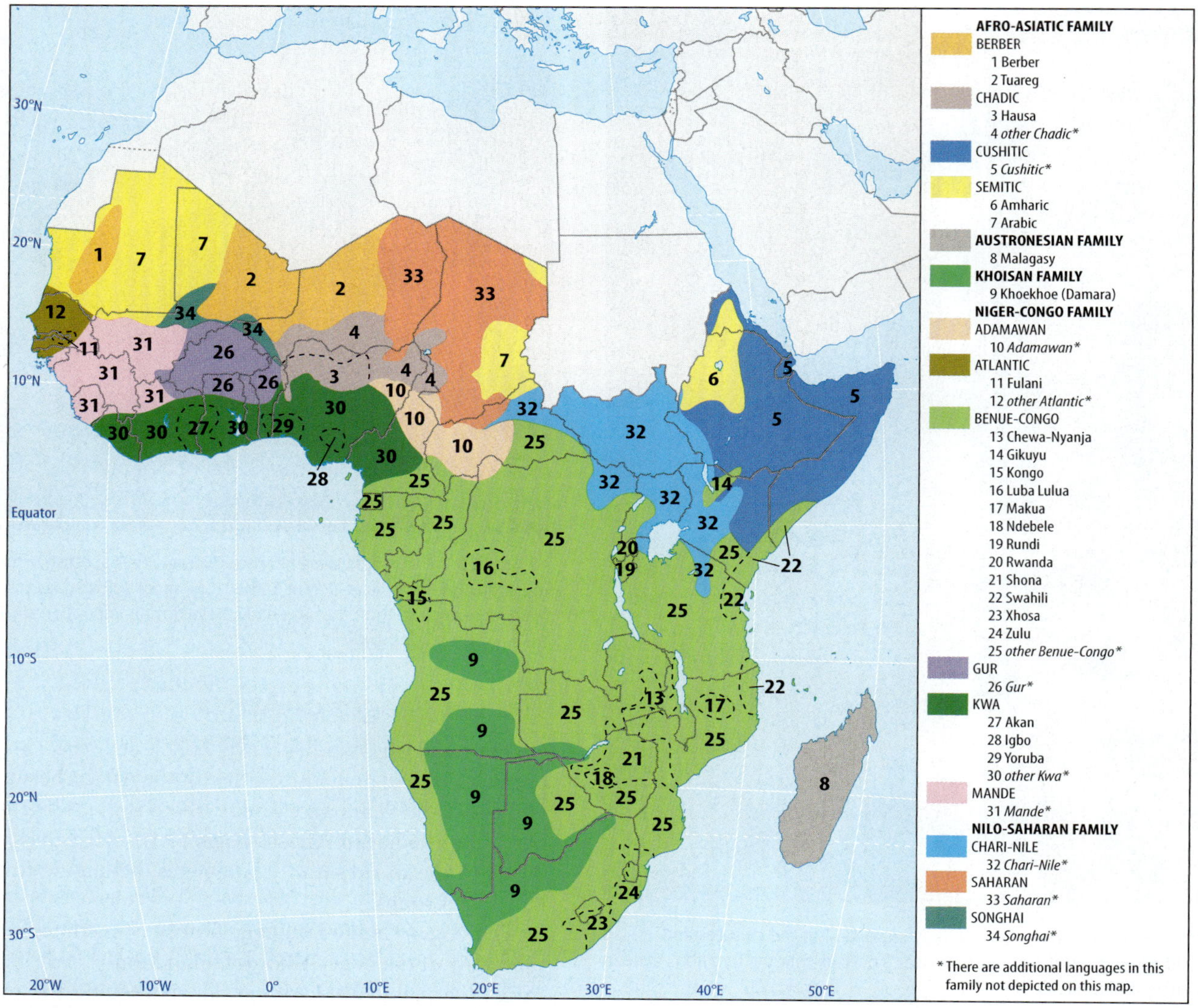

• Figure 9.11 Languages of Sub-Saharan Africa.
Source: Adapted with permission from Bernard Comrie, et al., The Atlas of Languages, Revised Edition (New York: Facts On File, 2003). Reprinted with permission of the Publisher.

largest, *Khoekhoe (Damara)*, is spoken by fewer than 200,000 people in Namibia. Recognizable by their prominent "clicking" sounds, these may be humankind's oldest languages.

Quite distinct from those four groups, the people of Madagascar speak a language without African roots. This is *Malagasy*, an
291 Austronesian tongue that originated in Southeast Asia.

Africa's list of lingua francas—the languages most likely to be recognized on the continent and those that Africans can use to speak to the wider world—is provided by the six official languages of the African Union, the continent's chief supranational
85 organization. They are English, French, Portuguese, Spanish, Swahili, and Arabic. Use of English is growing most rapidly, even in the formerly French-dominated "Francophone" countries. Countries that formerly were British colonies have a distinct advantage in the world of outsourced services: South Africa, Kenya, and Ghana, for example, have already opened call centers that they hope will rival those of India. Some Francophone countries also have call centers catering to French industries and clients.

The dominance of foreign languages in computers and other digital technologies could endanger African languages, many of which are already in peril because they are spoken by so few people. As elsewhere, many languages in Africa are threatened
by language death. Many linguists, however, see the computer 226
as a tool for *saving* the roughly 3,000 endangered languages of the world. They argue that the computer will make it much easier for people to learn and preserve their traditional tongues. Australian linguist James McElvenny uses software to revitalize vanishing languages (look up the Enduring Voices Project on the web). He gives young language learners portable electronic reference tools that give the definitions and sounds of words that are otherwise no longer spoken. Dr. Tucker Childs, a linguist working in the West African nation of Sierra Leone, is using a geographic information system (GIS) to plot locational variations of Kim and other endangered languages. With these findings, linguists can decide where to focus their efforts.

Africa's Belief Systems

The religious landscape of Africa is complex and fluid. Spiritualism is extremely strong, but spiritual affiliations and practices are more interwoven and flexible than in most other

INSIGHTS Africa's Greatest Conservationist

Not much larger than the common housefly, the tsetse fly of Sub-Saharan Africa packs a wallop. This insect carries two diseases, both known as **trypanosomiasis**, which are extremely debilitating to people and their domesticated animals. People contract **sleeping sickness** from the fly's bite, and cattle contract nagana. Since the 1950s, widespread efforts have been made to eradicate tsetse flies so that people can grow crops and herd animals in currently fly-infested wilderness areas (•**Figure 9.B**). Where the efforts have been successful, people have cleared, cultivated, and put livestock on the land. The results are mixed. Although people have been able to feed growing populations in the process of opening up these lands, they have also eliminated important wild resources and in many cases caused erosion, desertification, and salinization of the land. Where the tsetse fly has been eliminated, so has the wilderness. The diminutive tsetse fly thus may be characterized as a **keystone species**—one that affects many other organisms in an ecosystem. The loss of a keystone species—in this case, a fly that keeps out humans and cattle—can have a series of destructive impacts throughout the ecosystem. For its role in maintaining wilderness in Sub-Saharan Africa, some wildlife experts call the tsetse fly "Africa's greatest conservationist."

Joe Hobbs

• **Figure 9.B** Strenuous and successful efforts to eradicate tsetse flies have opened vast new areas of Africa to human use. This is a tsetse fly trap with two components: a liquid chemical called "simulated cow's breath" that attracts the fly and a pesticide-soaked tarp that kills it.

world regions. It is not uncommon for family members to follow different faiths or for an individual to change his or her religious beliefs and practices in the course of a lifetime.

Broadly, however, some dominant patterns of religious geography can be recognized (•**Figure 9.12**). Islam is the dominant religion in North Africa and the countries of the Sahel on the southern fringe of the Sahara. The **Ethiopian Orthodox Church**, closely related to the Coptic Christian faith of Egypt, makes Ethiopia an exception to the otherwise Islamic Horn of Africa region (•**Figure 9.13**). Islam is also the prevailing religion of the East African coast, where Arab traders introduced the faith. Muslims are a majority or strong minority in rural northern Nigeria and Tanzania, and there are minority Muslim populations in cities and towns across the continent. Christians are a majority in southern Nigeria, Uganda, Lesotho, and parts of South Africa. Both Christianity and Islam are strongest in the cities, whereas traditional religions prevail or overlap with these monotheistic faiths in rural areas.

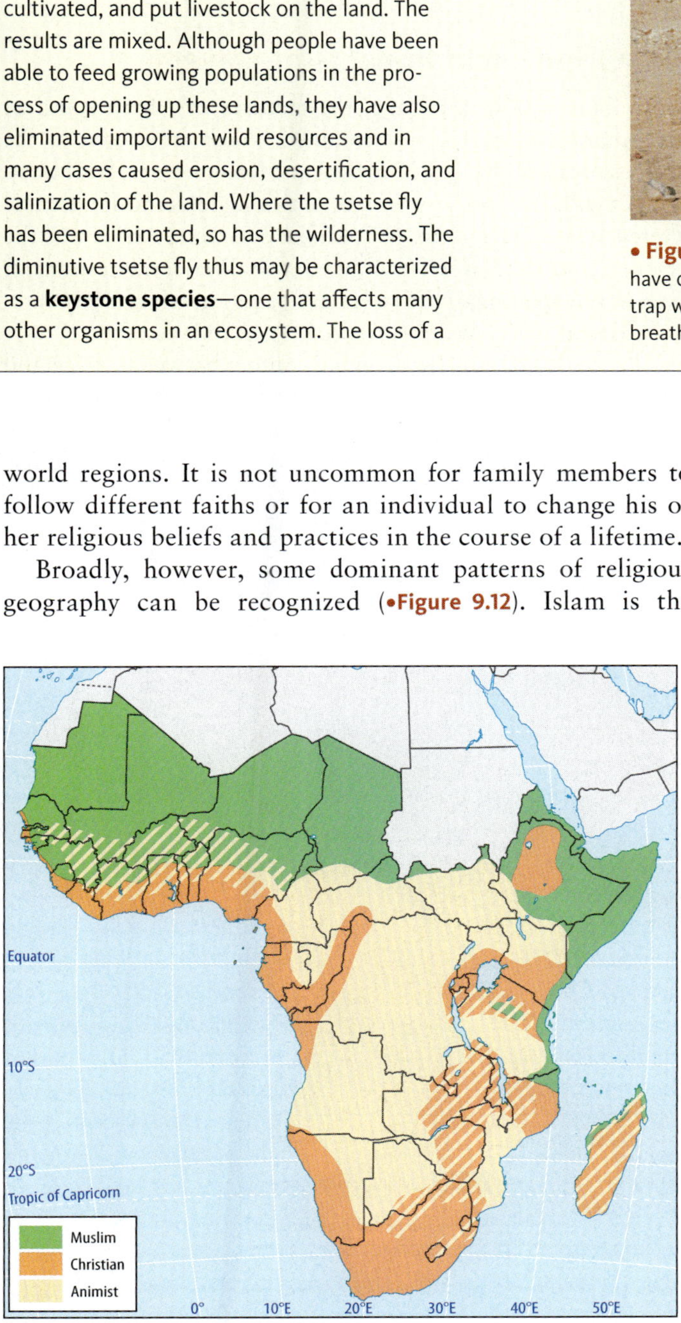

© Cengage Learning 2013

• **Figure 9.12** Religions of Sub-Saharan Africa.

Christians and Muslims have mounted efforts to dispose of "heretical" notions in African traditional belief systems, but these concepts are very resilient. Even where the large monotheistic faiths nominally prevail and their followers are devout, a strong substrate of indigenous beliefs has persisted and usually comfortably merged with the "official" faith—a syncretism also present in other cultures of the world. One example 228, 365 is the mingling of indigenous *vodun* ("voodoo") beliefs with Christianity and Islam in West Africa.

Indigenous African religions have many animistic elements, emphasizing that spiritual forces are manifested everywhere in the environment. Many spirits are tied to particular places in the landscape. Such beliefs contrast with the introduced Christian and Muslim faiths, which tend to see

Joe Hobbs

• **Figure 9.13** The Ethiopian Church has been the dominant religion in Ethiopia since the 4th Century. This priest's flock worships at the 12th Century Church of St. George in Lalibela. When my Mom and I visited Ethiopia in 1983 we were apparently the only tourists in the country! Ethiopia had a Marxist government that discouraged tourism and outlawed religion. The country's devout Christians and Muslims defied the ban.

nature as separate from God and people as apart from and superior to nature.[3] Across the spectrum of African cultures, the use of mediums to contact spirits of deceased ancestors, creator beings, or earth genies that can intercede on one's behalf is widespread (**see Perspectives from the Field, next page**). Reverence for ancestors is strong, reflected also in great respect for the elderly who are living.

Outsiders have tried, seldom successfully, to give accurate names to indigenous African belief systems, practices, and personnel: ancestor worship, animism, *force vitale* ("life force"), living dead, and witch doctor among them. Although it would be gratifying to have precise names and generalizations for African beliefs, they are unique and can best be understood by careful reading, observation, and discussion on a case-by-case basis.

The Origins and Impacts of Slavery

Until about 1,000 years ago, the cultures of Africa south of the Saharan desert barrier remained largely unknown to the peoples north of the desert. Egyptians, Romans, and Arabs had contact with the northern fringes of this region, and some trade filtered across the Sahara, but to most outsiders, Africa was the "Dark Continent," a self-contained, tribalized land of mystery. Even at the opening of the twentieth century, vast areas of interior tropical Africa were still little known to Westerners.

The tragic impetus for growing contact between Africa and the wider world was slavery (**•Figure 9.14**). Over a period of 12 centuries, as many as 25 million people from Sub-Saharan Africa were forced to become slaves, exported as merchandise

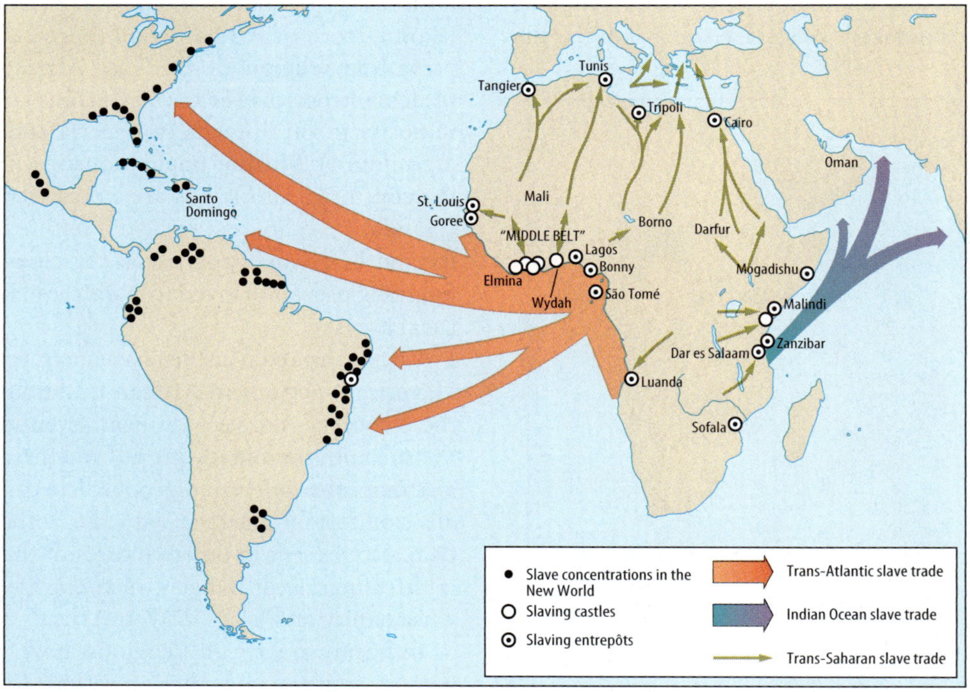

• **Figure 9.14** Slave export trade routes from Sub-Saharan Africa.

Source: Based on Stock, Robert, Africa South of the Sahara.

PERSPECTIVES FROM THE FIELD Under Madagascar

In the austral autumn of 2000, I went to Madagascar to investigate the relationship between people and caves on the island. It was a long journey: 42 hours from the time I left home in Columbia, Missouri, until I arrived in Antananarivo (Tana), Madagascar's capital. During a three-hour delay of the Air Madagascar flight from Paris, excess weight, including my baggage, was unloaded from the aircraft. Twelve hours after takeoff, I arrived exhausted and without provisions (except for 30 pounds of camera and video gear in my hand luggage) but very excited about being in this exceptional country.

Fortunately, once in Tana, I was in the capable hands of a dear friend and colleague, Steve Goodman. A zoologist based at Chicago's Field Museum, Steve is a pioneer in the natural history study of this extraordinary island, where nearly 90 percent of all plants and animals are endemic. Steve has lived here since 1989 and has virtually become Malagasy. His wife, Asmina, is of the island's coastal Sakalava culture. Steve, Asmina, and their son, Hisham, reside with an extended family network, and as soon as I arrived from the airport, I was immersed in special festivities surrounding the impending wedding of a family member. That afternoon, these included the *fumba fumba,* a ceremony in which the bride's and the groom's families seek approvals from their respective ancestors (*razana*) for the wedding. It is impossible to overstate the significance of the spirits of one's ancestors in this culture. The Malagasy peoples believe them to influence or control virtually every aspect of daily life.

When my baggage arrived two days later, Steve and I set out across the High Plateau, bound for the caves of northwestern Madagascar's region of Majunga. Not 3 hours into what should have been a 15-hour journey, Steve's vehicle developed a fuel line problem and began to hemorrhage gasoline. There was no choice but to return to Tana. At least this setback allowed me the opportunity to observe in more detail the environmental holocaust wrought by 2,000 years of human activity on the island (see Figure 9.31). Relentless clearing of forest for rice cultivation and cattle pasture has created a virtual moonscape. The rivers run blood-red with what is left of the island's precious topsoil.

Steve was drawn away by other obligations, but I had the good fortune to be joined by his sister-in-law, Patty Vavizara, who would be my French and Malagasy interpreter and caving companion for the next two weeks. We flew to Majunga, hired a four-wheel-drive vehicle, and made our way toward the vast and beautifully decorated caves of Anjohibe and Anjohikely.

My favorite cave is Andoboara, which we reached late at night. We had not set out until dusk, and our two local guides had difficulty getting oriented in the darkness. There was no road to follow, and their views were obscured by grasses nearly twice the height of the vehicle. But they finally located Andoboara Cave, small and uninteresting but for its human and spiritual associations and therefore of huge interest to me.

The people of the nearby village of Ambalakedi consider this cave sacred because on three separate occasions, most recently in 1998, grief-stricken parents whose children had wandered into the forest had recovered them alive here. The children had been abducted by a sometimes-malevolent spirit called *kalanoro,* who sought to rebuke the parents for not taking proper care of their children. The parents sought guidance from another spirit called *tromba,* who possessed a fellow villager and recited a laundry list of demands made by the *kalanoro* that would have to be met for the child to be found. Once the parents fulfilled their obligations by, for example, leaving honey and sacrificing zebu cattle in the place designated by the *tromba,* the *kalanoro* revealed via the *tromba* the lost child's location, and in each case the joyous reunion took place in the Andoboara Cave. As I was to learn in the coming weeks, there are countless caves in Madagascar bearing rich accounts of this kind.

Near the end of my stay, Patty and I took a five-day walk through the Ankarana Special Reserve near the island's northern tip. The park is dominated by a long wall of Jurassic-era limestone, eroded by wind and water into razor-sharp formations locally called *tsingy-tsingy* (meaning "ouch ouch," as in how your feet feel when you tread the stuff). The massif is perforated by scores of caves, some so enormous that even a powerful Petzl headlamp like mine could not illuminate their ceilings. Others are less grand in scale but graced with lovely calcite formations.

The Ankarana's sinkholes and the perimeter of the limestone massif contain protected remnants of dry tropical forest that are home to abundant wildlife. Each night, members of the primitive primate family of lemurs invaded our camp in search of food, calling so loudly that sleep came only through sheer exhaustion from the day's walking and caving. As a reptile lover, I was in paradise, indulging in the capture, photography, and release of boas and other snakes, riverine turtles, and—my favorites—the chameleons.

But the caves and their human connotations were of greatest interest. At one night's campfire, a man from a village just outside the park told me how his ancestors, members of the royal family of the Antakarana tribe, sought refuge in the caves for nearly three years during the 1800s. They were escaping would-be slavery at the hands of the Merina tribe of highland Madagascar, whose goal was to unite (or pacify and enslave) the island's 17 ethnic groups. The Antakarana had taken advantage of the Merina's hopeless disorientation in the underground labyrinth of the Ankarana massif and survived in its caves while hunting, foraging, and growing some crops in the hidden karst canyons and sinkholes. The first four Antakarana kings were buried in caves here and are still visited on a ceremonial pilgrimage once every five years by the current king, Issa, and his subjects. I met with him to seek his permission to enter the royal cave tombs, but he said it was *fady* (taboo) even for him to do so, except on the specified occasion; he would welcome me to take part in the next event.

One particular day in the Ankarana was among the most extraordinary in my life. With an excellent guide named Aurelian Toly, we walked through the massive Andrafiabe Cave for several hundred meters before reaching a giant sunlit chasm containing what botanists call a "sunken forest," essentially a tropical paradise surrounded by karst desert wilderness. Lemurs howled and parrots cried in this lost world. I thought, "Spielberg couldn't top this!" Crossing the forested ravine, we entered Cathedral Cave, whose entrance is littered with the hearths and pots used by the Antakarana in their time of refuge here. Well into the cave, we crossed a 600-foot-long (200 m) hill of guano built by a massive roost of insectivorous *Hipposideros* bats. Fresh deposits fell like raindrops on our helmets. Our noses were choked by the stench of ammonia and the dust of the

(continued)

PERSPECTIVES FROM THE FIELD (continued)

guano trodden underfoot. Our headlights shined upon seething masses of cockroaches and centipedes feasting on the treasures dropped by the bats. We couldn't help but crunch this fauna as we walked, and with the shrieking of the annoyed bats, I thought again of a Spielberg movie set.

My last excursion into underground Madagascar was in Crocodile Cave. It is appropriately named, for running through it is a river in which crocodiles seek habitat while the world outside is parched in the long dry season of April–October. On the cave's sand banks, we saw abundant crocodile tracks, some fresh, and we crossed the cave stream with trepidation. Instead of finding the great reptiles (some up to 19.7 ft [6 m] long have been seen here), about 3,300 feet (c. 1,000 m) into the cave we were astonished to come upon a dugout canoe beached on the bank. Next to the pirogue were three poles, each bearing on either end a strand of eight to ten cormorants and other waterfowl tied by their feet. Some had already died and others struggled for life. In the next stretch of cave passage, our headlamps revealed the three poachers responsible, and a confrontation ensued. The two park guides in my party of seven cried out, "We are from the national park service. Come forward. Drop your weapons because the warden has a gun and will use it against you!" I was surprised to learn that I was the warden, and the GPS pouch on my belt bore the gun. The ruse was enough to convince the poachers, who now pleaded for leniency. They said they had known hunting was forbidden in the park, but the need to feed their families had driven them to take this long and strange canoe voyage into the cave.

I still did not understand. Why were there waterfowl in a cave? The park rangers demanded that the hunters return immediately to release the birds where they had taken them, and marching behind the poachers 1.25 miles (2 km) further into the cave, I had my answer (**•Figure 9.C**). The cave opened up into another lost world, a great circular cavity in the Ankarana massif, this one filled not with forest but with marshes and lakes teeming with wildfowl. By this time, few of the birds were living, but the hunters did release them. The guides asked me (still feigning as warden) what to do with the criminals and their dead birds. "Would they do this again?" I asked. There was some discussion. Then the three poachers made what seemed a solemn vow that they would never again take wildlife from the park, lest the spirits of the ancestors of the Antakarana seek vengeance upon them. It was a particularly appropriate Malagasy way of dealing with a very understandable problem: the need of poor men to feed their families. For the geographer, the entire incident provided an especially fascinating insight into the relationship between people and caves in an extraordinary country.

• Figure 9.C The poachers (with birds on poles) after being apprehended by the Ankarana park staff, here escorting them out of the Crocodile Cave.

from their homelands. The trade began in the seventh century, with Arab merchants using trans-Saharan camel caravan routes to exchange guns, books, textiles, and beads from North Africa for slaves, gold, and ivory from Sub-Saharan Africa. As many as two-thirds of the estimated 9.5 million slaves exported between the years 650 and 1900 along this route were young women who became concubines and household servants in North Africa and Turkey. Male slaves usually became soldiers or court attendants (some of whom eventually assumed important political offices). From the eighth to nineteenth centuries, about 5 million more slaves were exported from East Africa to Arabia, Oman, Persia (modern Iran), India, and China. Again, most were women who became concubines and servants.

The cruel and lucrative traffic in slaves provided the initial motivation for European commerce along the African coasts, inaugurating the long era of European exploitation of Africa for profit and political advantage. The European-controlled slave trade was the largest by far, particularly with the development of plantations and mines in the New World. Between the sixteenth and nineteenth centuries, the capture, transport, and sale of slaves was the exclusive preoccupation of trade between the European world and West Africa. Portuguese and Spaniards began the trade in the fifteenth century, and within 100 years English, Danish, Dutch, Swedish, and French slavers were participating. The peak of the transatlantic slave trade was between 1700 and 1870, when about 80 percent of an estimated total of 10 million slaves made the crossing. In escape attempts, in transit, and in the famines and epidemics that followed slave raids in Africa, probably more than 10 million others died.

Slaves were a prized commodity in the **triangular trade** linking West Africa with Europe and the Americas. European ships carried guns, ammunition, rum, and manufactured goods to West Africa and exchanged them there for slaves. They then transported the slaves to the Americas, exchanging them for gold, silver, tobacco, sugar, cotton, rum, and tropical hardwoods to be carried back to Europe. As "raw material" and as the labor working the mines and plantations of Latin America, the West Indies, and Anglo America, slaves generated much of the wealth that made Europe prosperous and helped spark the 45, 32 Industrial Revolution.

Although Europeans carried out the triangular trade, their physical presence was limited to coastal shipping points. Africans were the intermediaries who actually raided inland communities to capture the slaves and assemble them at the coast for transit shipment. West African kingdoms initially acquired their own slaves in the course of waging local wars. As the demand for slaves grew, these kingdoms increasingly went to war for the sole purpose of capturing people for the trade. As the exports grew, so did the practice of Africans keeping African slaves. Even after Britain (in 1807) and the other European countries abolished slavery (1870), it continued to flourish in Africa. By the end of the century, slaves made up half the populations of many African states.

Slavery has not yet died out in the region (and indeed, a modern-day form of slavery exists in the form of human trafficking elsewhere in the world; see pages 145 and 395). In Mauritania, some Moors still enslave blacks, although the national government has outlawed this practice several times. Enslavement of children persists in West Africa. The typical pattern is that impoverished parents in one of the region's poorer countries, such as Benin, are approached by an intermediary who promises to take a child from their care and see that the boy or girl is properly educated and employed. This involves a fee (as little as $14) that, unbeknown to the parents, is a sale into slavery. The intermediary sells the child (for $250 to $400, on average) to a trafficker who sees that the child is transported to one of the region's richer countries, such as Gabon. There, the child ends up working without wages and under the threat of violence as a domestic servant, plantation worker (especially on cacao and cotton plantations), or prostitute. Benin is the leading slave supplier and trafficking center; Gabon, Ivory Coast, Cameroon, and Nigeria are the main buyers of slaves. Slavery also exists in Sudan. In the 1990s, well-intentioned church groups and other organizations in the West began paying hard cash to buy freedom for slaves in Sudan. Although many real slaves were freed, Sudanese profiteers also moved in with "counterfeit slaves," ordinary people hired to act as slaves until their "rescue" had been paid for.

The Impacts of Colonialism

In the sixteenth century, European colonialism began to overshadow and inhibit the growth of indigenous African civilizations. This is a diverse region, but one thing its peoples have in common is a sense that in the past they were humiliated and oppressed by outsiders.[4]

Portugal was the earliest colonial power to build an African empire. The epic voyage of Vasco da Gama to India in 1497–1499 via the Cape of Good Hope was the culmination of several decades of Portuguese exploration along Africa's western coasts. During the sixteenth century, Portugal controlled an extensive series of strong points and trading stations along both the Atlantic and Indian Ocean coasts of the continent. European penetration of the African interior began in 1850 with a series of journeys of exploration. Missionaries like David Livingstone, as well as traders, government officials, and adventurers and scientific explorers such as James Bruce, Richard Burton, and John Speke, undertook these expeditions. By 1881, when Africans still ruled about 90 percent of the region, foreign exploits had revealed the main outlines of inner African geography, and the European powers began to grab colonial territory in the interior.

Much of the "scramble for Africa" took place at the Conference of Berlin in 1884 and 1885, when the French, British, Germans, Belgians, Portuguese, Italians, and Spanish established their respective spheres of influence in the region. By 1900, only Ethiopia and Liberia had not been colonized. For more than half a century, Sub-Saharan Africa was a patchwork of European colonies (•**Figure 9.15**), and Europeans in these possessions were a privileged social and economic class.

At the outbreak of World War II in 1939, only three countries—South Africa, Egypt, and Liberia—were independent. The United Kingdom, France, Belgium, Italy, Portugal, and Spain controlled the rest. But after the war, mainly in the 1960s and 1970s, a sustained drive for independence was mounted. Ceasing to be a colonial region, Africa emerged with more than a quarter of the world's independent countries. This was a peaceful process in most instances, but bloodshed accompanied or followed independence in several countries.

Dependency theorists often point to Africa as a prime example of how colonialism created lasting disadvantages for the colonized. European colonization produced or perpetuated 44
many negative attributes of underdevelopment. These included the marginalization of subsistence farmers, notably those who colonial authorities, intent on cash crop production, displaced from quality soils to inferior land. In addition, European use 47
of indigenous labor to build railways and roads often took a high toll in human lives and disrupted countless families. Furthermore, the colonizers often corrupted traditional systems of political organization to suit their needs, sowing seeds of dissent and interethnic conflict. 338

The European colonial enterprise did have some positive impacts. The colonies, and the independent nations that succeeded them, were the beneficiaries of new cities and the transport links built with forced or cheap African labor; new medical and educational facilities (often developed through Christian missions); new crops and better agricultural techniques; employment and income provided by new mines and modern industries; new governmental institutions; and government-made maps useful for administration and planning. Such innovations were very helpful, but they were distributed unequally from one colony to another and were inadequate for the needs of modern societies when independence came.

One of the persistent problems in Africa's political geography is that many modern national boundaries do not correspond

• **Figure 9.15** Colonial rule in 1914. Germany lost its colonies to the British, French, and Belgians after World War I.

Source: Based on John Haywood, ed., Atlas of World History (Barnes and Noble Books, 1997).

with indigenous political or ethnic boundaries. In most cases, this is another legacy of colonialism: British, French, and other occupying powers created arbitrary administrative units that were transformed into countries as the colonial powers withdrew. Nigeria is a good example of an ethnically complex, unnaturally assembled nation. On the political map, it appears as an integral unit, but its boundaries have no logical basis in physical or cultural geography. Some Nigerians still refer to the **"mistake of 1914,"** when British colonial cartographers created the country despite its ethnic rifts (see Figure 9.22). The greatest divide is between Muslim north and Christian south, but there are at least 250 ethnic groups within the country. The British colonizers invested more in education and economic development in the south and built army ranks among northerners, whom they thought made better fighters; so the subsequent pattern is that the northern military leaders have tried to blunt the economic clout of southerners. Colonial administration and boundary drawing thus sowed seeds of modern conflict in Nigeria, a problem described in more detail in Section 9.6.

Although formal political colonialism has vanished, most countries still have important links with the colonial powers that formerly controlled them, and many foreign corporations that operated in colonial days still maintain a significant presence. France has a long history of post-independence intervention in the political and military affairs of its former African colonies. France is the only ex-colonial power to keep troops in Africa (with the highest numbers in Djibouti, Ivory Coast, Gabon, and Chad). France took steps to ensure that most of its former colonies trade almost exclusively with France and in turn supported national currencies with the French treasury. But France found this paternalistic approach to be extremely expensive and has been reducing its military presence and other costly assistance to its African clients.

9.4 Economic Geography

Sub-Saharan Africa ranks at the very bottom of every statistical indicator of global quality of life. Great poverty is characteristic of the region; all but two of the world's 20 poorest countries are there (see Table 9.1). About half of this region's people subsist on less than $1.25 per day, according to the World Bank. This is the only world region that has grown poorer in the last 25 years. It is the most difficult world region in which to conduct business, in part because of corruption and bureaucratic obstacles. A recent study of the world's civil wars since 1960 determined that there were three important risk factors for conflict: poverty, low economic growth, and high dependence on natural resources. Sub-Saharan Africa is richly endowed with these ingredients of war.

These are familiar statistics and connotations of Africa. But there is no denying that the region is changing for the better. Despite the global economic downturn, the region's economic growth of 5 percent annually between 2000 and 2010 was twice its pace in the previous two decades. Private companies are creating jobs. There is an emerging middle class; in other words, the number of people able to spend money on things other than life's bare necessities is growing. Roads, bridges, dams, and other infrastructure are improving, in many cases through Chinese investment. Governance is improving too, especially as corruption is being tackled. The number of significant conflicts was cut in half between the 1990s and 2000s. This attractive mix of factors has led to a surge in foreign direct investment. Some financial authorities believe that Sub-Saharan Africa as a group should be just as attractive for investment as the BRIC nations (Brazil, Russia, India, and China). Some even insist on expanding the acronym to BRICS, with the "S" standing for South Africa.

Let us take a closer look at the region's challenges and the efforts and opportunities that may overcome them.

Commodities: Boom!

Africa's place in the commercial world is mainly that of a producer of primary products, especially cash crops and raw materials (particularly minerals), for sale outside the region. In most nations, one or two products supply more than 40 percent of all exports—for example, in Angola oil and diamonds make up 99 percent of the country's exports, and in Kenya coffee and tea are overwhelmingly important (**•Figure 9.16**). Such countries are vulnerable to international oversupply of an export on which it is vitally dependent; oversupplies will cause prices to crash. But when demand for these exports is high, the money pours in. 45

The recent upward prices for commodities, driven especially by demand in China, India, and the older industrialized economies, have generally been a blessing for this resource-rich region. The challenge is to keep the blessing from becoming a mixed blessing, or worse yet a curse (**see A Closer Look, next page**).

Subsistence Agriculture

Famine is an all-too frequent visitor to Sub-Saharan Africa. Per capita food output in most of Sub-Saharan Africa has 331,
declined or remained flat since independence. The average 332

Joe Hobbs

• Figure 9.16 Kenya's economy is highly dependent on the export of coffee. Cash crops and other raw materials are typical exports of Sub-Saharan Africa.

A closer LOOK

The Resource Curse

Many countries in Africa have been described as suffering from the **resource curse**, also known as the "paradox of plenty." This is a paradox in that a country with a great abundance of a valuable natural resource often experiences lower economic growth than countries without such abundance. Nigeria provides a good example. Three hundred billion dollars in oil revenues have poured into the country in the past three decades, yet Nigerians have become poorer, with 70 percent of Nigerians living below the poverty line (see page 336 for more discussion of Nigeria's oil problems). Venezuela's oil minister in the 1960s was quoted famously as saying that oil was not black gold; it was "the devil's excrement."*

What causes the resource curse? There are economic reasons: it is difficult to plan when commodity prices (like Nigeria's oil) swing wildly up and down; oil prices, for example, can reach lofty heights and then plummet. These boom and bust cycles lead to risk-taking, debt, and overinvestment. With busts, there are budget crises that hurt the poor. Oil-fueled growth does not always create many jobs; although oil can account for as much as 80 percent of a country's revenues, it often employs fewer than 10 percent of the workforce. Against this backdrop of economic inequality, the revenue earned from oil crowds out other sectors, such as agriculture and manufacturing.

The resource curse is sometimes called the "Dutch Disease," because in the 1950s natural gas was discovered in the Netherlands' waters. As large foreign exchange inflows poured in, local costs went up, so imports undercut domestic manufacturing and agriculture. All domestic industries except the fossil fuel industry went into a tailspin.

There are also political reasons for and consequences of the resource curse. A glut of money from the resource provides incentives for corruption and repression of dissent. Officials and company executives get rich, and the poor get poorer. Oil-producing LDCs often spend as much as 10 times what poorer LDCs do on their militaries, and are more likely to get involved in conflicts.

*Quoted in Moises Naim, "Oil Can Be a Curse on Poor Nations." *Financial Times*, August 19, 2009, p. 7.

African consumes 10 percent fewer calories than two decades ago, and malnutrition afflicts almost half the region's children. Rapid population growth and drought are partly responsible for this trend. To support growing populations, farmers across Sub-Saharan Africa have shortened fallow periods and pressed their lands to yield more crops. The result has been an unprecedented degradation of the resource. Fully 75 percent of the region's farmland is severely low in the nutrients needed to grow crops, up from 40 percent a decade earlier. Studies suggest that at current rates, crop yields will fall as much as 30 percent by 2022. Most African farmers cannot afford fertilizers and are not familiar with the soil conservation techniques that would help reverse this ominous trend.

Over half of Africa's peoples practice subsistence agriculture (farming their own food but producing little surplus for sale) and pastoralism. Women do a large share of the farmwork—they produce 80 to 90 percent of Africa's food—in addition to household chores and the bearing and nurturing of children (Figure 9.8). Mechanization is rare, fertilizers are expensive, and so crop yields are low. In the steppe of the northern Sahel, both rainfall and cultivation are scarce. The more dependable rainfall of the southern Sahel creates a major area of rain-fed cropping, with unirrigated millet, sorghum, corn (maize), and peanuts the major subsistence crops. In the tropical savannas south of the equator, corn is a major subsistence crop in most areas, with manioc (cassava) and millet also widely grown. Corn, manioc, bananas, and yams are the major food crops of the rain forest areas.

Many peoples, particularly in the vast tropical grasslands both north and south of the equator, are pastoral. Herding of sheep and hardy breeds of cattle is especially important in the Sahel. As happened in Sudan's tragedy of Darfur, an increasing problem is that farmers often drive pastoralists from traditional grazing lands. Confined to smaller areas in which to browse and graze, the nomads' cattle, sheep, and goats often overgraze vegetation and compact the soil. Access to drinking water for livestock and people is also a major problem.

Most Africans who live by tilling the soil also keep some animals, even if only goats and poultry. Among African peoples such as the Maasai of Kenya and Tanzania and the Tutsi (Watusi) of Rwanda and Burundi, livestock not only contribute to the daily diet but are also an indispensable part of customary social, cultural, and economic arrangements. Cattle 339
are particularly important, with sheep and goats playing a smaller role. Traditional Maasai pastoralists are probably the most famous African example of close dependence on cattle. They milk and carefully bleed the animals for each day's food and tend them with great care. The Maasai give a name to each animal, and herds play a central role in the main Maasai social and economic events through the year.

Madagascar is a good example of an African country in which cattle represent status, wealth, and cultural identity. In order to enhance their social standing, Malagasy livestock owners tend to want larger numbers of animals even more than they want better quality animals (Figure 9.9). A Malagasy family practicing the ritual commemoration of deceased ancestors sacrifices a large number of zebu cows for fellow villagers, and similar feasts accompany other important ritual dates. Due to such demands, the population of zebu cattle on the island is about 11 million. Their forage needs have grave consequences for Madagascar's rain forests and other wild habitats. People clear the forests and repeatedly set fire to the cleared lands to provide a flush of green pasture for their livestock, causing a rapid retreat of the island's natural vegetation.

Commercial Agriculture and Marginalization

There is enormous potential to improve African agriculture. The region is a net food importer, yet 60 percent of the world's uncultivated arable land is in sub-Saharan Africa. Much of the land is inherently difficult to farm, but cultivation methods, roads, food storage, and other infrastructure can be improved.

Who will benefit from these improvements? The poor farmer, the wealthy landowners, or both?

Food shortages often result when governments promote the cultivation of cash crop commodities for export, instead of subsistence food crops for local people. Coffee, cacao, cotton, peanuts, and oil palm products in particular provide a means of gaining foreign exchange with which to buy foreign technology, industrial equipment, military armaments, and
47 consumption items for the elite. Secondary cash crops include sisal (grown for its fibers; see Figure 10.22), pyrethrum (used in insecticides), tea, tobacco, rubber, pineapples, bananas, cloves, vanilla, cane sugar, and cashew nuts.

The expansion of commercial agriculture requires more arable land, and it is fascinating and disturbing to see where this land is coming from. There is a new and distressing trend for small farmers whose land is owned by the government. In a classic example of marginalization, small farmers are being driven off their land as it is sold to investors from China, South Korea,
48 Libya, Saudi Arabia, and other countries. In some cases, there are well-established villages in what was sold to the investors as "vacant" land. The process is creating a new source of landless, unemployed poor in African cities. None of the income from rice and sugar cane and other products grown on these lands goes to local people. Sometimes the people push back. When Madagascar's government was about to sell about half of the country's arable land to a South Korean firm, opposition to Madagascar's president swelled and he was voted out of office.

Although cash crops are important sources of revenue in these poor countries, excessive dependence on them, as with other commodities, can be harmful to a country's economy. The income they bring often goes almost exclusively to already prosperous farmers and corporations. The prices they fetch are vulnerable to sudden losses amid changing world market conditions. The plants are susceptible to drought and disease. Finally, they are often grown instead of food crops, and cash crops do not feed hungry people.

Mineral Resources

Mining and mineral exports have had a strong impact on the physical and social geographies of Sub-Saharan Africa. This is a well-endowed region with, for example, at least 10 percent of the world's petroleum reserves, 60 percent of its diamonds, and a third of its cobalt (used in the manufacture of glass and ceramics). The three primary mineral source areas are South Africa and Namibia; the Democratic Republic of Congo–Zambia–Zimbabwe region; and West Africa, especially the areas near the Atlantic Ocean (•Figure 9.17). Notable

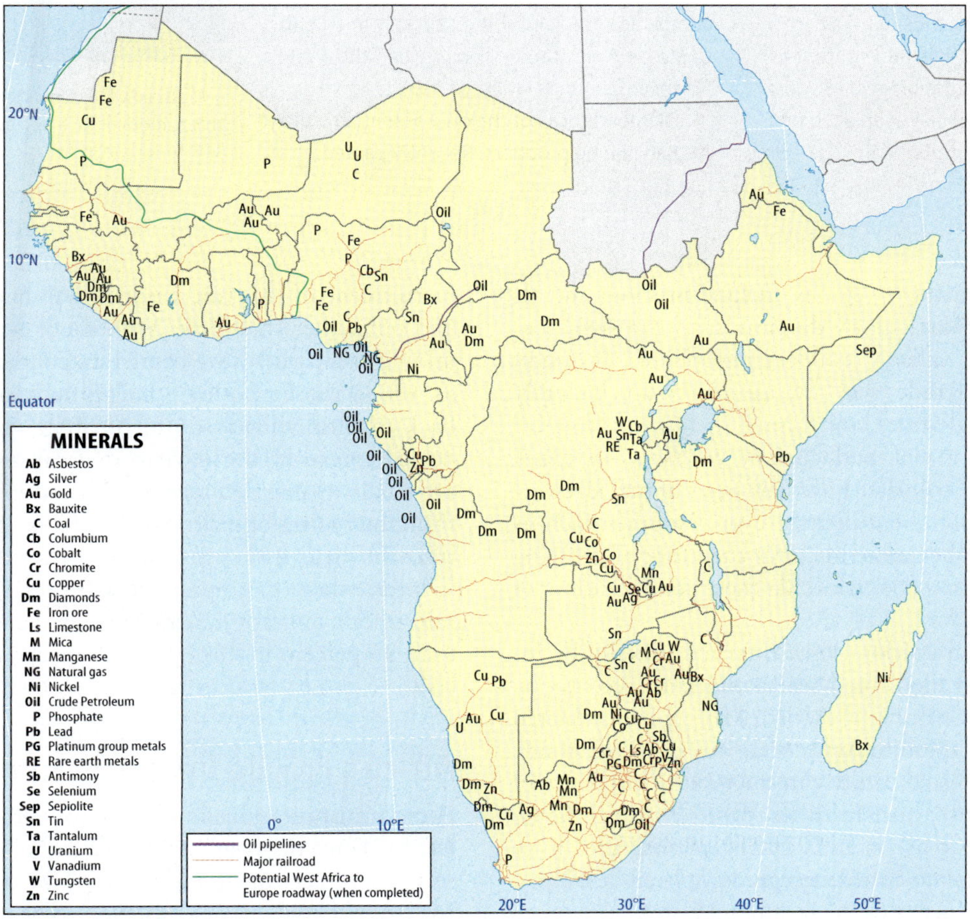

• Figure 9.17 Minerals, oil pipelines, and transportation links in Sub-Saharan Africa.
Source: Adapted from country mineral maps from USGS.

PROBLEM LANDSCAPE | Cleaning Up the Dirty Diamonds

"Diamonds are a girl's best friend," sang Marilyn Monroe. *Diamonds Are Forever*, proclaimed the title of a James Bond Film. Another film did not have such an endearing name: *Blood Diamond*. This film tells part of a sordid story about diamonds: as recently as 2004, as much as 15 percent of the world's annual production of rough diamonds was made up of **dirty diamonds** (also called **conflict diamonds** or **blood diamonds**), defined by the United Nations as "rough diamond used by rebel movements or their allies to finance conflict aimed at undermining legitimate governments." Gems of such questionable origin financed at least three African wars. But diplomatic, nongovernmental, and business efforts have largely stemmed the tide of Africa's dirty diamonds.

In 2002, following four years of negotiations, 45 countries endorsed a U.N.-backed certification plan, called the **Kimberley Process**, designed to ensure that only legally mined rough diamonds, untainted by violence, reach the market. Rough diamonds must be sent in tamper-proof containers with a certificate guaranteeing their origin and contents. The importing countries (the biggest of which are China and India, which have replaced Belgium and Israel as the world's leading diamond cutters and polishers) must certify that the shipments have arrived unopened and reject any shipments that do not meet the requirements. Only countries that subscribe to the Kimberley Process are allowed to trade in rough diamonds.

Corporate interests and consumer ethics have cleaned up the diamond trade. Diamond certification has been spearheaded by the world's leading diamond company (with about two-thirds of the market), the South African-based multinational named De Beers. De Beers was embarrassed by a report that it had bought $14 million worth of diamonds from Angolan rebels in a single year. Perceiving that a public relations debacle could lead to a business disaster, as happened with an organized boycott against fur products in the 1980s, De Beers seized the initiative. In the name of Africa's welfare, but certainly also as a means of increasing demand and profit, De Beers introduced **branded diamonds**, certified as coming from nonconflict areas. Other diamond companies followed suit, especially in an effort to please Americans—who buy half the world's finished diamonds (•**Figure 9.D**).

Diamonds are not the only potentially "dirty" African mining products. Recently passed American legislation requires that products made of tin, tungsten, tantalum, and gold, sold in the United States, be certified as free of conflict in the eastern Democratic Republic of the Congo (the labels read "DRC Conflict Free"). This requirement affects sales of airplanes, computers, cell phones, television sets, and other goods.

Joe Hobbs

• **Figure 9.D** Many consumers are unaware that this symbol of love and beauty may have had a sinister origin in Africa.

mineral exports from these regions include precious metals and precious stones (particularly diamonds; see Problem Landscape, this page), iron alloys, copper, phosphate, uranium, petroleum, and high-grade iron ore, all destined principally for Europe, the United States, and China.

Some of Africa's recent and current conflicts, in many cases fought by child soldiers, have not been about ideology or ethnicity but simply about control over resources such as diamonds, oil, and other minerals. It is therefore of growing interest in the consuming countries that these commodities not be tainted by violence.

Large multinational corporations, financed initially by investors in Europe and the United States and most recently by China, do most of the mining in Africa. Mining has attracted far more investment capital to Africa than any other economic activity. Money is invested directly in the mines, and many of the transportation lines, port facilities, power stations, housing and commercial areas, manufacturing plants, and other elements in the continent's infrastructure have been developed primarily to serve the needs of the mining industry.

Great numbers of workers in the mines are temporary migrants from rural areas, often hundreds of miles away. The recruitment of migrant workers has had important cultural and public health effects. Millions of Africans who work for mining companies have come into contact with Western ideas as well as those of other ethnic groups and have carried these back to their villages. Unfortunately, migrant labor has also helped spread HIV, the virus that causes AIDS. Most miners 316 are single or married men who spend extended periods away from their wives or girlfriends, and it is common that they contract the virus from prostitutes and then return home to their villages, where they spread it still further. The HIV infection rate among migrant miners working in South Africa was more than 25 percent in 2011—a high rate, but down from 40 percent just three years earlier.

Africa's Fragile Infrastructure

Poor transportation hinders development in Sub-Saharan Africa. Few countries can afford to build extensive new road or railroad networks, and much of the colonial infrastructure has deteriorated (•**Figures 9.17, 9.18**). The region critically needs a good international transportation network, together with a lowering of trade barriers, to enlarge market opportunities.

Poor transportation is also often a contributing factor to famine. In Ethiopia, for example, the road network is so poor that it is extremely difficult to get food from the western part of the country, which often has crop surpluses, to the eastern part, which has chronic food shortages. On more than one occasion, it has proved cheaper to ship food from the United States to eastern Ethiopia than to truck it across this African country! Similarly, a World Bank study showed that it cost $50 to ship a metric ton of corn from Iowa 8,500 miles (13,600 km) to the Kenyan port of Mombasa but $100 to move it from Mombasa 550 miles (880 km) inland to Kampala, Uganda. Transportation problems contribute to the high costs of agricultural inputs like fertilizers, which in this region cost two to three times what they do in Asia. Poor communication, especially in digital technology applications, also hinders development. But that is changing quickly, and for the better (see A Closer Look).

Cell phones are at the heart of a technological revolution in Africa's rural regions. Every phone owner needs to recharge the phone, but with most homes lacking electricity, that requires a trip to town and a recharge fee. The increasingly popular alterative is the purchase of an $80 Chinese-made photovoltaic power system that can recharge a phone and power a lamp. The investment pays for itself quickly. The solar panel is "green" in many ways, not only producing no greenhouse gases but also averting the need to buy candles, charcoal, batteries, wood, and kerosene. Perhaps the most impressive result of the purchase is a marked improvement in children's grades at school: they now have light for studying.

Whether by mobile phone or computer, in beginning to overcome the digital divide, African peoples are ramping up their
46 development prospects. The World Bank reported that for every 10 percentage points of increase in high-speed Internet access,

A closer LOOK

The Mobile Phone Revolution in Africa

There are more telephone land lines in the city of Tokyo than in all of Africa. So what, who needs land lines? Not Africans. Africa has the fastest growing mobile phone market worldwide—and of all the world regions, has been the most underserved by both mobile phone and Internet access. Mobile phone network coverage was available to just 10 percent of Sub-Saharan Africa's people in 1999, but that grew to 60 percent by 2008 and 85 percent in 2010, by which time there were more than 100 million users. There has also been explosive growth in broadband Internet coverage, made possible by recent projects laying tens of thousand of miles of undersea fiber optic cables.

Mobile phones are penetrating Africa far more quickly than the Internet, and African peoples are using them in many ways: to transfer money, buy and sell goods, and even turn on water wells. Without doubt, cell phones are a critical tool in rural development. In Uganda, for example, farmers in remote areas are communicating with plant scientists in the nation's capital to report the incidence and spread of a disease infecting bananas. The farmers take photos, mark Global Positioning System (GPS) locations, write remarks, and then transfer all these data by mobile phone to the scientists. The scientists then prepare an attack plan against the disease, and communicate disease treatment procedures back to the farmers. Using Google text messaging and an operator service, villagers are also able to learn more about farming practices and other critical issues like health care. Text messaging is improving farmers' access to markets. In time, web-enabled cell phones will make it much easier for millions of Africans to jump to the Internet without having to use computers.

Joe Hobbs

• **Figure 9.18** Many crucial African river crossings have no bridges; people must rely on ferries instead. Here, vehicles and people board a ferry to cross an estuary north of Mombasa on the Kenyan coast. Backup of traffic at these bottlenecks may take hours or even days to clear.

economic growth rises 1.3 percentage points. Internet connections are critical for the establishment of new businesses and the expansion of existing ones. Voice-over-Internet technology (VOIP, such as Skype) is making telephone calling more affordable for hundreds of millions of Africans and is also making African call centers more competitive with the established Indian market. Don't be surprised if one of your next customer service agents speaks with an African rather than Indian or Filipino accent.

In the meantime, please have a look at a valuable Internet tool created in Africa and used there mainly on cell phones: **Ushahidi** (ushahidi.com). Ushahidi (meaning "witness" or "testimony" in Swahili) is at the intersection of critical current events and their geospatial context. It originated as a means of tracking the violence accompanying the disputed 2008 presidential election in Kenya. Eyewitnesses sent text messages or wrote emails about what they saw, and each of these communications was "georeferenced" (placed on a known coordinate mapping system, in this case Google maps). It was therefore possible to see in real time where the hot spots of violence were. Authorities, supporters, opponents, and others were able to act or make plans with this information. This kind of information can be thought of as "crowdsourcing," or obtaining eyewitness
13 accounts from a large group of people.

There are no limits to the possible applications of Ushahidi. It is particularly valuable as a means of reporting the fallout from human and natural disasters such as the tsunami in Japan and the violence in eastern Congo described later, and as a means of calling people together for social action, as it was in
192, Occupy Wall Street and Arab Spring protests. Have a look at
414 the site to see what recent events Ushahidi has covered, and reflect on how inherently geographic this tool is. Having begun as "activist mapping," bringing together individual testimony with activist causes in their geospatial context, Ushahidi—like Twitter and Facebook—now has no bounds, with one caveat: Whatever Ushahidi is covering, it is doing so geospatially.

There is a widespread perception that Africa is always behind. Certainly much of the discussion in this chapter suggests that Africa has a lot of catching up to do. But Ushahidi is just one example of the surging innovation and creativity in African hands. It is as Pliny the Elder (23–79 C.E.) wrote: "There is always something new out of Africa."

Trade and Aid

One of the major obstacles to African economic development is that many countries outside the region have effectively closed their doors to African imports. These restrictions typically take the form of subsidies, high tariffs, or low quotas imposed on agricultural products (like cotton) or manufactured goods (like textiles). Potential importing countries, such as the United States, impose these restrictions to protect their own industries. For instance, the United States and African countries are the world's leading exporters of raw cotton. The U.S. government spends $2 billion each year to subsidize its 25,000 cotton farmers. This means higher U.S. production and exports, and consequently lower cotton prices and incomes for African farmers.

"Trade Not Aid!" is a growing mantra of African economics. An example of what reduced trade barriers might do for Africa is provided by the **African Growth and Opportunity Act (AGOA)**, passed by the U.S. Congress in 2000 and due to expire in 2017. AGOA reduced or ended tariffs and quotas on more than 1,800 manufactured, mineral, and food items that could be imported from Africa. African exports to the United States have surged, especially in the clothing industry. Apparel made in Lesotho, Kenya, Mauritius, South Africa, Madagascar, and elsewhere in Africa is now sold in Target and Walmart stores. The United States is considering extending AGOA's benefits with the group of countries comprising the **Southern African Customs Union (SACU)**: South Africa, Lesotho, Swaziland, Botswana, and Namibia. SACU is one of several trade pacts and organizations within Africa. These entities make it easier for international trade organizations to do business with Africa. In many cases, jobs and incomes for Africans have grown because of new opportunities created by trade pacts. AGOA, for example, helped tiny Lesotho establish a market niche in "sweatshop-free" garment production.

For decades, almost all of the countries of Sub-Saharan Africa were heavily in debt to foreign lenders. Young nations undertook costly development projects with borrowed money, particularly after the mid-1970s. Western financiers, planners, and contractors gave optimistic assessments of the benefits to be expected from such projects, and African leaders were ready to accept loans as a way to reap quick benefits from newly won independence. Countries soon found they had great difficulty in meeting even the interest payments on their debts, and their efforts to do so often resulted in further economic woes and destructive environmental effects. 46

Western donors began attaching more strings to loans as means of promoting democracy. African leaders were told that if they did not institute democratic reforms, hold fair elections, or improve bad human rights records, their countries would no longer be eligible for development aid or lending programs. Often the leaders made just enough concessions to win the aid without instituting real reform—a phenomenon known as **donor democracy** (•**Figure 9.19**).

In the West, increased foreign aid in the form of gifts or grants to Africa became controversial. Critics argued that foreign aid only increases dependence and deflects national attention away from problems underlying famine. International aid monies often free up funds from national treasuries to be spent on unfortunate uses. In the Horn of Africa, for example, when foreign money became available for famine relief, it allowed Ethiopia and Eritrea to spend more on weapons used in their war. Aid monies are often simply stolen, and monitoring by donors of how they are spent has been inadequate.

Since about 1990, donors have proceeded more cautiously, more often funneling money through private and religious groups working in health care, education, and small-business development using microcredit practices (**see Try It, next page**). Foreign aid has been very effective in advancing primary and secondary education in the region. However, many African governments view universities as potential hotbeds of dissent

and discourage investment in higher education. Africa's universities are crumbling, and the classic brain drain of African talent to Western institutions means there is that much less intellec-
56 tual capital left at home to invest in the region's development.

Africa's untenable debt situation, and the slow flow of aid to the region, began to change around 2005. Two major factors were at work. First, in the West, awareness of the humanitarian costs of not directing more resources to Africa was growing. Media, political, and corporate stars like the singer Bono, former U.S. President Bill Clinton, and Microsoft founder Bill Gates kept up a steady appeal for governments and citizens of the MDCs to be more generous to impoverished Africa. Secondly, just as it was becoming clear that the West was falling behind on its promised course to double aid to Africa, an Eastern giant stepped into the gap.

China Steps In

The second and more powerful shift in Africa's debt, trade, and aid picture came after 2006 with the meteoric rise of China's engagement with the region. China's trade with Sub-Saharan Africa increased 10-fold between 1999 and 2006, when China displaced Britain as the region's third most important trading partner (after the United States and France). By 2010, China was the region's leading trade partner. In its 2007 dialogue with the African Development Bank (AfDB), China pledged $20 billion in infrastructure and trade financing in the region.

Those funds, to be expended over a three-year period, represent a doubling of China's previous inputs and rival the value

Microcredit

Try It

A promising alternative to conventional development based on large, top-down projects funded with nationwide borrowing and debt is **microcredit** (or microlending, microfinance), the lending of small sums to poor people to set up or expand small businesses. Typically, poor people have no collateral—assets to pledge against loans they seek—and therefore cannot borrow from commercial banks. This drives them into the arms of local loan sharks, who may charge huge interest rates of up to 10 to 20 percent *per day*. But in the interest of fostering development, microlending organizations have begun to encourage poor borrowers to form a small group that will cross-guarantee all members' loans. One member of the group may take out a loan of $25, for example, to help start up a small business such as a roadside restaurant. Only when that individual has paid it back may the next person in the group receive a loan. Peer pressure minimizes the default rate, and the accumulated history of timely paybacks increases the entire group's creditworthiness.

Microlenders generally prefer to lend to women because women are more likely to use the money in ways that will feed, clothe, and otherwise benefit their children, whereas men may be more likely to spend the money on alcohol or other frivolous pursuits. One fear in Africa is that microloans will not be repaid if the borrower or someone in the borrower's family contracts HIV or some other debilitating illness. Therefore, the loans are often extended only after the borrower receives some amount of public health training, for example, in the use of birth control and in general food preparation hygiene.

One microlending project that has received widespread and generally positive attention is Kiva. Many people find it attractive because they can lend small amounts and track the progress of the individuals they are helping (see kiva.org). Want to try it? There is a chance you would lose some or all of your investment of $25 or more, but the Web site states that the repayment rate is 98.91 percent. You may want to at least explore the site to get a sense of how individuals and small groups around the world are seeking to improve their well-being through microborrowing.

© Chris McNulty/Alamy

• **Figure 9.19** In Scotland, an aid worker raises funds to fight poverty in Africa.

of those pledged by Africa's traditional financiers in the West. There is an important difference, however. Western institutions tend to have strings attached to loans and aid, insisting that African countries adopt democratic and other reforms to be eligible for the monies. China, in contrast, has a "no strings attached" policy. It is not insisting on political reforms, is not expressing concern about environmental or social responsibilities, and much to the chagrin of the West is even engaging with repressive governments that the West is trying to isolate. At the same time, Western governments and financial institutions
237, are increasingly alarmed that China is making rapid inroads
335, 367, into African markets at the expense of Western economies and
373 geopolitical interests.

Even Africans look with some trepidation of China's advance. Some complain about China's record on not hiring locals, instead bringing in Chinese nationals for virtually all labor and management positions. In other cases, Chinese firms have reportedly mistreated African workers and paid them what have been described as "slave wages." Some Chinese projects have been documented as having negative environmental impacts.

China is unabashed about its interests in Sub-Saharan Africa: it needs and wants the region's raw materials—especially oil, iron ore, copper, and cotton—to feed its insatiable industrial appetite. To many Africans, China is building the same kind of exploitive mercantile relationship that characterized the colonial West: it takes Africa's raw materials, transforms them into value-added products, and sells them mainly to its
367 own advantage. South Africa's president warned in 2007 that Africa risked becoming an economic colony of China. But desperate for assistance, even the wary are inclined to accept China's terms. The most typical arrangement is the "mines for infrastructure" deal, in which the African country grants mining concessions to the Chinese, and the Chinese build railways, roads, schools, and hospitals. There is especially great appreciation for the schools—there are simply not enough of them in many African countries.

China's model of state-led capitalism has proven attractive to many African governments. South African government officials go to Beijing to learn how to run state-owned companies more profitably. China is helping several African countries to build special economic zones, just as China did to spur its own
268 economic growth.

China and the West do not call all the shots. There are a growing number of African institutions designed to encourage economic growth and political stability. In 2002, the 53-member **African Union (AU)** was formed to replace the old and often inconsequential Organization of African Unity (OAU). One of the African Union's main articulated goals is to implement self-inspection or "peer review" to help promote democracy, good governance, human rights, gender equity, and development. It has formed an agency responsible for this monitoring, known as the **New Partnership for Africa's Development (NEPAD)**. NEPAD's premise is that improvements in such human affairs will improve the climate for international investment and assistance to Africa. The African Union has also created a 265-member Pan-African Parliament to make regional laws.

9.5 Geopolitical Issues

Sub-Saharan Africa has waxed and waned as a theater of geopolitical interest since the end of World War II. During much of the twentieth century, the great powers were interested in the region. To boost their competing aims during the Cold War years, the Soviet Union and the United States played African countries against one another, arming them with weapons with which to wage **proxy wars**. The superpowers also extended aid generously
to many African nations, but not simply for altruistic reasons. 178

Great power concerns about Africa changed with the end 24
of the Cold War around 1990, and the nature of conflict in Africa changed. The great powers withdrew support, but their weapons remained to prolong smoldering conflicts. The United States helped build the arsenals of eight of the nine countries involved in the recent Democratic Republic of Congo conflict, for
example. Cold War-era weapons like AK-47 assault rifles were 339
"dumped"—sold at low prices—in Africa, where they were not considered obsolete and where they fanned the flames of conflict. Increasingly, fighting began to spread across international borders. Previously, the United States and the Soviet Union had maintained a kind of security balance that kept warfare from becoming internationalized, but that restraint no longer exists.

The scenario of regional or even Africa-wide wars was realized with Africa's first "world war," centered on the Democratic Republic of Congo (see page 340). In the absence of regional or international powers to keep the peace, some countries (for example, Sierra Leone) simply imploded, fragmenting into fiefdoms run by factions that claimed to be revolutionaries but were essentially profiteers. Other national governments (as in Ethiopia) spent vast sums on expensive military aircraft while their people suffered from malnutrition.

The end of the Cold War constricted (and in the case of Russia, ruptured) aid pipelines. The United States cut its economic assistance to Sub-Saharan Africa by 30 percent between 1985 and 1992 and reduced it another 10 percent between 1992 and 2001. Why the waning concern? Former U.S. President George W. Bush was quoted as saying, "While Africa may be important, it doesn't fit into the national strategic interests, as far as I can see them."[5]

Then came 9/11. Al-Qa'ida's attack on the United States and the U.S. counterstrike against al-Qa'ida in Afghanistan raised fears that the Islamist organization might find new recruits and training grounds in Africa. Al-Qa'ida had already bombed U.S. embassies in Kenya and Tanzania in 1998 and in 2002 struck again in the Kenyan city of Mombasa. The United States maintains a focus on **terrorism hot spots** in the region, including Kenya, Somalia, Djibouti, Niger, Chad, and Mali, all of which are regarded as real or potential training grounds for al-Qa'ida-affiliated groups. As related on the next page, U.S. forces have made the tiny Horn of African country of Djibouti a major military center for potential action against nearby Somalia, Yemen, and other countries where Islamist militants are active. So far, U.S. interest in the hot spots has translated into military expenditures and partnerships with African governments, but not into wider development assistance.

In addition to the terrorism hot spots, and the Somali piracy that occasionally threatens commercial traffic in the region's vital sealanes, the United States has strong geopolitical interests in the region's oil-producing countries. These too became more critical after 9/11 and the U.S. war in Iraq, which heightened the sense that Middle Eastern oil supplies were
15 vulnerable. The United States is seeking more stable oil supplies and has pinned many hopes on Africa. Petroleum-rich Nigeria, Cameroon, Gabon, Congo, Angola, and Equatorial Guinea together in 2011 supplied 14 percent of the oil used by the United States, a figure projected by the U.S. Central Intelligence Agency to rise considerably in the coming years.

Increasingly, the United States will be competing with China for these dwindling finite fossil reserves, and given current trends, the geopolitical landscape of Sub-Saharan Africa will see Asia on the rise and American clout in decline. Countries that are not pleased with Washington's wishes or demands will find partners elsewhere, particularly in Asia. Some military analysts describe the emerging U.S.-China rivalry in Africa as a modern "Great Game", comparable to the British-Russian
241 contest over central Asia in the nineteenth century.

For many years following the deaths of U.S. service personnel in Somalia in 1993 (the "Blackhawk Down" incident), there was great reluctance to project U.S. military power in African trouble spots. Even Al-Qa'ida's simultaneous attacks on U.S. embassies in Kenya and Tanzania in 1998 did not change American military commitments in the region. The 9/11 attacks did, however. Since 2001, the United States has maintained a military base at Camp Lemonnier in Djibouti. Along with bases in Ethiopia and the Seychelles, it has been very useful in projecting U.S. power into Somalia and Yemen. Unmanned drones equipped with laser-guided weapons have been the weapons of choice against suspected al-Qa'ida (and in
183, Somalia, al-Qa'ida's affiliate al-Shabab) targets.
238 The United States views South Africa as the region's economic powerhouse and as the key to regional stability. South Africa's military strength, which included nuclear weapons until they were decommissioned in 1990, contributes to its global significance. It is strategically important for its frontages on both the Atlantic and Indian Oceans, its possession of Africa's finest transport network (vitally important to many African countries that trade with and through South Africa), and its diversified mineral wealth.

Finally, HIV/AIDS in the region is also a geopolitical concern. Despite Africa's apparent remoteness from the United States, the two are linked by hundreds of air traffic routes traveled by thousands of people each day. Hundreds of HIV-positive Africans, perhaps most of them unaware of their virus, pass through U.S. airport gateways daily. Some African strains 316
(subtypes) of the virus are more easily transmitted by heterosexual contact than the strains prevalent in the United States, so the risk to the population at large is greater. There are other dimensions of the potential burden on the United States. As long as the epidemic rages in Africa, there may be large flights of "medical refugees" seeking asylum in the United States. There could also be AIDS-related political instability or civil wars that invite U.S. military intervention in the region. How the world should respond to poverty and recurrent crises is one of the key questions facing the huge, resource-rich, problem-ridden but promising region of Sub-Saharan Africa. The following section examines some of these challenges in greater detail.

9.6 Regional Issues and Landscapes

The Sahel

Drought and Desertification in the Sahel

The Sahel region extends eastward from the Cape Verde Islands to the Atlantic shore nations of Mauritania, Senegal, and the Gambia, and inland to Mali, Burkina Faso ("Land of the Upright Men," formerly Upper Volta), Niger, Chad, and the new nation of South Sudan (•**Figure 9.20**).

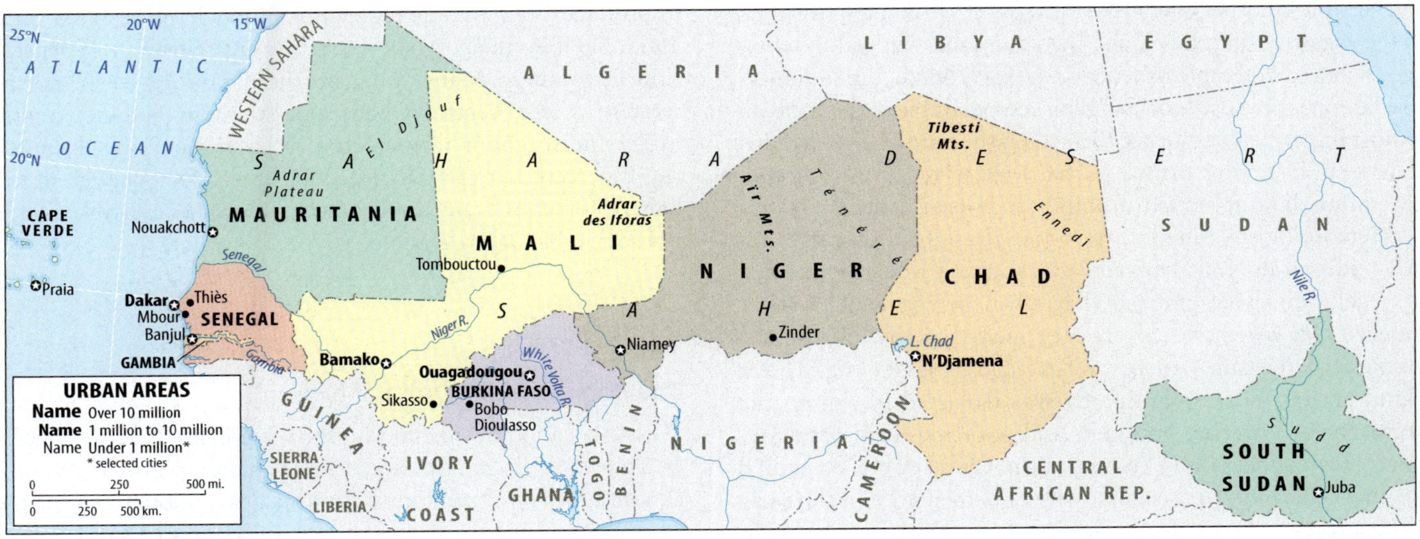

• **Figure 9.20** Principal features of the Sahel.

The Sahel region, like much of Sub-Saharan Africa, experiences periodic drought. Drought is a naturally occurring climatic event in which rain fails to fall over an area for an extended period, often years. When rain does return, the arid and semiarid ecosystems of the Sahel come to life with a profusion of flowering plants, insects, and herbivores like gazelles, whose populations climb when foods are abundant.

These ecosystems have high **resilience**, meaning they are able to recover from the stress of drought and have mechanisms to cope with a natural cycle that includes periods of dryness and rain.

Desertification is the destruction of that resilience and the biological potential of arid and semiarid ecosystems. It is an unnatural, human-induced condition that has afflicted the Sahel periodically and severely since the late 1960s, and thus it is different from the natural phenomenon of drought. However, the recent history of the Sahel suggests that drought can be the catalyst that initiates the process of desertification. During decades of good rains prior to the late 1960s, the Fulani (Fulbe), Tuareg, and other Sahelian pastoralists allowed their herds of cattle and goats to grow more numerous. The animals represent wealth, and people naturally wanted their numbers to increase beyond the immediate subsistence needs
160 of their families. However, due to declining death rates, the numbers and sizes of families keeping livestock were also very large. Thus, the unprecedented numbers of livestock that built up during the rainy years made the Sahelian ecosystem vulnerable to the impact of drought on an unprecedented scale.

Drought struck the region in 1968, persisted through 1973, and returned over many years through 1985. With the drought, annual plants failed to grow, so the large herds of cattle and goats turned to acacia trees and other perennial sources of fodder. They ate all of the edible vegetation. This destruction was a critical problem because plants play an important role in maintaining soil integrity by helping intercept moisture and funnel it downward through the root system. Plant litter and decomposer organisms working around the plant contribute to soil fertility and help stabilize the soil. With hungry livestock in the region eating the plants, the landscape changed. No longer anchored and replenished, good soils eroded. "Junk plants" like Sodom apple (*Calotropis*) replaced palatable plants, and an almost impermeable surface formed on the land. This degraded ecosystem lost its resilience so that even when rains returned, vegetation did not recover promptly as it usually does.

Meteorologists studying the Sahel drought discovered another important link between precipitation and the removal of vegetation. They noticed that when people and livestock reduced the vegetative cover, they increased the **albedo**, the amount of the sun's energy reflected by the ground. Fewer plants meant more solar energy was deflected back into the atmosphere, lowering humidity and reducing local precipitation. This connection is known as the **Charney effect**, named for Jules Charney, the scientist who documented it in the early 1970s. It was a tragic sequence: People responding to drought actually perpetuated further drought conditions.

The 1968–1973 drought devastated the great herds of the Sahel. An estimated 3.5 million cattle died. Two million pastoral nomads lost at least 50 percent of their herds, and many lost as much as 90 percent. The Niger and Senegal rivers dried up completely. The cost in human lives was estimated at 100,000 to 250,000. Environmental refugees poured into towns unprepared to deal with such a large and sudden influx. Many pastoral nomads gave up herding and settled down to become wage earners and, where possible, farmers.

This crisis resulted in an international effort to combat desertification, using recommendations developed by the United Nations in 1977. The U.N. report, however, soon became a case study in how *not* to deal with an environmental crisis in the developing world. Most of the recommendations were universal, high-technology solutions that proved impossible to implement in the villages and degraded pastures of the Sahel. For example, many international agencies attempted to relieve the suffering of Sahelian pastoralists by using techniques such as digging deep wells to water livestock. But these agencies failed to anticipate that these wells would act like magnets for great numbers of people and animals; although there was plenty to drink, many animals starved to death when they decimated what vegetation remained around the new water supplies.

Since the disappointment of the 1977 U.N. plan, and with the lessons relief organizations learned as a result, there has been greater focus on sustainable development, with its local
rather than universal solutions. In Niger, for example, hit by a 61
devastating drought as recently as 2005, farmers have ceased the traditional practice of clearing trees before planting crops and are now protecting trees and growing crops under and between them. They are also harvesting and selling many useful tree products. A key factor for the farmers' new concern for trees is that Niger's government released its long-held claim that all trees belonged to the state; now farmers recognize the trees as private property. New attitudes and practices like these, and a sustained period of good rains (which unfortunately have prompted periodic locust swarms), have combined to produce a remarkable regreening of much of the Sahel since 1985. Satellite imagery shows significant regrowth of vegetation in southern Mauritania, northern Burkina Faso, northwestern Niger, central Chad, and southern Sudan. Annual millet and sorghum harvests have risen in some cases as much as 70 percent above their pre-1985 levels. This change in the Sahel's fortunes is an important lesson that desertification is not inevitable or irreversible.

West Africa

The Poor, Oil-Rich Delta of Nigeria

West Africa extends from Guinea-Bissau eastward to Nigeria (•**Figure 9.21**). Its nine political units make up about 800,000 square miles (c. 2 million sq km), or nearly one-fourth of the area of the United States. These nine countries are Guinea-Bissau, Guinea, Sierra Leone, Liberia, Ivory Coast, Ghana,

• **Figure 9.21** Principal features of West Africa.

Togo, Benin, and Nigeria. The spatial, demographic, political, and economic giant of the region is Nigeria.

Oil is the economic lifeblood of populous Nigeria. Africa's largest oil producer, the country is well-endowed, ranking 10th in the world's proven reserves. Most of the production is concentrated in the Niger River Delta, home to about 12 million mostly Christian people (•**Figure 9.22**). They belong to several different ethnic groups who have one thing in common: they say they have derived few benefits and have suffered greatly from oil development in their homeland. They complain that for more than 30 years, thousands of oil spills have tainted their croplands and water, destroying their crops and fisheries, while the flaring off of natural gas has polluted their air and caused acid rain. They note that despite the enormous revenue generated from oil drilled on their land, little money has returned to the area; most goes to foreign oil companies like Shell and Chevron and to the Muslim-dominated government in the north. Meanwhile, most of them live in palm-roofed mud huts; half of the delta region lacks adequate roads, water supplies, and electricity. Schools have few books, and clinics have few medical supplies.

Where does Nigeria's oil money go? There is not enough transparency to answer this question. Incredibly, until June of 2011, it was illegal in Nigeria to publish official government data and statistics, including government accounts. Recently, $22 billion from the government's Excess Crude Account simply "went missing." An estimated $400 billion has been pilfered from Nigeria's treasury since the country became independent in 1960.

Since 1990, several ethnically based organizations have been agitating for change. Ethnic **Ogoni** activists founded the **Movement for the Survival of the Ogoni People (MOSOP)**, issuing a bill of rights in which they declared the right to a safe environment and more federal support of their people. One of the movement's leaders, a popular author, playwright, and television producer named Ken Saro-Wiwa, also called for self-determination for the Ogoni. Despite international condemnation, Nigeria's government executed him after a questionable murder trial.

Despite the return to civilian rule in Nigeria that came after Saro-Wiwa's death, dissent in the delta continued to grow. Ogoni defiance and contempt of government and the oil companies spread to the region's other major ethnic groups: the **Ibo (Igbo)**, **Yoruba**, and **Ijaw**. The Ijaw (who make up about 5% of Nigeria's population) have been especially strident in demanding that some of the oil money be used to fund roads, electricity, running water, and medical clinics. Ijaw activists have spoken of secession and self-determination if their demands are not met. They have periodically registered dissatisfaction by seizing onshore stations and offshore rigs belonging

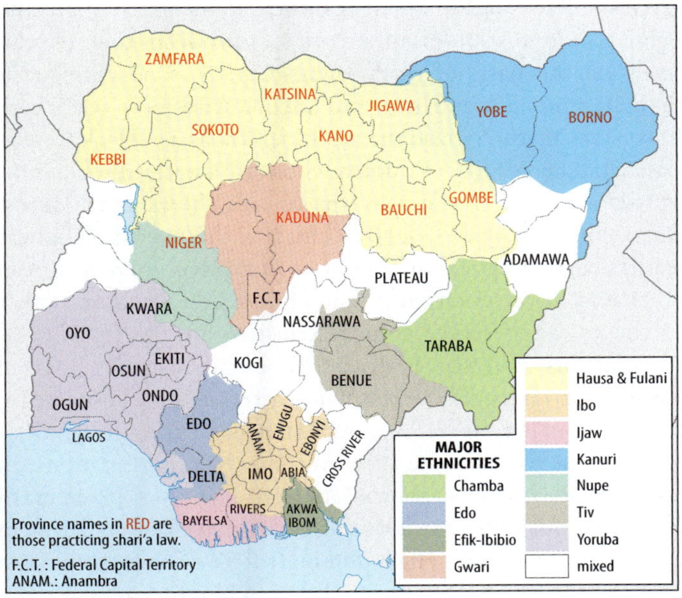

• **Figure 9.22** Nigeria's ethnic map.

to foreign oil companies, temporarily disrupting Nigeria's oil exports on both occasions. In 2006, a mainly Ijaw group calling itself the **Movement for the Emancipation of the Niger Delta (MEND)** escalated these tactics into frequent kidnappings of foreign oil workers. Opportunistic profiteers without well-established political views also joined in the action, and kidnapping became big business. An oil worker is typically freed unharmed upon payment of a ransom of several hundred thousand dollars.

This militancy in the delta has sent shockwaves through the world economy. Nigeria had touted itself as the safe alternative to Middle Eastern oil but has had to cut production by as much as 25 percent when tensions have been particularly high. The resulting shortages on world markets helped send oil and gasoline prices to record levels.

Questions of equity in a potentially rich country thus pose a threat to national, regional, and international stability. The peoples of the eight oil-producing states in the delta have been promised 13 percent of all petroleum revenues (compared with 5% previously), and have been told that the region's foul gas flares will be shut off. But locals say the promised revenues have been embezzled, while the flaring off of gas continues.

One of Nigeria's ironies is that in this oil-rich country, there are continuous shortages of gasoline and electricity. Neglect, mismanagement, corruption, and theft are all responsible. One of the cruelest and most repetitive Nigerian news stories is the immolation of scores and even hundreds of poor delta villagers who, in an activity known as "scooping" or "bunkering," illegally puncture gasoline or oil pipelines in an effort to collect and sell the fuels, only to inadvertently ignite the fires that consume them.

• **Figure 9.23** Principal features of East Africa.

East Africa

No More Divisionism?

Five countries make up East Africa: Kenya and Tanzania, which front the Indian Ocean, and landlocked Uganda, Rwanda, and Burundi (•**Figure 9.23**). Their total area is 703,000 square miles (1.8 million sq km), roughly the size of the U.S. state of Texas, with a total population in 2011 of 143 million (nearly half the population of the United States).

Episodes of bloodshed have punctuated East Africa's history. This section and the next section on west-central Africa tell a complex and tragic tale of genocide. These details are related here because the potential for renewed conflict is high and because the world community vowed not to forget what happened here.

Rwanda and Burundi have had tragic disputes between their majority and minority populations. About 85 percent of the population in Burundi and 90 percent in Rwanda are composed of the **Hutu (Bahutu)**, and most of the remainder of the two populations is **Tutsi (Watusi)**. These were not originally separate tribal or ethnic groups; the peoples speak the same language, share a common culture, and often intermarry. The distinction between Hutu and Tutsi was instead based on socioeconomic classification. Historically, the Tutsi were a ruling class that dominated the Hutu majority. Tutsi power and influence were measured especially by the vast numbers of cattle they owned. Tutsi who lost cattle and became poor came to be identified as Hutu, and Hutu who acquired cattle and wealth often became Tutsi.

Colonial rule in East Africa attempted to polarize these groups, increasing antagonism between them. German and 45, 325 Belgian administrators differentiated between them exclusively on the basis of cattle ownership: Anyone with fewer than 10 animals was Hutu, and anyone with more was Tutsi. Europeans mythologized the wealthier Tutsis as "black Caucasian" conquerors from Ethiopia who were a naturally superior, aristocratic race whose role was to rule the peasant Hutus. The colonists replaced all Hutu chiefs with Tutsi chiefs. These leaders carried out colonial policies that often imposed forced labor and heavy taxation on the Hutus. Education and other privileges were reserved almost exclusively for the Tutsis.

Ferocious violence between these two peoples marred Rwanda and Burundi intermittently after the states became independent in the early 1960s. After independence, the majority Hutus came to dominate the governments of both nations. In 1994, the death of Rwanda's Hutu president in a plane crash (in which Burundi's president also died) sparked civil war in Rwanda between the Hutu-dominated government and Tutsi rebels of the Rwandan Patriotic Front (RPF), based in Uganda. Hutu government paramilitary troops, still vengeful about decades of domination by Tutsis, systematically massacred Tutsi civilians with automatic weapons, grenades, machetes, and nail-studded clubs. Women, children, orphans, and hospital patients were not spared. An estimated 800,000 Tutsis and "moderate" Hutus (those who did not support the genocide) died. International media tracked these horrors, but outside nations did nothing to stop the fighting.

Well-organized and motivated Tutsi rebels seized control of most of the country and took power in the capital, Kigali, in 1994. That precipitated a flood of 2 million Hutu refugees mainly into neighboring Zaire, now the Democratic Republic of Congo (see page 340). Although the Tutsi-dominated government promised to work with the Hutus to build a new multiethnic democracy, Hutu insurgents, based mainly in eastern Zaire, continued to carry out attacks against Tutsi targets in Rwanda.

Rwanda now has a national unity policy aimed at reconciling Hutus and Tutsis; both occupy important government posts, and although the Tutsis still hold the political upper hand, they have encouraged the resettlement of Hutu refugees who fled the country during its troubles. In 2004, marking the 10-year anniversary of the genocide with a new policy meant to prevent its recurrence, Rwanda's government outlawed ethnicity. Now any Rwandan who speaks or writes of Hutus and Tutsis may be fined or imprisoned for practicing "divisionism." The country's past nevertheless continues to shape its future; for example, there is an ongoing defection of the faithful from Rwanda's majority Catholic Church to other religions, especially Islam but also smaller Christian denominations. Most of the country's worst massacres took place in Catholic churches, with the apparent knowledge and even participation of church leaders.

In Burundi, the Tutsi minority traditionally dominated politics, the army, and business. In 1993, there was a brief shift of power to the Hutu majority when Burundians elected the country's first Hutu president. Tutsis murdered him in October 1993, and a large-scale Hutu retaliatory slaughter of Tutsis ensued. An estimated 300,000 died in a decade of violence. Burundi's Tutsis and Hutus now have a power-sharing agreement in which the presidency rotates between them.

The peace has been maintained by the first-ever deployment of the African Union's **African Standby Force**. To help ensure the stability needed for economic and political progress throughout the region, the African Union established this continental military force. It takes orders from the AU's Peace and Security Council. The force will, in theory, deploy peacekeepers or peacemakers to ensure that the map of Africa, so filled with deadly conflicts in the past, has far fewer trouble spots in the future. The African Union has plans to create an African Court on Human and People's Rights and, like the European Union after which it is modeled, may also eventually seek a common currency.

Collective shame and embarrassment in the international community about the failure to stop the killing in Rwanda and Burundi led in subsequent years to numerous new anti-genocide measures. These include a United Nations "early warning" system to prevent genocide and a new forum for crimes against humanity in the **International Criminal Court**, established in 2002 in The Hague, Netherlands.

West Central Africa

The subregion of West Central Africa is flanked by Cameroon and the Central African Republic to the north and the Democratic Republic of Congo to the south (•Figure 9.24). It contains seven countries: Cameroon, Gabon, Central African Republic, Congo Republic (often known by the shorthand Congo-Brazzaville), Democratic Republic of Congo (known by the shorthand Congo-Kinshasa or DRC), São Tomé and Príncipe, and Equatorial Guinea.

Colonialism and Modern Struggles

During the last quarter of the nineteenth century, the Congo Basin was virtually a personal possession of Belgium's King Léopold II, whose agents ransacked it ruthlessly for wild rubber, ivory, and other tropical products gathered by Africans. This lawless era inspired Joseph Conrad's famous novel *Heart of Darkness* (1902), in which the trader Kurtz, at the point of death, evokes the ravaged Congo region with the cry, "The horror! The horror!"

In 1908, the Belgian government formally annexed the greater part of the Congo Basin, creating the colony of Belgian Congo. In 1960, the Congo colony became the independent Republic of the Congo. In 1971, it took the name Zaire, meaning "river." Following the overthrow of Zaire's government in 1997, the country was renamed Democratic Republic of Congo.

In recent decades, the region's troubles have focused on its major power, the Democratic Republic of Congo, and have been inseparable from events in neighboring East Africa.

• **Figure 9.24** Principal features of West Central Africa.

During his tenure in office from 1965 to 1997, then-Zaire's autocratic ruler, Mobutu Sese Seko, had neglected the country's development but acquired a massive fortune. Political opposition and public discontent with unemployment and inflation led to unrest in the early 1990s.

All-out war began after 1997, when Zairean army troops attempted to expel a force of indigenous Tutsis living in the country's far eastern region (for discussion of the region's Tutsis and Hutus, see page 338). The assault backfired when these Tutsis, supported by neighboring Tutsi-dominated governments in Rwanda and Burundi and by other sympathetic countries in the region, took up yet more arms and initiated fierce attacks on government forces.

The Tutsi counterattack quickly widened into an antigovernment revolt comprised of many ethnic factions and led by a non-Tutsi guerrilla named Laurent Kabila. Kabila's forces pushed westward virtually unopposed through the huge country and in May 1997 occupied Kinshasa with little loss of life. President Mobutu fled Zaire on the eve of Kabila's triumph. Kabila then declared himself president of the country he now called the Democratic Republic of Congo.

Like his predecessor Mobutu, he funneled much of the country's wealth and power into the hands of family and friends. He also obstructed international human rights investigations into the deaths and disappearances of thousands of Hutu refugees during the heady days of rebellion against Mobutu. And he failed to bring stability to the far east of his huge country, where an environment of chaos had given birth to his own bid for power.

In 1998, Hutu forces based in the eastern Congo-Kinshasa launched successive devastating attacks on Tutsi interests in Rwanda. Although the Tutsis of Rwanda had helped bring Kabila to power in Congo-Kinshasa, they now turned against him, forming the backbone of a rebel alliance bent on bringing down Kabila's young regime. The unrest led to something rather rare in the region, a broader fight dubbed **"Africa's First World War"** that involved nine countries and 20 rebel movements, and resulted in the deaths of more than 5 million people.

Although the fighting was sometime ferocious, it was not warfare that directly claimed most of the lives lost in this struggle between 1996 and 2001. Most of the casualties were civilians who died from starvation and disease or in widespread massacres directed against ethnic groups, usually Hutus killing Tutsis, but sometimes the reverse. The country appeared to be on the verge of dissolution when Laurent Kabila was assassinated by a bodyguard in 2001. He was succeeded by his son Joseph.

Conflict in Congo-Kinshasa has been enormously profitable for the warring factions, some foreign companies, and foreign governments (Uganda and Rwanda, for example, looted eastern Congo-Kinshasa of its gems, minerals, timber, agricultural produce, and wildlife, including elephant ivory from some of the country's national parks). Since 1998, most of the fighting has in fact been unabashedly about control over the Congo's vast mineral resources. Warlords and rebel proxies have been sustained by exploitation of minerals in the areas they controlled. Most of Congo's vast reserves of gold, diamonds, and copper are in the south center and west of the country, where the government is negotiating with the Chinese a vast minerals (coltan, tin, copper, diamonds, gold)-for-infrastructure deal (see page 334). In the east, where fighting to overthrow the government was centered, there is more gold. In this region, armed groups, now in concert with the Congo's army, control mines. There is a good chance that your cell phone contains coltan mined in this region and that it was not mined and exported "cleanly" as most diamonds are nowadays.

Although peace has been negotiated in this area, war could easily break out again. Now integrated into the Congolese army, ethnic Tutsi have been able to extend the area and the mineral wealth they control. Rival Rwandan Hutu militias, who control their own mineral wealth, have been carrying out reprisal attacks against Congolese civilians. Strife between Rwanda and the Congo is also enabled by expatriate sympathizers of the groups in Europe and the United States.

The Horn of Africa

The Galápagos Islands of Religion

In the extreme northeastern section of Sub-Saharan Africa, a great volcanic plateau rises steeply from the desert. This highland and adjacent areas occupy the greater part of the Horn of Africa, named for its projection from the continent into the Indian Ocean. This subregion includes the countries of Ethiopia, Eritrea, Somalia, and Djibouti (•Figure 9.25).

Of all these countries' peoples, Ethiopia's are the most ethnically and culturally diverse. About 45 percent, including the politically dominant **Amhara** peoples, practice **Ethiopian Orthodox Christianity**, an ancient branch of Coptic Christianity that came to Ethiopia in the fourth century from Egypt.

• **Figure 9.25** Principal features of the Horn of Africa.

Joe Hobbs

• **Figure 9.26** The churches of Lalibela, carved from volcanic rock in the twelfth and thirteenth centuries, are among Ethiopia's many Christian cultural treasures. For scale, note the men at left and right.

The entire area has had important cultural and historical links with Egypt, the Fertile Crescent, and Arabia. The Ethiopian monarchy based its origins and legitimacy on the union of the biblical King Solomon and the Queen of Sheba, who, tradition holds, gave birth to the first Ethiopian emperor, Menelik. Until a Marxist coup brought an end to the emperorship in the 1970s, Ethiopia's rulers were always Christian. Ethiopia has many outstanding Christian artistic and architectural treasures, including the 11 churches of Lalibela, carved from solid rock in the twelfth and thirteenth centuries (•Figure 9.26). Most of the rest of Ethiopia's people are either Muslims (who make up about 40% of the population) or members of Protestant, Evangelical, and Roman Catholic churches. There are still small numbers of **Falashas**, or **Ethiopian Jews**, in Ethiopia, a remnant of a very ancient and isolated Jewish population. The majority, about 100,000 people, have fled to Israel since the mid-1980s. Because this mountainous country has long served as an isolated refuge for such unique groups, it has been nicknamed the **Galápagos Islands of Religion**.

Perhaps the most unusual religion with roots in Ethiopia is that of Rasta, or the Rastafari movement of the Jah People (called **Rastafarianism** by non-Rasta). This faith, which originated in the 1930s in the West Indies island of Jamaica, has the central doctrine that the Ethiopian Emperor Haile Selaisse (1892–1975) was the earthly incarnation of Jah or Jehova (God). Ras Tafari was the Emperor's name before his coronation, and he was crowned on November 2, 1930, as the "King of Kings, Elect of God, and Conquering Lion of the Tribe of Judah (•Figure 9.27)."[6]

© Janine Wiedel Photolibrary/Alamy

• **Figure 9.27** The Lion of Judah.

Many poor Jamaicans gravitated to him for leadership and spiritual inspiration; he was the only black leader at the time to be recognized as legitimate in international circles. Their deep respect for him evolved quickly to reverence and worship, and the Rasta movement was born. The Rastafari faithful believe that blacks are the true children of Israel, and that on Judgment Day, Haile Selaisse will call them to come home to Zion, which they identify with Africa rather than Jerusalem. During his reign, Emperor Selaisse indulged the wishes of some Rastas to "repatriate" to Africa, allowing them to settle on his land in Ethiopia.

On April 21, 1966, Emperor Haile Selaisse's plane touched down on the airport tarmac in Kingston, Jamaica. For the 200,000 Rastafaris who had gathered, God Himself had come for a visit. He told the assembled that they should not return to Ethiopia until they first liberated Jamaica. Haile Selassie's short stay bestowed new legitimacy to the Rasta faith, and Rasta culture flowered and diffused rapidly after that. One of its principal carriers was Bob Marley, whose emotive cries for freedom thrust reggae onto the international music stage.

Southern Africa

In the southern part of Sub-Saharan Africa, five countries—Angola, Mozambique, Zimbabwe, Zambia, and Malawi—share the basin of the Zambezi River and have long had important relations with each other (•**Figure 9.28**). The first two were former Portuguese colonies, and the latter three were formerly British. Still farther south are South Africa and four countries whose geographic problems are closely linked to South Africa: the three former British dependencies of Botswana, Swaziland, and Lesotho, and the former German colony of Namibia.

Ethnicity, Colonialism, Strife, and Reconciliation in South Africa

Visitors from Western Europe and North America will find in South Africa (•**Figure 9.29**) most of the institutions and facilities to which they are accustomed. However, the Europeanized cultural landscape does not reflect the majority culture of this unusual country. Whites represent only about 10 percent of the total population, and blacks make up 79 percent. There are two other large racial groups: the so-called **coloreds**, of mixed origin (9%), and the Asians (2%), most of whom are Indians. Many peoples make up the black African majority. The **Zulu** and **Xhosa** (pronounced "*khoh*-suh") are both concentrated in hilly sections of eastern South Africa near the Indian Ocean and are the most populous of the nine officially recognized tribal groups.

A huge economic gulf separates South Africa's impoverished black Africans from about 5 million whites of European descent who dominate the economy and enjoy a lifestyle comparable to that of people in Western Europe and North America.

• **Figure 9.28** Principal features of southern Africa.

• **Figure 9.29** Principal features of South Africa.

South Africa's overall per capita GNI PPP of $10,050 is comparable to that of Brazil. For the unskilled and semiskilled labor needed to operate their mines, factories, farms, and services, the whites of South Africa have always relied mainly on low-paid black workers in most of the country and colored workers in Western Cape province. The white-dominated economy could not operate without them, and the Africans are in turn extremely dependent on white payrolls. Although the races are interlocked economically, relations between them have historically been very poor and have been structured to benefit the whites. A country whose people were legally segregated by race, South Africa was one of the world's most controversial nations until its new beginning in 1994. Following is a summary of how the country's racial structures—a critical issue in its modern geography—evolved.

Racial segregation characterized South African life from 1652 onward, but the **Afrikaners** (descendents of Dutch, German and French settlers) after 1948 systematized it with laws supporting the official policy of "separate development of the races," subsequently called "multinational development" and commonly known as **apartheid** (pronounced "uh-*pahr*-tayt"). The new laws imposed racially based restrictions and prohibitions on the entire population, but they weighed most heavily on black Africans and denied them political power. The laws fragmented and displaced families and ethnic groups, rearranging the country's cultural landscape, and culminated in the establishment of makeshift tribal states, or **homelands** (also known as **native reserves, Bantustans**, and **national states**), which were nominally intended to achieve "independence" as black nations. Whites ejected several million Africans not wanted in "white" South Africa from their homes and transferred them to the homelands, which were located on the most undesirable and least productive lands. Thus, great numbers of blacks were removed from cities and white farms and dumped in poverty-stricken, crowded, resource-poor areas, often far from where they had ever lived before.

The apartheid system drew furious criticism from many other nations. Most of the world community ostracized South Africa, and many countries, including the United States, imposed economic sanctions against it. Black unrest directed against apartheid and the general underdevelopment of the African majority became so widespread and violent between 1984 and 1986 that the government declared a state of emergency. But even after the state of emergency was lifted in 1990, violence continued. Much fighting took place between two large, tribally based rival factions: the Xhosa-dominated **African National Congress (ANC)**, led by Nelson Mandela, and the Zulu-dominated **Inkatha Freedom Party (IFP)**, led by Chief Mangosuthu Buthelezi.

With the country on the brink of anarchy, a complete official turnabout on the issue of apartheid resolved South Africa's ongoing racial crisis. It began in 1989 after Frederik W. de Klerk, an insider in the Afrikaner political establishment, became president. Urged on by white South African business leaders, antiapartheid activists, the powerful force of South African and international political opinion, the impact of international economic boycotts against South Africa, and his own stated convictions concerning the injustices and unworkability of apartheid, de Klerk launched a broad program to repeal the apartheid laws and put South Africa on the road to revolutionary governmental changes under a new constitution. Years of negotiations led to an all-race election in 1994, in which voters turned out in massive numbers to elect a new national parliament. The parliament then chose Nelson Mandela as president. (The white government had freed Mandela only four years earlier after a 27-year period of political imprisonment.) This historic election ended white European political control in the last bastion of European colonialism on the African continent. The apartheid laws became null and void, and the homelands were abolished.

Mandela, "South Africa's George Washington," retired in 1999 and was succeeded by a second black president, Thabo Mbeki, who resigned from office in 2008. The country's political landscape is remarkably stable, but challenges remain. The races are continuing on their long path of reconciliation. Some of the healing process has been formalized in the national **Truth and Reconciliation Commission**, which allows those who were party to racial violence during the apartheid era to confess their misdeeds and, in a sense, be absolved of them.

There is still a huge economic gulf between the haves and have-nots, but there are now more blacks—over 10 million—in the ranks of the middle and upper classes. Growing black wealth has boosted racial integration in neighborhoods. The black underclass continues to grow, however, and an overall unemployment rate of 25 percent fuels an ongoing epidemic of petty and violent crime, especially in the cities. Crime, the world's worst HIV/AIDS epidemic, a low level of education among workers, and relatively high labor costs have dampened foreign investment and economic growth in South Africa. No longer restrained by apartheid-era economic sanctions, South Africa has at the same time become a major investor in other African economies, buying banks, railways, cell phone networks, power plants, and breweries across the region.

There is a real need for land reform in South Africa because the white minority owns more than 70 percent of the productive land. A plan is in place to help reduce the imbalance by redistributing 30 percent of the country's farmland from white to black hands by 2014 on a willing-seller, willing-buyer basis. Progress is slow, however, and some landless blacks vow a takeover of white farms as occurred in neighboring Zimbabwe. A program is also under way to compensate the estimated 3.5 million blacks forcibly displaced by the government to the homelands between 1960 and 1982. More broadly, South Africa has an aggressive affirmative action program to help redress the economic imbalance between the races. The **Employment Equity Act** does not impose quotas but requires employers to move toward "demographic proportionality" based on the national proportions of race and gender. In another effort to secure more income for the black majority, the South African government assumed control of all the country's mineral resources in 2002. Mining companies now can exploit these resources only under state license and, presumably, with more supervision to ensure that profits are not drained excessively abroad or to whites.

The Indian Ocean Islands

Madagascar and the Theory of Island Biogeography

The islands and island groups off the Indian Ocean coast of Africa are unique in their cultures and natural histories. Madagascar, the Comoro Islands, Reunion, Mauritius, and the Seychelles have had African, Asian, Arab, European, and even Polynesian ethnic and cultural influences (•**Figure 9.30**). As

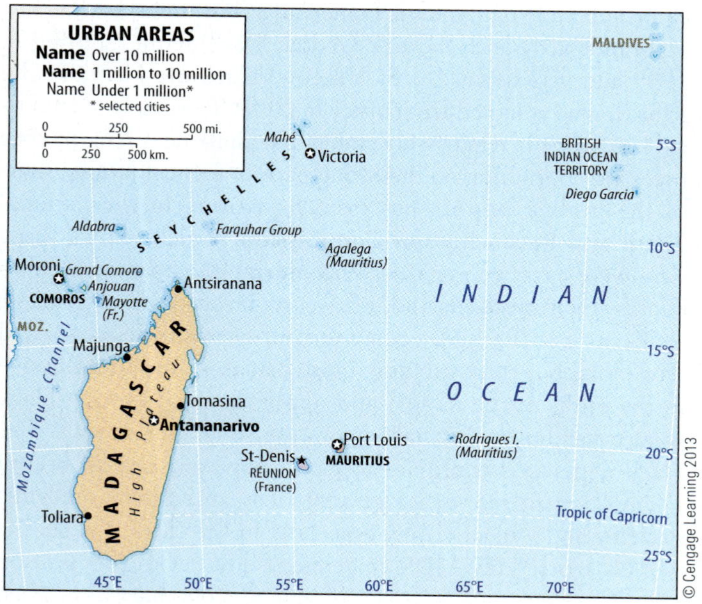

• **Figure 9.30** Principal features of the Indian Ocean islands.

island ecosystems, they are home to many endemic plant and animal species.

Madagascar (in French, *La Grande Ile*) is the fourth largest island in the world, nearly 1,000 miles (c.1,600 km) long and about 350 miles (c.560 km) wide. It lies off the southeast coast of Africa and has geological formations similar to those of the African mainland. Its distinctive flora and fauna include most of the world's lemur and chameleon species. These plants and animals are under tremendous pressure from people; Madagascar has 21 million inhabitants, many of them subsistence farmers who clear the island's forests to meet their needs.

The entomologist Edward Wilson and the biologist Paul Ehrlich have forecast that if tropical rain forests continue to be cut down at the present rate, a quarter of all of the plant and animal species on earth will become extinct by 2040.[7] They base their estimate on a model that correlates habitat area with the number of species living in the habitat. This **theory of island biogeography** emerged from observations of island ecosystems in the West Indies. It states that the number of species found on an individual island correlates with the island's area, with a 10-fold increase in area normally resulting in a doubling of the number of plant and animal [species]. If island A, for example, is 10 square miles in area and has 50 species, 100-square-mile island B may be expected to support 100 species.

What makes the theory useful in projecting species losses is the inverse of this equation: a 10-fold reduction in area will result in a halving of the number of species; therefore, 1-square-mile island C can be expected to hold only 25 species. In applying the model, ecologists treat habitat areas as if they were islands. Thus, if people cut down 90 percent of the tropical rain forest of the Amazon Basin, for example, the theory of island biogeography suggests that they would eliminate half of the species of that ecosystem. Scientists caution that the theory is only a tool meant to help in making rough estimates; the actual number of species lost with habitat removal may be higher or lower.

As a rough guideline, the theory of island biogeography is useful in projecting and attempting to slow the rate of extinction in the world's biodiversity hot spots, such as Madagascar. More than 90 percent of Madagascar's plant and animal species are endemic, occurring nowhere else on earth. Extinction 28
of species was well under way soon after people arrived on the island; the giant flightless elephant bird (*Aepyornis*) was among the early casualties. But human activities, particularly the clearing of forests to grow rice and provide pasture for zebu cattle, are eliminating habitat areas on the island at a faster rate than ever before. Meanwhile, scientists are anxious to learn whether some of Madagascar's remaining plants might be useful in fighting diseases such as AIDS and cancer. Already Madagascar's rosy periwinkle has yielded compounds effective against Hodgkin's disease and lymphocytic leukemia. Other species could become extinct before their useful properties ever become known.

How urgent is the task to study and attempt to protect plant and animal species in Madagascar? Scientists turn to the theory of island biogeography for an answer. Although people

• Figure 9.31 The subsistence needs of a growing human population have had ruinous effects on Madagascar's landscapes and wildlife. Before people came some 2,000 years ago, most of the island was forested, as in the Montagne d'Ambre National Park (left). Today, less than 10 percent of the island is forest, and a characteristic feature of its High Plateau is a barren gully, an erosion scar known locally as *lavaka* (right). The reddish color comes from the island's lateritic soil (see page 219)

have lived on Madagascar for less than 2,000 years, they have succeeded in removing 90 percent of the island's forest, setting the stage for some of the most ruinous erosion seen anywhere
290 on earth (**•Figure 9.31**). The theory of island biogeography suggests that in the process, they have caused the extinction of roughly half of the island's species. With Madagascar's human population on track to double in 25 years, and with pressure on the island's remaining wild habitats expected to increase accordingly, the task of conservation is extremely urgent.

Africa fits like a piece of a jigsaw puzzle into neighboring South America. From Africa, we cross the Atlantic to explore that neighboring piece and the greater region of Latin America.

summary

- Africa is the cradle of humankind, where hominids originated and from where they diffused. The region has seen many indigenous civilizations and empires and is ethnically and linguistically diverse.
- The majority of the region's people are rural. There are several clusters of dense population.
- HIV/AIDS is taking a large toll, dramatically lowering life expectancy and projections for population growth, especially in southern Africa. But progress is being made in the fight against the epidemic.
- Sub-Saharan Africa has a relatively low population density overall, but a majority of this region's people live in a small number of densely populated areas.
- Most of Africa consists of a series of plateau surfaces dissected by prominent river systems.
- One of the most spectacular features of Africa's physical geography is the Great Rift Valley, a broad, steep-walled trough. The feature marks the boundary of two crustal plates that are rifting, or tearing apart.
- Although about two-thirds of the region lies within the low latitudes and has tropical climates and vegetation, Sub-Saharan Africa contains a great diversity of climate patterns and biomes, some resulting from elevation rather than latitudinal position.
- Africa's diverse wildlife is often threatened by human population growth, urbanization, and agricultural expansion. Management of elephants and rhinoceroses involves international market demand and legal restrictions.
- In the four culture hearths of Sub-Saharan Africa, early indigenous people were responsible for several agricultural innovations, including the domestication of millet, sorghum, yams, cowpeas, okra, watermelons, coffee, and cotton.
- The four major language families of the region are Niger-Congo, Afro-Asiatic, Nilo-Saharan, and Khoisan. Malagasy, spoken only on Madagascar, is an Austronesian language. Islam and Christianity are major faiths, but indigenous belief systems are often mixed with them or exist on their own in some locales.
- The tragic impetus for growing contact between Africa and the wider world was slavery. Pockets of slavery still exist in the region.
- Most of Sub-Saharan Africa fell under European colonialism after the Conference of Berlin in 1884 and 1885.
- The majority of the people of Sub-Saharan Africa are poor, live in rural areas, and practice subsistence agriculture.

In most countries, per capita food output has declined or has not increased since independence. Export crops include coffee, cacao, cotton, peanuts, and oil palm products.

- Frequent droughts, lack of education, poor transportation, and serious public health issues have hindered development in Sub-Saharan Africa. Most countries are under-industrialized and overly dependent on the export of a few primary products.
- The export of minerals has had a particularly strong impact on the physical and social geography of Sub-Saharan Africa. The notorious trade in dirty diamonds has been largely cleansed.
- Most countries are heavily in debt to foreign lenders. Economic and humanitarian assistance to the region slowed considerably after the end of the Cold War, but it has picked up again since 9/11. China is the latest major economic power to engage in Africa and has a "no strings attached" policy when it comes to aid and trade with the region.
- Although many countries have been under authoritarian governments since independence, there has been some progress toward democracy. Serious political instability is characteristic of many African countries. Important links with the colonial powers that formerly controlled them remain strong in many countries of the region.
- The recently established African Union is a supranational organization dedicated to solving African problems without the intervention of outside powers.
- So often judged to be marginal in world affairs, Sub-Saharan Africa deserves and is receiving increased international attention because of its humanitarian problems, the global implications of its public health and environmental situations, problems in the management of its natural resource wealth, its oil reserves, and concerns about terrorism.
- Sub-Saharan Africa can be divided into seven subregions: the Sahel, West Africa, West Central Africa, East Africa, the Horn of Africa, southern Africa, and the Indian Ocean islands.
- The Sahel region extends eastward from the Cape Verde Islands to the Atlantic shore nations of Mauritania, Senegal, and the Gambia and inland to Mali, Burkina Faso, Niger, and Chad. The name Sahel in Arabic means "coast" or "shore," referring to the region as a front on the great desert "sea" of the Sahara.
- Between the late 1960s and 1985, the Sahel was subjected to severe droughts, which, in combination with increased human pressure on resources, prompted a process of desertification. Human and natural factors have helped restore the region recently.
- The most significant mining development in West Africa has been Nigeria's emergence as a producer and major exporter of oil. There are serious conflicts among ethnic, religious, and political groups in Nigeria, some resulting from the maldistribution of income from the country's oil wealth.
- The subregion of West Central Africa is flanked by Cameroon and the Central African Republic to the north and the Democratic Republic of Congo (DRC) to the south. The DRC was the epicenter of "Africa's First World War."
- East Africa includes the countries of Kenya, Tanzania, Uganda, Rwanda, and Burundi. Ethnic rivalries and conflicts have beset all of the East African countries. Warfare between the Hutus and Tutsis of Rwanda and Burundi in the 1990s was the most serious and was related to the violence in the Congo.
- Comprised of a great volcanic plateau and adjacent areas, the Horn of Africa includes the countries of Ethiopia, Eritrea, Somalia, and Djibouti. Ethiopia has been called the Galápagos Islands of religion.
- Southern Africa includes the countries of Angola, Mozambique, Zimbabwe, Zambia, Malawi, South Africa, Botswana, Swaziland, Lesotho, and Namibia.
- Racial segregation characterized South African life from 1652 onward, but it was systematized after 1948 under a body of laws known as apartheid.
- In 1994, Nelson Mandela was elected president of South Africa after an all-race election was held in the country. After years of conflict, this historic election ended apartheid and white European political control. Reconciliation between the races is continuing. The ranks of middle- and upper-class blacks are growing, but the gap between the haves and have-nots in South Africa is huge.
- The Indian Ocean islands of Madagascar, the Comoros, Réunion, Mauritius, and the Seychelles exhibit African, Asian, Arab, European, and even Polynesian ethnic and cultural influences. Madagascar has a wealth of endemic species under tremendous pressure by humans, who have cleared 90 percent of the original vegetation cover in the past 2,000 years.

Key Terms + Concepts

"Africa's First World War" (p. 340)
African Growth and Opportunity Act (AGOA) (p. 332)
African National Congress (ANC) (p. 343)
African Standby Force (p. 339)
African Union (AU) (p. 334)
Afrikaners (p. 343)
AIDS (acquired immunodeficiency syndrome) (p. 310)
albedo (p. 336)
Amhara (p. 340)
antiretroviral (ARV) drugs (p. 316)
apartheid (p. 343)
Bantustans (p. 343)
blood diamonds (p. 330)
branded diamonds (p. 330)
Chadic language subfamily (p. 319)
Charney effect (p. 336)
coloreds (p. 342)
conflict diamonds (p. 330)
Cushitic language subfamily (p. 319)

desertification (p. 336)
dirty diamonds (p. 330)
donor democracy (p. 332)
Employment Equity Act (p. 344)
Ethiopian Orthodox Christianity (p. 340)
Ethiopian Orthodox Church (p. 340)
Falashas (Ethiopian Jews) (p. 341)
"Galápagos Islands of Religion" (p. 341)
Great Rift Valley (p. 317)
homelands (p. 343)
human immunodeficiency virus (HIV) (p. 310)
Hutu (Bahutu) (p. 338)
Ibo (Igbo) (p. 337)
Ijaw (p. 337)
Inkatha Freedom Party (IFP) (p. 343)
International Criminal Court (p. 339)
keystone species (p. 321)
Khoisan language family (p. 319)
Kimberley Process (p. 330)
microcredit (p. 333)
"mistake of 1914" (p. 326)
Movement for the Emancipation of the Niger Delta (MEND) (p. 338)
Movement for the Survival of the Ogoni People (MOSOP) (p. 337)
nagana (p. 312)
national states (p. 343)
native reserves (p. 343)
New Partnership for Africa's Development (NEPAD) (p. 334)
Niger-Congo language family (p. 319)
Adamawan subfamily (p. 319)
Atlantic subfamily (p. 319)
Benue-Congo subfamily (p. 319)
Gur subfamily (p. 319)
Khordofanian subfamily (p. 319)
Kwa subfamily (p. 319)
Mandé subfamily (p. 319)
Nilo-Saharan language family (p. 319)
Chari-Nile subfamily (p. 319)
Saharan subfamily (p. 319)
Songhai (p. 319)
Ogoni (p. 337)
pandemic (p. 316)
PEPFAR (p. 316)
proxy war (p. 334)
Rastafarianism (p. 341)
resilience (p. 336)
resource curse (p. 328)
San (p. 319)
sleeping sickness (p. 321)
Southern African Customs Union (SACU) (p. 332)
terrorism hot spot (p. 334)
theory of island biogeography (p. 344)
triangular trade (p. 324)
Truth and Reconciliation Commission (p. 343)
trypanosomiasis (p. 321)
Tutsi (Watusi) (p. 338)
Ushahidi (p. 332)
Xhosa (p. 342)
Yoruba (p. 337)
Zulu (p. 342)

Review Questions

1. Where are the five principal areas of population concentration in Sub-Saharan Africa?
2. What factors account for the high HIV infection rates in Sub-Saharan Africa? What can be done to fight the epidemic? What is HIV/AIDS doing to life expectancy and the age structure profiles in countries where it is epidemic?
3. What are the principal climatic zones and biomes of Sub-Saharan Africa?
4. Where were the major slave trade routes in and from Sub-Saharan Africa?
5. Where were the major colonial possessions of various European powers in Sub-Saharan Africa?
6. What is the significance of cattle in many cultures of Sub-Saharan Africa?
7. What are the region's major export crops? Why does dependence on them make a country vulnerable?
8. What is a keystone species? Why is the tsetse fly considered a keystone species in Sub-Saharan Africa?
9. How have cellphones revolutionized African communications, and what can their potential benefits include?
10. What do the terms *donor democracy* and *donor fatigue* mean, and what do they reveal about Africa's political geography?
11. What impacts did the end of the Cold War have on many African countries?
12. How is China's ascendancy affecting Africa's economies and geopolitical status?
13. What are the seven subregions of Sub-Saharan Africa? In the case studies of section 9.6, what are some of the most significant countries in each of these subregions?
14. In what ways is the Sahel both a physical and cultural boundary zone in Africa?
15. What country has the greatest demographic, political, and economic clout in West Africa? What unique difficulties does this country face with respect to its oil wealth?
16. What was the original distinction between the Hutus and the Tutsis? What issues brought them into conflict in the 1990s?
17. Why is Ethiopia known as the "Galápagos Islands of Religion"?
18. What tensions emerged between blacks and whites over the issue of land tenure in Zimbabwe? What effects has government policy had on land reform?
19. What are the main ethnic groups in South Africa? What were their traditional social and economic roles? How have these changed since the end of apartheid?
20. How are deforestation and species loss measured in Madagascar?

Notes

1. Quoted in "Rebranding Africa," *New York Times*, July 9, 2009, p. A21.
2. Lydia Polgreen, "Misery Loves Optimism in Africa," *New York Times,* March 5, 2006, p. WK-1.
3. Robert Stock, *Africa South of the Sahara*, 2nd ed. (New York: Guilford Press, 2004), p. 41.
4. Ibid.
5. Quoted in Ian Fischer, "Africans Ask If Washington's Sun Will Shine on Them," *New York Times,* February 8, 2001, p. A3.
6. This information comes from an excellent online description of the Rasta, http://en.wikipedia.org/wiki/Rastafarian#Haile_Selassie_and_the_Bible (accessed September 9, 2005). "The Galapagos Islands of Religion" was the title of a Marjorie Coeyman's *Christian Science Monitor* article on Ethiopia's religions, on March 30, 2000.
7. Edward O. Wilson, "Threats to Biodiversity," *Scientific American, 261*(3), 1989, pp. 60–66.

Online Resources

CourseMate: Make the most of your study time by accessing everything you need to succeed in one place. Read your textbook, take notes, review flashcards, watch videos, complete activities, take practice quizzes, and more—online with CourseMate. Log in at **www.cengagebrain.com**.

Glossary

1948–1949 Arab-Israeli War A war fought between Israel and the Arab countries of Egypt, Iraq, Syria, Lebanon, and Transjordan after Israel declared its independence as a Jewish country. Hundreds of thousands of Arab Palestinians chose or were forced to flee Israel during this war, many settling in the Gaza Strip and the West Bank.

1973 Arab-Israeli War A war launched against Israel by Egypt and Syria in an attempt to reverse the Arab losses of the 1967 war. Egypt succeeded in winning back the eastern side of the Suez Canal.

Abkhaz-Adyghean language family A collection of languages spoken in the Caucasus region.

Aboriginal languages Indigenous languages to Australia that are not clearly related to any languages outside the continent.

Aborigines People indigenous to Australia that likely arrived in a single major migration 50,000 years ago. Europeans colonizing Australia in the eighteenth century slaughtered large numbers of Aborigines and drove them into marginal areas of the continent. Aborigines have long been Australia's most disadvantaged group and were not allowed to vote until 1962.

absolute (mathematical) location Determined by the intersection of lines, such as latitude and longitude, providing an exact point expressed in degrees, minutes, and seconds.

Adamawan A branch of Bantu languages spoken in West Africa.

adaptation Measures designed to cope with and reduce the unavoidable impacts of climate change.

adaptive reuse New uses for older buildings and stores, often accompanied by a shift from decline to steady renewal in an urban neighborhood.

Adena A mound-builder Native American culture that lived in what is now the northeastern United States.

Africa's First World War A widespread war involving nine African countries and 20 rebel movements between 1996 and 2001 that resulted in the deaths of more than 5 million people.

African Americans Americans of African ancestry.

African Development Bank (AfDB) A multi-government-owned bank, based in the Ivory Coast and dedicated to social and economic development (especially in loans for infrastructure) in African countries.

African Growth and Opportunity Act (AGOA) A bill passed by the U.S. Congress in 2000 that reduced or ended tariffs and quotas on many manufactured, mineral, and food items that could be imported from Africa.

African National Congress (ANC) The governing political party in South Africa since 1994. During the apartheid era, the ANC was dedicated to the nonviolent resolution of conflicts between ruling whites and the black majority. The ANC was banned until 1990. ANC leader Nelson Mandela became South Africa's first post-apartheid president in 1994.

African Standby Force A continental military force established by the African Union.

African Union (AU) An organization of most African countries formed to promote democracy, good governance, human rights, gender equity, and development.

Afrikaners Dutch-speaking descendants of Dutch, French Huguenot, and German settlers in South Africa. They are the largest white population in South Africa; also known as Boers.

Afro-Americans See *African Americans.*

Afro-Asiatic language family A language family that includes Semitic languages such as Arabic and Hebrew.

Age of Discovery (Age of Exploration) The three to four centuries of European exploration, colonization, and global resource exploitation and trading led largely by European mercantile powers, beginning with Columbus at the end of the fifteenth century and continuing into the nineteenth century.

Age of Exploration See *Age of Discovery.*

age structure diagram (population pyramid) The graphic representation of a country's population by gender and 5-year age increments.

Agricultural (Neolithic or New Stone Age or Food-Producing) Revolution The domestication of plants and animals that began about 10,000 years ago.

Agricultural Triangle See *Slavic core.*

AIDS (acquired immunodeficiency syndrome) The disease caused by HIV. AIDS is a global pandemic that has killed over 25 million people since the first case was reported in Africa in 1959.

Ainu An indigenous people and language to Japan.

Al-Aqsa *Intifada* A Palestinian uprising, beginning in 2000, triggered by Israeli Prime Minister Ariel Sharon's visit to the Temple Mount in Jerusalem.

Al-Asqa Mosque A sacred Islamic site on the Temple Mount in Jerusalem.

al-Haraam ash-Shariif See "Noble Sanctuary."

Al-Qa'ida A terrorist group formerly led by Osama bin Laden that was behind several attacks against U.S. interests, including the September 11, 2001, attacks.

Al-Qa'ida in Iraq An insurgent group arising in Iraq and affiliated with al-Qa'ida during the Iraq War, led by Abu Mus'ab al-Zarqawi until 2006.

Alawite A minority Shiite Muslim group.

albedo The amount of the sun's energy reflected by the ground. Less vegetation cover correlates with high albedo, and vice versa.

Aleuts An indigenous people of the Aleutian Islands in Alaska.

Algic language family A formerly widespread Native American language family across what is now Canada and the United States.

Allah God in Islam.

Alliance of Small Island States A group of 39 island countries arguing for significant reductions in global greenhouse gas emissions.

Altaic languages A language family that includes Turkish, Korean, and Japanese.

alternative development An overall effort to enhance the education, health, and livelihoods of Latin American villages in an attempt to reduce the amount of cropland devoted to coca and poppies.

American Dream The concept of upward mobility in American society, associated in part with material possessions such as homes and cars.

American Invasion Mexican term for the Mexican War.

Amerindians See *Native Americans.*

Amhara The politically dominant ethnic group of Ethiopia.

amnesty In the U.S. immigration debate, not persecuting illegal immigrants if they meet certain conditions and granting them permanent resident status.

Anasazi A Native American civilization living in what is now the southwestern United States up until about the year 1500. The

Anasazi gradually abandoned the pueblo dwellings they inhabited for unknown reasons.

ancestor veneration A widespread custom in eastern Asia, with shrines and offerings dedicated to preserving the memory of one's ancestors.

anchor baby A baby born in the United States to foreign parents.

animism The belief that natural processes and objects possess souls.

Attila Line See *Green Line* (Cyprus).

Antarctic Circle A line of longitude (65.56° S) commonly used to divide the southern middle latitudes to the north from the southern high latitudes to the south.

Antarctic Treaty An agreement signed by 45 countries forbidding any exploitation of Antarctica's natural resources until 2048.

Anthropocene Epoch A proposed new name for the present geological epoch, currently termed the Holocene.

antiretroviral (ARV) drugs A medication for treatment of HIV infection.

anti-Semitism Anti-Jewish sentiments and activities.

apartheid The Republic of South Africa's former official policy of "separate development of the races," designed to ensure the racial integrity and political supremacy of the white minority.

aquaculture The cultivation of aquatic organisms such as fish for food.

Arab A person of Semitic Arab ethnicity whose ancestral language is Arabic.

Arab-Israeli War of 1956 See *Suez Crisis*.

archipelago A chain or group of islands.

Arctic Bridge A potential shipping route between Churchill, Canada, and Murmansk, Russia, across the Arctic Ocean.

Arctic Circle A line of longitude (65.56° N) commonly used to divide the northern middle latitudes to the south from the northern high latitudes to the north.

Arctic National Wildlife Refuge (ANWR) A wildlife refuge in Alaska that contains 3 to 8 billion barrels of oil.

Armenian genocide The deaths of an estimated 1.5 million Armenians between 1915 and 1918. Turkey and Azerbaijan have not accepted responsibility for their roles in the genocide.

arms race Usually associated with the competition between the United States and the Soviet Union, it refers to rival and potential enemy powers increasing their military arsenals—each in an open-ended effort to stay ahead of the other.

Ashkenazi Jews Jewish immigrants to Israel from Europe after Israel's independence.

Asia-Pacific Economic Cooperation (APEC) group A 21-member organization (including China, Australia, and the United States) that has agreed to establish free and open trade and investment among member countries by 2020.

Asian Americans Americans of Asian ancestry.

Asian Century A term for the twenty-first century, anticipating the large role many Asian countries may play in changing the globe.

Asian Tigers Several Asian regions including South Korea, Taiwan, Hong Kong, and Singapore that had rapidly growing economies in the late twentieth century.

Association of Southeast Asian Nations (ASEAN) A regional economic association of ten Southeast Asian countries.

asylum People given permission to immigrate on the grounds that they would be harmed or persecuted in their country of origin.

atmosphere The layer of gases surrounding the Earth.

atoll Low islands made of coral and usually having an irregular ring shape around a lagoon.

Attila Line See *Green Line* (Cyprus).

Australia New Zealand United States (ANZUS) Security Alliance A defense and security agreement among the three countries, created in 1951. New Zealand has not been an active member since 1987 because of the country's anti-nuclear stance.

Austric language family A language family of Southeast Asia.

Austronesian subfamily The family of languages of most indigenous peoples of the Pacific Islands.

Austronesians Original settlers of many locations in Southeast Asia and the Pacific beginning after 5000 B.C.E. Many of today's Micronesians and Polynesians are their descendants.

autonomous regions A region of a country granted a measure of self-rule without being fully independent of that country.

autonomy Self-rule, generally with reference to Palestinians' rights to run their own civil (and some security) affairs in portions of the West Bank and Gaza Strip allocated to them in the 1993–2000 peace agreements.

ayatollah A revered and powerful cleric at the head of the religious establishment in Iran.

azimuthal A type of map projection in which the developable surface is based on a geometric plane.

Aztec-Tanoan language family A language family that included the Nahuatl language the Aztecs spoke.

Aztecs A Native American people that created an empire across central Mexico in the fifteenth century, which was destroyed by the Spanish in 1521.

baby bounty A financial incentive provided by the Italian government to families to have more than one child, to combat declining population.

Balkanization The fragmentation of a political area into many smaller independent units, as in former Yugoslavia in the Balkan Peninsula.

balloon effect A term referring to the fact that if a balloon is pressed in one place, the air shifts elsewhere. This term is often used to describe areas where drugs are produced and shipped through; when drug supply or shipment is eradicated in one area, it will tend to surge in previously marginal locations.

Baltic languages A language group of Europe that includes Latvian and Lithuanian.

banana war Reference to the incidence in the late 1990s, when the United States wanted to sell bananas grown in Latin America by U.S. corporations to countries of the European Union, but the EU members refused to buy them, as they wanted to protect their investments in banana farming in former European colonies in Africa and the Pacific.

bank In physical geography, an elevated area of the sea bottom.

Bantustans See *homelands*.

barrio A densely settled neighborhood in city space that is characteristically inhabited by migrants of Latino or Hispanic origin.

barter The exchange of goods or services in the absence of cash.

Basques A people, living primarily in Spain and France, who are ethnically and culturally unrelated to those of their host country majorities.

Bay of Pigs Invasion A failed attempt by the John F. Kennedy administration to overthrow Fidel Castro by military force.

***bazaar* (*suq*)** The central market of the traditional Middle Eastern city, characterized by twisting, close-set lanes and merchant stalls.

Benue-Congo A language family found in central and southern Africa.

Benelux The name used to collectively refer to Belgium, the Netherlands, and Luxembourg.

Berber An ethno-linguistic group of people living in northern Africa.

big bang The addition of ten primarily Eastern European countries to the European Union in 2004.

biodiversity See *biological diversity.*

biodiversity hot spots A ranked list of places scientists believe deserve immediate attention for flora and fauna study and conservation.

biogeography The distribution of living things on Earth.

biological diversity (biodiversity) The number of plant and animal species and the variety of genetic materials these organisms contain.

biomass The collective dried weight of organisms in an ecosystem.

biome A terrestrial ecosystem type categorized by a dominant type of natural vegetation.

biosphere The global ecological system, including all the relationships played out among the lithosphere, hydrosphere, and atmosphere. Also called the ecosphere.

birth dearth The declining population of many European countries from low birth rates.

birth rate The annual number of live births per 1,000 people in a population.

birth rate solution Intentionally lowering birth rates to prevent an overpopulation crisis.

birth tourism A practice in which an expectant mother from an LDC comes to the United States for the purpose of having her baby on U.S. soil, making the baby an automatic U.S. citizen.

black-earth belt An important area of crop and livestock production spanning parts of Russia, Ukraine, Moldova, and Kazakhstan. The main soils of this belt are mollisols.

blacks See *African Americans.*

blizzard A severe winter storm with heavy snowfall and high winds.

blood diamonds See *dirty diamonds.*

blowback Term associated with attacks on U.S. forces by former *mujahadiin* armed and trained by the United States to fight the Soviet occupation in Afghanistan.

Bolivarian Revolution Hugo Chávez's term for his policy agendas after taking control of Venezuela.

Bolshevik Revolution The seizure of the Russian government by Vladimir Lenin's faction of the Communist Party in 1917.

boreal forest A type of forest (also called taiga) primarily composed of coniferous trees that thrive in subarctic climates.

borough In New York City, one of the five administrative divisions: Manhattan, Brooklyn, Queens, Staten Island, or the Bronx.

Boswash A term for the megalopolis stretching from Boston to Washington, D.C.

bourgeoisie In Marxist doctrine, the capitalist class.

brain drain The exodus of educated or skilled persons from a poor to a rich country or from a poor to a rich region within a country.

branded diamonds Diamonds produced by the De Beers company, certified as coming from nonconflict areas.

break-of-bulk point A classic geographic term describing a point in transit when bulk goods must be removed from one mode of transport and installed on another. A trainload of grain carried to a port for transshipment on cargo boats or barges is a common example.

BRIC economy An acronym for Brazil, Russia, India, and China, four countries that may have the world's largest economies by 2050.

British–North America Act An act of the British Parliament in 1867 uniting several British North American colonies to create the independent country of Canada.

bubble economy See *economic bubble.*

Buddhism A religion based on the life and teachings of Siddhartha Gautama, known as the Buddha (or Enlightened One), who lived in the sixth century B.C.E. in India.

Burmese Way to Socialism Rigid state control of the economy in Myanmar.

California Gold Rush The discovery of gold in California, leading to a population boom in that state.

California Trail One of the major westward routes that American pioneers took in the middle of the nineteenth century, leading to what is now Sacramento, California.

Camp David Accords A 1979 agreement between Israel and Egypt that returned the Israeli-occupied Sinai Peninsula to Egypt in exchange for Egypt's recognition of Israel as a sovereign state.

Canada-U.S. Free Trade Agreement A free trade agreement between Canada and the United States in 1988 that was the forerunner to NAFTA.

Canadian Shield The ancient geological core of North America.

Cao Dai A faith originating in Vietnam that blends elements of Buddhism, Daoism, Confucianism, Judaism, Christianity, and Islam.

cap-and-trade system An international system designed to reduce carbon emissions by having richer countries that produce lots of greenhouse gases pay poorer, energy-inefficient countries to cut their emissions. This allows the richer countries to offset their carbon emissions with the cuts in emissions made by the poorer countries. The poorer country is obligated to use the income from cap-and-trade to invest in energy-efficient and nonpolluting technologies.

carbon dioxide (CO_2) A greenhouse gas, the human production of which is very likely the cause of global warming.

carbon sequestration The capture and long-term storage of carbon dioxide. Usually refers to natural processes, but there are also artificial means.

carbon sink An area such as forest, farmland, or ocean that sequesters a large amount of carbon dioxide.

cargo cults Belief systems that appeared in New Guinea and on various Pacific islands during World War II when "Stone Age" peoples were exposed for the first time to Western technology and consumer goods.

carrying capacity The size of a population of any organism that an ecosystem can support.

Carter Doctrine President Jimmy Carter's declaration, following the Soviet Union's invasion of Afghanistan in 1979, that the United States would use any means necessary to defend its vital interests in the Persian Gulf Region. The "vital interests" were interpreted to mean oil, and "any means necessary" interpreted to mean that the United States would go to war with the Soviet Union if oil supplies were threatened.

cartography The craft of designing and making maps, the basic language of geography. In recent years, this traditional manual art has been changed profoundly through the use of computers and geographic information systems (GIS) and through major improvements in machine capacity to produce detailed, colored map products.

cash (commercial) crops Crops produced generally for export.

caste The hierarchy in the Hindu religion that determines a person's social rank. It is established by birth and cannot be changed.

cataracts Areas along the Nile River where its valley narrows and rapids form.

Catholic Republicans One side of "The Troubles" in Northern Ireland, those who wished the British Army to leave the area and for Ireland to be unified.

Celtic Tiger A nickname for Ireland in the 1990s for the country's rapidly growing economy.

Central Arizona Project (CAP) A canal network bringing water from the Colorado River to the populous cities of Arizona.

Chadic A family of languages spoken mainly in northern Nigeria.

chaebols A selection of small companies in South Korea that were heavily funded by the government, eventually growing into giant conglomerates.

Chang Jiang Water Transfer Project A massive engineering project in China that will move water from China's wet south into its populated but drier north.

Chari-Nile A language family spoken in east-central Africa.

Charney Effect Observed by an atmospheric scientist named Charney, this states that the less plant cover there is on the ground, the higher the albedo—solar energy deflected back into the atmosphere—and therefore the lower the humidity and precipitation.

Chernobyl The site in the Ukraine where, in April 1986, the worst nuclear power plant accident in history occurred. It is thought that approximately 5,000 people died, and a zone with a 20-mile (32 km) radius is still virtually uninhabitable; 116,000 people were moved from the area, and cleanup continues to this day.

chernozem A Russian term meaning "black earth." It is a grassland soil that is exceptionally thick, productive, and durable.

Chibcha A small, pre-Columbian civilization in South America renowned for their exquisite gold works.

chokepoint A strategic narrow passageway on land or sea that may be closed off by force or threat of force.

choropleth maps Maps that are drawn to show the differing distribution of goods or geographic characteristics (including population) across a broad area. Such maps are good for generalizations but often mask significant local variations in the presence of the item being mapped.

Christianity One of the world's major faiths, predominant in Europe and many former European colonial areas. Christianity has several major denominations, including Roman Catholicism, Protestantism and Eastern Orthodox.

Chukotko-Kamchatkan A language family of far northeastern Siberia.

Church of the Holy Sepulcher Located in Jerusalem, this is the center of the Christian world, containing the places where tradition says Jesus Christ was crucified and buried.

City of Light A nickname for Paris, France.

civilization The complex culture of urban life.

climate The average weather conditions, including temperature, precipitation, and winds, of an area over an extended period of time.

Closer Economic Relations Agreement A free-trade pact between Australia and New Zealand.

coal A fossil fuel, used for creating heat and electricity, which has many industrial uses. The Industrial Revolution began in European locations situated near large deposits of coal.

coconut civilization A term characterizing the importance of coconuts in the economies and diets of several low-island Pacific countries.

coke Coal with most of its volatile constituents burned off.

Cold War The tense but generally peaceful political and military competition between the United States and the Soviet Union, and their respective allies, from the end of World War II until the collapse of the Soviet Union in 1991.

collective farm *(kolkhoz)* A large-scale farm in the former Soviet Union that usually incorporated several villages. Workers received shares of the income after the obligations of the collective had been met.

collectivization The process of forming collective farms in Communist countries.

collectivized agriculture See *collective farm.*

colonization The European pattern of establishing dependencies abroad to enhance economic development in the home country.

Colorado River Compact A 1922 agreement between seven western U.S. states that the Colorado River flows through, which determined how much river water each state may take.

Coloreds A South African term for persons of mixed racial ancestry.

Columbian Exchange The transfer of plants and animals between Europe and the Americas after Christopher Columbus's voyages.

comanagement A joint system of national park management in Australia between the Aborigines and the Australian government.

command economy A centrally planned economy typical of the Soviet Union and its Communist allies, in which the government rather than free enterprise determines the production, distribution, and sale of economic goods and services.

commercial crops See *cash crops.*

Common Market An earlier name given to the (current) 25 countries that make up the European Union. In 1957, an initial six countries combined to form the European Economic Community (EEC), and this supranational community has grown to have considerable economic and political importance in Europe. See *European Union.*

Commonwealth of Independent States (CIS) A loose political and economic organization of former Soviet republics (except for Estonia, Latvia, and Lithuania) formed in 1991.

Commonwealth of Nations A voluntary association of 54 countries around the world, most of them former British colonies.

communal violence Term used in India to refer to sectarian violence.

communism A type of national economy characterized by abolition of trade unions, abolition of private ownership in fields such as manufacturing and commerce, varying degrees of state ownership of agriculture, centrally planned and directed economies, and one-party dictatorial government.

Compact of Free Association A 1986 agreement between the United States and the Marshall Islands and the Federated States of Micronesia granting independence and subsidies to the two Pacific countries in exchange for renting their military sovereignty to the United States.

Comprehensive Test Ban Treaty A treaty signed by 149 nations in 1996 that prohibits all nuclear tests.

compromise projections A map projection that does not preserve any one metric but are designed to distort all metrics roughly equally to achieve a map that "looks right."

conflict diamonds See *dirty diamonds.*

conformal projections A map projection that preserves shape.

Confucianism A Chinese sociopolitical philosophy based upon the writings of Confucius, either on its own or blended with other religions.

conic A type of map projection in which the developable surface is based on a cone.

coniferous forest Needleleaf evergreen trees; most bear seed cones.

consumption overpopulation The concept that a few persons, each using a large quantity of natural resources from ecosystems across the world, add up to too many people for the environment to support.

continental drift See *plate tectonics.*

continental islands Once attached to nearby continents, these are the islands north and northeast of Australia, including New Guinea.

contingent sovereignty An agreement in Myanmar between the government and several minority ethnic groups, granting them more civil rights and economic opportunities.

Convention on the Law of the Sea A 1970s United Nations treaty permitting a sovereign power to have greater access to surrounding marine resources.

coordinate system A grid consisting of horizontal and vertical lines used to establish absolute location. On the Earth's surface, latitude and longitude lines form the coordinate system.

Copts A Christian sect in Egypt.

core location An area of the world with significant importance due to a central location relative to others.

Cornucopians People who argue that humans have always been able to conquer food shortages in the past, thanks to technological innovations, and therefore will be able to in the future as well. Also called *technocentrists.*

Corporate Average Fuel Economy (CAFÉ) standard A 1975 U.S. law stating that passenger cars produced by all auto companies must achieve an average of 27.5 miles per gallon.

Cossacks Peasant soldiers of the Russian steppes who were among the first Russian settlers in Siberia in the sixteenth century.

credit crisis A financial crisis originating in the United States in 2008, after the housing bubble burst, that quickly spread worldwide, causing global selloffs in stock exchanges and a widespread economic recession.

Creole Tongues that developed among black slaves and indentured servants in the Americas, consisting mainly of vocabularies from the European colonizers' languages.

Creole languages See *Creole.*

crop substitution Efforts to get Latin American farmers to switch from growing illegal coca or poppies to legal flowers or coffee.

Crusades A series of European Christian military campaigns between the eleventh and fourteenth centuries aimed at recapturing Jerusalem and the rest of the Holy Land from the Muslims.

Cuban Americans Cubans who fled Castro's regime for the United States, and their descendants.

Cuban Missile Crisis The debacle between the United States and the Union of Soviet Socialist Republics in 1962 that put both countries on the brink of nuclear war. Soviet rocket launch facilities were being built in Cuba, presumably for possible nuclear sites on the United States. President Kennedy demanded that the Soviets withdraw the missiles; they did.

cultural landscape The landscape modified by human transformation, thereby reflecting the cultural patterns of the resident culture.

culture The values, beliefs, aspirations, modes of behavior, social institutions, knowledge, and skills that are transmitted and learned within a group of people.

culture hearth An area where innovations develop, with subsequent diffusion to other areas.

Cushitic A family of languages spoken in Ethiopia and Somalia.

cylindrical A type of map projection in which the developable surface is based on a cylinder.

Dakota A Native American people living on the Great Plains.

Dalai Lama The spiritual and political leader of Tibetan Buddhism (Lamaism).

Dalits The bottom of the social hierarchy in India, not part of the caste system. Once known as "untouchables."

Daoism (Taoism) A Chinese school of thought, second only to Confucianism, originating in the *Tao-te Ching* in the sixth century B.C.E. Daoism philosophy is commonly blended with elements of Buddhism.

Dayton Accord A peace agreement creating a joint multiethnic, democratic national government in Bosnia and Herzegovina, responsible for foreign and economic policies, while also recognizing a second tier of government charged with overseeing internal functions composed of the Federation of Bosnia and Herzegovina and the Republika Srpska.

death rate The annual number of deaths per 1,000 people in a population.

death rate solution A catastrophic increase in death rates, seen as "nature's solution" to overpopulation.

debt-for-nature swap An arrangement in which a certain portion of international debt is forgiven in return for the borrower's pledge to invest that amount in nature conservation.

decertification A country's loss of foreign aid from the United States if that country is deemed to be uncooperative in the war on drugs.

decommissioning (of dams) A major modification or removal of a dam primarily for environmental reasons.

deforestation The removal of trees by people or their livestock.

degrees Unit of measurement for longitude and latitude. There are 180 degrees of latitude (90 degrees north of the equator, and 90 degrees south) and 360 degrees of longitude (180 degrees east of the Prime Meridian, 180 degrees west).

deindustrialization An economy shifting away from industries requiring large amounts of energy and labor toward an economy based on services and the production of high-tech goods.

delta The usually triangular-shaped alluvial area at the mouth of a river.

demilitarized zone (DMZ) The border between North and South Korea.

demographic transition A model describing population change within a country. The country initially has a high birth rate, a high death rate, and a low rate of natural increase; then moves through a middle stage of high birth rate, low death rate, and high rate of natural increase; and ultimately reaches a third stage of low birth rate, low or medium death rate, and low or negative rate of population increase.

demography The study of population.

denying the countryside to the enemy The use of explosives and defoliants such as Agent Orange by U.S. troops to create enormous damage to the farms and forests of Vietnam.

dependency theory A theory arguing that the world's more developed countries continue to prosper by dominating their former colonies, the now independent less developed countries.

desert An area too dry to support a continuous cover of trees or grass. A desert generally receives less than 10 inches (25 cm) of precipitation per year.

desert shrub vegetation Scant, bushy plant life occurring in deserts of the middle and low latitudes where there is not enough rain for trees or grasslands. The plants are generally xerophytic.

desertification The expansion of a desert brought about by changing environmental conditions or unwise human use.

development A process of improvement in the material conditions of people often linked to the diffusion of knowledge and technology.

devolution The process by which a sovereign country releases or loses more political and economic control to its constituent elements, such as states and provinces.

Diaspora The scattering of the Jews outside Palestine beginning in the Roman Era.

digital divide The divide between the handful of countries that are the technology innovators and users and the majority of nations that have little ability to create, purchase, or use new technologies.

direct rule The policy of administering Northern Ireland directly by the central government of the United Kingdom.

dirty diamonds Rough diamonds used by rebel movements or their allies to finance conflict aimed at undermining legitimate governments.

distance-decay relationship The geographical concept that the interactions between places (and the people that occupy them) decline as the distance between them grows.

divine rule by clerics Iranian principle of the Guardian Theologian, an infallible supreme leader whose authority is based on Shiite interpretations of Islam.

Dome of the Rock A Muslim shrine built on the Temple Mount in Jerusalem, on the site where two Jewish temples had previously stood.

domestication The controlled breeding and cultivation of plants and animals.

Dominican Republic-Central America Free Trade Agreement (DR-CAFTA) A free-trade organization created in 2005, comprised of the United States, the Dominican Republic, and five Central American countries.

domino effect In political terms, the Cold War concept that if one country (e.g. Vietnam) were to fall to communism, neighboring countries also would.

donor democracy A government that makes just enough concessions in holding elections or human rights to win foreign aid, without introducing any real reforms.

dot-com bubble See *tech bubble.*

dot map Type of map that uses dots to represent a stated amount of some phenomenon within a political unit.

doubling time The number of years required for the human population of a given area to double.

downstream countries A country that lies along a river that has already passed through at least one other country.

Dravidian language family A language family primarily found in southern India.

Dreamtime In Australian Aboriginal myth, the creation of the Earth and all things on it, created by the "dreams" of humanity's ancestors during the Dreamtime.

drought avoidance Adaptations of desert plants and animals to evade dry conditions by migrating (animals) or being active only when wet conditions occur (plants and animals).

drought endurance Adaptations of desert plants and animals to tolerate dry conditions through water storage and heat loss mechanisms.

Druze Members of a small offshoot of Shiite Islam, living mainly in the Golan Heights.

dry farming Planting and harvesting according to the seasonal rainfall cycle.

dumping Selling a product for less than the cost to produce it.

Dust Bowl An eight-year drought in the 1930s that ruined crops and livelihoods in the Great Plains, spurring many people to move west to California.

east longitude See *longitude.*

Eastern hemisphere See *hemisphere.*

ecodevelopment See *sustainable development.*

ecological bankruptcy The exhaustion of environmental capital, leading to potential political and social crises.

ecological footprint The amount of biologically productive land needed to sustain a person's consumption and absorb wastes.

ecologically dominant species A species that competes more successfully than others for nutrition and other essentials of life.

economic bubble The sharply increasing valuations of companies or entire economic sectors to lofty heights that cannot be sustained, often leading to recessions when these bubbles "burst."

economic shock therapy Russia's economic transformation in the early 1990s from a command economy to a free-market economy. Overseen by Boris Yeltsin, this transformation was difficult for a country accustomed to government direction in all economic matters, thus the "shock."

El Nino Southern Oscillation (ENSO; El Nino) A reversal of normal climatic patterns that occurs every few years, when the waters of the eastern Pacific Ocean become much warmer than usual, which can lead to global climatic disruptions.

emigrant A person moving away from a place.

emission-trading system See *cap-and-trade.*

Employment Equity Act A South African law that requires employers to move toward "demographic proportionality" in the workforce, based on national proportions of race and gender.

endemic species A species of plant or animal found exclusively in one area.

energy crisis The petroleum shortages and price surges sparked by the 1973 oil embargo.

energy *maquiladoras* Power generation plants built by American firms just inside Mexico that provide electricity to the U.S. grid but are subject to less stringent pollution controls.

enosis The desire of the Greek majority in Cyprus for political unification with Greece.

environmental determinism The concept that the physical environment has played a sovereign role in the cultural development of a people or landscape. Also known as *environmentalism.*

environmental perception The concept that how people view the world and its landscapes and resources influences their uses of the earth and therefore the condition of the earth.

equal-area projection A type of map projection that preserves area.

equator A line or latitude on the Earth's surface (0 degrees) halfway between the North and South Poles. The dividing line between the northern and southern hemispheres.

equidistant projection A type of map projection that preserves distance from one specific point only to all other points.

Eskimo-Aleut language family A Native American language family found in extreme northern North America.

Eskimos See *Inuit.*

ETA (Basque Homeland and Liberty) A militant Basque group that carried out many terrorist attacks in Spain from the 1960s until 2011.

ethanol A fuel made primarily from corn.

Ethiopian Orthodox Christianity See *Ethiopian Orthodox Church.*

Ethiopian Orthodox Church The Christian church of Ethiopia, related to the Coptic church of Egypt.

ethnic cleansing The relocation or killing of members of one ethnic group by another to achieve some demographic, political, or military objective.

Eurabia A term indicating the growing Muslim population in Europe.

euro Part of the authority of the European Union has been the institution of a new currency that has become "coin of the realm" since early 2002. Not all EU nations accepted the euro.

European core Primarily the countries of Western Europe, including Germany, France and the United Kingdom.

European debt crisis A financial crisis beginning in 2010, when it appeared that Greece might default on its loans. Greece did not want wealthier European countries dictating its spending and taxation policies, whereas wealthier countries such as Germany did not want to spend their own taxes to bail out Greece and other heavy borrowers.

European Economic Community (EEC) An economic organization designed to secure the benefits of large-scale production by pooling resources and markets. The name has been changed to Economic Union. See also *Common Market.*

European Greenbelt A mosaic of national parks and other protected areas stretching across 18 countries.

European periphery countries located in Northern, Southern, and Eastern Europe.

European Union (EU) The current organization, begun in the 1950s as the Common Market. It now is made up of 27 nations (see Chapter 3).

Eurozone The 17 countries that use the Euro as currency.

Evil Empire President Ronald Reagan's term for the Soviet Union in 1983.

exclusive economic zone (EEC) An area up to 200 miles (320 km) offshore of a country, in which only that country can conduct economic activities such as fishing.

Exodus The departure of the Israelites, led by Moses, from Egypt around 1200 B.C.E.

exotic species A nonnative species introduced into a new area.

extensive land use A livelihood, such as hunting and gathering, that requires the use of large land areas.

external costs (externalities) Consequences of goods and services that are not priced into the initial cost of those goods and services.

extinction capital of the world A term used for the Hawaiian Islands because of the irreversible impacts that exotic species, population growth, and development have had on the native wildlife.

extraction reserve Areas of tropical rainforest in Brazil where valuable resources such as rubber are allowed to be commercially extracted.

exurbanization The development of the distant outer suburbs of a city.

facts on the ground The purpose of the Israeli settlements in the West Bank, to make the Israeli presence so entrenched that its withdrawal would be very unlikely.

fair trade movement The selling of commodities such as coffee through cooperatives that provide a living wage for the people in poorer countries growing that product, instead of through commercial producers.

Falashas An ancient Jewish group in Ethiopia, living in isolation for thousands of years.

fall line A zone of transition in the eastern United States where rivers flow from the harder rocks of the Piedmont to the softer rocks of the Atlantic and Gulf Coastal Plain. Falls and/or rapids are characteristic features.

fallow A period of a decade or more in which tropical areas that have been farmed must be rested in order to revert to wild vegetation, thus restoring the soil's fertility.

Falung Gong A quasi-Buddhist movement that is forbidden by the Chinese government.

fault A break in a rock mass along which movement has occurred. A break due to rock masses being pulled apart is a tensional, or "normal," fault, whereas a break due to rocks being pushed together until one mass rides over the other is a compressional fault. The processes of faulting create these breaks.

faulting The processes of rock crowding together or pulling apart along tectonic plate boundaries.

favelas Shantytowns in Brazil.

feng shui A traditional Chinese theory for selecting favorable sites for building homes, cities, and other urban or architectural structures.

feral animals Domesticated animals that have abandoned their dependence on people to resume life in the wild.

Fertile Crescent The arc-shaped area stretching from southern Iraq through northern Iraq, southern Turkey, Syria, Lebanon, Israel, and western Jordan, where plants and animals were domesticated beginning about 10,000 years ago.

Fertile Triangle See "Slavic Coreland."

final status issues Issues deferred to the end of the Oslo Peace Process between Israel and the PLO. Finally dealt with at Camp David in 2000, they included the status of Jerusalem, the fate of Palestinian refugees, Palestinian statehood, and borders between Israel and a new Palestinian state. The negotiations broke down over these final status issues.

Finno-Ugric A subfamily of the Uralic language family, which includes Finnish and Hungarian.

First Nations A term used by Native American groups in Canada to refer to themselves, except for the Inuit.

First Peoples A Canadian term for Native Americans, including the Inuit.

First Temple A Jewish temple built by King Solomon on the Foundation Stone in Jerusalem that was later destroyed by Babylonians.

fish farming See *aquaculture.*

Five Themes of Geography A list created by the National Council for Geographic Education and the Association of American Geographers, summarizing what geography is all about: location, place, human–environment interaction, movement, and region.

flow map A type of map that uses arrows to indicate movement of people or goods from one place to another.

food fights A type of trade war between the United States and the European Union, involving the sales of foods, especially genetically modified foods.

food chain The sequence through which energy, in the form of food, passes through an ecosystem.

Food-Producing Revolution See *Agricultural Revolution.*

Footprints of the Ancestors In Australian Aboriginal creation myth, a term for the paths humanity's ancestors walked.

foraging See *hunting and gathering.*

formal region See *region.*

Formosan A subfamily of the Austronesian language family.

Foundation Stone The great rock in Jerusalem that the First Temple was built upon.

Forty-Niners Gold prospectors making their way to California during the 1849 gold rush.

fracking See *hydraulic fracturing.*

frankenfoods See *genetically modified foods.*

Free Aceh Movement (GAM) A secessionist movement in the Indonesian region of Aceh.

Free Trade Area of the Americas (FTAA) A proposed free-trade area encompassing most countries of the Western Hemisphere.

fuelwood crisis Deforestation in the less developed countries, caused by subsistence needs.

functional region See *region.*

G-21 (Group of 21) An economic group of 21 LDCs strongly involved in discussions of subsidies and tariffs of agricultural goods at World Trade Organization meetings.

Gadsden Purchase A strip of land in southern Arizona that the United States purchased from Mexico in 1853.

Galapágos Islands of Religion A term for the religious diversity of Ethiopia.

gaming industry The development of casinos and an economy dependent upon gambling for many Native American tribes.

Gaza-Jericho Accord See *Oslo I Accord.*

Gazprom A state-owned Russian natural gas company.

genetically modified (GM) foods Products such as rice, corn, and tomatoes, produced in the United States (and banned in Europe), that have been bio-engineered to be more productive and resilient.

genetically modified organism (GMO) foods See *genetically modified foods.*

gentrification The social and physical process of change in an urban neighborhood by the return of young, often professional, populations to the urban core. These people are often attracted by the substantial nature of the original building stock of the place and the proximity to the city center, which generally continues to have major professional opportunities. Although this process brings an urban landscape back into a primary role as a tax base, it does dispossess a considerable number of minority peoples.

genuine autonomy The Dalai Lama's request to China for Tibet to have authority over most internal Tibetan matters except foreign affairs and defense.

geographic information systems (GIS) The growing field of computer-assisted geographic analysis and graphic representation of spatial data. It is based on superimposing various data layers that may include everything from soils to hydrology, to transportation networks, to elevation. Computer software and hardware are steadily improving, enabling GIS to produce ever more detailed and exact output.

geography The study of the spatial order and associations of things. Also defined as the study of places, the study of relationships between people and environment, and the study of spatial organization.

geologic hot spot A small area of Earth's mantle where molten magma is relatively close to the crust. Hot spots are associated with island chains and thermal features.

geomancy The arrangement of homes and villages for maximum harmony with natural and spiritual environments.

geospatial A term pertaining to the geographic location and characteristics of natural or constructed features and boundaries on, above, or below the Earth's surface, especially referring to data that is geographic and spatial in nature.

Germanic languages A language family of northern and central Europe including German, Dutch, Swedish, and English.

ghetto Impoverished areas of central cities where there are many problems, including high crime and lack of jobs.

ghurba The worldwide diaspora of Palestinians.

gigantomania A Soviet preoccupation with fulfilling huge quotas or implementing grandiose schemes.

glacial deposition In the process of continental and valley glaciation, the deposition of moraines that become lateral or terminal in the act of glacial retreat. This same process also leads to glacial scouring as moving ice picks up loose rock and reshapes the landscape as the glacier moves forward or retreats.

glacial scouring See *glacial deposition.*

glaciation A geologic and climatologic process in which great ice sheets form in the Arctic and Antarctic regions and advance toward the equator.

glasnost Policies introduced by Russian leader Mikhail Gorbachev in the 1980s to allow a more democratic political system, more freedom of expression, and a more market-oriented economy.

globalization The spread of free trade, free markets, investments, and ideas across borders and the political and cultural adjustments that accompany this diffusion.

Global Warming Solutions Act of 2006 A California law requiring the state's greenhouse gas emissions to be cut to 1990 levels by 2020.

Gold Rush of 1849 See *California Gold Rush.*

Golden Crescent A term for portions of Iran, Afghanistan, and Pakistan, referring to the large amounts of the world's opium supply being produced there.

Golden Horde Russian term for the Kazan Tatars in the thirteenth century.

Golden Triangle A region in the hinterlands of Myanmar, Laos, and Thailand that is the center of Southeast Asia's illegal drug production.

Good Friday Agreement An agreement between the Unionists and the IRA in 1998 to replace direct rule of Northern Ireland with the power-sharing Northern Ireland Assembly.

Gosplan (Committee for State Planning) The agency in Moscow that formulated the Soviet Union's national economic plans.

GNI PPP Per capita gross national income purchasing power parity (per capita GNI PPP), a method of comparing the real value of output between countries' economies by combining gross national income (GNI) and purchasing power parity (PPP).

graduated symbol map A type of map that uses simple symbols such as circles or bar graphs scaled proportionally to the quantity of the attribute being mapped.

Great Depression The worst economic downturn in U.S. history, beginning in 1929 and lasting through most of the 1930s.

Great Game A rivalry between Russia and the United Kingdom in the nineteenth and early twentieth centuries over strategic influence in Afghanistan and other areas of Central Asia.

Great Migration The migration of about 6 million blacks from the rural South to the cities of the Northeast and Midwest between 1915 and 1970.

Great Rift Valley As the result of tectonic processes, a broad, steep-walled trough extending from the Zambezi Valley in southern Africa northward to the Red Sea and the valley of the Jordan River in southwestern Asia.

Great Volga Scheme The transformation of the Volga River in Russia in the twentieth century to make the river more navigable and to supply hydroelectric power and water for irrigation.

Great Wall of Louisiana Several projects in the planning stage to minimize flooding in Louisiana, including rebuilding the eroded barrier islands in the Gulf of Mexico, restoring the wetlands along the coast, and building a semipermeable levee system.

Greek The ethnolinguistic group inhabiting Greece.

Green Line (Cyprus) A buffer zone dividing Cyprus between the Greek Cypriots and the Turkish Cypriots.

Green Line (Israel) A term referring to Israel's border with the West Bank as of 1967.

Green Revolution The introduction and transfer of high-yielding seeds, mechanization, irrigation, and massive application of chemical fertilizers to areas where traditional agriculture has been practiced.

greenhouse effect The observation that increased concentrations of carbon dioxide and other gases in Earth's atmosphere cause a warmer atmosphere.

greenhouse gases Gases such as carbon dioxide and methane that trap heat from the Sun in the atmosphere, increasing the Earth's temperature.

Greenwich Meridien See *prime meridien.*

gross domestic product (GDP) The value of goods and services produced in a country in a given year. It does not include net income earned outside the country. The value is normally given in current prices for the stated year.

gross national income (GNI) A measure of a country's wealth, counting gross domestic production plus income from abroad such as rents, profits and labor.

gross national product (GNP) The value of goods and services produced internally in a given country during a stated year plus the value resulting from transactions abroad. The value is normally expressed in current prices of the stated year. Such data must be used with caution in regard to developing countries because of the broad variance in patterns of data collection and the fact that many people consume a large share of what they produce.

Group of Eight (G-8) The world's eight most economically powerful and politically influential countries: Canada, France, Germany, Italy, Japan, Russia, the United Kingdom, and the United States.

GUAM A regional grouping of Georgia, Ukraine, Azerbaijan, and Moldova, promoting economic integration, democratic reforms, and orientation toward Europe and away from Russia.

guano Seabird excrement that is also found in phosphate deposits.

guest workers Migrants (mainly young and male) from less developed countries, who are employed, sometimes illegally, in more developed countries.

Gulf Stream The strong ocean current originating in the tropical Atlantic Ocean that skirts the eastern shore of the United States, curves eastward, and reaches Europe as a part of the broader current called the North Atlantic Drift.

Gulf War The 1990–1991 confrontation between a large military coalition, spearheaded by the United States, and the Iraqi forces that had invaded Kuwait. The allies successfully drove Iraqi forces from Kuwait. Earlier, the Gulf War had referred to the 1980–1988 war between Iraq and Iran.

Gwich'in A Native American people in Alaska opposed to oil development in ANWR.

Gypsies See *Roma.*

hacienda A Spanish term for large rural estates owned by the aristocracy in Latin America.

hajj Pilgrimage to Mecca.

Hamas Palestinian Islamist party long dedicated to the destruction of Israel and opposed to the Palestinian-Israeli peace process. A major rival to the moderate al-Fatah party within the Palestine Liberation Organization.

Han Chinese The original Chinese peoples who settled in North China on the margins of the Yellow River (Huang He) and who were central to the development of Chinese culture. Han Chinese make up more than 90 percent of China's population.

Hanification Efforts by the Chinese government to increase ethnic Han presence in China's outlying areas, particularly Tibet.

Hapa The mixed-race offspring between Asian immigrants to the United States and Canada and other ethnic groups.

Hawaiian language The native Hawaiian language, a member of the Malayo-Polynesian subfamily of the Austronesian language family.

Hebrews A term for Jews in Biblical times.

hemisphere Half of the Earth's surface.

Hermit Kingdom A term for North Korea.

hero projects The construction of dams, railways, factories, and the like, encouraged by the Soviet government.

high islands Generally the result of volcanic eruptions, these are the higher and more agriculturally productive and densely populated islands of the Pacific World.

high latitudes See *latitude.*

Hinduism A regionally varied, complex belief system that lacks a definite creed or theology but includes a belief in reincarnation, the reverence of many living things, an unlimited pantheon of deities, and an infinite range of types of permissible worship.

hinterland The region of a country that lies away from the capital and largest cities and is most often rural or even unsettled; it is often seen in the minds of economic planners as an area with development potential.

Hizbullah Iranian-inspired and supported Shiite Lebanese political party holding many seats in Lebanon's Parliament. Broadly regarded as a terrorist organization, and opposed to the existence of Israel.

Hmong-Mien subfamily A language family of Southeast Asia and southern China.

Hokan-Siouan language family A Native American language family in North America.

Holocaust Nazi Germany's attempted extermination of Jews, Roma (Gypsies), homosexuals, and other minorities during World War II.

homelands Ten former territorial units in South Africa reserved for native Africans (blacks). The homelands had elected African governments, and some were designated as "independent" republics, although they were not recognized outside South Africa. Formerly known as *Bantustans*, they were abolished in 1994. Also known as *native reserves* and *national states.*

Homestead Act An 1862 U.S. law that opened up much of the interior of the country for settlement. A pioneer family could claim up to 160 acres of farming land for a $10 fee.

Hopewell A mound-builder Native American civilization that lived in what is now the U.S. Midwest.

homogenous region See *region.*

Human Development Index (HDI) A United Nations–devised ranked index of countries' development that evaluates quality of life issues (such as gender equality, literacy, and human rights) in addition to economic performance.

human–environment interaction Study of the ways human beings use and change the natural environment.

human geography One of the two major branches of geography, primarily studying the world's peoples, their spatial relationships, and their usage of the natural environment.

human immunodeficiency virus (HIV) The virus that causes AIDS.

human trafficking Movement of people against their will or intention, largely for the sex trade and manual labor.

humid continental A climate with cold winters, warm to hot summers, and sufficient rainfall for agriculture, with the greater part of the precipitation in the summer half-year.

humid subtropical A climate that generally occupies the southeastern margins of continents, with hot summers, mild to cool winters, and ample precipitation for agriculture.

humus Decomposed organic soil material. Grasslands characteristically provide more humus than forests do.

hunting and gathering A mode of livelihood, based on collecting wild plants and hunting wild animals, generally practiced by preagricultural peoples. Also known as *foraging.*

hurricane Large tropical storms that form over the warm waters of the Atlantic and Pacific Oceans.

Hutu (Bahutu) The majority population of Rwanda and Burundi. Ethnically, culturally, and linguistically the same as the minority Tutsi; the differences between the groups are socioeconomic, with the Tutsi a ruling class that dominated the Hutu majority.

hydraulic fracturing (fracking) A method of extracting natural gas by pumping water, sand and chemicals underground at high pressure, creating fissures in the rock that allow the natural gas to rise to the surface.

hydrologic cycle The process by which the sun evaporates seawater into water vapor that is later released as freshwater precipitation.

hydropolitics Political leverage and control over water.

hydrosphere All of the world's water features.

Ibo (Igbo) A major ethnic group in southern Nigeria.

Ice Age A period of widespread glaciation from about 2 million to 10,000 years ago.

ice cap A climate and biome type characterized by permanent ice cover on the ground, no vegetation (except where limited melting occurs), and a severely long, cold winter. Summers are short and cool.

Ijaw A major ethnic group in southern Nigeria.

illegal alien A migrant moving to another country illegally.

immigrant A person moving into a place.

Inca An indigenous people of South America who created an empire running from modern-day Ecuador to Chile. The Inca Empire was crushed by the Spanish in 1532.

Indians See *Native Americans.*

Indo-European languages A major language group made up of smaller language groups such as Germanic, Romance, and Slavic.

Indo-Fijians Residents of Fiji who are descended from indentured laborers from India, brought to the islands by the British in the early 1900s.

Indo-Iranian subfamily A language family that contains many of the languages spoken in India, Pakistan, Afghanistan, and Iran.

Indus Waters Treaty A 1960 treaty between India and Pakistan, describing how much water from the Indus River each country can use.

Industrial Revolution A period beginning in mid-eighteenth-century Britain that saw rapid advances in technology and the use of inanimate power.

informal settlements Settlements by people who do not hold legal claims to the land their homes are built upon.

information technology (IT) The Internet, wireless telephones, fiber optics, and other technologies characteristic of more developed countries. Generally seen as beneficial for a country's economic prospects, IT is also spreading in the less developed countries.

Inkatha Freedom Party (IFP) South African political party with mainly Zulu membership. Founded by Zulu chief Mangosuthu Buthelezi, the IFP is a rival of African National Congress (ANC), long led by Nelson Mandela.

insolation Heat from the sun.

insourcing Keeping certain jobs within a country instead of offshoring those jobs to another country.

intensive land use A livelihood such as farming, requiring use of small land areas.

intensive subsistence agriculture Agricultural practices built around the growing of cereals (especially rice) that require a steady output of arduous manual labor.

Intergovernmental Panel on Climate Change (IPCC) A United Nations panel, composed of 2,500 atmospheric scientists from 130 countries, that has concluded that the evidence for global warming is "unequivocal."

internally displaced persons (IDPs) People dislodged and impoverished by strife in their home countries but who have little prospect of emigrating from that country.

International Atomic Energy Agency (IAEA) A United Nations agency that carries out inspections of nuclear facilities.

International Criminal Court A United Nations forum for crimes against humanity, located in the Netherlands.

International Date Line A line roughly concurrent with the 180-degree line of longitude (with several deviations for political and practical reasons) where the beginning of one day and the end of another day meet. The date west of the International Date Line is one day ahead of the date east of the line.

interstate highway system A massive infrastructure development in the United States that started in the 1950s and serves as the primary network for trucking cargo across the country.

Intifada Palestinian uprising against Israeli occupation.

Inuit A people living in Greenland and far northern Canada and Alaska, sometimes known as Eskimos to others.

Inupiat A Native American people in favor of oil production in ANWR.

Iranian A subgroup of the Indo-European language family. Persian and Tajik are Iranian languages.

Irish Republican Army (IRA) A terrorist group based in Northern Ireland that launched many attacks in the United Kingdom, which were designed to drive out the British army from Northern Ireland.

Iron Curtain A term used by Winston Churchill to describe the dividing line in Europe between capitalist democracies to the west and Communist authoritarian regimes to the east.

irredentism Movement by an ethnic group in one country to revive or reinforce kindred ethnicity in another country—often in an effort to promote succession there.

irrigation The artificial placement of water to produce crops, generally in arid locations.

isarithmic map A type of map that uses lines or bands of color to join the points of equal value of some phenomenon across a mapped area.

Islam A monotheistic faith built upon the foundations of Judaism and Christianity. One of the world's major religions, it originated in the seventh century in what is now Saudi Arabia.

Islamic law (sharia) Islamic interpretations of divine intention in legal matters, based on the Qur'an, the Hadith, and Islamic scholars' opinions in matters of jurisprudence.

Islamic Movement of Uzbekistan (IMU) A group dedicated to the overflow of Uzbekistan's president and the creation of an Islamic state in the Fergana Valley of Central Asia.

Islamists Islamic militants.

Islamophobia A fear of Islam and the Muslims that practice it.

isolationism The tendency of a country to stay out of the conflicts of other countries.

isostatic rebound The slow uplift of land that had been pressed down by the great weight of glaciers during the Ice Age.

Israelites A term for Jews in Biblical times.

Jainism A small but influential religion of India renowned for its respect for geographic features and animal life.

Jain A practitioner of Jainism.

Jasmine Revolution The uprising in Tunisia against that country's leader in early 2011.

Jemaah Islamiah (JI, or the Islamic Group) A militant Islamist group in Southeast Asia.

Jewish settlements Israeli citizens settling in the West Bank to strengthen Israel's claims to the area.

Jews Practitioners of Judaism.

jihad A term that for Muslims can have a range of meanings, from expressing one's commitment to a personal conviction to all-out war against nonbelievers.

juche North Korean principle of self-reliance.

Judaism The first significant monotheistic faith, originating at least 4,000 years ago. It does not have a fixed creed or doctrine, and it is not a proselytizing religion.

Judea and Samaria Biblical kingdoms that are now part of the West Bank.

Ka'aba A Muslim shrine located in Mecca.

Kanaka Maoli The term native Hawaiians use for themselves.

karma An important Buddhist concept of a person's acts and the consequences of those acts.

karoshi Japanese term for death by overwork.

Kartvelian (South Caucasian) language A collection of languages spoken in the Caucasus region.

keystone species A species that affects many other organisms in an ecosystem.

Khalistan A desired homeland for India's Sikhs.

Khoisan language family The languages of the San people in southern Africa. These languages, nearly extinct, are among the world's oldest.

Kimberley Process A United Nations–backed certification plan for ensuring that only legally mined rough diamonds, untainted by violence, reach the market.

kleptocracy An economic and political system based on crime.

knowledge economy An economy based on innovation and services.

Kurdistan Workers Party (PKK) The main Kurdish resistance to Turkish rule.

Kurds The largest minority group in Turkey and the largest non-Arab minority in Iraq. Kurds are the world's largest ethnic group without a country.

Kyoto Protocol A treaty on climate change signed by 160 countries in Kyoto, Japan, in 1997 and put into force in 2005. It requires MDCs to reduce their greenhouse gas emissions by more than 5 percent below their 1990 levels by the year 2012.

Labor Party The liberal political party of Israel.

Lakota Native American people living on the Great Plains.

land empire The establishment of colonies in a continental hinterland as opposed to overseas.

Land for Peace Part of the peace process between Israel and Arabs, with Israel swapping Arab lands it occupied in 1967 in exchange for peace with its neighbors.

land hemisphere See *hemisphere*.

land reform In Latin America, the reallocation of land owned by the wealthy to small farmers.

landscape A portion of the earth's land surface. Geographers are interested in the transformation of natural landscapes into cultural landscapes.

landscape perspective The understanding of the changes people have made on the natural environment over time.

language death When the last speaker of a certain language dies

lapse rate The decrease in temperature over a corresponding increase in elevation. The lapse rate is (on average) about 3.6°F (2.0°C) for each increase of 1,000 feet (305 m) in elevation.

large-scale map A map constructed to show considerable detail in a small area.

Laskar Jihad A militant Islamist group in Indonesia.

laterite A material found in tropical regions with highly leached soils; it is composed mostly of iron and aluminum oxides that harden when exposed and make cultivation difficult.

lateritic soils A soil characteristic of many tropical areas. Excessive exposure to the sun and repeated wetting and drying turns the soil into a bricklike substance nearly impossible to cultivate.

***latifundia* (sing. *latifundio*)** Large agricultural Latin American estates with strong commercial orientations.

Latino An American of Hispanic background.

latitude A measurement that denotes position with respect to the equator and the poles. Latitude is measured in degrees, minutes, and seconds, which are described as parallels. Places near the equator are said to be in low latitudes; places near the poles are in high latitudes. The Tropic of Cancer and the Tropic of Capricorn, at 23.5°N and 23.5°S, respectively, and the Arctic Circle and Antarctic Circle, at 66.5°N and 66.5°S, respectively, form the most commonly recognized boundaries of the low and high latitudes. Places occupying an intermediate position with respect to the poles and the equator are said to be in the middle latitudes. There are no universally accepted definitions for the boundaries of the high, middle, and low latitudes.

Law of Return An Israeli law permitting all Jews living in Israel to have Israeli citizenship.

less developed countries (LDCs) The world's poorer countries.

levee A raised riverbank designed to prevent flooding.

liberal autonomy See *autonomous regions*.

Liberal Democratic Party (LDP) The dominant political party of Japan since 1952.

lifeboat ethics Ecologist Garrett Hardin's argument that, for ecological reasons, rich countries should not assist poor countries.

Liga Veneta An Italian organization wishing to create an autonomous region centered on the city of Venice.

Likud Party The conservative political party in Israel.

lithosphere The rocky portion of the Earth's surface.

location A concept central to all geographic analysis. Where something is relates to all manner of influences, from climate to migration routes, and is a crucial component in trying to understand patterns of historic and economic development.

lock Devices that allow ships to move from one water level to another in a canal.

loess A fine-grained material that has been picked up, transported, and deposited in its present location by wind; it forms an unusually productive soil.

longitude A measurement that denotes a position east or west of the prime meridian (Greenwich, England). Longitude is measured in degrees, minutes, and seconds, and meridians of longitude extend from pole to pole and intersect parallels of latitude.

loose nukes Stolen radioactive materials.

Louisiana Purchase A large purchase of land by the United States from France in 1803 that significantly added to the size of the United States.

low islands Made of coral, these are the generally flat, drier, less agriculturally productive, and less densely populated islands of the Pacific World.

low latitudes See *latitude*.

Ma'adan (Marsh Arabs) People living on marshlands in Iraq, creating a unique culture found nowhere else in the Middle East. Most of their habitat was destroyed by Saddam Hussein after the first Gulf War.

Mahayana Buddhism One of the two branches of Buddhism, originating in India and spreading northward to Tibet and China.

mainstream European American culture A culture based on the English language and British educational principles brought by British colonists combined with core American values that developed on U.S. soil.

major cluster of continuous settlement An area where all habitations lie no more than 3 miles (5 km) from other habitations in at least six different directions, and roads or railroads lie no more than 10–20 miles (16–32 km) away in at least three directions.

Malayo-Polynesian language subfamilies A subfamily of the Austronesian language family.

Mandé A language family of west Africa.

Malthusian scenario The model forecasting that human population growth will outpace growth in food and other resources, with a resulting population die-off.

Manifest Destiny The opening and settlement of new frontiers across the United States from the Atlantic to the Pacific Oceans.

map A representation of various phenomena over all or a part of the Earth, either on a globe or a flat surface.

map projection A way to minimize distortion in one or more properties of a map (direction, distance, shape, or area).

maquiladoras Operations dedicated to the assembly of manufactured goods, generally in Mexico and Central America, from components initially produced in the United States or other places. With the enactment of NAFTA in 1994, there was a massive expansion of *maquiladora* operations.

marginalization A process by which poor subsistence farmers are pushed onto fragile, inferior, or marginal lands that cannot support crops for long and that are degraded by cultivation.

Maria Lionza A syncretic faith in Venezuela that blends Catholic, West African, and indigenous beliefs.

marine west coast A climate occupying the western sides of continents in the higher middle latitudes; it is greatly moderated by the effects of ocean currents that are warm in winter and cool in summer relative to the land.

Maronites A Christian sect in Lebanon.

Marshall Plan The plan designed largely by the United States, after the conclusion of World War II, by which U.S. aid was focused on the rebuilding of the very Germany that had been its enemy in the war just concluded. Secretary of State George Marshall (who had been Chief of Staff of U.S. Army from 1939–1945) was central to the plan's design and implementation.

Masdar A carbon-neutral city under construction in the United Arab Emirates.

mathematical location See *absolute location*.

Maya A Native American ethnic group inhabiting southern Mexico and northern Central America.

Mayan language family A language family of southern Mexico and Guatemala.

medina An urban pattern typical of the Middle Eastern city before the twentieth century.

Mediterranean (dry-summer subtropical) A climate that occupies an intermediate location between a marine west coast climate on the poleward side and a steppe or desert climate on the equatorward side. During the high-sun period, it is rainless; in the low-sun period, it receives precipitation of cyclonic or orographic origin.

Mediterranean scrub forest The xeroyphytic vegetation typical of hot, dry summer, Mediterranean climate regions. Local names for this vegetation type include *maquis* and *chaparral*.

megacity A city with more than 10 million inhabitants.

megadiversity country A country containing a large percentage of the world's animal and plant species.

megalanguage One of the world's most-spoken languages, including Chinese, English, Spanish, Hindi, and Arabic.

megalopolis A chain of metropolitan areas with a population over 10 million.

Mekong River Commission A coalition of Vietnam, Laos, Cambodia, and Thailand, to determine rules to ensure water quality and quantity of the Mekong River.

Melanesia (black islands) A group of relatively large islands in the Pacific Ocean bordering Australia.

mental map A term used to define the geographies encompassed in every individual's mind as a series of locations, access routes, physical and cultural characteristics of places, and often a general sense of the good or bad of locales.

mercantile colonialism The historical pattern by which Europeans extracted primary products from colonies abroad, particularly in the tropics.

Mercator Projection A cylindrical, conformal map projection intended to be used for navigation developed by Gerardus Mercator in 1569.

Mercosur Acronym for Southern Cone Common Market, an economic association among Brazil, Argentina, Paraguay, and Uruguay.

meridian See *longitude*.

mestizo In Latin America, a person of mixed European and Native American ancestry.

Methane Greenhouse gas resulting from human activities (from rice paddies and the guts of ruminating animals like cattle).

Mexican War A war between the United States and Mexico in the 1840s, in which Mexico lost a significant portion of its territory to the United States.

Mezzogiorno A term for southern Italy.

microcredit The lending of small sums to poor people to set up or expand small businesses.

Micronesia (tiny islands) Thousands of small and scattered islands in the central and western Pacific Ocean, mainly north of the equator.

microstate A political entity that is tiny in area and population and is independent or semi-independent.

Middle Eastern ecological trilogy The model of mostly symbiotic relations among villagers, pastoral nomads, and urbanites in the Middle East.

middle latitudes See *latitude*.

migration A temporary, periodic, or permanent move to a new location.

millenarian movements See *cargo cults*.

minifundia **(sing. *minifundio*)** Small Latin American agricultural landholdings, usually with a strong subsistence component.

minutes A unit of measurement equal to 1/60th of a degree.

Mississippian A mound-builder Native American culture that lived in what is now the southeastern United States.

misdeveloped country A term sometimes applied to post-Soviet Russia because of the rampant criminality and corruption in its society and its demographic trends of high death rates and low birth rates.

Mistake of 1914 Nigerian term for the creation of the country's modern borders by the British.

mitigation Measures taken to avoid the adverse impacts of climate change in the long term.

Mogollon A Native American civilization that existed in what is now the southwestern United States.

Mohajirs Muslim immigrants to Pakistan from India after the two countries were partitioned in 1947.

mollisol soils Thick, productive, and durable soils, such as the *chernozem*, whose fertility comes from abundant humus in the top layer.

money laundering Ways of turning money earned in illegal ways into money that looks legal.

Mongols Conquerors of most of Russia in the thirteenth century.

Mon-Khmer A branch of the Austro-Asiatic language subfamily that includes the Khmer and Vietnamese languages.

monoculture The single-species cultivation of food or tree crops, usually very economical and productive but threatening to natural diversity and change.

Monroe Doctrine A policy of U.S. President James Monroe stating the United States would prevent European countries from establishing any new colonies in the Western Hemisphere.

monsoon A current of air blowing fairly steadily from a given direction for several weeks or months at a time. Characteristics of a monsoonal climate are a seasonal reversal of wind direction, a strong summer maximum of rainfall, and a long dry season lasting for most or all of the winter months.

Montreal Protocol A 1989 international treaty to ban chlorofluorocarbons.

moor A rainy, deforested upland, covered with grass or heather and often underlain by water-soaked peat. Also, Muslim inhabitants of Spain.

more developed countries (MDCs) The world's wealthier countries.

Mosan language family A Native American language family along the Pacific coast.

Mound Builder civilizations A collection of Native American culture groups named for their creation of large earthen mounds. They created extensive trading networks and urban settlement patterns.

Movement for the Emancipation of the Niger Delta (MEND) A mainly ethnically Ijaw group in Nigeria that has expressed dissatisfaction with the redistribution of Nigeria's oil wealth by kidnapping foreign oil workers.

Movement for the Survival of the Ogoni People (MOSOP) An ethnic Ogoni organization in Nigeria that petitions the government for a safe environment and more federal support for their people.

mujahidin Anti-Soviet rebel bands in Afghanistan during the Soviet invasion of the 1980s.

mulatto A person of mixed European and black ancestry.

mullah Religious leaders in Iran.

Multiculturalism Act A 1988 Canadian law that officially established Canada as a multicultural society, recognizing and encouraging traditional identities.

multi-Latina companies Companies, established in Latin America by Latin American workers who gained experience in the United States, that invest in the United States.

multinational companies Companies that operate, at least in part, outside their home countries.

Muslims Adherents of Islam.

mutually assured destruction (MAD) The concept of nuclear deterrence: if one side in a conflict were to launch nuclear weapons, the other would. Both countries would be obliterated.

Na-Dene language family A Native American language family of western North America, including Apache and Navajo.

nagana See *trypanosomiasis.*

Nakh-Dagestanian A collection of languages spoken in the Caucasus region.

narcoterrorist organization Armed insurgent groups that are funded by the sale of illegal drugs.

National Petroleum Reserve An area along the North Slope of Alaska estimated to have more petroleum resources than ANWR.

nationalism The drive to expand the identity and strength of a political unit that serves as home to a population interested in greater cohesion and often expanded political power.

national minorities A substantial population of ethnic groups originating in other nations but now living in foreign countries.

Native Americans The indigenous peoples of North and South America who originally migrated to the New World from Asia over a land bridge during the Ice Age. Christopher Columbus misidentified them as "Indians" in 1492.

native reserves See *homelands.*

Native Title Bill Legislation passed by the Australian government in 1993, addressing many issues relating to Aborigines' rights to claim and buy land.

natural replacement rate The highest rate at which a renewable resource can be used without decreasing its potential for renewal. Also known as *sustainable yield.*

natural landscape An environment without (or before) transformations caused by humans.

natural resource A product of the natural environment that can be used to benefit people. Resources are human appraisals.

Naxalites Maoist insurgents in India that claim to champion the causes of the rural poor.

Nazca An extinct ethnic group from Peru that left massive carvings in the desert two thousand years ago.

Near Abroad Russia's name for the now-independent former republics, other than Russia, of the Union of Soviet Socialist Republics.

Near North Australian term for neighboring Asian countries.

Negroes An archaic term for African Americans.

neocolonialism The perpetuation of a colonial economic pattern in which developing countries export raw materials to, and buy finished goods from, developed countries. This relationship is more profitable for the developed countries.

neo-Europes Areas colonized by Europeans in lands similar climatically and geographically to Europe, creating "new Europes" with cultural similarities and ties to Europe.

Neolithic Revolution See *Agricultural Revolution.*

Neo-Malthusians Supporters of forecasts that resources will not be able to keep pace with the needs of growing human populations.

new American imperialism The belief that post-9/11 actions by the United States marked the beginning of a new era of aggressive global involvement.

New Asian Tigers A collection of Southeast Asian countries that have had rapid economic growth.

new lands See *virgin and idle lands.*

New Partnership for Africa's Development (NEPAD) An agency of the African Union responsible for monitoring improvements in human rights across Africa.

New Rome A term referring to Constantinople as a center of Christendom in the late Roman period.

newly industrializing countries (NICs) The more prosperous of the world's less developed countries.

Niger-Congo language family A language family of central Africa.

Night Journey In Islam, Muhammad's brief ascension into heaven from the great rock beneath the Dome of the Rock.

Nile Water Agreement A treaty guaranteeing Egypt 75 percent of the Nile River's flow; it forbids any projects downstream that might threaten the volume of water reaching Egypt.

Nilo-Saharan language family A language family spoken in parts of Africa around the Sahara Desert.

NIMBY An acronym for "not in my back yard." Refers to the opposition of local residents to new developments in their area.

nirvana In Buddhism, a transcendent state in which one is able to escape the cycle of birth and rebirth.

Nitrous oxide Greenhouse gas resulting from human activities (from the breakdown of nitrogen fertilizers).

Noble Sanctuary (al-Haraam ash-Shariif) What Muslims call the Temple Mount.

nodal region See *region.*

Non-Pama-Nyungan A branch of the Aboriginal language family spoken in northern Australia.

nonselective migrants People responding to push factors.

nontimber forest products (NTFPs) Goods, other than trees, made from products found in forests, to prevent deforestation.

North American Free Trade Agreement (NAFTA) A free-trade agreement among Canada, the United States, and Mexico.

North Atlantic Drift A warm current, originating in tropical parts of the Atlantic Ocean, that drifts north and east, moderating temperatures of Western Europe. Also known as the *Gulf Stream.*

North Atlantic Treaty Organization (NATO) A military alliance formed in 1949 that included the United States, Canada, many European nations, and Turkey.

North European Plain The level to rolling lowlands that extend from the low countries on the west, through Germany to Poland. These are areas rich in agricultural development, dense human settlements, and a number of major industrial centers. The plain is broken by a number of rivers flowing from the Alps and other mountain systems in central Europe into the North and Baltic Seas.

north latitude See *latitude.*

North Pole A point at 90 degrees north latitude. The highest latitude in the northern hemisphere.

Northern Alliance The main domestic opponent of the Taliban in Afghanistan.

Northern Hemisphere See *hemisphere.*

Northern League An Italian organization calling for the creation of Padania as an autonomous area or independent country.

Northern Sea Route A waterway developed by the Soviets to provide a connection between the Pacific and Arctic oceans.

Northwest Passage A shipping channel through Canada's Arctic islands if global warming decreases sea ice enough to make a reliable route.

Nuclear Nonproliferation Treaty (NPT) See *Comprehensive Test Ban Treaty.*

Nunavut A Canadian territory created in 1999 to be the home of the Inuit.

Oceania A term collectively referring to Australia, New Zealand, and the islands of the Pacific Ocean.

Occupied Territories The territories captured by Israel in the 1967 War: the Gaza Strip and West Bank, captured, respectively, from Egypt and Jordan, and the Golan Heights, captured from Syria.

official state sponsors of terrorism A U.S. list of countries that sponsor terrorism.

offshore outsourcing (offshoring) See *outsourcing.*

Ogoni An ethnic group in Nigeria.

oil embargo The 1973 embargo on oil exports imposed by Arab members of the Organization of Petroleum Exporting Countries (OPEC) against the United States and the Netherlands.

oligarchs Literally "rule of the few," where a small number of wealthy businessmen control much of a state's production.

one-child campaign See *one-child policy.*

one-child policy Official policy of the Chinese government, limiting families to one child.

Operation Barbarossa A German onslaught against the Soviet Union during World War II.

Operation Crossroads A series of 67 nuclear experiments by the United States in the Marshall Islands of the Pacific Ocean after World War II.

Operation Enduring Freedom The U.S. military strike against Afghanistan that began in October 2001.

Operation Iraqi Freedom The U.S.-led ground assault on Iraq that began in March 2003.

Orange Revolution The 2005 Ukrainian presidential election that brought pro-Western Victor Yushchenko to power.

Oregon Trail One of the major routes leading westward that American pioneers took in the middle of the nineteenth century, leading to what is now the Willamette Valley of Oregon.

Organization of American States (OAS) A political organization of every country in North and South America, except Cuba.

Organization of Petroleum Exporting Countries (OPEC) An international organization created to get higher profits from oil sales to its member countries.

orientation The relationship between direction on a map and the corresponding compass direction in reality.

original affluent society A description of hunter-gatherers, who are thought by some scholars to have enjoyed harmony with the natural world and suffered few social and psychological problems.

Oslo I Accord An agreement signed by Israeli prime minister Yitzhak Rabin and Palestinian leader Yassir Arafat, designed to be a pathway to peace between Israel and the Palestinians.

Oslo II Accord Another agreement, not yet fully implemented, between Israel and the Palestinian Authority that would transfer more of the West Bank to Palestinian control in exchange for increased Palestinian efforts to crack down on terrorists and guarantee Israel's security.

Oto-Manguean language family A family of Native American languages spoken in northern Mexico.

outsourcing The flight of jobs and technology from countries with high manufacturing and service costs to countries with lower costs.

overseas foreign workers (OFW) A term referring to the one-quarter of the workforce of the Philippines that works outside that country and sends back remittances that make up more than 10 percent of the Philippines' GDP.

Padania A term for northern Italy.

paddy (padi) The field in which rice is grown.

Palestine Liberation Organization (PLO) The Palestinian military and civilian organization created in the 1960s to resist Israel and recognized in the 1990s as the sole legitimate organization representing official Palestinian interests internationally.

Palestinian Authority (PA) The Palestinian government created as a result of the Oslo I Accord.

Palestinians Arabs who formed the majority of inhabitants of Palestine prior to the creation of the modern country of Israel.

pampa Subtropical grasslands in South America.

Pama-Nyungan branch of the Aboriginal language family spoken over most of Australia.

Pan-African Parliament A lawmaking body of the African Union.

pan-Turkism The goal of uniting all Turkic peoples of Asia.

Pancasila A pan-Indonesian nationalist ideology designed to neutralize all ethnic identities.

pandemic A very widespread epidemic.

Papuan languages A collection of the 750 languages spoken on New Guinea and surrounding islands.

Papuans People of mostly Melanesian origin that live on New Guinea.

parallel A latitude line running parallel to the equator.

páramos Alpine meadows in Latin America.

Parsis An ethnic group in India that practices Zoroastrianism.

Parti Quebecois (PQ) A separatist political party in Quebec.

pastoral nomads Desert peoples of the Middle East and northern Africa who migrate through arid lands with their livestock, following patterns of vegetation and rainfall.

Penutian language family A Native American language family of Mexico and California.

People of the Book Term used by Muslims to refer to Jews and Christians.

people overpopulation The concept that many persons, each using a small quantity of natural resources to sustain life, add up to too many people for the environment to support.

PEPFAR An acronym for President's Emergency Plan for AIDS Relief, in which the George W. Bush administration pledged over $45 billion to combat AIDS in Africa.

per capita GDP A measure of economic well-being found by dividing a country's gross domestic product by its population.

per capita gross national income A measure of economic well-being found by dividing a country's gross domestic product plus any foreign output by domestically-owned producers by its population.

per capita gross domestic product purchasing power parity (GDP PPP) In this figure, annual per capita gross domestic product (GDP)—the total output of goods and services a country produces for home use in a year—is divided by the country's population. For comparative purposes, that figure is adjusted for purchasing power parity (PPP), which involves the use of standardized international dollar price weights that are applied to the quantities of final goods and services produced in a given economy. The resulting measure, per capita GDP PPP, provides the best available starting point for comparisons of economic strength and well-being between countries.

perceptual region See *region.*

perestroika See *glasnost.*

peripheral location An area of the world with less importance due to a distant location relative to others.

permafrost Permanently frozen subsoil.

Permanent Fund A government surplus in Alaska, created from oil revenues.

Persian The language spoken by the majority of Iranians.

physical geography The subdiscipline of geography most concerned with the climate, landforms, soils, and physiography of the earth's surface.

Pidgin The official language of Papua New Guinea, a combination of English and foreign words mixed with indigenous vocabulary and grammar.

PIGS An acronym for Portgual, Ireland, Greece, and Spain, used in reference to the weaker economies of those countries as opposed to those of France and Germany.

Pillars of Islam The five fundamental tenets of the faith of Islam, which are *profession of faith, prayer, almsgiving, fasting,* and the *hajj.*

pivotal countries Those countries whose collapse would cause international refugee migration, war, pollution, disease epidemics, or other international security problems.

place In geographic analysis of a given locale, the nature of place identity becomes a means of understanding people's response to that particular place. Determination of the environmental and cultural characteristics that are most frequently associated with a certain place helps to establish that "sense of place" for a given location.

Plan Colombia A U.S.-funded effort to reduce the amount of coca production in Colombia.

Plan Patriot An operation in Colombia involving U.S. military personnel and civilian contractors targeting FARC to reduce drugs and secure oil reserves.

plates Giant sections of rock the lithosphere is split into, which slowly drift over time.

plate tectonics The dominant force in the creation of the continents, mountain systems, and ocean deeps. The steady, but slow, movement of these massive plates of the earth's mantle and crust has created the positions of the continents and major patterns of volcanic and seismic activity. Areas where plates are being pulled under other plates are called subduction zones.

platform economy Common in high-tech industries, with a corporation headquartered in the United States but most of the work on its products done overseas.

Pleistocene overkill hypothesis A hypothesis stating that hunters and gatherers of the Pleistocene Era hunted many species to extinction.

plinthite The bricklike substance into which lateritic soils can turn.

podzol Soil with a grayish, bleached appearance when plowed, lacking in well-decomposed organic matter, poorly structured, and very low in natural fertility. Podzols are the dominant soils of the taiga.

Polynesia (many islands) A Pacific island region that roughly resembles a triangle with its corners at New Zealand, the Hawaiian Islands, and Easter Island.

population change rate The birth rate minus the death rate in a population.

population explosion The surge in Earth's human population that has occurred since the beginning of the Industrial Revolution.

population implosion A dramatic loss of population, caused by low fertility rates or death rates that are much higher than birth rates.

population pyramid See *age structure diagram.*

population replacement level The number of new births required to keep a population steady. Generally calculated as 2.1 children per woman in MDCs.

postindustrial Description of an economy or society characterized by the transformation from manufacturing to information management, financial services, and the service sector.

potato famine The starvation of 10 percent of Ireland's population in the late 1840s, caused by a crop disease called blight.

potlatch A Native American system in which rank and status were determined by the quantity and quality of material goods one could give away.

Poverty Point A mound-builder Native American culture that lived in what is now the southeastern United States.

power projection hub In geopolitical terms, a strategic outpost from which a country could launch naval, air, and other forces against an enemy.

prairie An area of tall grass in the middle latitudes, composed of rich soils that have been cleared for agriculture. The original lack of trees may have been due to repeated burnings or periodic drought conditions.

Pre-1967 borders The borders of Israel after 1949 and before the Six-Day War of 1967 in which Israel captured and occupied the Gaza Strip, West Bank, and Golan Heights.

precautionary principle The notion that is it appropriate to attempt to reduce or eliminate any risky practice.

Pre-Columbian times North and South America before Christopher Columbus's first voyage in 1492.

precipitation Water falling from the sky to the surface in the form of rain, snow, sleet or hail.

preemptive engagement A U.S. policy developed after 9/11 that allowed the country to take unilateral military action if necessary to defuse threats to its security whenever and wherever those threats occur.

primate city A city that dominates a country's urban scene and is usually defined as being larger than the country's second and third largest cities combined. Primate cities are generally found in developing countries, although some developed countries, such as France, have them.

prime meridian The line of zero degrees longitude, which passes through Greenwich, England. It separates the Western Hemisphere from the Eastern.

privatization The shift of ownership of economic activities from the state to nongovernmental.

proletariat In Marxism, the industrial working class.

Promised Land Canaan, for the Israelites in Biblical times.

Protestant Reformation The schism in Western Christianity that began in 1517 when Martin Luther registered his objections to much of the dogma, rituals, and political organization of the Roman Catholic Church.

Protestant Unionists One side of "The Troubles" in Northern Ireland, those who wished Northern Ireland to remain unified with the United Kingdom and not unify with the rest of Ireland.

Protestant work ethic In the United States, the emphasis on work, productivity, and economic success in an individual's life. Despite its name, this ethic is shared by people of many faiths and ethnicities.

Protestantism One of the denominations of Christianity, which began in the Protestant Reformation of the sixteenth century.

Proto-Asiatic A family of languages spoken largely in northeastern Russia.

proxy war During the Cold War, the situation created when the United States and the Union of Soviet Socialist Republics played African countries against one another, arming them with weapons, to boost their competing aims.

pueblo An interconnected series of residences and ceremonial centers composed of adobe, used by the Hohokam culture of New Mexico.

purchasing power parity (PPP) A method of comparing the real value of output between different countries' economies, considering factors such as differences in relative prices of goods and services.

push and pull forces of migration Emigration caused by so-called push factors, as when hunger or lack of land "pushes" peasants out of rural areas into cities, or by pull factors, as when an educated villager responds to a job opportunity in the city. People responding to push factors are often referred to as nonselective migrants, whereas those reacting to pull factors are called selective migrants. Both push and pull forces are behind the rural-to-urban migration that is characteristic of most countries.

Quechua See *Quechua-Aymaran language family.*

Quechua-Aymaran language family A language family of central South America, to which the language of the Inca belonged. Many residents of Peru and Bolivia still speak Quechua as a first language.

Rastafarianism A religion originating in Jamaica in 1930 that holds that Ethiopian emperor Haile Selaisse was the earthly incarnation of God.

Red-Dead Peace Conduit A joint venture between Israel and Jordan that will carry seawater from the Gulf of Aqaba northward to replenish the shrinking Dead Sea.

Reducing Emissions from Deforestation and Forest Degradation (REDD) A United Nations effort to create a financial value for the carbon stored in forests and offer incentives for developing countries to reduce carbon emissions from forested areas and invest in low-carbon development.

reference map A type of map primarily depicting the locations of various features on the Earth's surface.

refugees Victims of severe push factors such as persecution, political repression, and war.

region A "human construct" that is often of considerable size, that has substantial internal unity or homogeneity, and that differs in significant respects from adjoining areas. Regions can be classed as formal (homogeneous), functional, or vernacular. The formal region, also known as a uniform region, has a unitary quality that derives from a homogeneous characteristic. The United States is an example of a formal region. The functional region, also called the nodal region, is a coherent structure of areal units organized into a functioning system by lines of movement or influence that converge on a central node or trunk. A major example would be the trading territory served by a large city and bound together by the flow of people, goods, and information over an organized network of transportation and communication lines. Vernacular or perceptual regions are areas that possess regional identity, such as the Sun Belt, but share less objective criteria in the use of this regional name. General regions, such as the major world regions in this text, are recognized on the basis of overall distinctiveness.

Regional Initiative for the Infrastructure Integration of South America An international effort to improve the economies of South American countries by linking them together by road networks and other infrastructure projects.

relative location The location of a place defined by relationship to other places.

remittances Money earned by workers in a foreign country sent back to that worker's home country.

remote sensing Through the use of aerial and satellite imagery, geographers and other scientists have been able to get vast amounts of data describing places all over the face of the earth. Remote sensing is the science of acquiring and analyzing data without being in contact with the subject. It is used in the study of patterns of land use, seasonal change, agricultural activity, and even human movement along transport lines. This process relates closely to GIS. See also *geographic information systems (GIS).*

renationalization A process of reversing the trend toward private ownership of economic production in Russia during the presidency of Vladimir Putin.

renewable resource A resource, such as timber, that is grown or renewed so that a continual supply is available. A finite resource is one that, once consumed, cannot be easily used again. Petroleum products are a good example of such a resource and because it takes too much time to go through the process of creation, they are not seen as renewable.

reservations Areas of land reserved for Native American tribes by the U.S. government.

resilience An ecosystem that can recover from the stress of drought and has mechanisms to cope with a natural cycle that includes extended periods of dryness and rain.

resource curse A paradox of plenty, when a country with a great abundance of a valuable natural resource experiences lower economic growth than countries without such abundance.

Richter scale A system for measuring the strength of earthquakes.

right of return The Palestinian Arab principle that refugees (and their descendants) displaced from Israel and the Occupied Territories in the 1948 and 1967 wars be allowed to return to the region.

risk minimization The exploitation of multiple resources so some resources will support a population if others fail.

River Sea A term referring to the enormity of the Amazon River.

road map for peace An effort led by the United States, the European Union, Russia, and the United Nations to reach a peace settlement between Israel and the Palestinians, ultimately leading to an independent Palestinian state.

Roma (Gypsies) Originating in India, they are one of Europe's largest ethnic minorities. Commonly referred to as Gypsies, they often move from place to place in caravans and have been discriminated against for centuries.

Roman Catholic Church The largest denomination of Christianity, led by the Pope in Vatican City.

Romance languages A language family primarily of southern Europe, including Italian, Spanish and French.

Roosevelt Corollary A 1904 policy in which the United States declared it had the right to supervise the internal affairs of Latin American countries to ensure U.S. national security.

Rose Revolution The peaceful transition of power in the 2003 presidential election in Georgia.

rules of origin A NAFTA requirement that half or more of all the components of any manufactured good must originate in Canada, the United States, or Mexico.

rural-to-urban migration See *urbanization.*

Rus Viking adventurers from Scandinavia that settled in Slavic regions during the Middle Ages.

Russian cross The dramatic rise in death rates and fall of birth rates in Russia after the breakup of the Soviet Union.

Russian Revolution The overthrow of Czar Nicholas II of Russia in 1917.

Russification The effort, particularly under the Soviets, to implant Russian culture in non-Russian regions of the former Soviet Union and its Eastern European neighbors.

Rust Belt An area across the Midwest United States that was home to many heavy industries.

sacred space (sacred place) Any locale that people hold in reverence, such as places of worship, cemeteries, and battlefields.

Saharan A subfamily of the Nilo-Saharan language family spoken in north-central Africa.

salinization The deposition of salts on, and subsequent fertility loss in, soils experiencing a combination of overwatering and high evaporation.

salmon war A dispute regarding salmon fishing between the United States and Canada until it was resolved in 1999.

San The proper name for Bushmen, a major group of the Khoisan ethnolinguistic cultures living mainly in Botswana and Namibia. They were among the last people on Earth to practice hunting and gathering.

sand sea A virtual "ocean" of sand characteristic of parts of the Middle East, where people, plants, and animals are all but nonexistent.

Santa Fe Trail One of the major routes leading westward that American pioneers took in the middle of the nineteenth century, leading to what is now New Mexico.

Santeria A syncretic religion originating in the Caribbean, blending Catholicism, indigenous, and Yoruba beliefs.

savanna A low-latitude grassland in an area with marked wet and dry seasons.

scale The size ratio represented by a map; for example, a map with a scale of 1:12,500 is portrayed as 1/12,500 of the actual size.

Schengen Agreement A framework among signatory countries allowing for free movement of people without need for visas, passports, or other border controls.

Schengenland A term for the countries participating in the Schengen Agreement.

scorched earth The wartime practice of destroying one's own assets to prevent them from falling into enemy hands.

seafloor spreading The creation of new crust from molten material rising from the Earth's mantle through ridges on the ocean floor.

seamount An underwater volcanic mountain.

Second Kuwait A term describing the Russian island of Sakhalin and the potential large reserves of oil and natural gas it has offshore.

second law of thermodynamics A natural law stating that high-quality, concentrated energy is increasingly degraded as it passes through the food chain.

Second Russian Revolution The collapse of the Soviet Union in 1991.

Second Temple The second Jewish temple to stand on the Temple Mount, which was destroyed by the Romans.

seconds A unit of measurement equal to 1/60 of a minute, or 1/3,600 of a degree.

sectarian violence Conflict between people of different religions.

security fence A physical barrier constructed by Israel in the West Bank designed to prevent Palestinian suicide bombers from reaching Israel.

sedentarization Voluntary or coerced settling down, particularly by pastoral nomads in the Middle East.

segregation A policy of separation of races.

seismic activity Usually refers to tectonic forces that result in earthquakes.

selective migrants People reacting to pull factors.

semiarid A mainly dry climate, with sparse precipitation.

Semitic language A language family that includes Hebrew and Arabic.

Sephardic Jews Jews who had inhabited Palestine and other parts of the Middle East.

settler colonization The historical pattern by which Europeans sought to create new or "neo-Europes" abroad.

Sephardic Jews Jews who had inhabited Palestine and other parts of the Middle East since early times.

Shanghai Cooperation Organization (SCO) An organization composed of China, Russia, and several Central Asian countries designed to combat terrorism, separatism, and extremism in the region.

shantytown Areas of often squalid slums on the outskirts of large cities throughout the developing world.

shatter belt A large, strategically located region composed of conflicting states caught between the conflicting interests of great powers.

shell banks A bank existing only on paper that specializes in money laundering.

Shia, or Shiite The branch of Islam regarding male descendants of Ali, the cousin and son-in-law of the Prophet Muhammad, as the only rightful successors to the Prophet Muhammad.

shifting cultivation A cycle of land use between crop and fallow years that is needed to work around the infertility of tropical soils. After clearing tropical forest, a farmer may get only a few years of crops before there is no fertility in the soil and must move on to clear more land. In the meantime, forest reclaims the previously farmed plots. Where new lands are not available, fallow periods on old fields are reduced or eliminated, resulting in soil deterioration. Also known as *swidden* or *slash-and-burn cultivation.*

shifting ranching A destructive process in which ranchers raise cattle on areas of former rainforest originally cleared by farmers. When the land is exhausted, ranchers move their cattle on to the next plot of cleared land.

Shintoism The strongly nationalistic religion of Japan, a blend of nature reverence, Japanese folklore, and Chinese ritual.

Silicon Valley A region of the San Francisco Bay area known for its concentration of high-tech companies and workers.

Silk Road A mostly overland trade route that stretched from Venice to Xi'an in China.

Sinhalese The majority ethnic group of Sri Lanka, predominantly Buddhist.

Sinn Fein The political counterpart to the Irish Republican Army.

Sino-Tibetan language family A major language family of Asia, including Chinese and its variants.

Sioux Native American people living on the Great Plains.

site The specific geographic location of a given place.

situation The accessibility of that site and the nature of the economic and population characteristics of that locale.

Six-Day War A 1967 war between Egypt, Syria, and Jordan against Israel, resulting in a dramatic Israeli victory and the Israeli occupation of the Gaza Strip, West Bank, and Golan Heights.

six essential elements of geography 1. The world in spatial terms; 2. Places and regions; 3. Physical systems; 4. Human systems; 5. Environment and society; 6. Uses of geography.

slash-and-burn cultivation See *shifting cultivation.*

Slavic Core (Fertile Triangle or Agricultural Triangle) The large area of the western former Soviet region containing most of the region's cities, industries, and cultivated lands.

Slavic languages A language family in eastern Europe including Russian, Polish and Czech.

sleeping sickness See *trypanosomiasis.*

small-scale map A map constructed to give a highly generalized view of a large area.

socialization State ownership of economic production.

Solidarity An independent trade union in Poland, the creation of which was one of the conditions resulting in the withdrawal of Soviet control in Poland in 1989.

songlines See *Footprints of the Ancestors.*

south latitude See *latitude.*

South Pole A point at 90 degrees south latitude. The highest latitude in the southern hemisphere.

Southeast Anatolia Project (GAP) An agricultural effort in Turkey aiming to double the amount of the country's irrigable farmland.

Southern African Customs Union (SACU) A trade organization including South Africa, Lesotho, Swaziland, Botswana, and Namibia.

Southern Cone Common Market (Mercosur) See *Mercosur.*

Southern Hemisphere See *hemisphere.*

Soviet satellites Communist countries in Eastern Europe between World War II and 1989.

space In geography, the exact placement of locations on the face of the Earth.

Spanglish A hybrid tongue of Spanish and English.

spatial Geographers recognize spatial distributions and patterns in Earth's physical and human characteristics. The term *spatial* comes from the noun *space*, and it relates to the distribution of various phenomena on Earth's surface. Geographers portray spatial data cartographically—that is, with maps.

special economic zones (SEZs) In China, the urban areas designated after the 1970s to attract investment and boost production through tax breaks and other incentives.

spodosols Acidic soils that have a grayish, bleached appearance when plowed, lack well-decomposed organic matter, and are low in natural fertility. Also known as *podzols.*

state farm (*sovkhoz*) A type of collectivized state-owned agricultural unit in the former Soviet Union; workers receive cash wages in the same manner as industrial workers.

State Law and Order Restoration Council (SLORC) The military government that seized control of Burma in 1988 and changed the country's name to Myanmar.

steppe (temperate grassland) A biome composed mainly of short grasses. It occurs in areas of steppe climate, which is a transitional zone between very arid deserts and humid areas.

storm surge A powerful wave in the ocean that is associated with landfalling hurricanes.

strategic hamlet Fortified communities in Vietnam that the U.S. military forced millions of South Vietnamese farming families into, in a failed attempt to prevent Viet Cong forces from infiltrating South Vietnamese villages.

String of Pearls China's strategy of gaining critical allies along the Indian Ocean shoreline.

subarctic A high-latitude climate characterized by short, mild summers and long, severe winters.

subduction The process of one tectonic plate descending below another.

subduction zone See *plate tectonics.*

subsidies Payments made by governments to domestic producers.

Suez Crisis A British, French, and Israeli invasion of Egypt in 1956 after Egyptian president Nasser nationalized the Suez Canal. International pressure caused the withdrawal of the invading forces, and the canal stayed in Egyptian control.

summer monsoon See "Monsoon."

Sunni The branch of Islam regarding successorship to the Prophet Muhammad as a matter of consensus among religious elders.

superpower A country with a bigger economy, stronger military, and pervasive popular culture than most other countries. Currently the United States is the only global superpower.

supranational organization An international organization in which member countries are united beyond the authority of any single national government and are planned and controlled by a group of nations.

suq A large commercial zone adjoining the ceremonial and administrative heart of a Middle Eastern city. See *bazaar.*

sustainable development (ecodevelopment) Concepts and efforts to improve the quality of human life while living within the carrying capacity of supporting ecosystems.

sustainable yield (natural replacement rate) The highest rate at which a renewable resource can be used without decreasing its potential for renewal.

sweatshops Factories in the LDCs infamous for low wages and long working hours. Part of the dark side of globalization: Chinese and other Asian companies in particular can undercut competition by underpaying and overworking employees.

swidden cultivation See *shifting cultivation.*

syncretism The incorporation of some animistic beliefs into mainstream religions.

taiga A northern coniferous (needleleaf) forest.

Tai-Kadai sufamily

Taliban One of the mujahidiin factions in Afghanistan fighting the Soviet invasion in the 1980s. The Taliban gained control over most of the country in the 1990s and instituted a very strict code of Islamic law. The Taliban were driven from power during the U.S.-led Operation Enduring Freedom but remain a significant insurgent force.

Tamils The minority ethnic group of Sri Lanka, primarily Hindu.

Tamil Tigers Tamil insurgents in Sri Lanka that wanted to establish an independent Tamil homeland. They were defeated by the Sri Lanka government in 2009.

Taoism See *Daoism.*

tar sands Mineral deposits in which liquid petroleum is mixed with sand and clay. Extracting the oil from tar sands is expensive and consumes large amounts of energy and water.

tariffs Tax penalties imposed on imports.

Tatars Nomads of Central Asia that conquered much of Russia in the thirteenth century along with the Mongols.

tech bubble An economic bubble in the United States in the late 1990s, centering on high-tech products and services.

technocentrists (cornucopians) Supporters of forecasts that resources will keep pace with or exceed the needs of growing human populations.

tectonic forces Processes that derive their energy from within the earth's crust and serve to create landforms by elevating, disrupting, and roughening the earth's surface.

temperate grasslands See *steppe.*

temperate mixed forest Dominant forest type in areas with humid subtropical *and* humid continental climates.

Temple Mount A site in Jerusalem sacred to both Jews and Muslims. Two Jewish temples stood on the Temple Mount, and currently the Muslim Dome of the Rock occupies the site.

Teotihuacanos A Native American ethnic group that lived in central Mexico two thousand years ago, who were later replaced by the Aztecs.

terra nullius A legal doctrine declaring a location "no one's land," allowing for claims to that land.

terraces Carefully constructed rice paddies arranged in a stair-step fashion.

terrorism Premeditated, politically motivated violence perpetrated against noncombatant targets by subnational groups or clandestine agents, usually intended to influence an audience.

terrorism hot spot An area regarded by the United States as real or potential training grounds for al-Qa'ida-affiliated groups.

thematic map A type of map showing the spatial distribution of one or more attributes across a given area.

theory of Himalayan environmental degradation The concept that overpopulation on steep slopes in the middle and high elevations of the Himalaya leads to excessive deforestation and erosion through shifting cultivation. This has many knock-on effects downstream, including flooding and siltation of hydroelectric reservoirs.

theory of island biogeography A theoretical calculation of the relationship between habitat loss and natural species loss, in which a 90 percent loss of natural forest cover results in a loss of half of the resident species.

Theravada Buddhism One of the two branches of Buddhism, found largely in Sri Lanka and Southeast Asia. Therevada Buddhists claim to follow the true teachings and practices of the Buddha.

Third Revolution The belief that sustainable development will bring about a marked shift in human ways of interacting with the natural environment, so dramatic that it will be compared with the origins of agriculture and industry.

Third Rome A term used to describe Moscow's importance in Christian affairs (after Rome and Constantinople) in the eleventh century.

Thirteenth Amendment An amendment to the U.S. Constitution in 1865, ending the practice of slavery.

Tibeto-Burman subfamily A language subfamily spoken in Southeast Asia and western China.

tierra caliente A Latin American climatic zone reaching from sea level upward to approximately 3,000 feet (914 m). Crops such as rice, sugarcane, and cacao grow in this hot, wet environment.

tierra fría A Latin American climatic zone found between 6,000 feet (1,800 m) and 12,000 feet (3,600 m) above sea level. Frost occurs and the upper limit of agriculture and tree growth is reached.

tierra helada A Latin American climatic zone above 10,000 feet (3,048 m) that has little vegetation and frequent snow cover.

tierra templada A Latin American climatic zone extending from approximately 3,000 to 6,000 feet (914–1,829 m) above sea level. It is a prominent zone of European-induced settlement and commercial agriculture such as coffee growing.

Tiger Cubs See *New Asian Tigers.*

tipping point In climate change, a time beyond which some changes are irreversible, likely bringing rapid acceleration in temperature and its related effects.

Tornado Alley An area of the Midwestern United States where tornado activity is very common.

Torres Strait Islanders A distinct indigenous people of Australia with origins in Papua New Guinea.

tortilla wars Increases in the cost of corn in Mexico from U.S. imports that threatened to make it too expensive for the poor to buy. Also refers to the controversial appearance of genetically modified corn in Mexican fields at the same time.

Tory In the American Revolution, colonists who wished to remain politically connected with the British government.

Totonac language family A small language family of eastern Mexico.

trade barriers Restrictions on international trade, especially those created by MDCs to protect their industries from competition from cheaper goods produced in LDCs.

trade wars The use of barriers to restrict trade between countries to protect their interests and industries.

Trading with the Enemies Act A U.S. law forbidding Americans from spending more than $300 in Cuba.

Trans-Amazon Highway A highway linking the Atlantic coast of Brazil with the Pacific coast of Peru.

Trans-Pacific Partnership A free-trade initiative between the United States, Australia, New Zealand, and several countries in Asia and South America.

tree line The upper limit (or northernmost limit) of natural tree growth.

Trench Deep linear Earth feature (for example, the Mariana Trench off Japan).

triangular trade The sixteenth- to ninteenth-century trading links between West Africa, Europe, and the Americas, involving guns, alcohol, and manufactured goods from Europe to West Africa exchanged for slaves. Slaves brought to the Americans were exchanged for the gold, silver, tobacco, sugar, and rum carried back to Europe.

Tropic of Cancer A line of latitude located at 23.44°N. The general dividing line in the northern hemisphere between the tropics to the south and the mid-latitudes to the north.

Tropic of Capricorn A line of latitude located at 23.44°S. The general dividing line in the southern hemisphere between the tropics to the north and the mid-latitudes to the south.

tropical deciduous forest The vegetation typical of some tropical areas with a dry season. Here, broadleaf trees lose their leaves and are dormant during the dry season and then add foliage and resume their growth during the wet season.

tropical rain forest A low-latitude broadleaf evergreen forest found where heat and moisture are continuously, or almost continuously, available.

tropical savanna A relatively moist low-latitude climate that has a pronounced dry season.

The Troubles A period of strife in Northern Ireland marked by violence between Catholic Republicans and Protestant Unionists.

Truth and Reconciliation Commission A South African commission that allows those who were party to racial violence during the apartheid era to confess their misdeeds and, in a sense, be absolved of them.

trypanosomiasis A disease spread by the tsetse fly of Africa. Known as sleeping sickness in humans, and nagana in domestic cattle.

tsunami Large waves in the ocean created by undersea earthquakes.

Tuareg An ethnolinguistic group in northern Africa.

tundra A region with a long, cold winter, when moisture is unobtainable because it is frozen, and a very short, cool summer. Vegetation includes mosses, lichens, shrubs, dwarf trees, and some grass.

Turkic A subfamily of the Altaic language family, spoken mainly in western and central Asia.

Turks The predominant ethnic group of Turkey.

Tutsi (Watusi) See *Hutu.*

two-state solution A 1947 United Nations proposal to establish separate Jewish and Arab states in Palestine.

underemployment The underutilization of labor.

underground economy The "black market" typical of the former Soviet region and many LDCs.

undifferentiated highland A climate that varies with latitude, altitude, and exposure to the sun and moisture-bearing winds.

undocumented workers People who are in the United States illegally but have obtained temporary or long-term employment.

uniform region See *region.*

union state The integration of economic, political, and defense systems between two independent countries.

United Nations Resolutions 242 and 338 Resolutions passed by the United Nations after the Middle Eastern wars of 1967 and 1973, respectively, calling on Israel to withdraw from the Occupied Territories.

untouchables See *Dalits.*

upper limit of agriculture The highest elevation at which agricultural activities can be performed, at about 12,000 feet (3,600 m).

upstream countries Countries where a major river originates that have a large amount of control over that river's water, at the expense of countries farther down the river's course.

Uralic languages A language group that includes Finnish, Hungarian, and many Siberian languages.

urbanites In the Middle Eastern ecological trilogy, the educated people who provide manufactured goods and innovations, secular and religious educational instruction, and cultural amenities.

urbanization The population growth of cities, mostly from the movement of people from rural regions to built-up areas.

Ushahidi A nonprofit software firm and platform founded in Kenya. The software uses "crowdsourcing" for activist mapping of political, environmental, and other problems. It is a fine example of local people in an LDC developing a geospatial tool kit to address problems in development.

usufruct Nondestructive mutual use of resources by multiple clans or tribes located within the territory of another.

value-added manufacturing The process of refining and fabricating more valuable goods from raw or semi-processed materials.

value-added products Finished products, worth much more than the raw materials they are created from.

Varangians See *Rus.*

vernacular region See *region.*

Viet Cong An insurrectionist Communist force in South Vietnam, supported by the North Vietnamese government, that aided the North's conquest of the South and reunification of the country during the Vietnam War.

villagers In the Middle Eastern ecological trilogy, the peasant farmers that form the cornerstone of the trilogy.

virgin and idle lands (new lands) Steppe areas of Kazakhstan and Siberia brought into grain production in the 1950s.

virtual fence A high-tech surveillance system for parts of the U.S.-Mexico border.

volcanism Movement of molten material from the Earth's mantle, usually released through volcanoes.

wa A Japanese concept of harmony based upon the principle of economic equality.

walkabout A ritual journey for Australian Aborigines in which a boy on the verge of adulthood follows the pathway of his creator-ancestors.

Wallace's Line A marked biogeographic division located in Indonesia.

Walmart effect The loss of independent or small retail establishments in towns after being unable to compete with lower prices at Walmart and similar large stores.

War of 1812 A war between the United States and the United Kingdom, fought largely as a result of the United States attempting to annex Canada.

war on drugs A long-running U.S. policy of reducing drug usage within the United States, mostly by attempting to eradicate drugs at their sources, especially in South America.

War on Terror An open-ended, unconventional war declared by President George W. Bush in 2001 against the Islamist groups that carried out terrorist attacks against the United States.

Warsaw Pact A military alliance, now dissolved, consisting of the Soviet Union and the European countries of Poland, East Germany, Czechoslovakia, Hungary, Romania, and Bulgaria.

Washington Consensus A political-economic philosophy by which the United States would pressure democratic governments of Latin America to open their markets to international trade, reduce tarriffs, sell off state companies to private owners, and make other economic changes to bring prosperity to the region.

water hemisphere See *hemisphere.*

Water Towers of Asia The glaciers and snowpack of the Himalaya, Karakorum, and other great mountain ranges.

Way of the Law See *Footprints of the Ancestors.*

weapons of mass destruction (WMD) Weapons designed to kill large numbers of people by chemical, biological, or nuclear means.

weather The atmospheric conditions prevailing at one time and place.

welfare state A national government that provides generous social services to citizens, paid for by high taxation rates.

west longitude See *longitude.*

westerly winds An airstream located in the middle latitudes that blows from west to east. Also known as *westerlies.*

Western Big Development Project A Chinese project to improve infrastructure and the economic livelihoods of residents of China's autonomous regions.

Western Hemisphere See *hemisphere.*

Western Wall The most sacred site in the world accessible to Jews, the only remnant of the Second Temple.

wet rice cultivation Rice harvests producing two or three crops per year that can sustain large populations over long periods of time.

wheat war An economic dispute between the United States and Canada in the 1990s, as wheat yields were high and prices were very low.

White Australia policy An official Australian immigration policy until 1972 that excluded Asian and black immigrants.

white gold A term indicating how valuable cotton in Central Asia was to the Soviet economy.

Wik case A 1996 Australian high court ruling that Aborigines can claim title to public lands held by farmers and ranchers under long-term pastoral leases granted by state governments.

winter monsoon See "Monsoon."

Window on the West Russia's city of St. Petersburg, located on the Baltic Sea, giving Russia marine access to Europe.

world city A city that is a center of economic and political power in the global economy.

Xhosa One of the most populous ethnic groups of South Africa.

yang In *feng shui*, the masculine, bright, and positive.

yin In *feng shui*, the feminine, dark, and negative.

Yoruba A major ethnic group in southern Nigeria.

youth bulge A demographic situation in which a large percentage of a country's population is young.

zero population growth (ZPG) The condition of equal birth rates and death rates in a population.

zero-sum game A term referring to a gain by one party resulting in an equivalent loss to another.

Zion A synonym for Jerusalem.

Zionist movement

Zomia A term for the upland areas of Southeast Asia.

Zoroastrianism Religion practiced by the Parsis of India.

Zulu One of the most populous ethnic groups of South Africa.

Index

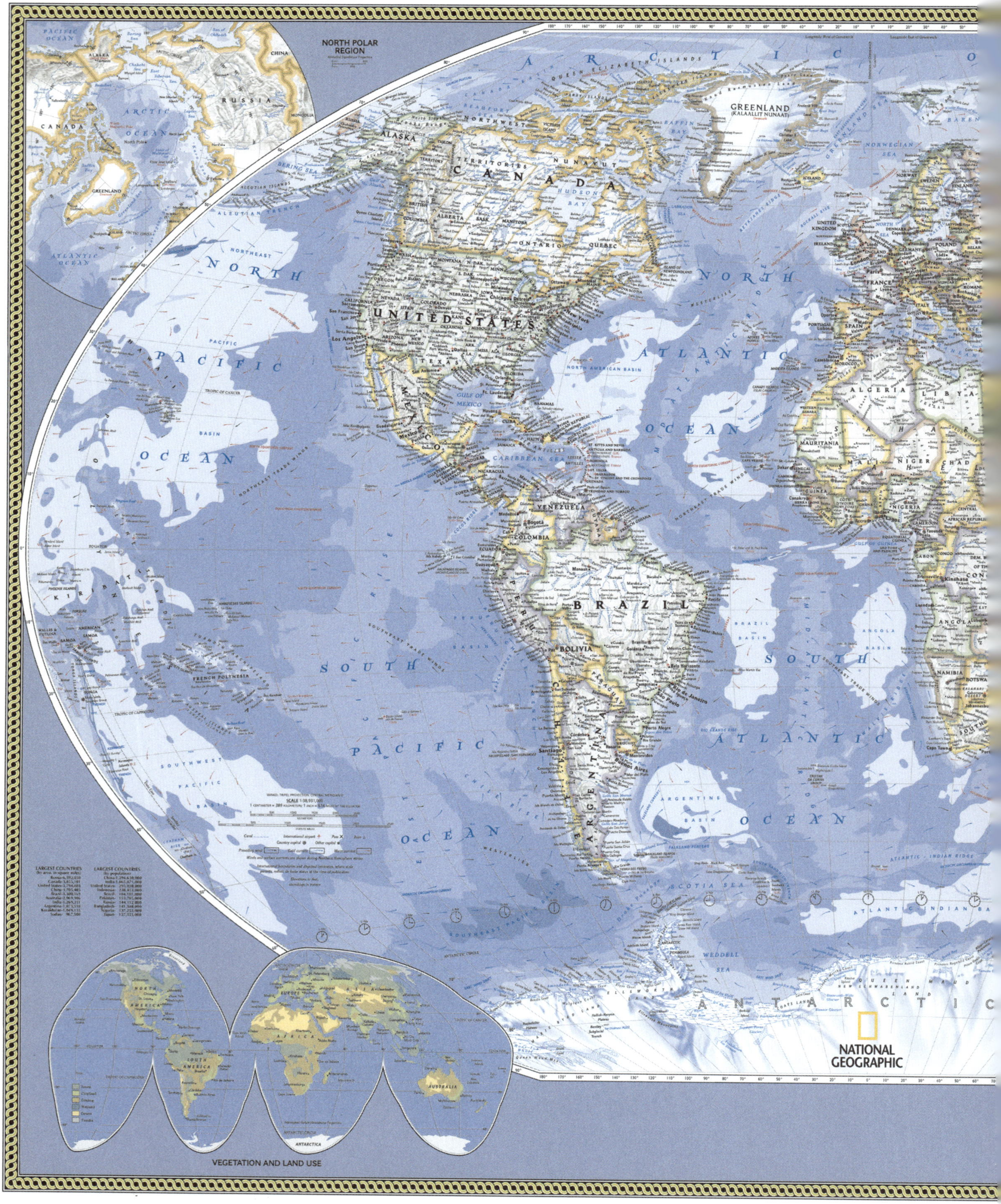
NORTH POLAR REGION
PACIFIC OCEAN
ALASKA
RUSSIA
CHINA
MONGOLIA
CANADA
ARCTIC OCEAN
North Pole
GREENLAND
ATLANTIC OCEAN
A R C T I C
QUEEN ELIZABETH ISLANDS
GREENLAND (KALAALLIT NUNAAT)
NORTHWEST TERRITORIES
NUNAVUT
C A N A D A
HUDSON BAY
ALBERTA
ONTARIO
QUEBEC
NORTH PACIFIC
NORTH ATLANTIC
UNITED STATES
MEXICO
GULF OF MEXICO
CARIBBEAN SEA
OCEAN
VENEZUELA
COLOMBIA
PERU
BRAZIL
BOLIVIA
ARGENTINA
KIRIBATI
FRENCH POLYNESIA
SOUTH PACIFIC OCEAN
SOUTH ATLANTIC OCEAN
ARGENTINE BASIN
SCOTIA SEA
WEDDELL SEA
ANTARCTIC
UNITED KINGDOM
IRELAND
FRANCE
SPAIN
PORTUGAL
MOROCCO
ALGERIA
LIBYA
MAURITANIA
MALI
NIGER
CHAD
NIGERIA
ANGOLA
NAMIBIA
NORWAY
ICELAND
VEGETATION AND LAND USE
NORTH AMERICA
SOUTH AMERICA
EUROPE
AFRICA
AUSTRALIA
ANTARCTICA
NATIONAL GEOGRAPHIC